H alpr

W9-AUG-129

Advances in
ORGANOMETALLIC CHEMISTRY

VOLUME 54

Advances in Organometallic Chemistry

EDITED BY

ROBERT WEST

ORGANOSILICON RESEARCH CENTER
DEPARTMENT OF CHEMISTRY
UNIVERSITY OF WISCONSIN
MADISON, WISCONSIN USA

ANTHONY F. HILL

RESEARCH SCHOOL OF CHEMISTRY
INSTITUTE OF ADVANCED STUDIES
AUSTRALIAN NATIONAL UNIVERSITY
CANBERRA, ACT, AUSTRALIA

FOUNDING EDITOR

F. GORDON A. STONE

VOLUME 54

Amsterdam • Boston • Heidelberg • London • New York • Oxford • Paris
San Diego • San Francisco • Singapore • Sydney • Tokyo
Academic Press is an imprint of Elsevier

Academic Press is an imprint of Elsevier
84 Theobald's Road, London WC1X 8RR, UK
Radarweg 29, PO Box 211, 1000 AE Amsterdam, The Netherlands
The Boulevard, Langford Lane, Kidlington, Oxford OX5 1GB, UK
30 Corporate Drive, Suite 400, Burlington, MA 01803, USA
525 B Street, Suite 1900, San Diego, CA 92101-4495, USA

First edition 2006

Copyright © 2006 Elsevier Inc. All rights reserved

No part of this publication may be reproduced, stored in a retrieval system or transmitted in any form or by any means electronic, mechanical, photocopying, recording or otherwise without the prior written permission of the publisher

Permissions may be sought directly from Elsevier's Science & Technology Rights Department in Oxford, UK: phone (+44) (0) 1865 843830; fax (+44) (0) 1865 853333; email: permissions@elsevier.com. Alternatively you can submit your request online by visiting the Elsevier web site at http://elsevier.com/locate/permissions, and selecting *Obtaining permission to use Elsevier material*

Notice
No responsibility is assumed by the publisher for any injury and/or damage to persons or property as a matter of products liability, negligence or otherwise, or from any use or operation of any methods, products, instructions or ideas contained in the material herein. Because of rapid advances in the medical sciences, in particular, independent verification of diagnoses and drug dosages should be made

ISBN-13: 978-0-12-031154-5
ISBN-10: 0-12-031154-2
ISSN: 0065-3055

For information on all Academic Press publications
visit our website at books.elsevier.com

Printed and bound in The Netherlands

06 07 08 09 10 10 9 8 7 6 5 4 3 2 1

Working together to grow
libraries in developing countries
www.elsevier.com | www.bookaid.org | www.sabre.org
ELSEVIER BOOK AID International Sabre Foundation

Contents

Intramolecular Interconversions in Cyclosilazane Chemistry: A Joint Experimental–Theoretical Study

UWE KLINGEBIEL, NINA HELMOLD and STEFAN SCHMATZ

Molecular Alumo-Siloxanes and Base Adducts

MICHAEL VEITH

Progress in the Chemistry of Stable Disilenes

MITSUO KIRA and TAKEAKI IWAMOTO

Metallacycloalkanes – Synthesis, Structure and Reactivity of Medium to Large Ring Compounds

BURGERT BLOM, HADLEY CLAYTON, MAIRI KILKENNY and JOHN R. MOSS

The Chemistry of Cyclometallated Gold(III) Complexes with C,N-Donor Ligands

WILLIAM HENDERSON

Sterically Demanding Phosphinimides: Ligands for Unique Main Group and Transition Metal Chemistry

DOUGLAS W. STEPHAN

Transition Metal Complexes of Tethered Arenes

JOANNE R. ADAMS and MARTIN A. BENNETT

Contributors

Numbers in parentheses indicated the pages on which the authors' contributions begin.

JOANNE R. ADAMS (293), Research School of Chemistry, Institute of Advanced Studies Australian National University, Canberra, ACT 0200, Australia

MARTIN A. BENNETT (293), Research School of Chemistry, Institute of Advanced Studies Australian National University, Canberra, ACT 0200, Australia

BURGET BLOM (149), Department Of Chemistry, University Of Cape Town, Rondebosch 7701, South Africa

HADLEY CLAYTON (149), Department Of Chemistry, University Of Cape Town, Rondebosch 7701, South Africa

NINA HELMOLD (1), Institut für Anorganische Chemie, Tammannstrasse 4, D-37077 Göttingen, Germany

WILLIAM HENDERSON (207), Department of Chemistry, The University of Waikato, Private Bag 3105, Hamilton, New Zealand

TAKEAKI IWAMOTO (73), PRESTO, Japan Science and Technology Agency, 4-1-8 Honcho Kawaguchi, Saitama 332-0012, Japan

MAIRI KILKENNY (149), Department Of Chemistry, University Of Cape Town, Rondebosch 7701, South Africa

MITSUO KIRA (73), Department of Chemistry, Graduate School of Science, Tohoku University, Aoba-ku, Sendai 980-8578, Japan

UWE KLINGEBIEL (1), Institut für Anorganische Chemie, Tammannstrasse 4, D-37077 Göttingen, Germany

JOHN R. MOSS (149), Department Of Chemistry, University Of Cape Town, Rondebosch 7701, South Africa

STEFAN SCHMATZ (1), Institut für Physikalische Chemie, Tammannstrasse 6, D-37077 Göttingen, Germany

DOUGLAS W. STEPHAN (267), Chemistry and Biochemistry, University of Windsor, Windsor, ON Canada N9B 3P4

MICHAEL VEITH (49), Institut für Anorganische Chemie, Universität des Saarlandes, Im Stadtwald, D-66401 Saarbrücken, Germany

Intramolecular Interconversions in Cyclosilazane Chemistry: A Joint Experimental–Theoretical Study

UWE KLINGEBIEL,[a,*] NINA HELMOLD[a] and STEFAN SCHMATZ[b,*]

[a]*Institut für Anorganische Chemie, Tammannstrasse 4, D – 37077 Göttingen, Germany*
[b]*Institut für Physikalische Chemie, Tammannstrasse 6, D – 37077 Göttingen, Germany*

I
SUMMARY

The chemistry of cyclosilazanes and borazines is a topic of high current interest in view of their importance for preceramic polymers. On pyrolysis these compounds yield various useful ceramic materials, e.g., silicon and boron carbide, silicon and boron nitride and ternary or quaternary mixtures of such materials.[1] These ceramic materials show high stability against corrosion and thermal shock, high creep resistance, low electrical conductivity and a low coefficient of thermal expansion, all of which are potentially useful properties. One of the best application prospects is the use of the preceramic polymers as a protective coating material for carbon fibers.

As far as we know, rearrangement of cyclosilazanes is the first step of polymerization. To understand the polymerization processes of cyclosilazanes, fundamental research on the chemistry of this class of silicon rings is highly desirable.

In comparison with other groups, silyl groups show great mobility. Silyl group migration depends on the propensity of the silicon atom to be bonded to the atom with the largest negative partial charge. Essentially three types of interconversions are known:

*Corresponding author.
E-mail: uklinge@gwdg.de (U. Klingebiel), sschmat@gwdg.de (S. Schmatz).

ADVANCES IN ORGANOMETALLIC CHEMISTRY
VOLUME 54 ISSN 0065-3055/DOI 10.1016/S0065-3055(05)54001-2

© 2006 Elsevier Inc.
All rights reserved.

(a) Keto-enol tautomery,[2,3] which was used in the synthesis of the first phoshaethenes and silaethenes, (b) Anionic isomerization and (c) Neutral isomerization,[4,5] which lead, e.g., to the preparation of isomeric silylhydroxylamines[6,7] and silylhydrazines.[8–10] In ring systems, silyl migrations can lead to ring contractions or expansions.[10–12] Of all the isomerization processes in inorganic ring systems, the interconversions in cyclosilazane chemistry are those which are now best understood in terms of the mechanisms involved.[11,12] Therefore, it seems to be an appropriate time to summarize the experimental results and to compare them with theoretical data in a comprehensive review.

II
INTRODUCTION

A. *Formation of Four-, Six- and Eight-Membered* ($\rangle$Si–NH)-*Rings*

The six-membered cyclotri- and the eight-membered cyclotetrasilazanes were the first well-defined silicon–nitrogen rings. They were prepared by Brewer and Haber by ammonolysis of dimethyldichlorosilane in 1948[13] (Scheme 1).

Since 1950, ammonolysis reactions of many dichlorosilanes have been studied. Usually cyclic trimers and tetramers were formed. But very little was known about the reaction mechanisms.[14]

The reaction is a complex process involving substitution reactions and/or homo- and heterofunctional condensations (Scheme 2).

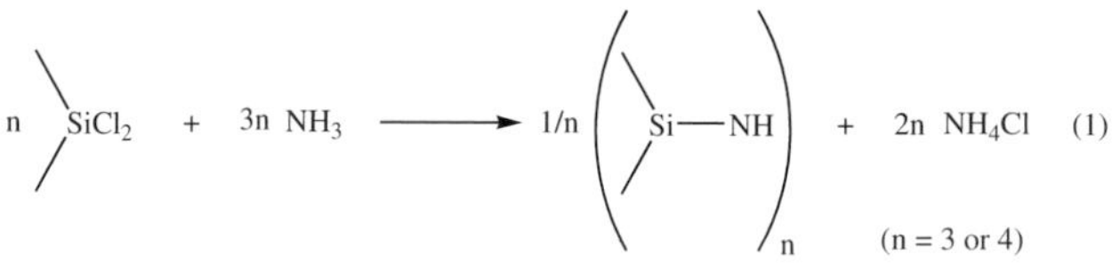

$$n \; \rangle SiCl_2 \; + \; 3n\; NH_3 \longrightarrow 1/n \left(\rangle Si{-}NH \right)_n \; + \; 2n\; NH_4Cl \quad (1)$$

$$(n = 3 \text{ or } 4)$$

SCHEME 1.

$$-\!\!\!\overset{|}{\underset{|}{Si}}\!-Cl \; + \; NH_3 \longrightarrow -\!\!\!\overset{|}{\underset{|}{Si}}\!-NH_2 \; + \; HCl$$

$$-\!\!\!\overset{|}{\underset{|}{Si}}\!-NH_2 \; + \; H_2N-\overset{|}{\underset{|}{Si}}\!- \longrightarrow -\!\!\!\overset{|}{\underset{|}{Si}}\!-\overset{H}{N}-\overset{|}{\underset{|}{Si}}\!- \; + \; NH_3 \quad (2)$$

$$-\!\!\!\overset{|}{\underset{|}{Si}}\!-NH_2 \; + \; Cl-\overset{|}{\underset{|}{Si}}\!- \longrightarrow -\!\!\!\overset{|}{\underset{|}{Si}}\!-\overset{H}{N}-\overset{|}{\underset{|}{Si}}\!- \; + \; HCl$$

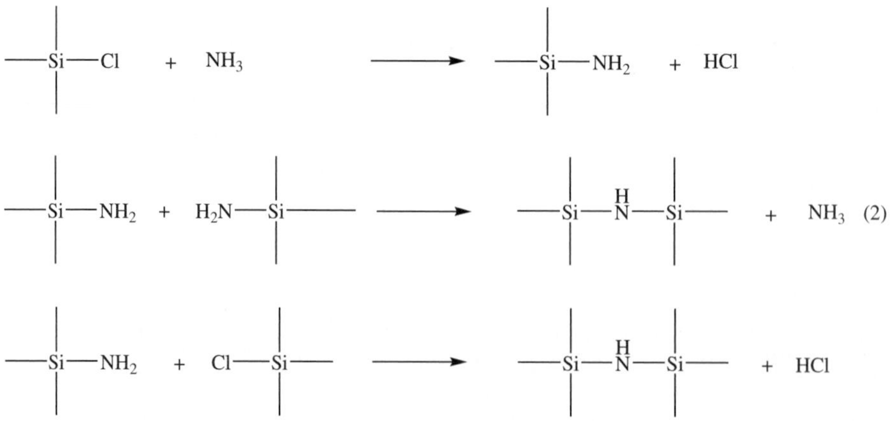

SCHEME 2.

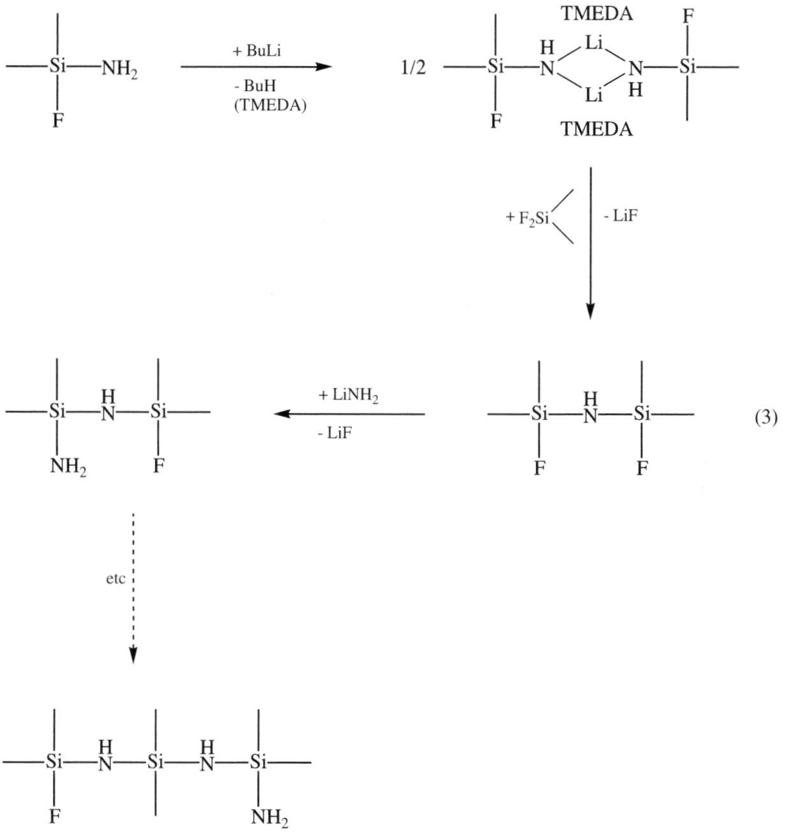

SCHEME 3.

Many intermediates can be postulated in the course of the ammonolysis of dichlorosilanes, but only some of them have been isolated, which indicates the complexity of the process.

Primary products of such condensations can be synthesized in the reaction of fluorosilanes with lithium amide (Scheme 3) so that the formation mechanisms of the ring compounds can be studied.[15]

In contrast to the ammonolysis of chlorosilanes, the four-membered ring was formed along with the six- and eight-membered rings by heating lithium fluorosilylamide (Scheme 4).[16]

Four-membered (SiNH) rings were previously unknown.[14] However, rings can also be synthesized in a stepwise manner *via* various acyclic compounds. The NH_2-substituted compounds can be lithiated and LiF leads to the formation of rings.[15]

Hexa-tert-butylcyclotrisilazane and octa-tert-butylcyclotetrasilazane are the bulkiest known cyclosilazanes. Hexa-tert-butylcyclotrisilazane is a planar six-membered

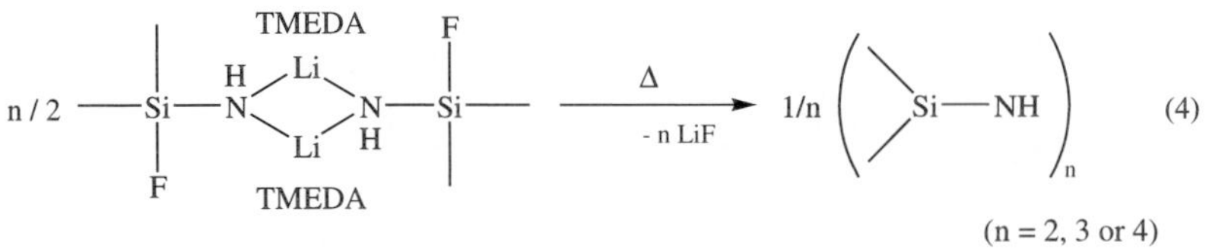

$$n/2\ \left[\mathrm{F{-}Si{-}N(H)}\,\mathrm{Li_2(TMEDA)_2}\,\mathrm{N(H){-}Si{-}F}\right] \xrightarrow[\text{- n LiF}]{\Delta} 1/n\left(\mathrm{{>}Si{-}NH}\right)_n \quad (4)$$

(n = 2, 3 or 4)

SCHEME 4.

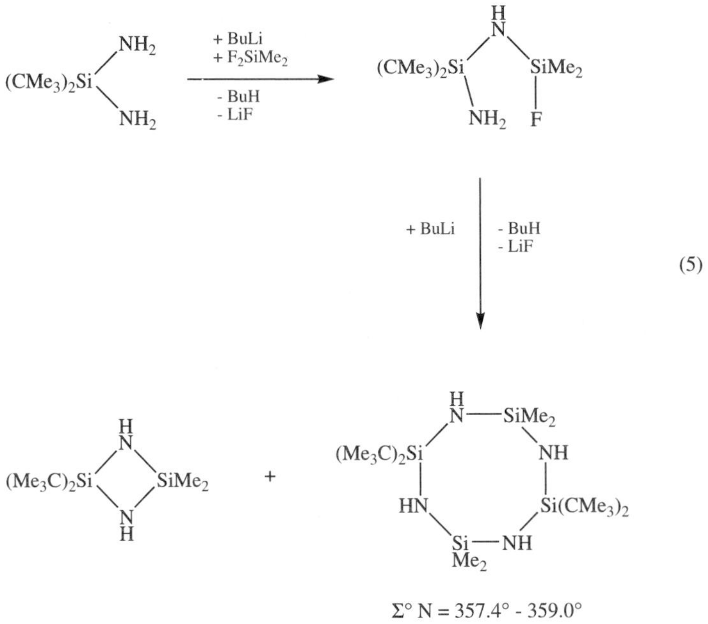

$$(CMe_3)_2Si(NH_2)_2 \xrightarrow[\text{- BuH, - LiF}]{\text{+ BuLi, + }F_2SiMe_2} (CMe_3)_2Si(NH_2){-}NH{-}SiMe_2F$$

$$\xrightarrow[\text{- BuH, - LiF}]{\text{+ BuLi}} (Me_3C)_2Si(NH)_2SiMe_2 \;+\; [(Me_3C)_2Si{-}NH{-}SiMe_2{-}NH]_2 \quad (5)$$

$\Sigma^\circ N = 357.4^\circ - 359.0^\circ$

SCHEME 5.

ring,[17] and octa-tert-butylcyclotetrasilazane is the most planar eight-membered ring known so far.[16] The deviation of the ring atoms from the plane is only 26 pm.

Another possible reaction pathway starts from stable diaminosilanes; it allows for different organyl substituents at the silicon atoms in the cyclosilazane products (Scheme 5).

The four-membered ring is planar[18] with the NSiN angles being smaller than 90°, while the SiNSi angles are larger than 90°. Until now the eight-membered ring is the only one that has a saddle conformation with the Si atoms being approximately coplanar and the N atoms alternatively located above and below this plane.[18]

1. Dimerization of Iminosilenes

Four-membered rings are obtained when alkyl-, aryl- or silylaminofluorosilanes are used as starting material.[19] Lithiated aminofluorosilanes are often very stable and can be purified by distillation or sublimation (Scheme 6).

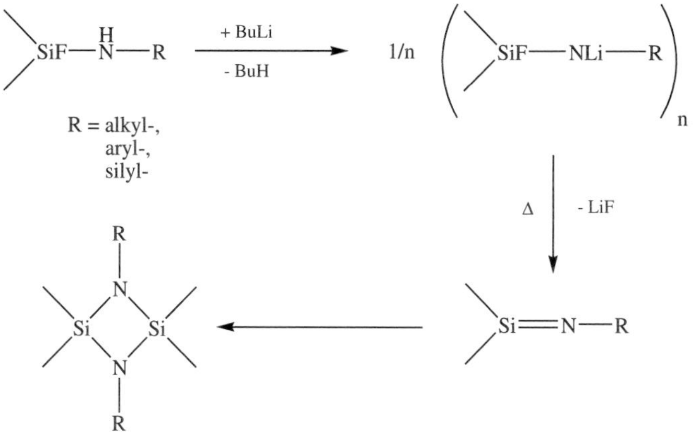

$$>SiF-\overset{H}{N}-R \xrightarrow[-\,BuH]{+\,BuLi} 1/n\left(>SiF-NLi-R\right)_n \xrightarrow[-\,LiF]{\Delta} >Si{=}N-R \longrightarrow (>Si-NR)_2 \quad (6)$$

R = alkyl-, aryl-, silyl-

SCHEME 6.

Several interesting mechanistic features of silicon chemistry have been found. Attempts to eliminate LiF led to dimerization in a head to tail (2 + 2) cycloaddition or rearrangement of the formed iminosilene.[19] The limits of dimerization are reached with the dimer diisopropyl-(tri-*tert*-butylphenylimino)silene, for which the crystal structure analysis shows severe steric distortions.[20]

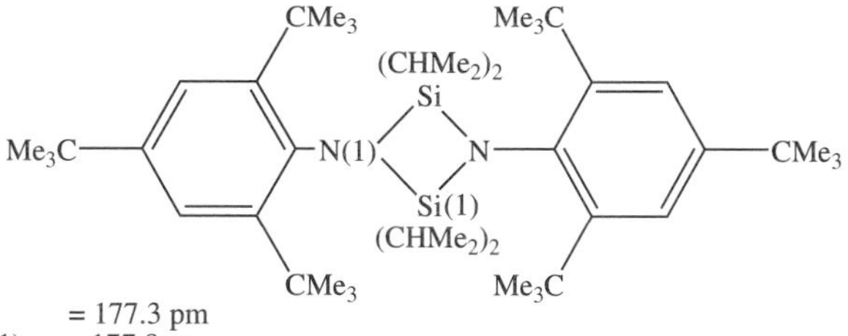

Si - N = 177.3 pm
Si - N(1) = 177.8 pm
N - Si(1) = 179.6 pm
Si(1) - N(1) = 180.1 pm

a. Electrophilic 1,3-Silyl Group Migration

If the iminosilene is too bulky to dimerize, it often exhibits unusual ways of attaining the coordination number four for the silicon atom. For example, bis(silyl)amino-iminosilenes cyclize with simultaneous electrophilic 1,3-migration of the silyl group from the amine to the more negatively charged imine nitrogen[3] (Scheme 7).

^{15}N-NMR measurements prove that the electron density at the nitrogen atom is smaller in silylimines than in organylimines. Silyliminosilenes are usually obtained in dimeric form.[19,21,22]

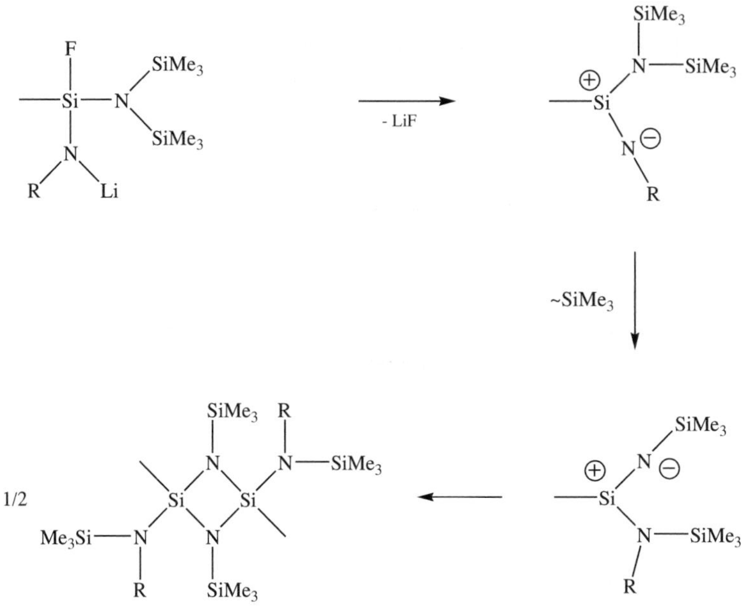

(7)

SCHEME 7.

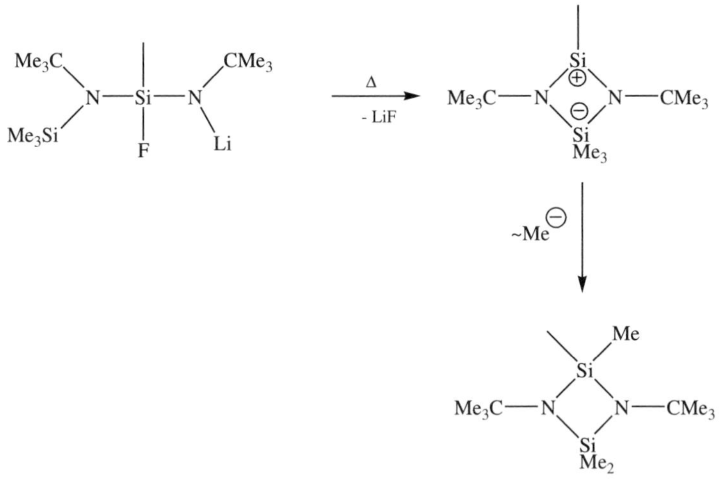

(8)

SCHEME 8.

b. Nucleophilic 1,3-Methanide-Ion Migration

Another surprising ring closure occurs in a molecule with organic substituents bonded to the amine and the imine nitrogen, e.g., tert-butyl groups. Electrophilic silyl group migration is not complete because the two nitrogen atoms are symmetrically substituted. Instead (Scheme 8), a cyclic ylide is formed. It is stabilized by an

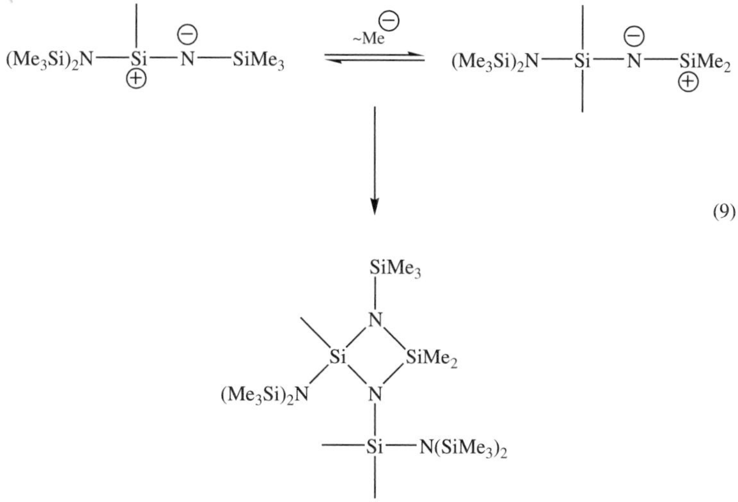

(9)

SCHEME 9.

intramolecular nucleophilic 1,3-rearrangement of a methanide ion from one silicon to the other;[15,23,24] a four-membered ring is obtained.

If the central silicon carries two silylamino groups, only half an equivalent of the iminosilene undergoes migration of a methanide ion. The result is a cross dimer of the two ylides,[25] as shown in Scheme 9.[15]

The two diastereomers could be isolated and separated by recrystallization.[25]

From this, two general rules can be formulated:

(1) Silyl groups migrate to more negatively charged atoms or anions. The isomerization is electrophilic and follows an S_N2 mechanism.
(2) The existence of two differently coordinated silicon atoms in the same molecule or anion leads to nucleophilic migration of a methanide ion.

B. *Cyclodisilazane Anions*

Cyclodisilazane anions and cyclodisilazanes in *cis*-conformation are found in the reaction of dilithiated bis(silylamino)fluorosilanes with chlorotrimethylsilane. The dilithium salt of the corresponding bis(silylamino)chlorosilane is obtained. LiCl elimination in the presence of THF leads to the formation of silaamidides (which are isolated as dimers),[19,26] four-membered cyclodisilazane anions and $(thf)_3Li–Cl–Li(thf)_3^+$ or $Li(thf)_4^+$ cations (Scheme 10).[19,27] Hydrolysis of dimeric silaamidides is a facile synthesis leading to cyclodisilazanes in the *cis*-conformation. Examples are shown in Figs. 1 and 2 (Scheme 11).[19,27]

$$+ Me_3SiCl \quad - Me_3SiF \quad THF$$

1-3

$- LiCl$ $\quad \Delta$

$Li^{\oplus}(THF)_4$ (10)

4, 6

$[(THF)_3Li{-}Cl{-}Li(THF)_3]^{\oplus}$

5

	R	R'
1, 4	$Si(CMe_3)_2Me$	F
2, 5	$SiMe_2CMe_3$	CMe_3
3, 6	$SiMe_2CMe_3$	C_6H_5

SCHEME 10.

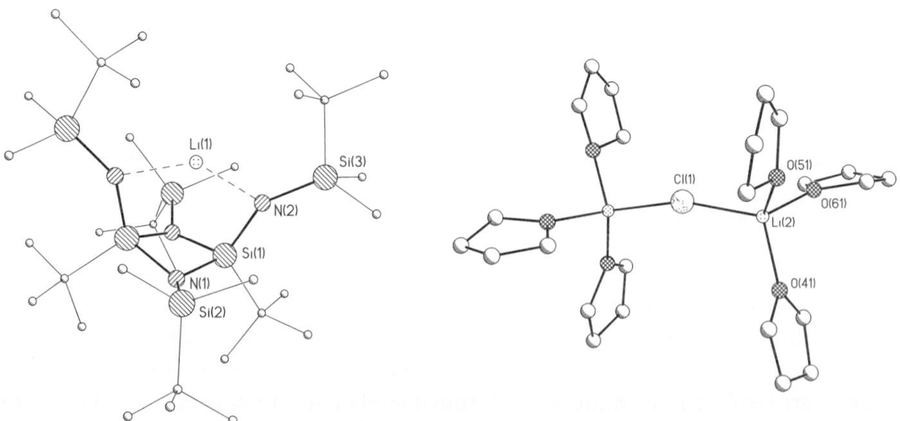

FIG. 1. Crystal structure of the anionic and cationic part of **5**.

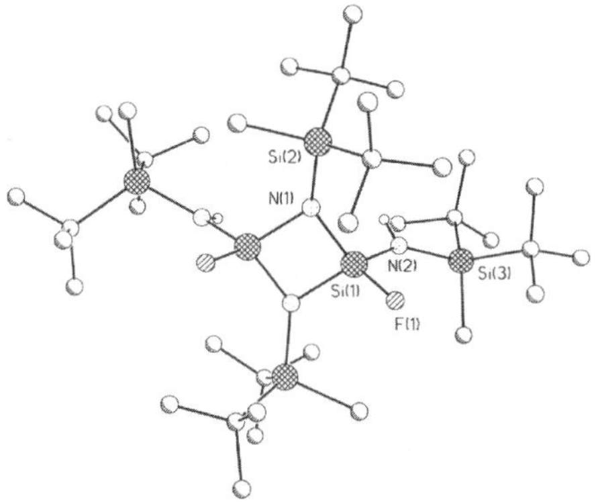

7

Si(1) - F(1)	159.0(2) pm	Si(1) - N(1)	173.7(3) pm
N(1) - Si(2)	175.9(3) pm	N(2) - Si(3)	176.3(3) pm
Si(1) - N(2)	169.3(3) pm		

FIG. 2. Crystal structure of **7**.

Si(1)–N(1)	178.3(4)	Si(3)–N(2)–Si(1)	146.2(3)
N(2)–Si(3)	166.4(4)	N(2)–Li(1)–N(2)#1	137.2(6)
N(2)–Li(1)	200.3(6)	Li(2)–Cl(1)–Li(2)#2	163.5(7)
Li(2)–Cl(1)	223.7(1)		

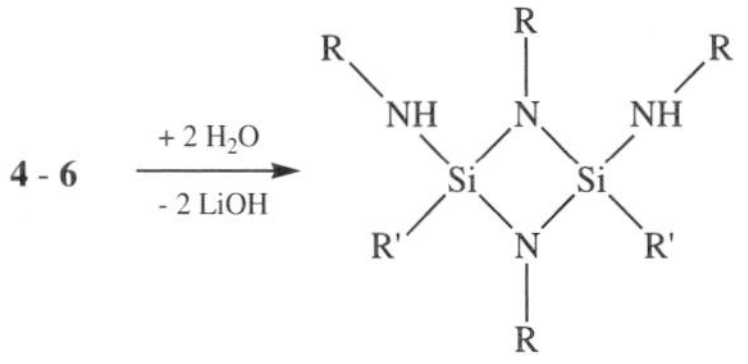

	R	R′
4, 7	$Si(CMe_3)_2Me$	F
5, 8	$SiMe_2CMe_3$	CMe_3
6, 9	$SiMe_2CMe_3$	C_6H_5

(11)

SCHEME 11.

C. *Cyclodisilazane Cations*

The nitrogen atom in silicon–nitrogen compounds is normally in a structurally planar environment, and rarely exhibits basic properties. However, when cyclodisilazanes with bulky substituents, e.g., $[(Me_2HC)_2Si–NCMe_3]_2$, are heated in *n*-hexane/CH_2Cl_2, mono- and diprotonated cyclodisilazane cations are formed with $AlCl_4^-$ and/or $^-Cl_3Al–O–AlCl_3^-$ as anionic counterparts (Scheme 12).[28]

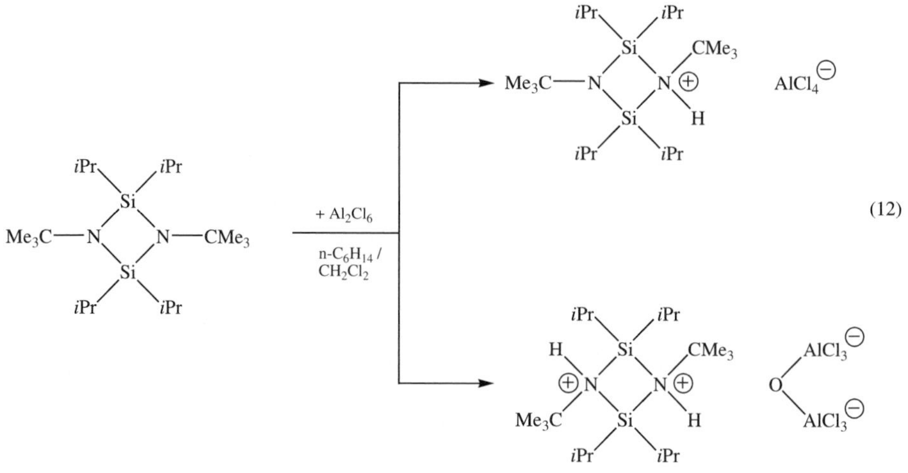

(12)

SCHEME 12.

The Si–N bonds in the cyclodisilazane (174.7 pm) are significantly shorter than in the dication (179.3 pm), i.e., they are weakened in the course of the reaction because of the sp^3-hybridization of the nitrogen atoms in the resulting cations.

III
2,2,4,4,6,6-HEXAMETHYLCYCLOTRISILAZANE

A. *Isomeric Cyclotri- and Cyclodisilazanes, DFT Calculations*

A possible isomer of the cyclotrisilazane $(Me_2Si\text{-}NH)_3$, **A**, is a cyclodisilazane $HN(SiMe_2)_2NSiMe_2NH_2$, **B**, with an exocyclic $SiMe_2NH_2$ group.

Density functional theory (DFT) calculations have been carried out to elucidate the structure and energetics of the various isomeric (Si–N) rings as well as the corresponding anions and dianions. The local minima on the potential energy surfaces were verified by computation of the eigenvalues of the respective Hessian matrices. From these, harmonic vibrational frequencies and the zero-point vibration corrected energetics were calculated. The calculations were carried out for isolated molecules in the gas phase. The theoretical results are expected to be reliable for molecules in non-polar or weakly aprotic polar solvents.

Using the B3LYP hybrid functional of Becke (B3LYP)[29] and a 6–31G(d) basis set comprising 234 contracted Gaussian-type orbitals (cGTOs), the energetic difference between the six-membered $(Me_2Si\text{-}NH)_3$ (HMCTS), **A**, and its four-membered isomer **B** is calculated to be 16.7 kcal/mol. If zero-point vibrational effects are included, this value slightly reduces to 16.6 kcal/mol (Fig. 3). The six-membered ring exhibits a boat-conformation, where the atoms N(1), Si(2), Si(3) and N(3) form a plane and N(2) and Si(1) are located above it. The hydrogens at N(1) and N(3) point downwards, while that at N(2) point upwards. The molecule has C_s symmetry. The endocyclic angles at

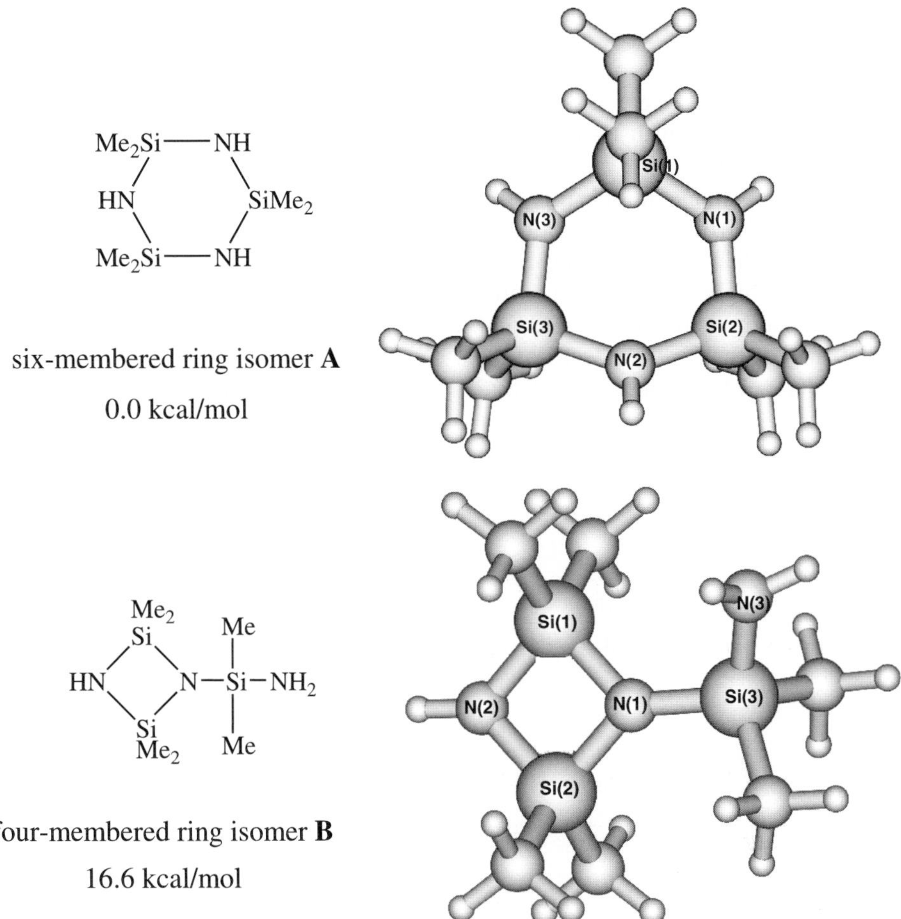

FIG. 3. Calculated structures of the six- and four-membered ring isomers [B3LYP/6–31G(d)].

the silicon atoms are approximately tetrahedral, while those at the nitrogens are much wider. All Si–N-bond distances are very similar. The most important structural parameters are r(Si(2)–N(2) = 175.0 pm, r(N(1)–Si(2)) = 174.5 pm, r(Si(1)–N(1)) = 175.5 pm; α(N(1)–Si(1)–N(3)) = 109.0°, α(Si(1)–N(1)–Si(2)) = 125.1°, α(N(1)–Si(2)–N(2)) = 104.7°, α(Si(2)–N(2)–Si(3)) = 130.7°.

The four-membered ring is planar (sum of angles 360°), but not symmetric owing to the exocyclic $SiMe_2NH_2$ group. The angles at the nitrogens are about 6° wider than those at the silicons: α(Si(2)–N(2)–Si(1)) = 93.7°, α(N(1)–Si(2)–N(2)) = 87.1°, α(Si(1)–N(1)–Si(2)) = 92.5° and α(N(2)–Si(1)–N(1)) = 86.7°. As was found for the six-membered ring, the four Si–N distances in the ring are very similar: r(N(2)–Si(2)) = 175.0 pm, r(Si(2)–N(1)) = 176.7 pm, r(N(1)–Si(1)) = 177.3 pm and r(Si(1)–N(2)) = 175.6 pm. The transannular Si···Si interatomic distance is calculated to be 255.7 pm.

B. *Isomeric Cyclotri- and Cyclodisilazane-Anions and -Dianions, DFT Calculations*

DFT calculations were also performed for the anionic isomers of HMCTS (Fig. 4). Additional diffuse functions were included in the basis set (6–31 + G(d)), resulting in 280 cGTOs.

While only a single anionic structure is possible for the six-membered ring, two anions can be formed starting from the four-membered species: The lithiation can take place in the ring or at the chain (Fig. 4). In the former case, the anion is 22.9 kcal/mol higher in energy than the six-membered anion, while in the latter case, the anion energy is higher by 19.2 kcal/mol (Fig. 5).

The ring of the six-membered-anionic species **A′** is planar with bond distances $r(\text{Si(1)–N(1)}) = 173.3\,\text{pm}$, $r(\text{N(1)–Si(2)}) = 179.7$, $r(\text{Si(2)–N(2)}) = 165.8\,\text{pm}$ and

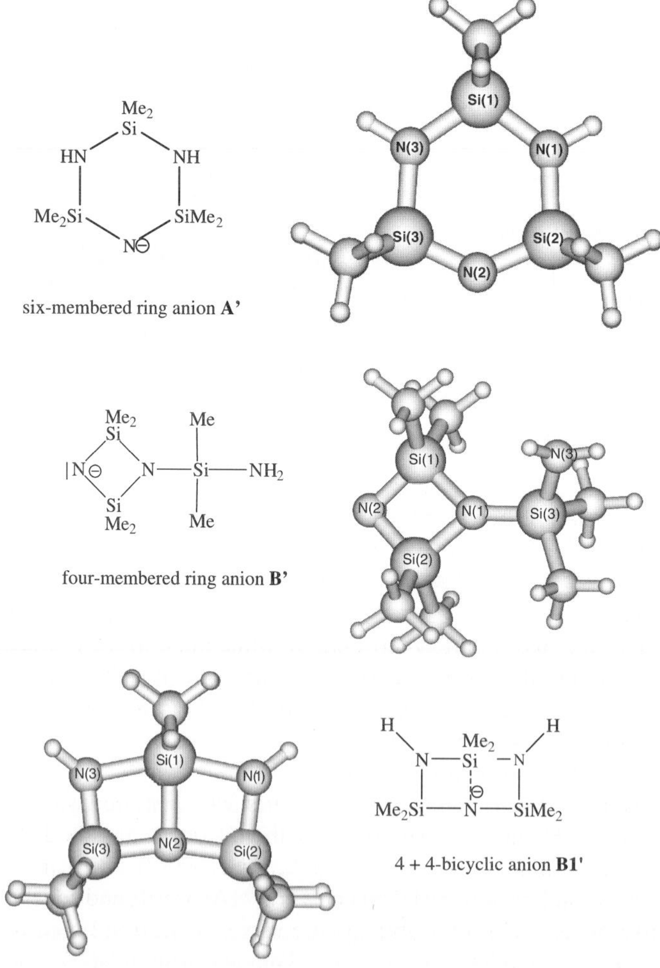

FIG. 4. Calculated structures of the anions of HMCTS [B3LYP/631 + G(d)].

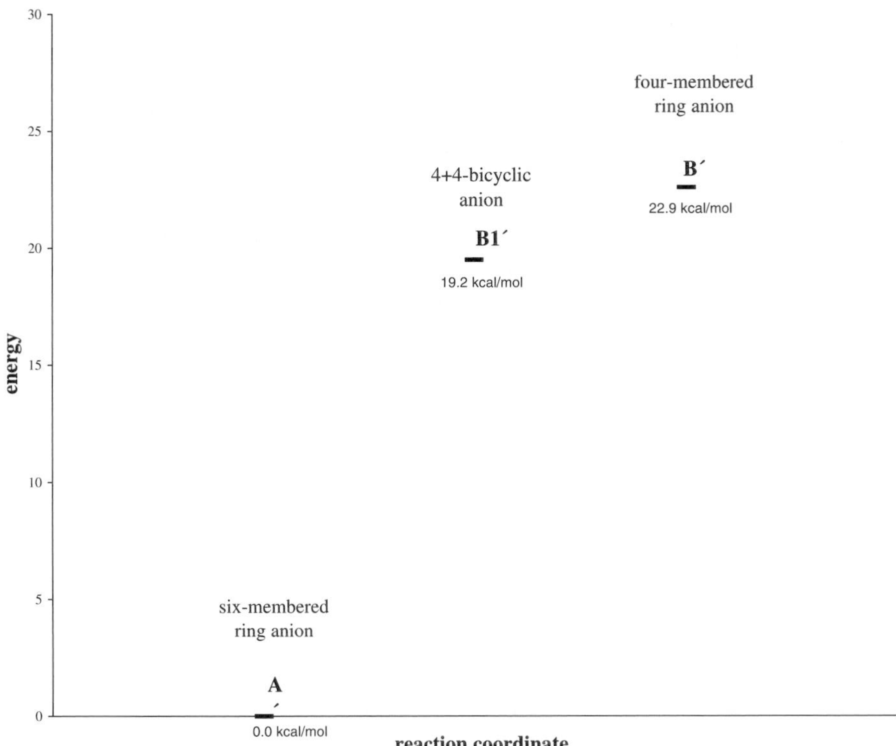

FIG. 5. Energy diagram of the anionic isomers **A**′–**C**′ of HMCTS. The energies refer to the most stable isomer **A**′ and include zero-point energy contributions.

angles α(N(3)–Si(1)–N(1)) = 109.1°, α(Si(1)–N(1)–Si(2)) = 128.5°, α(N(1)–Si(2)–N(2)) = 111.5° and α(Si(2)–N(2)–Si(3)) = 130.9°.

Endocyclic lithiation of the four-membered neutral species yields a ring **B**′ with considerably more acute angles at the nitrogens and more obtuse angles at the silicon atoms: α(Si(1)–N(2)–Si(2)) = 90.4°, α(N(2)–Si(2)–N(1)) = 93.4°, α(Si(2)–N(1)–Si(1)) = 83.1° and α(N(1)–Si(1)–N(2)) = 93.2°. The Si–N bonds at the anionic center are more than 10 pm shorter than the other two endocyclic bonds: *r*(N(2)–Si(2)) = 170.0 pm, *r*(Si(2)–N(1)) = 181.8 pm, *r*(N(1)–Si(1)) = 182.1 pm, *r*(Si(1)–N(2)) = 170.1 pm. The transannular Si···Si distance is calculated to be 241.3 pm and thus more than 14 pm shorter than in the neutral species **B**.

Exocyclic lithiation, however, results in a bicyclic structure **B1**′. The HN group attacks the γ-silicon atom of the ring. The structure contains two four-membered rings and a five-coordinated silicon atom. The endocylic Si(3)–N(2)–Si(2) angle is found to be 194.1°. The two NH units are equivalent, so that one cannot distinguish between the endo- and exocylic NH in the starting material. The bicycle is almost, but not exactly planar. The two HN hydrogens show *trans* configuration and the molecule exhibits C_2 symmetry with the C_2-axis through N(2) and Si(1). The most important structural parameters are as follows: *r*(Si(1)–N(3)) = 193.8 pm,

r(N(3)–Si(3)) = 171.2 pm, r(Si(3)–N(2)) = 171.7 pm, r(Si(1)–N(2)) = 187.6 pm; α(Si(1)–N(3)–Si(3)) = 94.9°, α(N(3)–Si(1)–N(1)) = 156.9°, α(N(3)–Si(3)–N(2)) = 89.4°. While the four Si–N bonds containing the $SiMe_2$ groups at the corners of the species are relatively short, the bond that connects the two rings and, in particular, the bonds Si(1)–N(3) and Si(1)–N(1) are unusually long. Thus, the system **B1′** can easily undergo rearrangements either towards the more stable six-membered ring anion **A′** (by breaking the Si(1)–N(2) bond) or towards the energetically slightly higher four-membered ring **B′** (by opening one of the rings of the bicycle).

Figure 5 graphically displays the high difference in the energies of the six-membered anion **A′** and the four-membered-isomeric anions **B′** or **B1′**.

While theoretically the anionic HMCTS (**A′**) has two possible cyclodisilazane isomers (**B′**, **B1′**), experimental results give evidence that only the isomers **A′** and **B1′** are of importance. For example, it was found that the anion of HMCTS (**A′**) always reacts with Me_3SiCl, fluorosilanes and -boranes in a molar ratio 1:1 with retention of the ring size.

We have also calculated the structures of the dianions of HMCTS (Fig. 6).

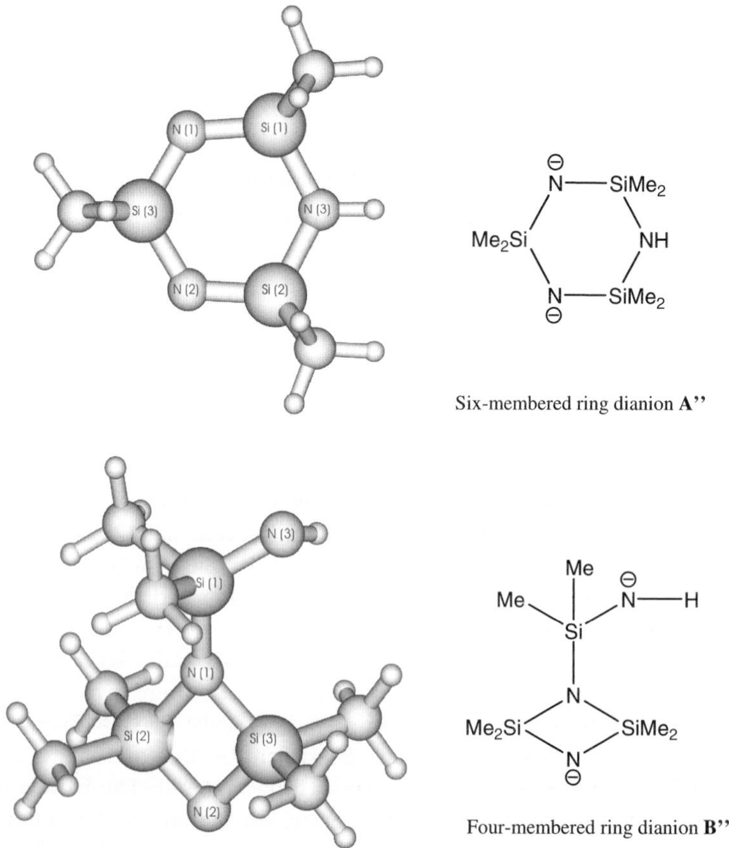

FIG. 6. Calculated structures of the dianions of HMCTS [B3LYP/6–31G+G(d)].

The six-membered-ring dianion **A″** is found to be 22.6 kcal/mol more stable than the four-membered-ring anion **B″**. This result is confirmed by experiments, which show that substitution reactions occur at the six-membered-ring system.

The calculated structure of the six-membered dianion exhibits C_{2v} symmetry and a perfectly planar Si–N skeleton. The endocyclic angles amount to α(Si–NH–Si) = 124.0°, α(NH–Si–N) = 114.3°, α(Si–N$^-$–Si) = 124.6° and α(N$^-$–Si–N$^-$) = 118.2°. The bond lengths are r(NH–Si) = 178.1 pm, r(NHSi–N$^-$) = 166.0 pm and r(N–SiN$^-$) = 170.9 pm.

C. *Substitution Reactions*

Because of the weakness of the Si–N bond, ring cleavage of cyclosilazanes occurs upon treatment with halides of main group elements. For more than 20 years after the synthesis of the six-membered ring $(Me_2SiNH)_3$, only the substitution reaction of the N-bonded hydrogen by one trimethylsilyl group was known.[30]

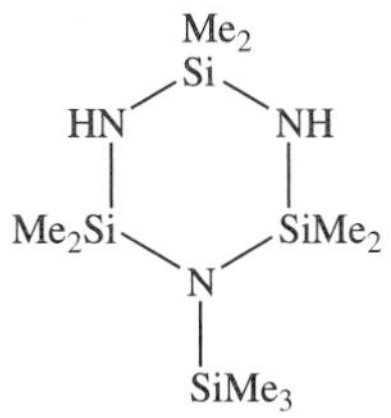

Me_2
Si
HN NH
Me_2Si $SiMe_2$
N
$SiMe_3$

Reactions of the ring or its lithium salt with di-, tri- or tetrachlorosilane destroyed the ring moiety.[14] However, mono-, di- and trisubstitution of the lithium derivatives of the six-membered ring employing fluorosilanes and -boranes is possible without any problems, even at high temperatures.[31–37] Generally, it was found that in absence of strong steric or electronic restraints the dilithium salt of HMCTS also reacts with fluoroboranes and -silanes to give substituted six-membered rings (Scheme 13).

In a one-pot reaction, di- and trisubstituted compounds are found as well. The reason for this is that the hydrogen atom in the NH group of the substituted ring is more acidic compared with that in the initial cyclotrisilazane salt (**A′**). The first product can be lithiated by the lithium salt of HMCTS (**A′**) and reacts with other fluorosilanes or -boranes in the reaction mixture, e.g., coupling of the cyclotrisilazane with borazines occurs in the reaction of lithiated HMCTS and F_2B–NR (R = Me, Et) in molar ratios 1:1, 1:2 and 1:3. The crystal structure of a trisubstituted compound is $[(Me_2Si)N]_3[B_3F_2(NEt)_3]_3$, **10** (Scheme 14, Fig. 7).[37]

The six-membered (Si–N) ring crystallizes in a boat conformation.[37] The nitrogen atoms exhibit a planar environment, $\Sigma° N = 359.6°$. The Si–N bonds have a length of approximately 174 pm.

Lithium salts of fluorosilyl-substituted rings are thermally stable and can be isolated without any problems, as shown in Scheme 15.[37]

The lithium salt shown in Fig. 8 crystallizes from THF as a dimer in a boat conformation with a planar $(Li–N)_2$ ring. The partially anionic Li–N bond causes increasing electron density at the nitrogen atom N(1) and therefore a shortening of the N–Si bonds in its neighborhood.

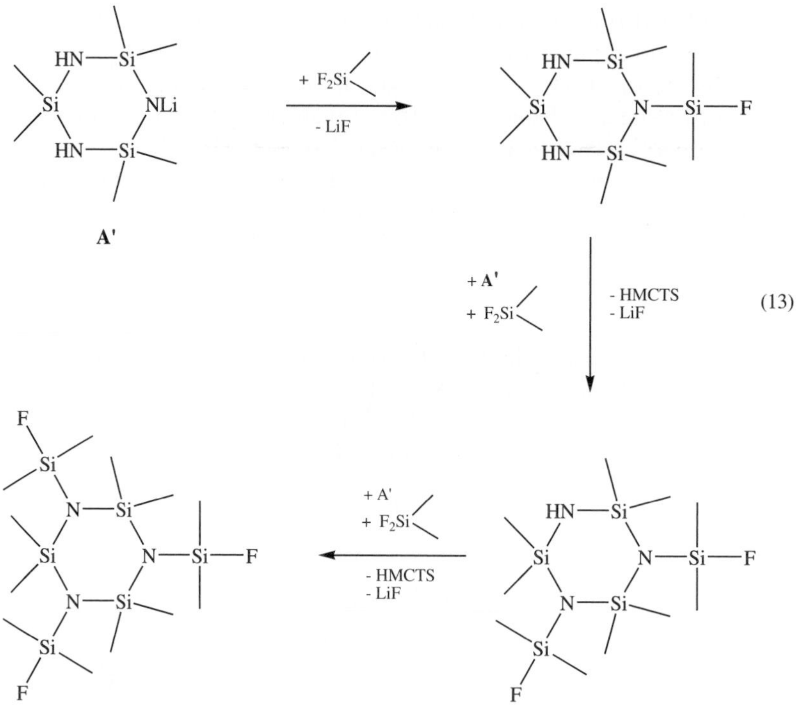

(13)

SCHEME 13.

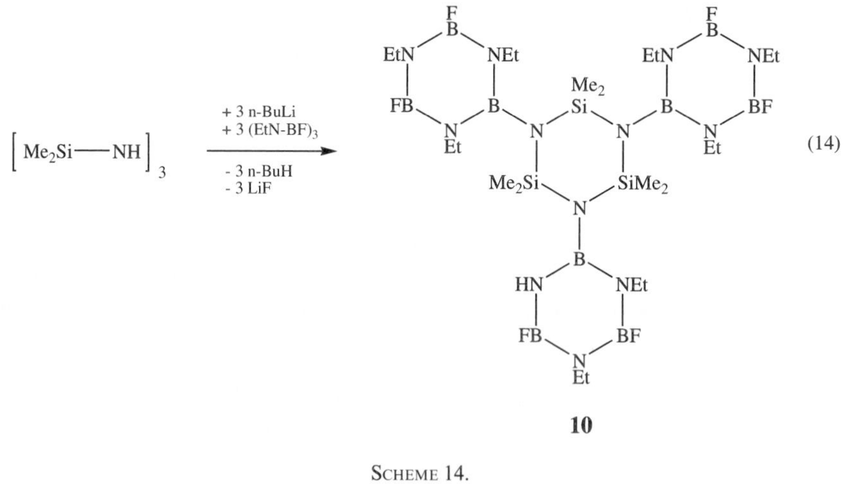

(14)

SCHEME 14.

Starting with this salt, it is also possible to substitute the six-membered ring with different silyl or boryl groups, as shown in Scheme 16. The structure of one such product, **12**, is shown in Fig. 9.

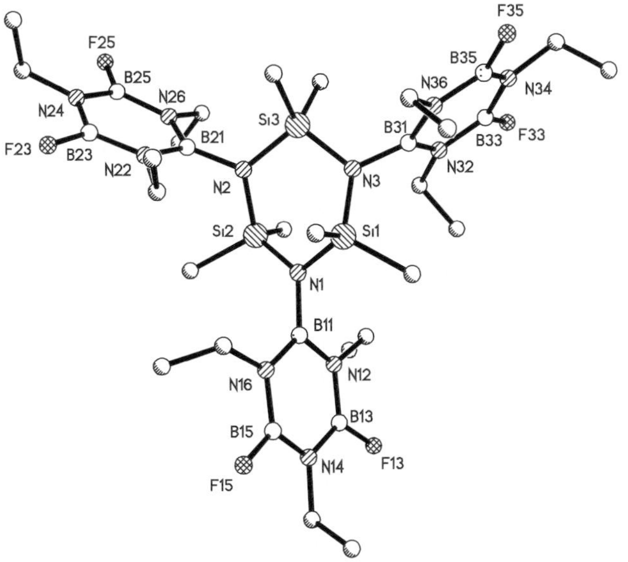

FIG. 7. Crystal structure of $[(Me_2Si)N]_3[B_3F_2(Net)_3]_3$, (**10**).

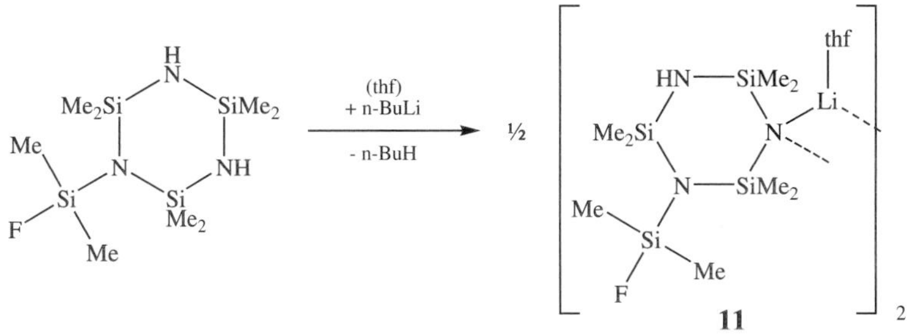

(15)

SCHEME 15.

Ring couplings occur without any difficulties as well, e.g.,[38,39]

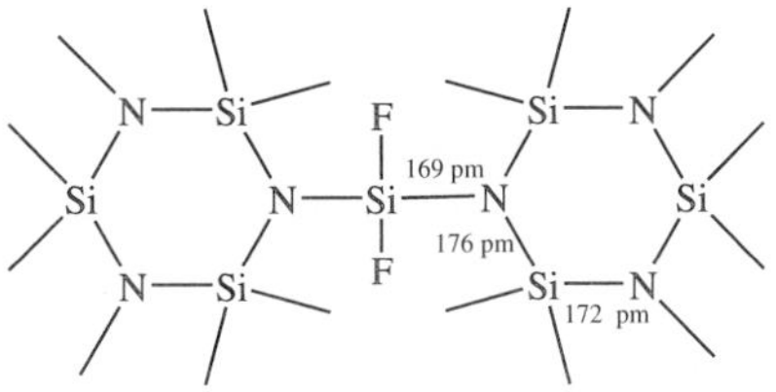

Like all unsymmetrically substituted rings, the SiF_2-coupled molecule is far from being planar.[38] The ring torsion angles do not correspond to any ideal conformation of six-membered rings. The electron-withdrawing fluorine atoms have a considerable effect on the geometry of the silicon–nitrogen skeleton, causing shortening of the exocyclic Si–N bonds and lengthening of the nearest ring bonds compared to

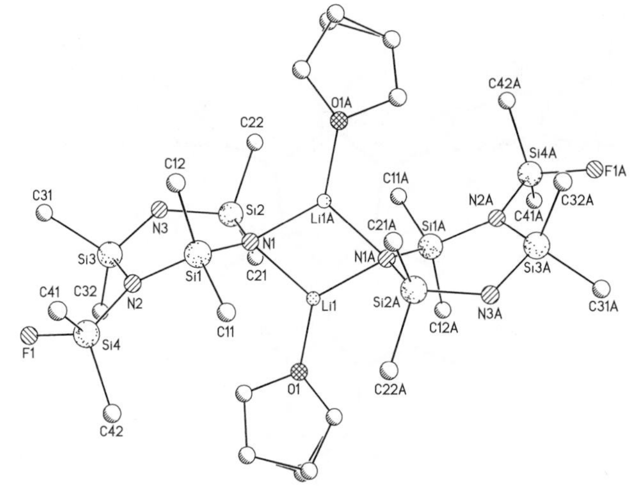

11

FIG. 8. Crystal structure of **11**.

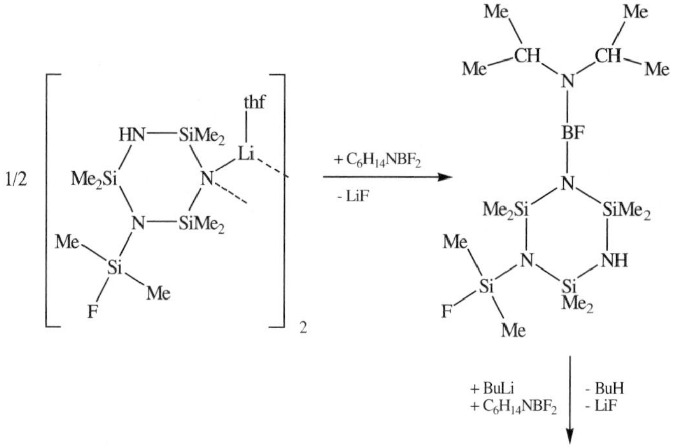

(16)

SCHEME 16.

the other endocyclic bonds. The weakening of these Si–N bonds in the rings is important for interconversion reactions as will be shown in the next section.

D. *Interconversion Reactions*

1. Equilibrium between Silyl-Substituted Cyclotri- and Isomeric Cyclodisilazane-Anions

Depending on the reaction conditions and the first substituent, the situation is much more complicated in the case of lithium salts of monosilyl-substituted cyclotrisilazanes.

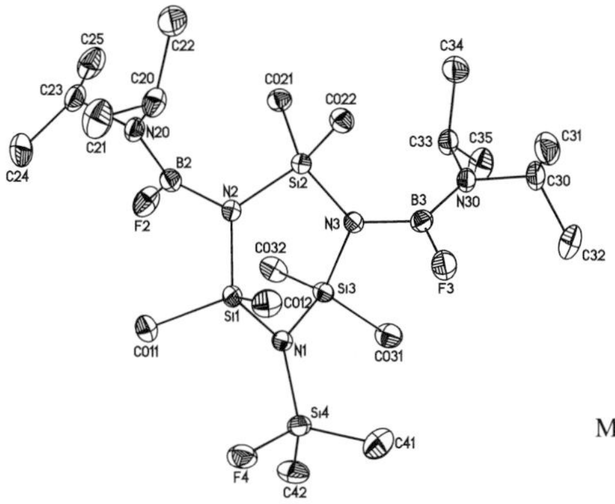

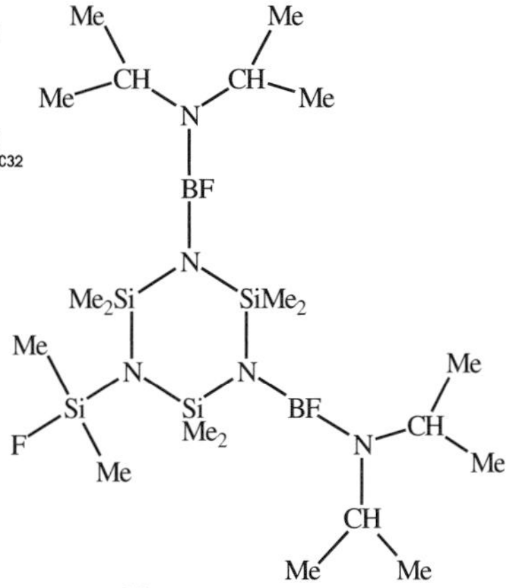

Si(1) – N(2)	173.5 pm
Si(4) – N(2)	170.5 pm
N(1) – B(1)	144.3 pm
N(10) – B(1)	138.5 pm

12

FIG. 9. Crystal structure of **12**.

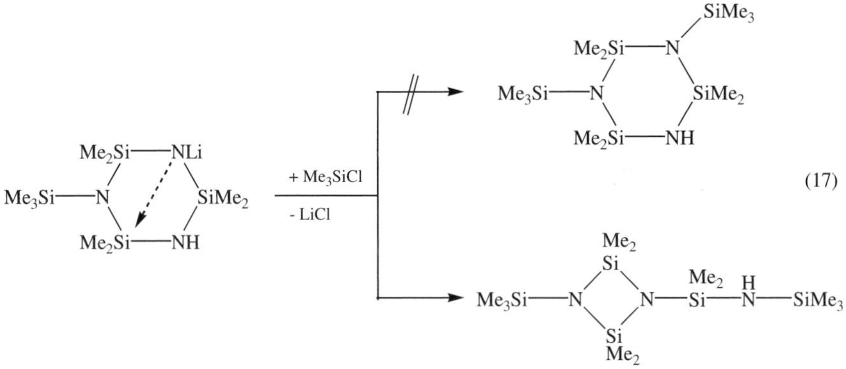

(17)

SCHEME 17.

In 1969, Fink found that a compound that had been described as a bissilyl-substituted six-membered ring was in fact an isomeric cyclodisilazane (Scheme 17).[40,41]

Later, he found equilibrium between the six- and the four-membered-ring anion, which depends very strongly on the temperature at which the reaction is conducted.[41] Higher temperatures lead to a better Si→N contact across the ring and therefore isomerization is possible. Substitution is preferred at low temperatures. We have studied the anionic rearrangement in reactions with fluorosilanes and boranes and found three more factors that influence the equilibrium between the six- and four-membered rings illustrated in Scheme 18.[42,43]

(1) Kinetic effects of the substituents:[32,42–44] Bulky groups increase the tendency of ring contraction. The comparison between the lithium salt of the

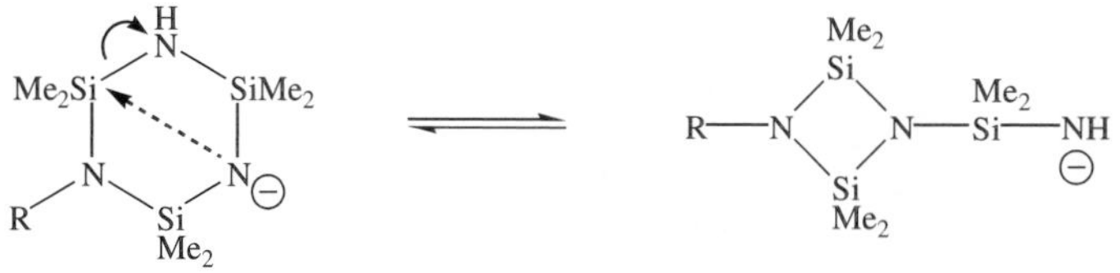

SCHEME 18.

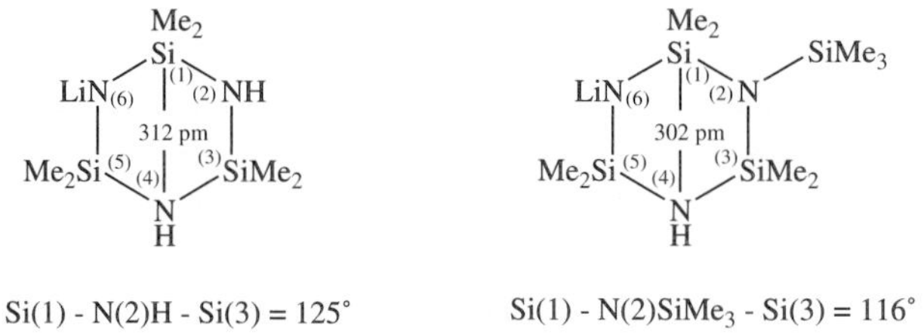

$Si(1) - N(2)H - Si(3) = 125°$ $Si(1) - N(2)SiMe_3 - Si(3) = 116°$

SCHEME 19.

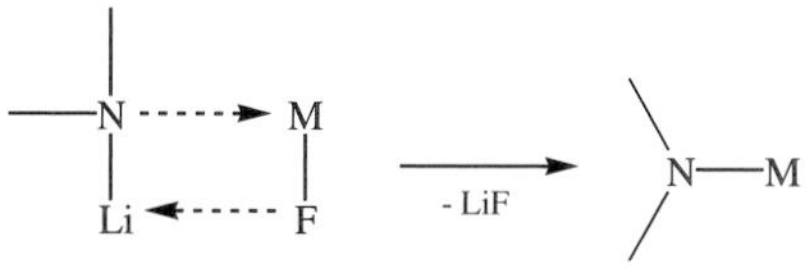

SCHEME 20.

unsubstituted and Me_3Si-substituted cyclotrisilazane (Scheme 19), which both crystallize from THF as dimers in boat conformation, shows that the bulky Me_3Si group decreases the Si(1)–N(2)–Si(3) angle and therefore the distance of the Si(1) and N(4) atoms above the ring.[32,44]

(2) Thermodynamic effects of the substituents:[45,46] Substituents that increase the basic character of the nitrogen raise the tendency to ring contraction and vice versa.

This explains why di- and trisubstitution of the six-membered ring with fluorosilanes and -boranes is possible. These groups decrease the basic character of the ring and its tendency to ring contraction.

(3) Properties of the attacking group:[45,46] If the attacking ligand is a Lewis acid, the basic character of the nitrogen is decreased, as shown by the four center mechanism in Scheme 20.

The ring size is retained. For example, the isomeric four-membered ring is formed easily in reactions of the lithiated cyclotrisilazane with aminofluorosilanes, but the ring size is retained with the Lewis acid aminofluoroborane. Starting with the isomeric four-membered ring, ring substitution is observed with aminofluorosilane and ring expansion with aminofluoroborane.[45,46]

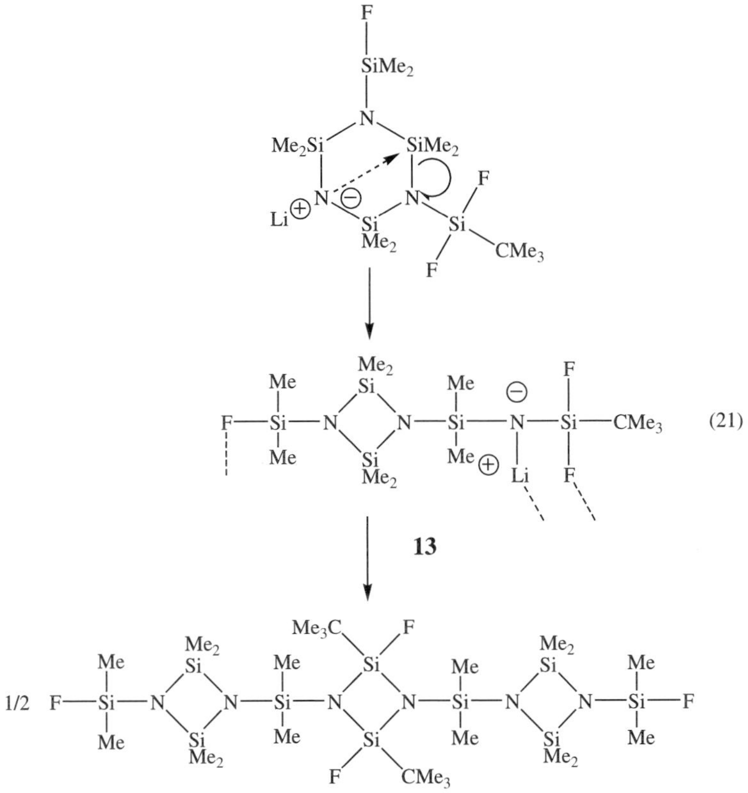

(21)

13

SCHEME 21.

2. Position of the Si–N-Bond Cleavage, Formation of the Silyl-Bridged Cyclodisilazanes

Experimental studies allow a prediction of the bond that will be cleaved and the isomer of the four-membered ring that will be obtained.

As discussed above, the Si–N bond in six-membered-ring compounds is longer and weaker in the neighborhood of an SiF_2 group (because of the electron-withdrawing effect) than the one in the neighborhood of an SiF or $SiMe_3$ group. The weakest bond is broken as shown in Scheme 21.

The lithium salt of the Me_3CSiF_2- and Me_2SiF-substituted six-membered ring crystallizes as an isomeric four-membered ring.[34,47] In the solid state the lithium compound forms a polymeric lattice *via* Li···F contacts (Fig. 10).

The Li–F distances (185 pm) are about 7 pm shorter than the Li–N contacts (192 pm). The lithium ion and the SiF_2 group shorten the N(1)–Si(1)-bond length to 162.9 pm. LiF can be eliminated thermally, and silyl-bridged four-membered rings are isolated. Such compounds have recently been used as precursors for Si_3N_4 ceramics.[1,16]

Another approach to silyl-coupled cyclodisilazanes starts with a fluorosilyl-substituted cyclotrisilazane and half an equivalent of BuLi. In this case, only half of

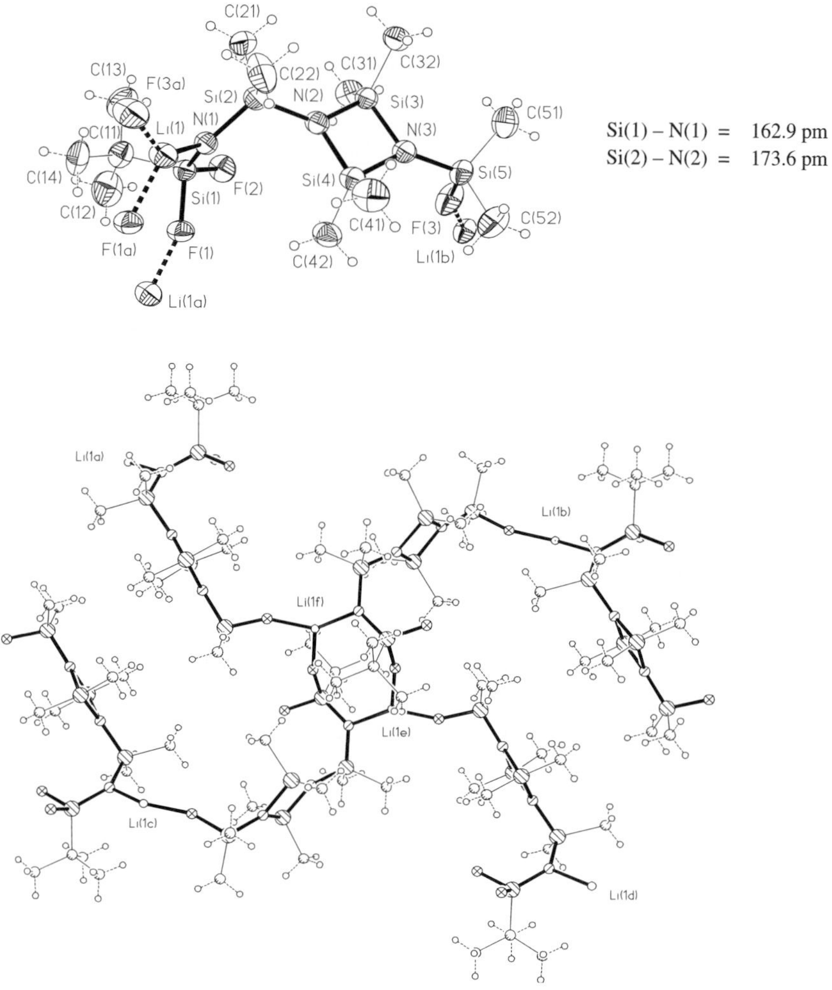

FIG. 10. Crystal structure of **13**.

the six-membered-ring molecules rearrange to the four-membered-isomeric anion and react with the remaining Me_2SiF-substituted cyclotrisilazane. This ring contracts with the other equivalent of BuLi. LiF is eliminated thermally and the central four-membered ring is formed.[41,47,48] (Scheme 22, Fig. 11)

a. Disubstitution of HMCTS with Ring Contraction

As shown in Schemes 13, 14, the synthesis of coupled cyclotrisilazanes and borazines is possible in the ratios 1:1, 1:2 and 1:3.[33,37]

Using bulkier groups and higher reaction temperatures, the lithium salt of 1-borazinyl-cyclotrisilazane contracts to give the four-membered cyclodisilazane **15**

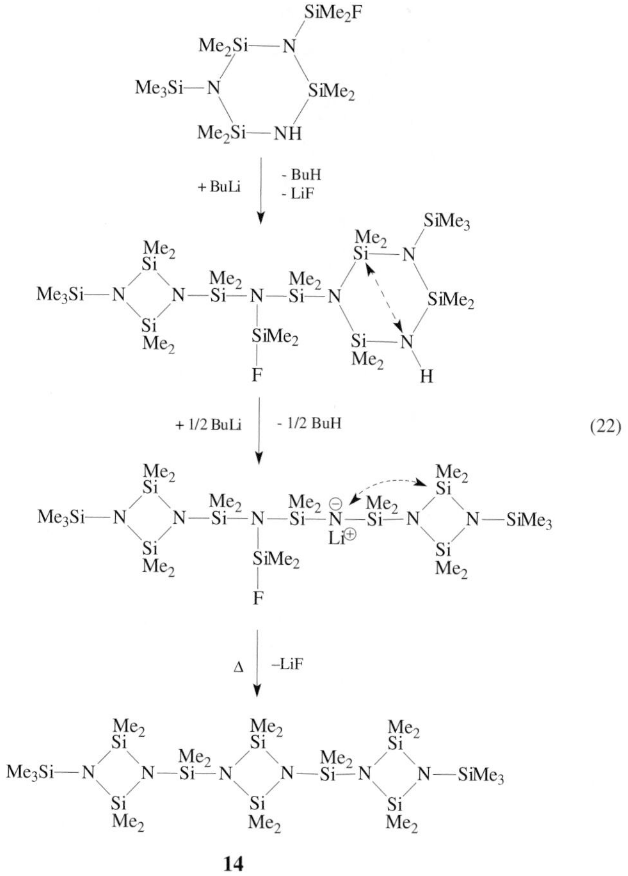

(22)

14

SCHEME 22.

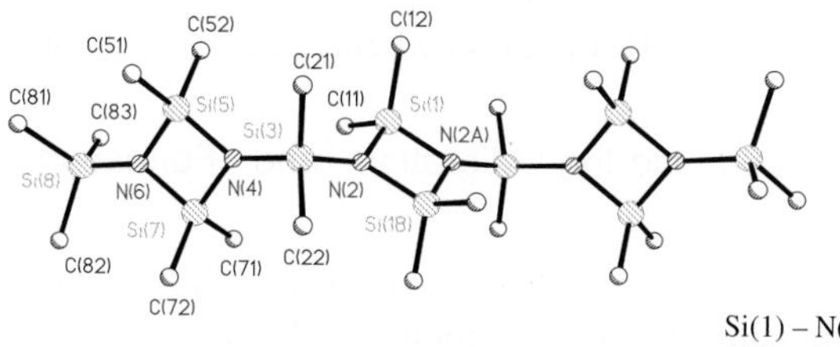

Si(1) – N(2) = 174.8 pm
Si(3) – N(2) = 172.7 pm
Si(8) – N(6) = 171.1 pm

FIG. 11. Crystal structure of **14**.

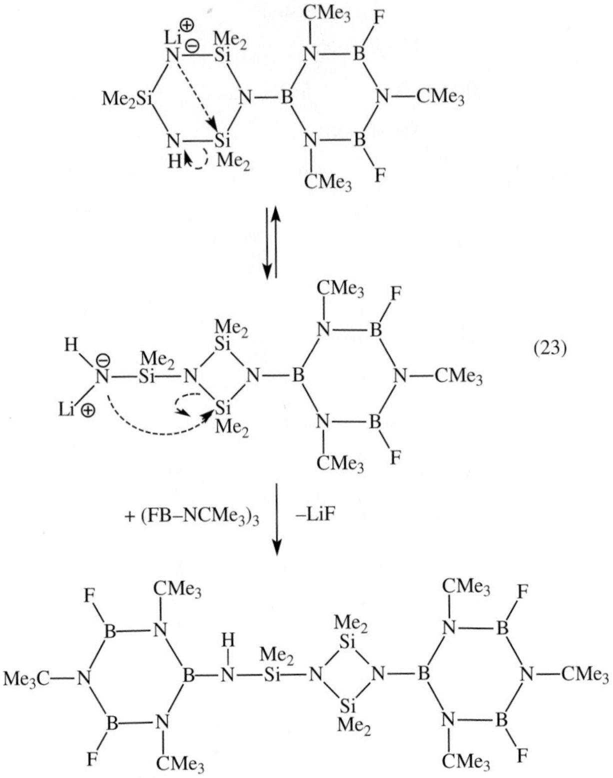

(23)

$+ (FB\text{–}NCMe_3)_3$ $-LiF$

15

SCHEME 23.

(Scheme 23, Fig. 12). In a second step the borazine is substituted at the resulting nitrogen ion, which is not part of a ring.[31,36,49]

The cyclodisilazane is not precisely planar. The sum of the angles around the nitrogen atom N(1) is 352.1°. Both of the borazine rings have a boat conformation.[49]

b. Ring Contraction, Methanide-Ion Migration and Formation of Bicyclic Compounds

The reaction of SiF-coupled six-membered rings with lithium organyls first produces ring contraction and subsequently eliminates LiF. As discussed in Section II, we have now obtained an intermediate ylide that contains silicon atoms with the coordination numbers three and five. This gives rise to a nucleophilic migration of a methanide ion. A fused bicyclic compound is formed (Scheme 24). This compound has also been characterized by an X-ray analysis and the structure is shown in Fig. 13. Both the four-membered rings are planar.[42,46,50]

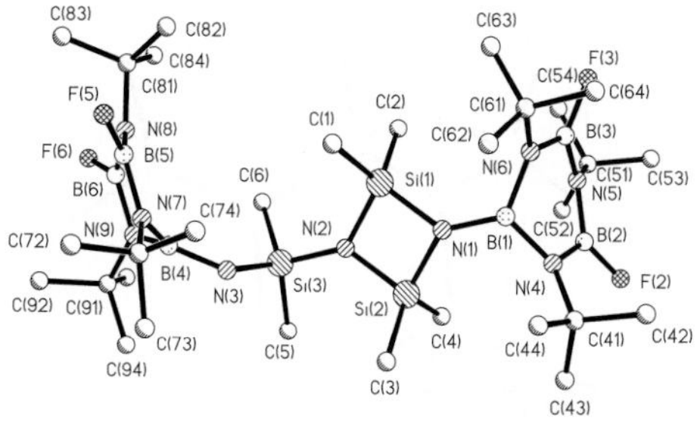

N(1) – B(1)	=	141.7 pm
N(6) – B(1)	=	148.4 pm
Si(1) – N(1)	=	177.0 pm
Si(3) – N(3)	=	174.9 pm
N(3) – B(4)	=	140.5 pm
Si(3) – N(2)	=	171.9 pm
Σ° N(1)	=	352°

FIG. 12. Crystal structure of **15**.

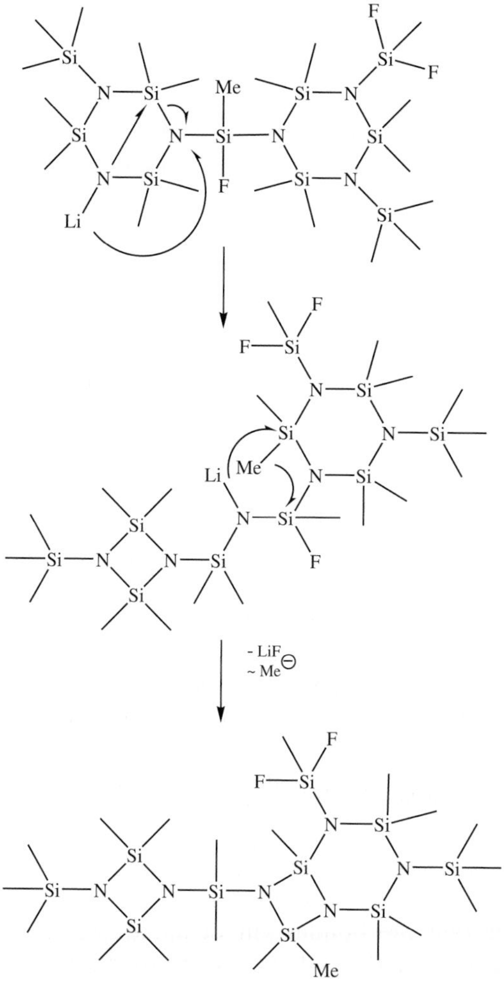

(24)

- LiF
~ Me⊖

16

SCHEME 24.

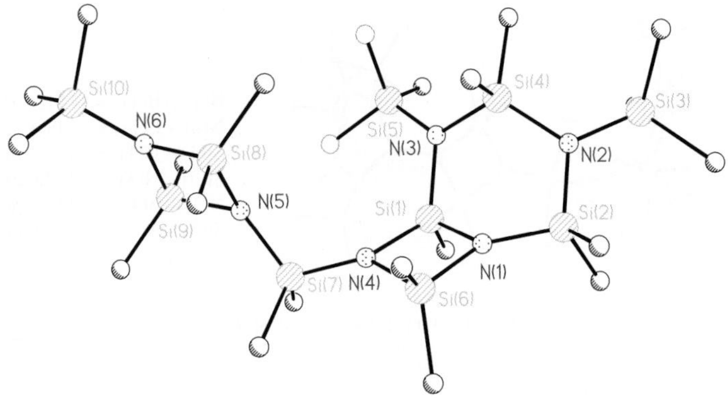

FIG. 13. Crystal structure of **16**.[50]

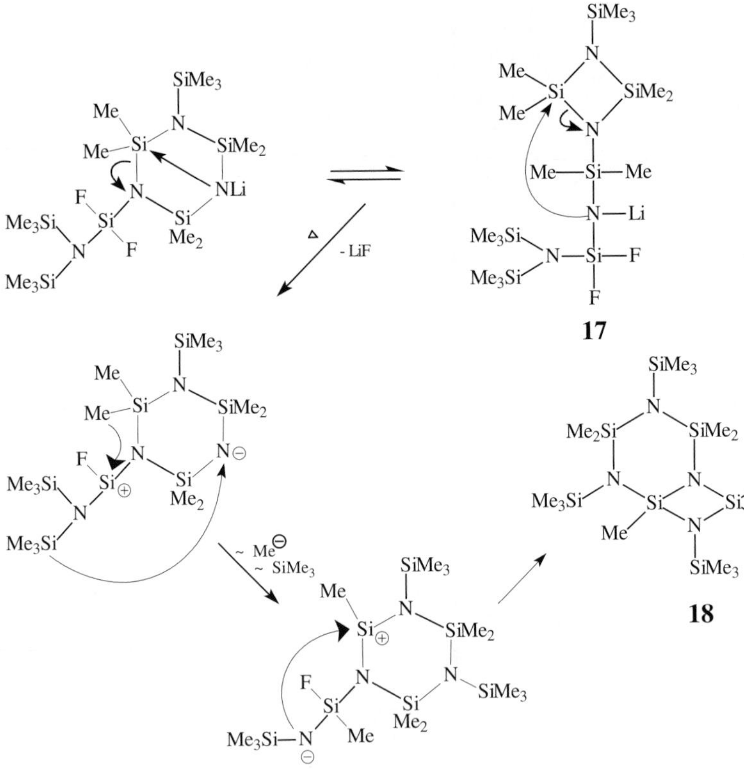

SCHEME 25.

(25)

It is understandable that the lithium salt of an Me_3Si- and $(Me_3Si)_2N$–SiF_2-substituted six-membered ring **17** crystallizes as an isomeric four-membered ring (Scheme 25, Fig. 14).

Elimination of LiF leads to an intermediate ylide containing silicon atoms with the coordination numbers three and four as well as nitrogen atoms with the

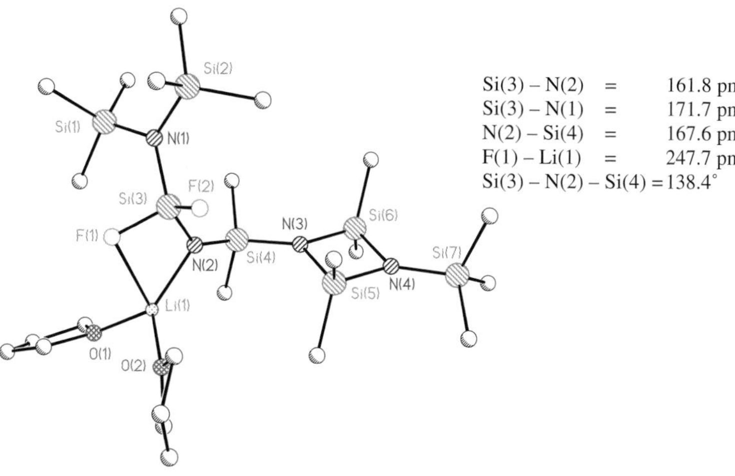

FIG. 14. Crystal structure of **17**.

coordination numbers two and three. Under these conditions both a nucleophilic migration of a methanide ion and an electrophilic migration of an $SiMe_3$ group can occur, as discussed in Section II. Accordingly, a bicyclic compound **18** is formed and characterized by X-ray.[51,52]

The Si(3)–N(2) bond of the four-membered cyclodisilazane isomer has the length of an Si–N double bond.

IV

2,2,4,4,6,6,8,8-OCTAMETHYLCYCLOTETRASILAZANE, OMCTS

A. *Conformers and Isomeric Compounds of OMCTS, DFT Calculations*

We have carried out DFT (B3LYP/6–31G(d)) calculations (the basis set comprises 312 cGTOs) in order to establish the energetic order of the different possible isomers of $(Me_2Si–NH)_4$, OMCTS (Fig. 15). At the local minima on the potential energy surfaces, the Hessian matrices were computed. Harmonic vibrational frequencies were used to calculate the zero-point vibration-corrected energetics. (Results are collected in Table I and Fig. 16.)

The most stable isomer is the twist conformation **A** where two opposite NH units are both above or below the approximate ring plane defined by the four silicon atoms (the four Si atoms are not exactly located in one plane). This species has no symmetry elements and consequently belongs to the point group C_1. The chair conformation **A1** is energetically located at 2.1 kcal/mol above the twist form. At first glance, the highly symmetric boat conformation **A2**, which is 6.2 kcal/mol higher in energy than the most stable isomer **A**, exhibits C_{4v} symmetry. In fact, all Si–N-bond distances are equal in length (175.7 pm), and the four N–Si–N angles measure 115.7°, while the Si–N–Si angles are calculated to be 134.9°. However, a closer examination reveals that the bond lengths show differences in the order of

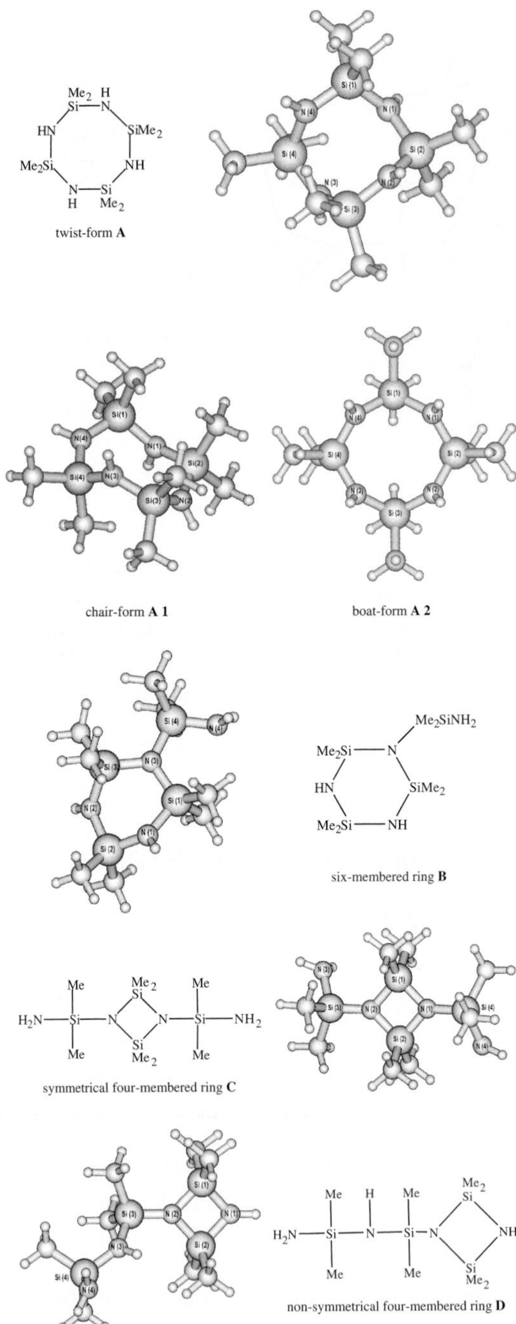

FIG. 15. Calculated structures of the isomers of OMCTS [B3LYP/6–31G + (d)].

TABLE I

RELATIVE ENERGIES OF THE DIFFERENT ISOMERS AND CONFORMERS OF OMCTS (B3LYP/6–31G(d) RESULTS)

	E (kcal/mol)	E_0 (kcal/mol)
A	0.0	0.0
A1	1.7	2.1
A2	5.1	6.2
B	5.9	6.3
C	12.0	11.9
D	16.7	16.6

Note: E denotes the electronic energy while E_0 includes zero-point vibrational effects.

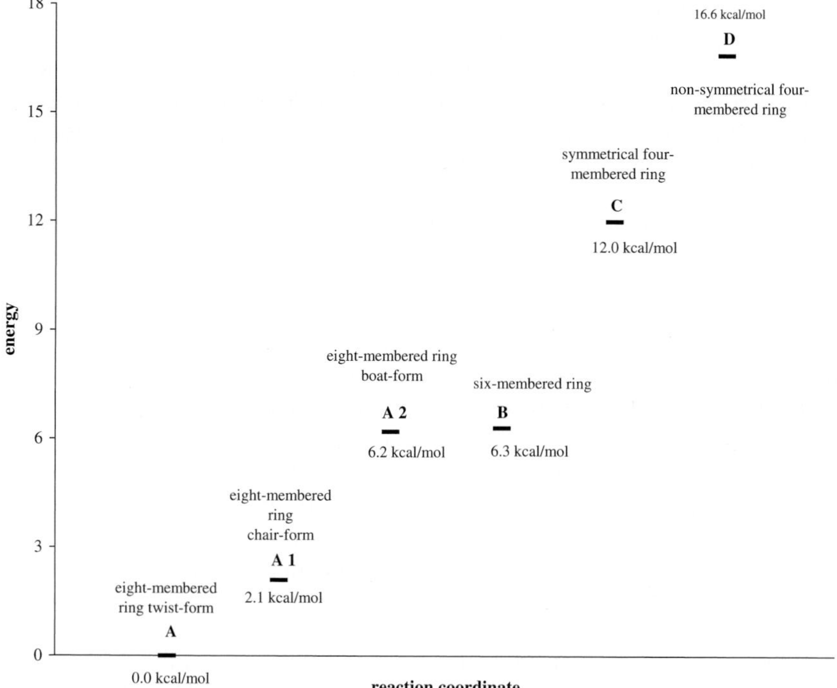

FIG. 16. Energy diagram of the neutral isomers **A**–**D** of OMCTS. The energies refer to the most stable isomer **A** and include zero-point energy contributions.

0.05 pm and the torsional angles differ strongly. With only one two-fold symmetry axis and two symmetry planes, isomer **A2** exhibits C_{2v} symmetry.

In single crystals of OMCTS,[53] two conformers – the chair-(**A1**) and the boat-form (**A2**) – are found (see Fig. 15). The calculated most stable twist conformation **A** appears only in the unsymmetrical $[(Me_3C)_2Si–NH–SiMe_2NH]_2$.

In addition to the three conformers **A**, **A1** and **A2** of the eight-membered-ring compounds, OMCTS can exist in the following structures: a six-membered (**B**) and a symmetrical (**C**) and a non-symmetrical (**D**) four-membered ring with exocyclic $SiMe_2NH_2$ groups.

B is only slightly higher in energy than the symmetric eight-membered ring **A2**. When zero-point energy effects are included, the latter two isomers are almost isoenergetic. The tertiary nitrogen atom in **B** is surrounded by two coplanar silicons. The six-membered ring is not planar; one of the NH nitrogen atoms is above the approximate ring plane, while the other points below it.

The planar four-membered ring **C** with two exocyclic Si–N units is much less stable. The two terminal NH_2 groups are found at the same side of the ring, but in *trans* position. Thus, the point group of this species is C_2.

Isomer **D** with the exocyclic $SiMe_2$–NH–$SiMe_2$–NH_2 chain exhibits an almost planar four-membered ring (sum of angles 179°). The species is by far the least stable isomer of OMCTS discussed in this work.

The energy diagram in Fig. 16 explains that in the reaction of Me_2SiCl_2 and NH_3 apart from the six-membered ring, HMCTS, only the eight-membered ring, OMCTS, is formed. In a single crystal structure determination of OMCTS, the chair- and the boat-form (**A1** and **A2**) are found in a ratio of 1:1.[14]

B. *Substitution Reactions, Synthesis of Mono- and Disubstituted Eight-Membered Rings and Bicyclic Compounds*

Until 1988, 40 years after its first synthesis, no reaction of OMCTS with retention of the ring size had been published. Investigations of the six-membered $(Me_2Si\text{-}NH)_3$ frequently showed rearrangements in reactions of anionic cyclosilazanes. Therefore, it seemed natural to investigate if ring size is retained in the alkaline derivatives of the eight-membered ring. It was found that octamethylcyclotetrasilazane reacts with *n*-butyllithium[54] or alkaline elements[55] to give alkali salts, which crystallize as dimeric THF adducts. An example is shown in Scheme 26. In the dimers, two eight-membered rings are connected by a planar alkali nitrogen four-membered ring. Lithium is tricoordinated, sodium tetracoordinated and potassium penta- and hexacoordinated.[55] The coordinatively bonded THF in the lithium compound can be exchanged with the Lewis-base TMEDA.[55]

Both lithium salts of HMCTS and mono- and dilithium salts of OMCTS were used to substitute the ring with fluoroboryl,[36,49,56,57] -silyl[58–61] and -phosphoryl[62] groups to couple cyclosilazanes *via* BF, SiF and PF units and to bridge the eight-membered ring with SiFR groups (Scheme 27).[37]

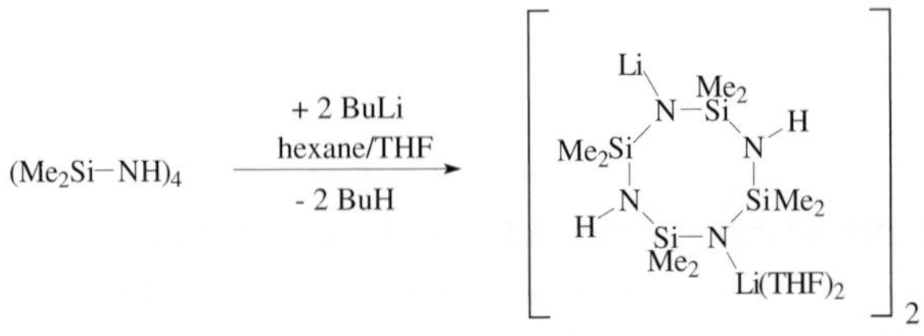

$$(Me_2Si{-}NH)_4 \xrightarrow[\text{- 2 BuH}]{\text{+ 2 BuLi, hexane/THF}} \left[\text{Li–N–SiMe}_2\text{–NH–SiMe}_2\text{–N(Li(THF)}_2\text{)–SiMe}_2\text{–NH–SiMe}_2 \right]_2 \quad (26)$$

SCHEME 26.

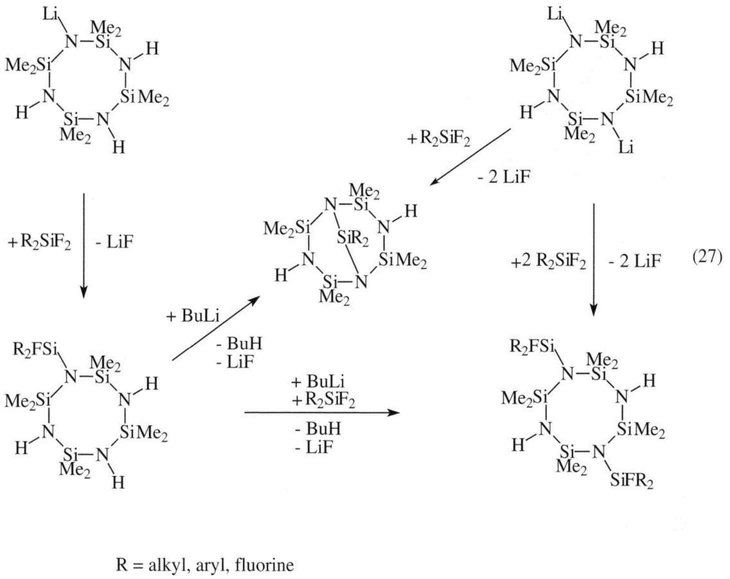

+R_2SiF_2 / - LiF

+ BuLi / - BuH / - LiF

+R_2SiF_2 / - 2 LiF

+2 R_2SiF_2 / - 2 LiF (27)

+ BuLi / +R_2SiF_2 / - BuH / - LiF

R = alkyl, aryl, fluorine

Scheme 27.

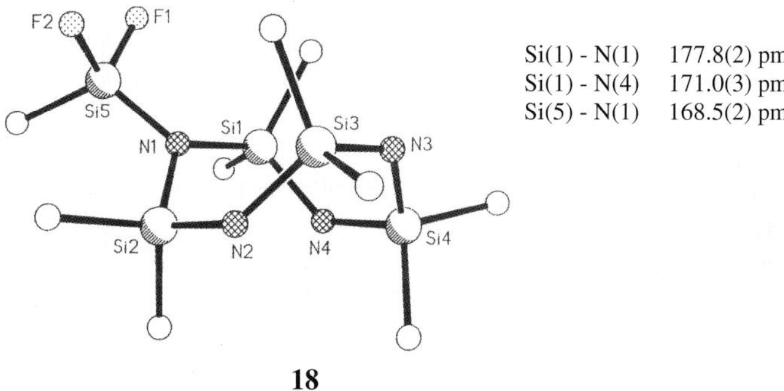

18

Fig. 17. Crystal structure of $MeSiF_2N(Me_2SiNHSiMe_2)_2NH$, **18**.

The X-ray structure of the FSiPh-bridged compound indicates that the nitrogen atoms of the Si_3N units have a pyramidal environment.[54,61]

Monosubstituted eight-membered rings crystallize in a boat conformation (see Fig. 17). The molecules form chains *via* H–F bridges in the crystals.[61]

Disubstituted rings crystallize in a chair conformation (see Fig. 18). The molecules contain three-dimensional H–F bridges. The rings are perpendicular to each other.[61]

C. *Interconversion Reactions*

1. Structural Isomers of the Mono-Anion of OMCTS, DFT Calculations

The most likely isomeric mono-anions of OMCTS are shown in Scheme 28.

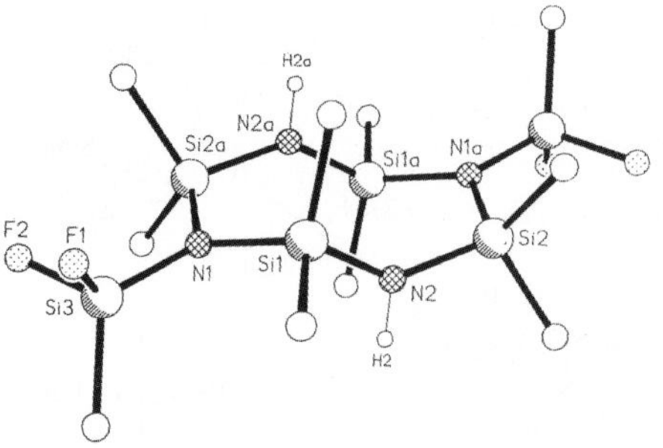

Si(1) - N(1) 176.4(3) pm
Si(1) - N(2) 171.9(3) pm
Si(3) - N(1) 169.5(2) pm

19

FIG. 18. Crystal structure of $(MeSiF_2\text{-}N(SiMe_2)_2NH)_2$, **19**.

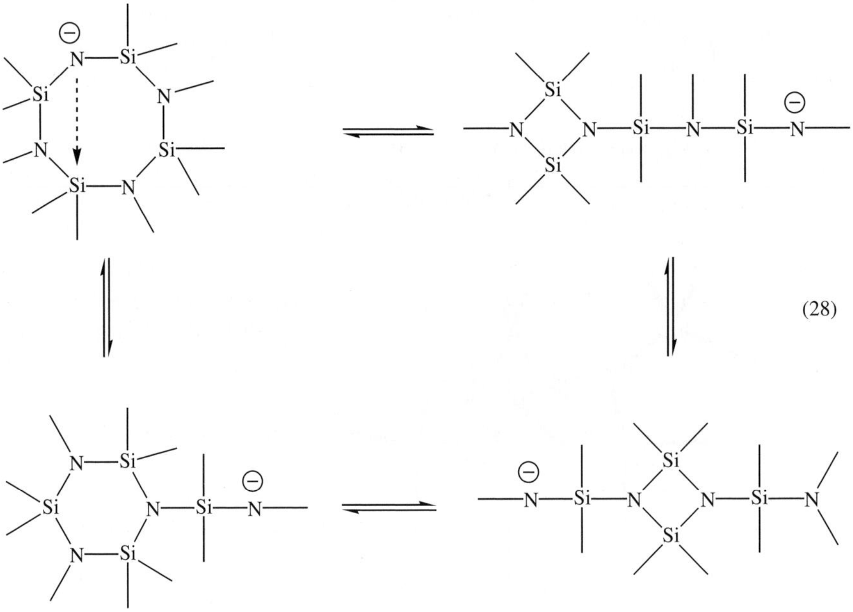

(28)

SCHEME 28.

We have carried out calculations to establish the structures and energies of possible mono-anions of OMCTS. To account for the anionic character of the species, additional diffuse functions were included in the basis set. The B3LYP/6–31G + (d) calculations (374 cGTOs) are reported in Table II and in Figs. 19 and 20. The zero-point corrected energies are also given in this table. In addition, single-point calculations with the large 6–311 + G(2d,p) basis set (626 cGTOs) were carried out (E_{big} in Table II). The results from the latter calculations give evidence that the large basis set is responsible for a decrease in all energies with respect to isomer **A′** (up to 2 kcal/mol), but that, on the other hand, the energetic order of the different isomers

TABLE II

RELATIVE ENERGIES OF THE DIFFERENT ISOMERS AND CONFORMERS OF THE ANION OF OMCTS (B3LYP/6–31 + G(d) RESULTS)

	E (kcal/mol)	E_0 (kcal/mol)	E_{big} (kcal/mol)
A′	0.0	0.0	0.0
B′	6.03	6.13	4.24
B1′	12.55	12.24	11.79
C′	15.73	15.19	15.47
D′	13.16	12.34	12.50
D1′	24.83	24.42	23.38
D2′	32.87	31.53	30.85

Note: E denotes the electronic energy while E_0 includes zero-point vibrational effects. E_{big} is obtained from calculations with a large basis set and without zero-point correction.

is preserved. We are thus confident that the theoretical results presented in this work are in accord with DFT data at the basis set limit.

The most stable isomer is the eight-membered ring **A′**, as was found for the neutral system. There is only one possible isomer because the four NH groups in **A** are equivalent. The Si(4)–N(4)–Si(1) angle in **A′** is calculated to be 153.7° and exists only in the twist conformation.

The six-membered ring **B** can be lithiated at two different positions. Endocyclic lithiation yields structure **B′**, which is 6.1 kcal/mol above the eight-membered anion **A′**, comparable to the neutral system. The Si–N–Si angle at the anionic center measures 130.4°; it is more than 23° smaller than in **A′** because of the smaller ring size.

Lithiation at the exocyclic NH_2 group yields a completely different structure **B1′**. Here, the HN unit attacks a silicon atom of the ring, forming a 6 + 4 bicyclus. The four-membered ring shows only 5° deviation from planarity, while the six-membered ring exhibits a pronounced twist structure. The tertiary nitrogen atom is not exactly planar. The transannular Si···Si distance in the four-membered ring is calculated to be 269.9 pm. The bicyclic structure **B1′** is energetically located at 12.2 kcal/mol above species **A′**.

Lithiation of the symmetrical OMCTS isomer **C** yields only one isomer of the anion. Here, again a bicyclic structure is formed that is relatively close in energy to **B1′**.

The non-symmetrical four-membered ring species of OMCTS, **D**, can be lithiated in three different positions. Removal of the proton at the endocyclic NH unit results in a structure **D1′** that is 24.4 kcal/mol above **A′**. The transannular Si···Si distance is calculated to be 241.2 pm and the Si–N–Si angle at the anionic site is 90.4°.

Lithiation of the secondary nitrogen atom of the exocyclic chain yields a relatively stable structure **D′** (12.3 kcal/mol above **A′**). The four-membered ring is almost planar and the Si–N–Si angle is calculated to be 161.0°. The transannular Si···Si distance is larger than in **D1′**, 254.1 pm. This species can easily arrange to the less stable compound **C′** by nucleophilic attack of N(3) on Si(2).

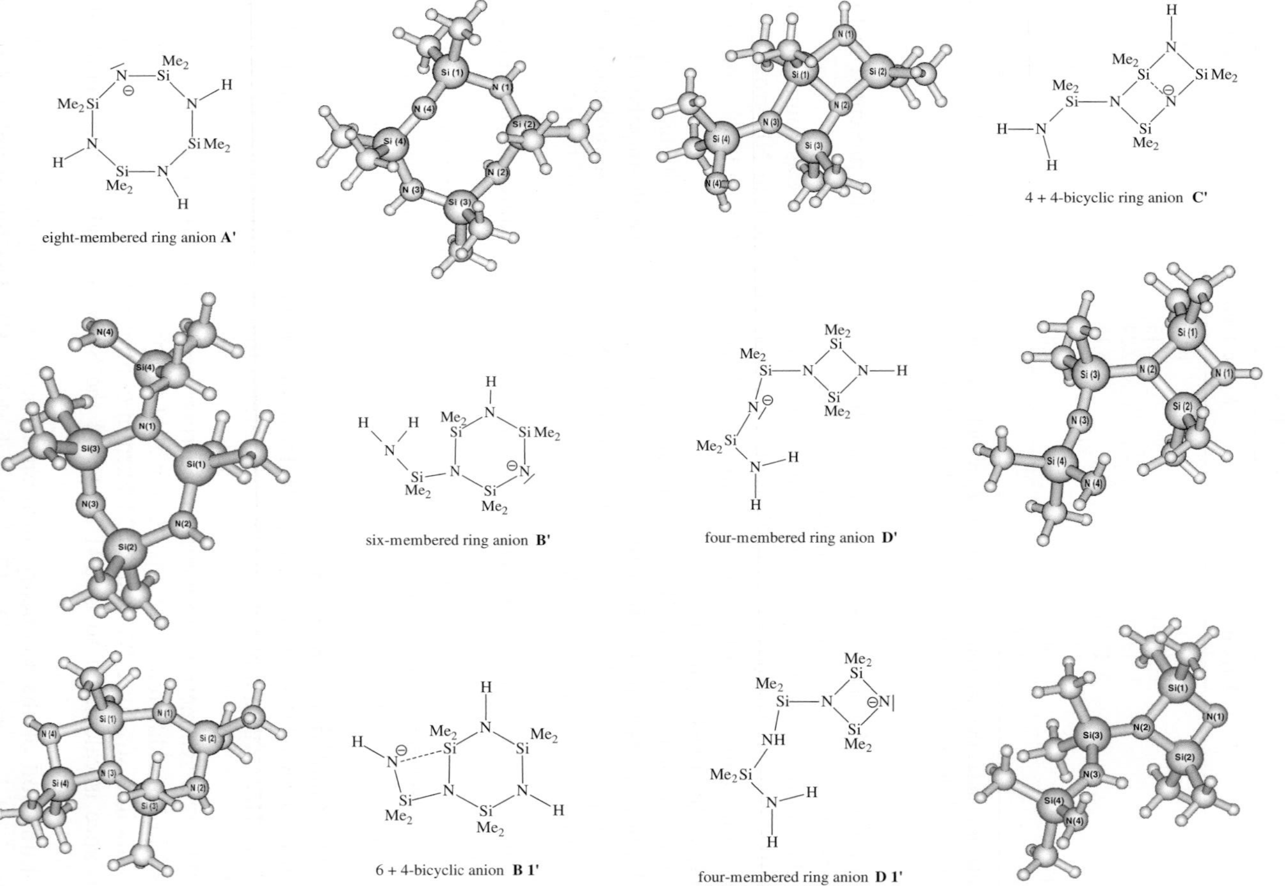

FIG. 19. Calculated structures of the mono-anionic isomers of OMCTS [B3LYP/6–31 + G(d)].

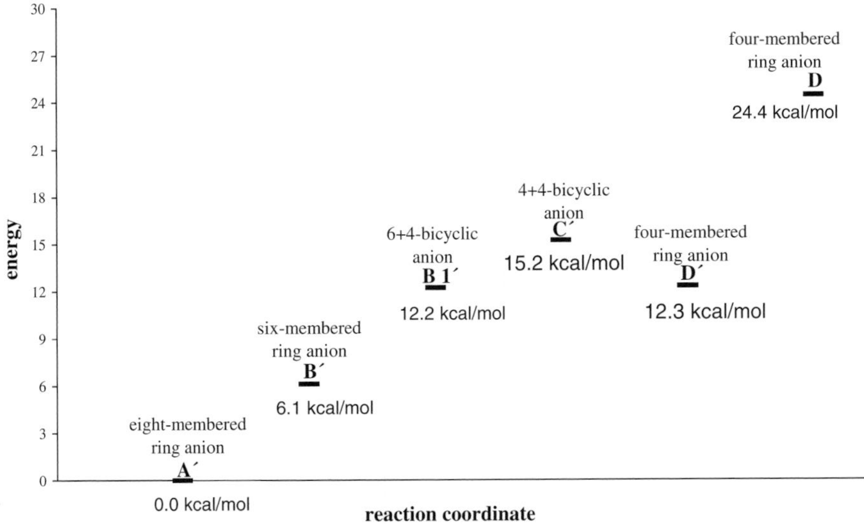

FIG. 20. Energy diagram of the anionic isomers **A′**–**D 1′** of OMCTS. The energies refer to the most stable isomer **A'** and include zero-point energy contributions.

Lithiation of the terminal NH_2 group yields a very unstable species **D2′** (32.9 kcal/mol above **A′**; not shown in Fig. 19), which, however, can stabilize itself by forming the 6 + 4 bicycle **B1′**. Similar to **D′**, the Si···Si distance is 254.8 pm.

The energy diagram Fig. 20 explains that mono-substitution of OMCTS with fluoroboranes, -phophanes and -silanes is possible and that a symmetrically substituted four-membered ring is formed in stepwise substitution (**C′**).

2. Experimental Results

a. Ring Contractions of the Mono-Anion of OMCTS with Triorganyl Chlorosilanes and Borazines

In contrast to substitution reactions of lithium salts of OMCTS with fluoroboranes, -phosphanes and -silanes no mono- or disubstituted OMCTS with a triorganylsilyl group is known. According to chemical experiments, the anion of the OMCTS contracts in reactions with Me_3SiCl and forms a silyl-substituted six-membered-ring isomer 20,[31,56–58,60] (Scheme 29).

No substitution of **B 1′**–**D 1′** (Figs. 19 and 20) is known.

The mono-anion of the eight-membered ring contracts forming a six-membered ring according to Scheme 29. The lower reactivity of chlorotrimethylsilane compared to fluorosilanes requires a higher reaction temperature and leads to a ring contraction of the mono-anion of OMCTS. The six-membered isomer **B 1′** is formed and substituted at the exocyclic $–SiMe_2–NHLi$ unit with a trimethylsilyl group (**20**[59,66]). This six-membered ring can be substituted or contracted to form a symmetrically disubstituted four-membered ring[31,56,63–65] (Scheme 30).

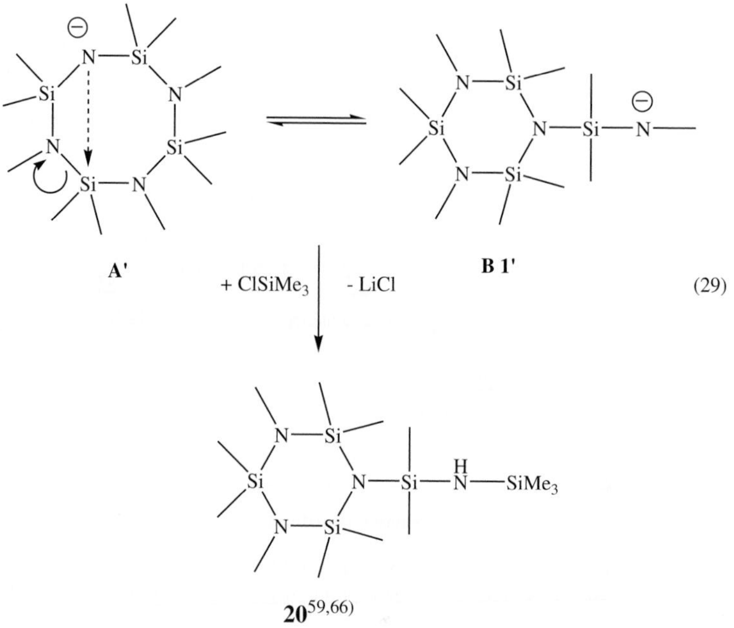

SCHEME 29.

These experimental results are in good agreement with DFT calculations, which show that the lithium salt **A′** is 6.1 kcal/mol more stable than the lithium salt **B′**.

Bulky fluoroborazines, like tri-*tert*-butyl-trifluoroborazine, show the same reaction course with the mono-lithium salt of OMCTS, shown in Scheme 31. Smaller fluoroborazines like triethyl-trifluoroborazine substitute the eight-membered ring Fig. 21.[37]

b. DFT Calculations, Structural Isomers of the Dianion of OMCTS

We have performed further calculations on dianionic structures of OMCTS isomers. It was found that the dianion of OMCTS has seven other structural isomers, another eight-membered ring (**A1″**), two six-membered rings (**B″**, **B1″**), one symmetrically substituted (**C″**) and three unsymmetrically substituted four-membered rings (**D″**–**D2″**) (Fig. 22). The energy diagram of the dianionic isomers of OMCTS (**A″**–**D2″**) is shown in Fig. 23 (see also Table III).

There are two possible isomers of the eight-membered ring **A** in which the two negative charges are either opposite to each other across the ring (**A′**) or in alpha position to each other (**A1″**). Clearly, the C_2 symmetric structure **A′** is more stable by 9.6 kcal/mol compared to isomer **A1″**. The Si–N–Si angles are calculated to be 158.7°. In the energetically higher species, one of the Si–N–Si angles is slightly smaller (156.7°), while the other is larger (162.7°).

The six-membered OMCTS **B** can be dilithiated at two endocyclic positions, yielding an isomer **B1″**, 20.5 kcal/mol above the most stable species, and at one

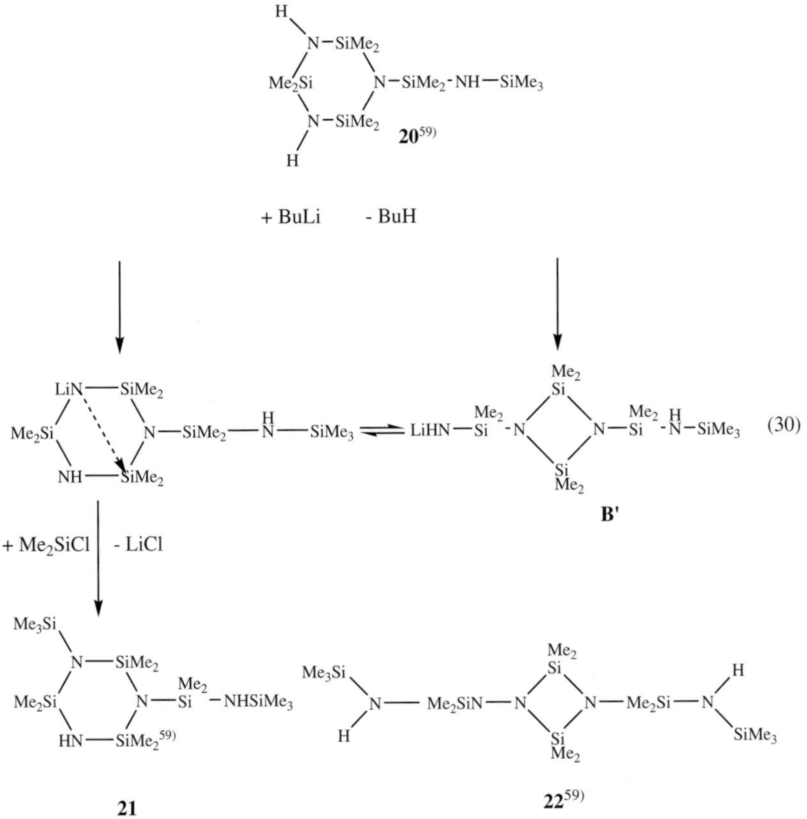

H
N–SiMe$_2$
Me$_2$Si N–SiMe$_2$-NH–SiMe$_3$
N–SiMe$_2$
H
20[59)]

+ BuLi - BuH

LiN–SiMe$_2$
Me$_2$Si N–SiMe$_2$–$\overset{H}{N}$–SiMe$_3$ ⇌ LiHN–Si(Me$_2$)-N(Si(Me$_2$))$_2$N–Si(Me$_2$)-$\overset{H}{N}$–SiMe$_3$ (30)
NH–SiMe$_2$

B'

+ Me$_2$SiCl - LiCl

Me$_3$Si
N–SiMe$_2$
Me$_2$Si N–Si(Me$_2$)–NHSiMe$_3$
HN–SiMe$_2$[59)]

21

Me$_3$Si(H)N–Me$_2$SiN–N(Si(Me$_2$))$_2$N–Me$_2$Si–N(H)SiMe$_3$

22[59)]

SCHEME 30.

endo- and one exocyclic position, resulting in a structure **B**″ that is 4.9 kcal/mol more stable. While in **B**″ one charge is localized at the less favorable primary amino group, the larger distance between the two charges makes this structure more stable. Formation of a 6+4 bicyclus would result in a much less stable conformation owing to the repulsion of the two negative centers.

OMCTS isomer **C** can only yield one dianion **C**″ (lithiation of the two exocyclic amino groups). The structure is 26.4 kcal/mol higher than the most stable species **A**″.

The non-symmetrical four-membered ring **D** can result in three different dianions. Lithiation of the secondary exocyclic and the endocyclic structure yields a dianion **D**″ that is comparable in energy (17.7 kcal/mol) with the two six-membered-ring dianions **B**″ and **B1**″.

Lithiation of the endo- and the primary exocylic NH group ends up in a structure **D1**″ at 24.0 kcal/mol.

Lithiation of both exocyclic HN units yields a very unstable species. A 4+4 bicyclus **D2**″ is formed at 33.5 kcal/mol above **A**″. The Si–N bond that connects the

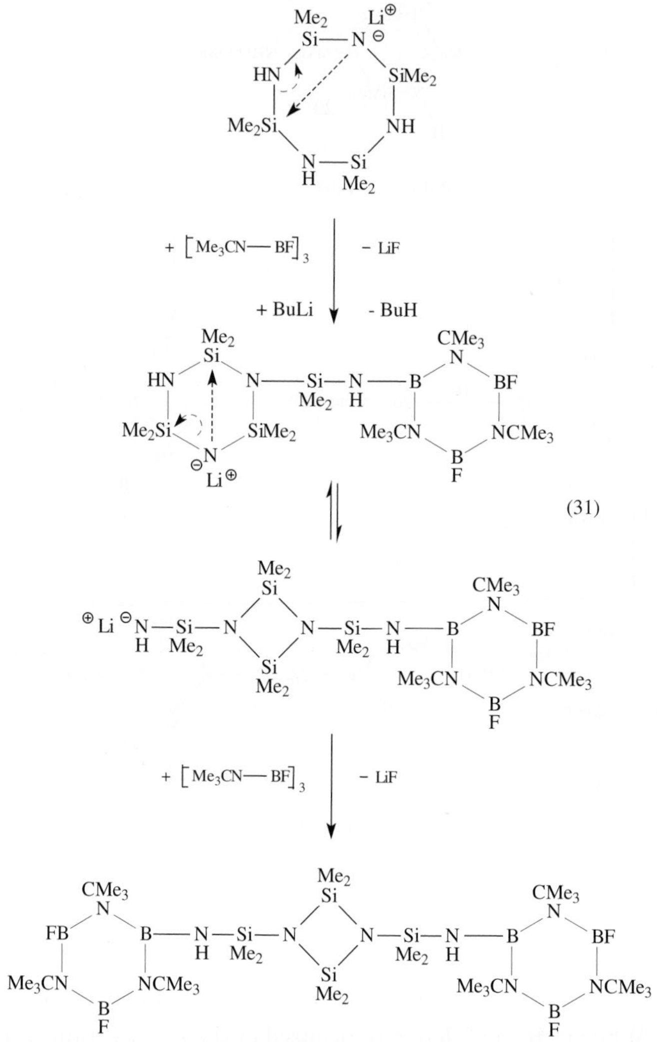

$+ [Me_3CN\text{—}BF]_3$ $- LiF$

$+ BuLi$ $- BuH$

(31)

$+ [Me_3CN\text{—}BF]_3$ $- LiF$

23[37)]

SCHEME 31.

two rings measures 185.9 pm and the two transannular Si···Si distances are calculated to be 272.3 pm and 263.4 pm.

The energy diagram Fig. 23 shows that disubstitution of the eight-membered ring is possible and that ring contraction – in contrast to the reaction course of the mono-anion – should lead to the formation of a non-symmetrically substituted four-membered ring (**D**″).

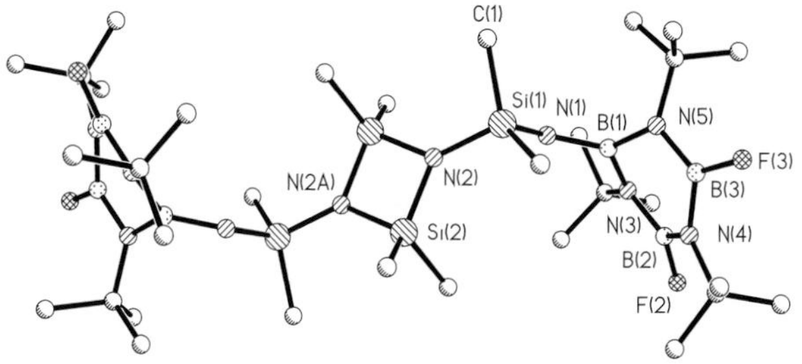

Si(1)–N(2)	171,8 pm
N(1)–B(1)	141,6 pm
B(1)–N(5)	147,0 pm
Si(1)–N(1)	175,0 pm
Si(2)–N(2)–Si(2A)	90,76°
B(3)–N(5)–B(1)	114,0°
N(2)–Si(2)–N(2A)	89,24°

FIG. 21. Crystal structure of **23**.[37]

3. Experimental Results

a. Ring Contraction and Formation of Unsymmetrically Substituted Four-Membered Rings

Depending on the reaction conditions and the attacking agents, the dilithiated eight-membered ring contracts (Scheme 28, Fig. 22, **D**″, III.4.1) forming an unsymmetrical dianionic four-membered ring (**D**″, Fig. 22) that reacts with fluorosilanes according to Scheme 32 to give compounds of type **24**[59,60].

So far, no tris(silyl)-substituted cyclotetrasilazane is known. Starting with the isomeric unsymmetrically substituted cyclodisilazane, the next substituent R′ is bonded exocyclically at the four-membered ring yielding **25** (Scheme 33).[31,56,63,64,66]

Difunctional halides react spontaneously with the dilithium derivatives, quantitatively forming a second four-membered ring **26**[31,66] from the unsymmetrical four-membered ring. A spirocyclic compound with a high molecular weight is obtained starting with the dilithium salt of the unsymmetrically substituted cyclodisilazane with SiF_4 (Scheme 34 Fig. 24).[66]

A further equilibrium, which can be controlled by the reaction temperature, exists between the trissilyl-substituted asymmetrical four-membered, six-membered and the symmetrical four-membered-ring anions (Scheme 35).[52]

Perhaps for sterical reasons, it has not been possible so far to prepare a tetrakis-silylsubstituted asymmetrical four-membered-ring isomer.

Higher reaction temperatures lead to the formation of the four-membered and lower reaction temperatures to the six-membered-ring isomer (Scheme 35).[52]

As the experiments show, the same principles that were found for the equilibrium of the six- and four-membered-ring anions can also be applied to the eight-, six- and four-membered systems.

As already shown in Scheme 18, bulky groups kinetically promote the tendency of ring contraction. A comparison of the crystal structures of $(Me_2Si–NH)_4$ and the bis(tert-butyldifluorosilyl)-substituted eight-membered ring shows the following significant aspects:

(1) The electron-withdrawing effect of the fluorine atoms leads to a shortening of the exocyclic Si–N bond and lengthening of the nearest Si–N-ring bonds.
(2) The bulky SiF_2CMe_3 group decreases the Si–N–Si-ring angle.
(3) Both effects together decrease the non-bonding N–Si distance across the ring and increase the ring contraction.

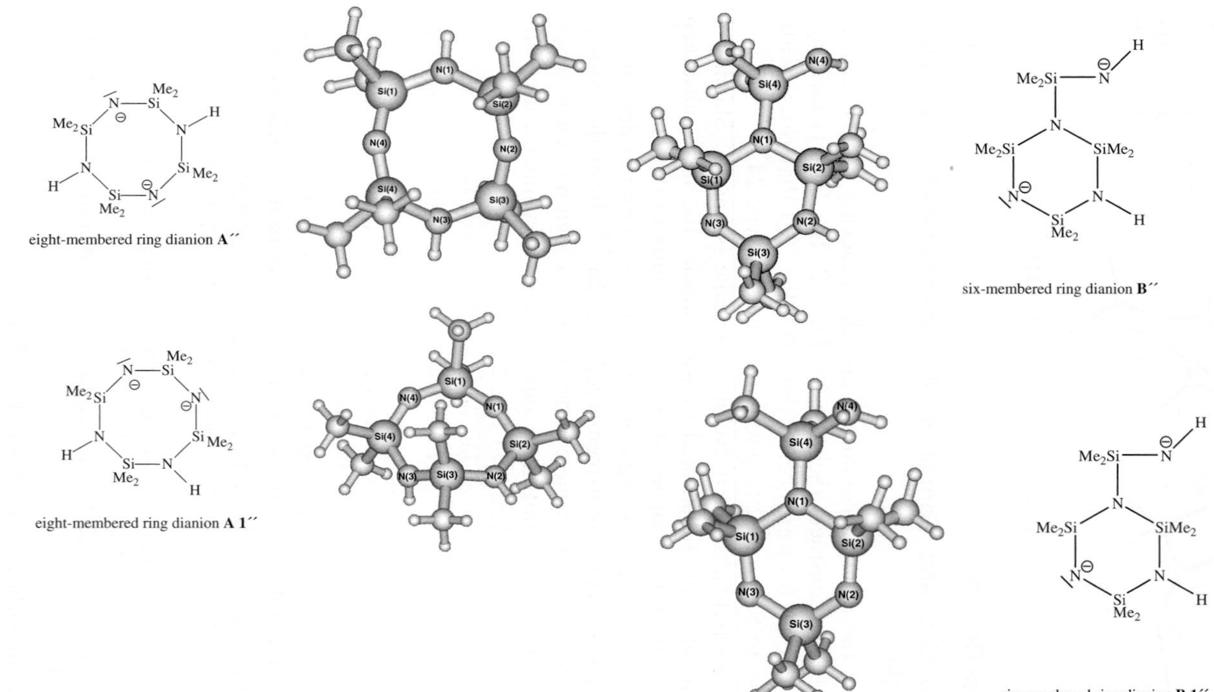

FIG. 22. Calculated structures of the dianionic isomers of OMCTS [B3LYP/6–31 + G(d)].

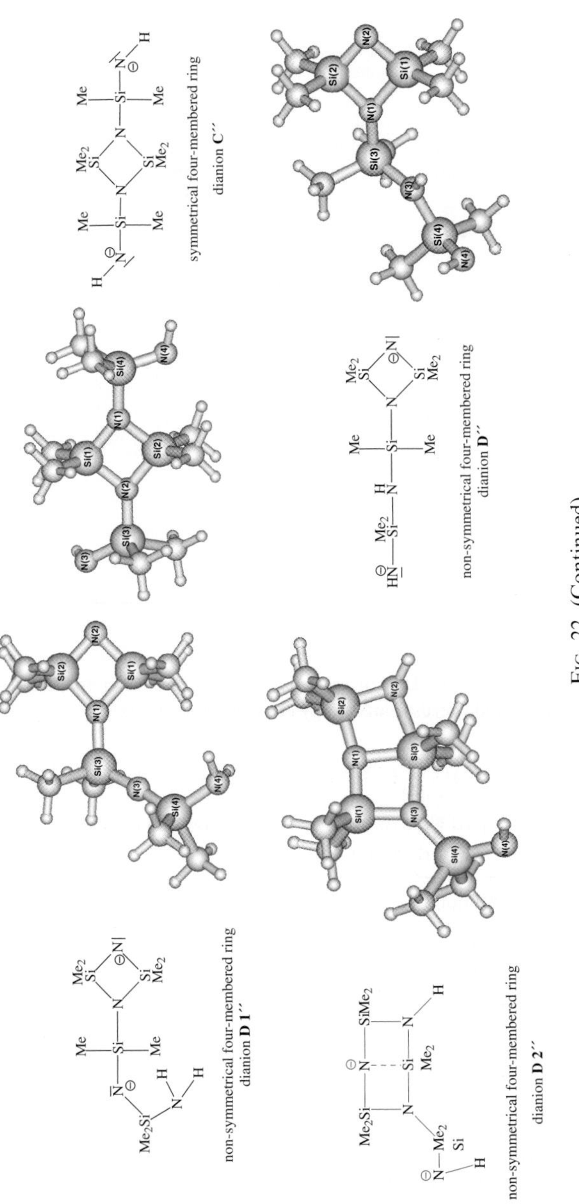
non-symmetrical four-membered ring
dianion **D 1′′**
non-symmetrical four-membered ring
dianion **D 2′′**
non-symmetrical four-membered ring
dianion **D′′**
symmetrical four-membered ring
dianion **C′′**

FIG. 22. (Continued)

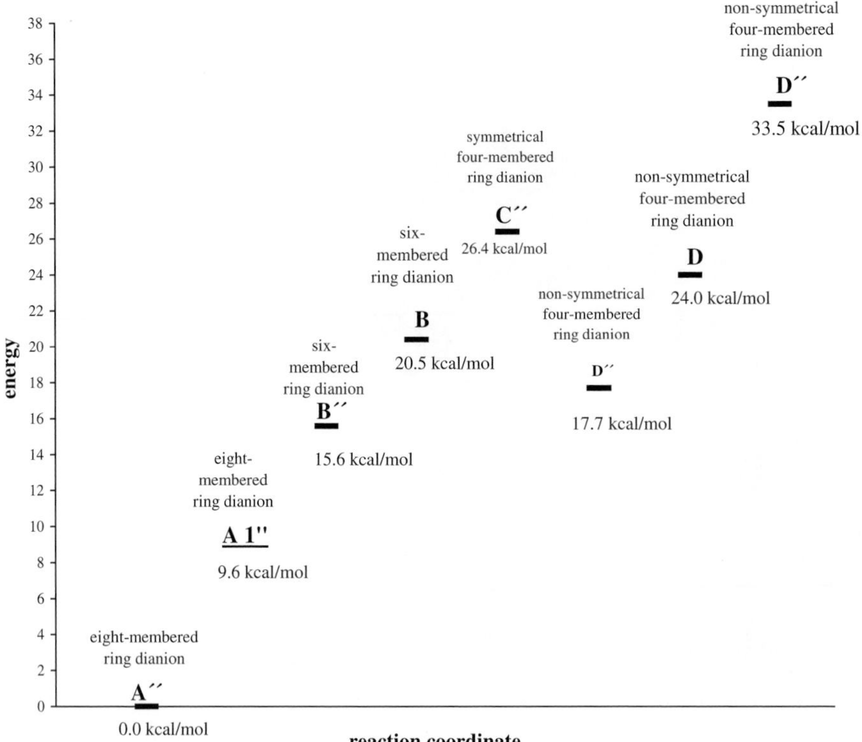

FIG. 23. Energy diagram of the dianionic isomers **A**″–**D2**″ of OMCTS. The energies refer to the most stable isomer **A**″ and include zero-point energy contributions.

TABLE III

RELATIVE ENERGIES OF THE DIFFERENT ISOMERS AND CONFORMERS OF THE DIANION OF OMCTS (B3LYP/6–31 + G(D) RESULTS)

	E (kcal/mol)	E_0 (kcal/mol)
A″	0.0	0.0
A1″	9.4	9.6
B″	15.2	15.6
B1″	18.9	20.5
C″	27.3	26.4
D″	17.3	17.7
D1″	23.7	24.0
D2	33.9	33.5

Note: E denotes the electronic energy while E_0 includes zero-point vibrational effects.

We discussed isomerization reactions of HMCTS in Section III and of OMCTS in this chapter. Again, there is the question which rearrangement will occur in the anion of the bis(difluorosilyl)-substituted eight-membered ring, e.g., in the bis(tert-butyldifluorosilyl)-substituted-ring system shown in Scheme 36. Assuming that

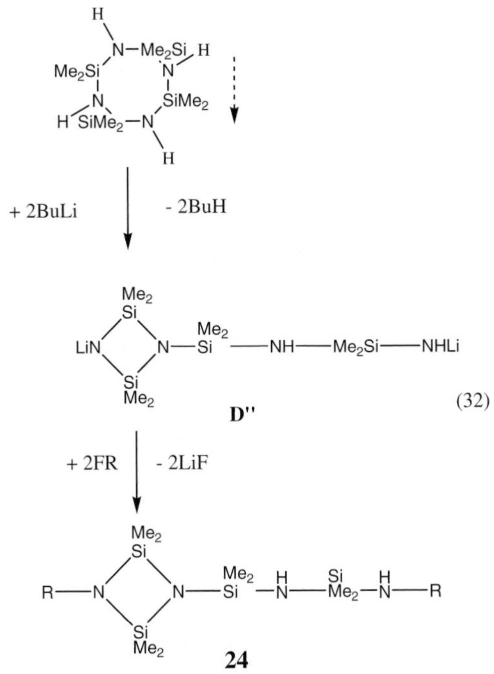

+ 2BuLi - 2BuH

D''

(32)

+ 2FR - 2LiF

24

R: $SiMe_3$, F_2SiCMe_3[59)]

SCHEME 32.

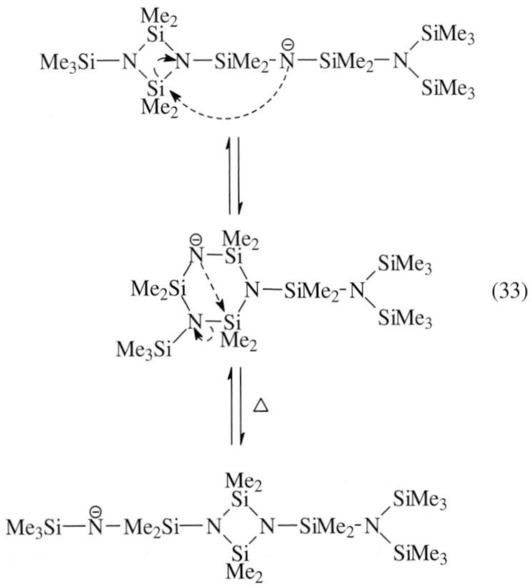

(33)

SCHEME 33.

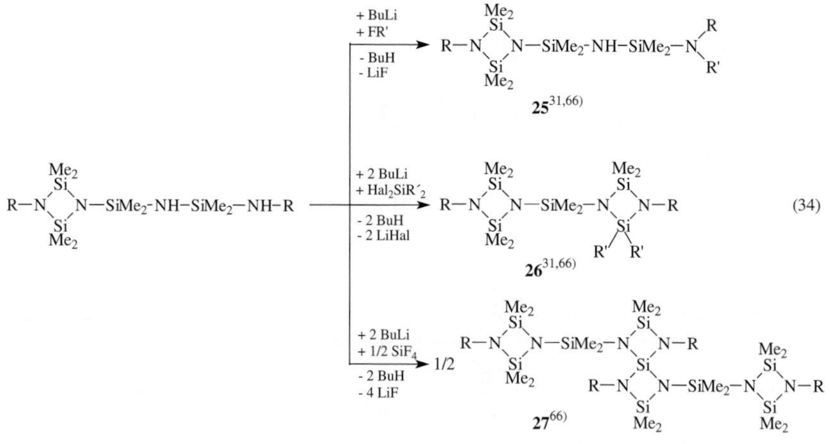

(34)

$R,R' = SiMe_3$[66], $FSiMe_2$, $HSiMe_2$[60]

SCHEME 34.

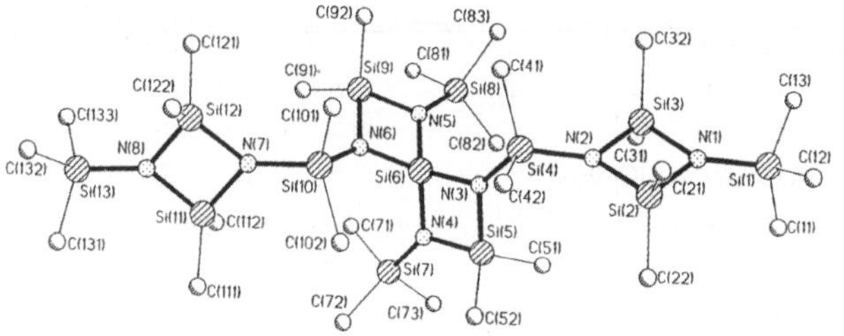

N(1) - Si(2) 173.5(7) pm Si(1) - N(1) 171.6(7) pm

FIG. 24. Crystal structure of **27**.

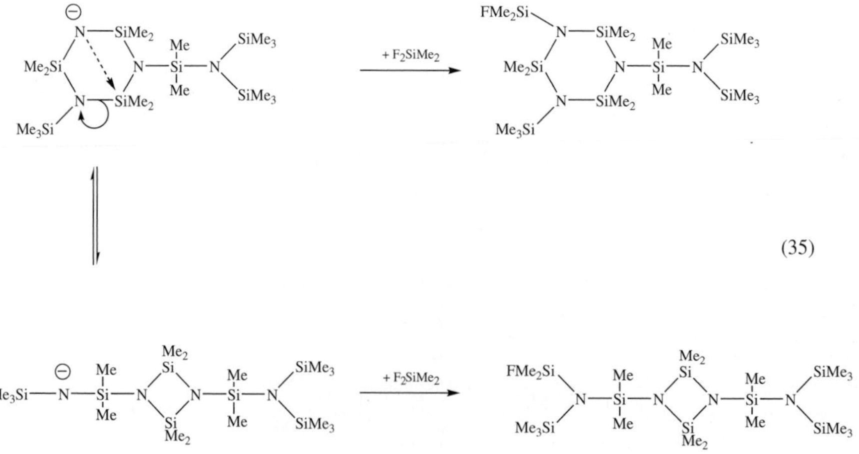

(35)

SCHEME 35.

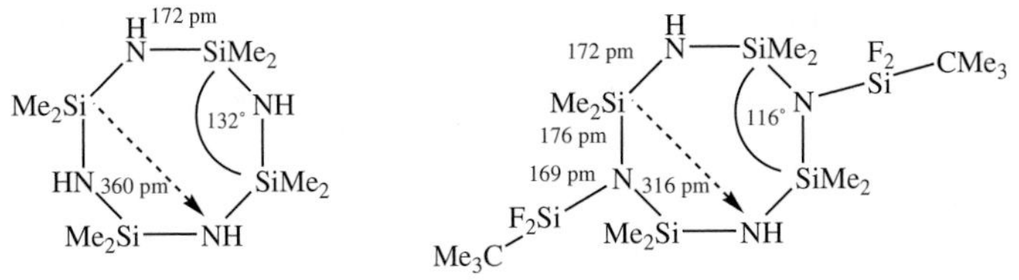

(36)

SCHEME 36.

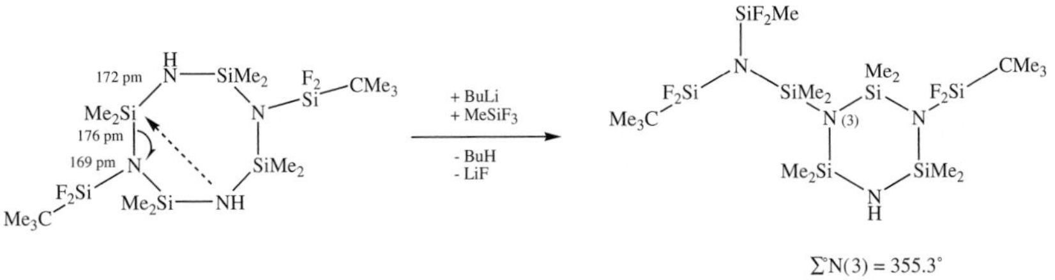

(37)

SCHEME 37.

eight-membered rings isomerize following the same rules as the six-membered Si–N rings, a six-membered-ring system will be formed in an interconversion reaction. The Si–N-ring bond in the neighborhood of the SiF_2 substituent is longer and therefore weaker than in the neighborhood of a hydrogen atom or a Me_3Si group. Indeed, as shown in Scheme 37, the weakest bond breaks and a six-membered ring is obtained.[63,65]

V

OUTLOOK

As was shown in this contribution, interconversion reactions in (Si–N)-ring chemistry are quite well understood. The knowledge of the rearrangements will surely lead to further interesting results in Si–N chemistry and help to understand polymerization and pyrolysis reactions. The reactivity of cyclosilazanes allows the preparation of large molecules, which already have properties of polymers. Starting from OMCTS, pure Si_2N_2NH, α-Si_3N_3 and β-Si_3N_4 were synthesized.[49,67] The insertion of boron atoms will possibly allow the synthesis of ternary SiNB ceramic powders in a pyrolytic reaction in an atmosphere of NH_3.

REFERENCES

(1) Jaschke, B.; Klingebiel, U.; Riedel, R.; Doslik, N.; Gadow, R. *Appl. Organomet. Chem.* **2000**, *14*, 671.
(2) Brook, A. G.; Brook, M. A. *Adv. Organomet. Chem.* **1996**, *39*, 71.
(3) Becker, G. Z. *Anorg. Allg. Chem.* **1976**, *423*, 242.
(4) West, R.; Boudjouk, P. *J. Am. Chem. Soc.* **1973**, *95*, 3987.
(5) Wolfgramm, R.; Müller, T.; Klingebiel, U. *Organometallics* **1998**, *17*, 3222.
(6) Schmatz, S.; Ebker, C.; Labahn, T.; Stoll, H.; Klingebiel, U. *Organometallics* **2003**, *22*, 490.
(7) Boudjouk, P.; West, R. *Intra-Sci. Chem. Rep.* **1973**, *7*, 65.
(8) West, R.; Ishikawa, M.; Bailey, R. E. *J. Am. Chem. Soc.* **1967**, *89*, 4981.
(9) Bode, K.; Klingebiel, U. *Adv. Organomet. Chem.* **1996**, *40*, 1.

(10) Klingebiel, U.; Schmatz, S.; Gellermann, E.; Drost, C.; Noltemeyer, M. *Mh. Chem.* **2001**, *132*, 1105.
(11) Klingebiel, U. *Phosphorus, Sulfur, Silicon* **1989**, *41*, 361.
(12) Egert, E.; Kliebisch, U.; Klingebiel, U.; Schmidt, D. Z. *Naturforsch* **1987**, *42b*, 23.
(13) Brewer, S. D.; Haber, C. P. J. *J. Am. Chem. Soc.* **1948**, *70*, 3888.
(14) Haiduc, J. *The Chemistry of Inorganic Ring Systems*, Wiley-Interscience, London, **1970**, 370.
(15) Klingebiel, U.; Vater, N. *Chem. Ber.* **1983**, *116*, 3282.
(16) Werner, E.; Klingebiel, U.; Pauer, F.; Stalke, D.; Riedel, R.; Schaible, S. Z. *Anorg. Allg. Chem.* **1991**, *596*, 35.
(17) Clegg, W.; Sheldrick, G. M.; Stalke, D. *Acta Crystallogr. Sect. C* **1984**, *40*, 433.
(18) Rakebrandt, H.-J.; Klingebiel, U.; Noltemeyer, M.; Wenzel, U.; Mootz, D. *J. Organomet. Chem.* **1996**, *524*, 237.
(19) Hemme, I.; Klingebiel, U. *Adv. Organomet. Chem.* **1996**, *39*, 159.
(20) Stalke, D.; Keweloh, N.; Klingebiel, U.; Noltemeyer, M.; Sheldrick, G. M. Z. *Naturforsch* **1987**, *42b*, 237.
(21) Clegg, W.; Klingebiel, U.; Krampe, C.; Sheldrick, G. M. Z. *Naturforsch* **1980**, *35b*, 275.
(22) Walter, S.; Klingebiel, U. *Coord. Chem. Revs.* **1994**, *130*, 481.
(23) Klingebiel, U.; Meller, A. Z. *Anorg. Allg. Chem.* **1977**, *428*, 27.
(24) Klingebiel, U.; Meller, A. *Angew. Chem.* **1976**, *88*, 647.
(25) Clegg, W.; Klingebiel, U.; Sheldrick, G. M. Z. *Naturforsch* **1982**, *37b*, 423.
(26) Underiner, G. E.; Tan, R. P.; Powell, D. R.; West, R. *J. Am. Chem. Soc.* **1991**, *113*, 8437.
(27) Klingebiel, U.; Tecklenburg, B.; Noltemeyer, M.; Schmidt-Bäse, D.; Herbst-Irmer, R. Z. *Naturforsch* **1998**, *53b*, 355.
(28) Kliebisch, U.; Klingebiel, U.; Stalke, D.; Sheldrick, G. M. *Angew. Chem.* **1986**, *98*, 921.
(29) The B3LYP hybrid method of Becke includes Becke's exchange functional (Becke, A. D. *J. Chem. Phys.* **1993**, *98*, 5648) and the Lee, Yang and Parr non-local correlation functional (Lee, C.; Yang, W.; Parr, R. G. *Phys. Rev. B* **1988**, *37*, 785). The GAUSSIAN98 suite of programmes was used throughout (Frisch, M. J.; Trucks, G. W.; Schlegel, H. B.; Scuseria, G. E.; Robb, M. A.; Cheeseman, J. R.; Zakrzewski, V. G.; Montgomery, J. A.; Stratmann, R. E.; Burant, J. C., Dapprich, S.; Millam, J. M.; Daniels, A. D.; Kudin, K. N.; Strain, M. C.; Farkas, O.; Tomasi, J.; Barone, V.; Cossi, M.; Cammi, R.; Menucci, B.; Pomelli, C.; Adamo, C.; Clifford, S.; Ochterski, J.; Petersson, G. A.; Ayala, P. Y.; Cui, Q.; Morokuma, K.; Malick, D. K.; Rabuck, A. D.; Raghavachari, K.; Foresmann, J. B.; Cioslowski, J.; Ortiz, J. V.; Baboul, a. g.; Stefanov, B. b.; Liu, G.; Liashenko, A.; Piskorz, P.; Komaromi, I.; Gomberts, R.; Martin, R. L.; Fox, d. J.; Keith, T.; Al-Laham, M. A.; Peng, C. Y.; Nanayakkara, A.; Gonzales, C.; Challocombe, M.; Gill, P. M. W.; Johnson, B.; Chen, W.; Wong, M. W.; Andres, J. L.; Gonzales, C.; Head-Gordon, M.; Replogle, E. S.; Pople, J. A. *GAUSSIAN98*, Revisison A.7; GAUSSIAN, Inc.: Pittsburgh, PA, **1998**).
(30) Fink, W. *Helv. Chim. Acta* **1962**, *45*, 1081.
(31) Neugebauer, P.; Jaschke, B.; Klingebiel, U. (Z. Rappoport, Y. Apeloig, Eds.), *Supplement Si: The Chemistry of Organic Silicon Chemistry* Vol. 3, **2001**, Interscience Publishers, John Wiley & Sons, Ltd., p. 429.
(32) Egert, E.; Kliebisch, U.; Klingebiel, U.; Schmidt, D. Z. *Anorg. Allg. Chem.* **1987**, *548*, 89.
(33) Werner, E.; Klingebiel, U. *Phosphorus, Sulfur, Silicon* **1993**, *83*, 9.
(34) Klingebiel, U.; Noltemeyer, M.; Rakebrandt, H.-J. Z. *anorg. allg. Chem.* **1997**, *623*, 281.
(35) Frenzel, A.; Herbst-Irmer, R.; Klingebiel, U.; Schäfer, M. *Phosphorus, Sulfur, Silicon* **1996**, *112*, 155.
(36) Schaible, S.; Riedel, R.; Boese, R.; Werner, E.; Klingebiel, U.; Haase, M. *Appl. Organomet. Chem.* **1994**, *8*, 491.
(37) Jaschke, B.; Helmold, N.; Müller, I.; Pape, T.; Noltemeyer, M.; Herbst-Irmer, R.; Klingebiel, U. Z. *anorg. allg. Chem.* **2002**, *628*, 2071.
(38) Clegg, W.; Noltemeyer, M.; Sheldrick, G. M.; Vater, N. *Acta Crystallogr. Sect. B* **1981**, *37*, 986.
(39) Hesse, M.; Klingebiel, U.; Skoda, L. *Chem. Ber.* **1981**, *114*, 2287.
(40) Breed, L. W. *Inorg. Chem.* **1968**, *7*, 1940.
(41) Fink, W. *Helv. Chim. Acta* **1969**, *52*, 2261.
(42) Klingebiel, U. *Inorg. Reac. Met.* **1990**, *17*, 116.
(43) Klingebiel, U. *Phosphorus, Sulfur, Silicon* **1989**, *41*, 361.
(44) Haase, M.; Sheldrick, G. M. *Acta Crystallogr. Sect. C* **1986**, *42*, 1009.

(45) Klingebiel, U.; Skoda, L. Z. *Naturforsch* **1985**, *40b*, 913.
(46) Klingebiel, U. *Nachr. Chem. Tech. Lab.* **1987**, *35*, 1042.
(47) Werner, E.; Klingebiel, U.; Pauer, F.; Stalke, D. Z. *anorg. allg. Chem.* **1991**, *596*, 35.
(48) Clegg, W.; Hesse, M.; Klingebiel, U.; Sheldrick, G. M.; Skoda, L. Z. *Naturforsch* **1980**, *35b*, 1359.
(49) Jaschke, B.; Klingebiel, U.; Riedel, R.; Doslik, N.; Gadow, R. *Appl. Organomet. Chem.* **2000**, *14*, 671.
(50) Clegg, W.; Klingebiel, U.; Sheldrick, G. M.; Skoda, L.; Vater, N. Z. *Naturforsch* **1980**, *35b*, 1503.
(51) Clegg, W. *Acta Crystallogr. Sect. B* **1980**, *36*, 2830.
(52) Dippel, K.; Werner, E.; Klingebiel, U. *Phosphorus, Sulfur, Silicon* **1992**, *64*, 15.
(53) Smith, G. S.; Alexander, L. E. *Acta Crystallogr.* **1963**, *16*, 1015.
(54) Dippel, K.; Klingebiel, U.; Noltemeyer, M.; Pauer, F.; Sheldrick, G. M. *Angew. Chem.* **1988**, *100*, 1093.
(55) Dippel, K.; Klingebiel, U.; Kottke, T.; Pauer, F.; Sheldrick, G. M.; Stalke, D. *Chem. Ber.* **1990**, *123*, 237.
(56) Werner, E.; Klingebiel, U.; Dielkus, S.; Herbst-Irmer, R.; Schaible, S.; Riedel, R. Z. *anorg. allg. Chem.* **1994**, *620*, 1093.
(57) Helmold, N.; Jaschke, B.; Klingebiel, U. *Phosphorus, Sulfur, Silicon* **2001**, *169*, 245.
(58) Tecklenburg, B.; Klingebiel, U.; Noltemeyer, M.; Schmidt-Bäse, D. Z. *Naturforsch* **1992**, *47b*, 855.
(59) Dippel, K.; Klingebiel, U. Z. *Naturforsch* **1990**, *45b*, 1147.
(60) Werner, E.; Klingebiel, U. *J. Organomet. Chem.* **1994**, *470*, 35.
(61) Dippel, K.; Klingebiel, U.; Kottke, T.; Pauer, F.; Sheldrick, G. M.; Stalke, D. Z. *anorg. allg. Chem.* **1990**, *584*, 87.
(62) Klingebiel, U.; Noltemeyer, M.; Rakebrandt, H.-J. Z. *Naturforsch* **1997**, *52b*, 775.
(63) Großkopf, D.; Klingebiel, U. Z. *anorg. allg. Chem.* **1993**, *619*, 1857.
(64) Dippel, K.; Klingebiel, U. *Chem. Ber.* **1990**, *123*, 1817.
(65) Dippel, K.; Klingebiel, U.; Pauer, F.; Sheldrick, G. M.; Stalke, D. *Chem. Ber.* **1990**, *123*, 779.
(66) Dippel, K.; Klingebiel, U.; Marcus, L.; Schmidt-Bäse, D. Z. *Anorg. Allg. Chem.* **1992**, *612*, 130.
(67) Schaible, S.; Riedel, R.; Aldinger, F.; Klingebiel, U. (M. J. Hoffmann, P. F. Becher, Eds.), *Key Engineering Materials*, Petzow, **1993**, p. 81.

Molecular Alumo-Siloxanes and Base Adducts

MICHAEL VEITH*

Institut für Anorganische Chemie, Universität des Saarlandes and Leibniz-Institut für Neue Materialien, Im Stadtwald, D-66401 Saarbrücken, Germany

I

INTRODUCTION

The following review is concerned with the synthetic and structural chemistry of molecular alumo-siloxanes, which combine in a molecular entity the elements aluminum and silicon connected by oxygen. They may be regarded as molecular counterparts of alumo-silicates, which have attracted considerable attention owing to their solid-state cage structures (see for example: zeolites).[1–3] Numerous applications have been found for these solid-state materials: for instance the holes and pores can be used in different separation techniques.[4,5] Recently the channel and pore structures of zeolites and other porous materials have been used as templates for nano-structured materials and for catalytical purposes.[6–9]

The large scale synthesis and modification of alumo-silicates has frequently been developed using hydrothermal methods.[1] These methods have the advantages that they are cheap and simple with respect to the starting materials, but suffer generally from restrictions to the diversification of the desired compounds. Although much progress has been made by using template methods, there are still a lot of parameters to be considered to obtain an alumo-silicate of high purity and of a precise channel or pore size.[6–10]

Alumo-phosphates hasve been known to have some structural and chemical similarities to alumo-silicates. Indeed, formally they may be derived from SiO_2 by substituting the element silicon by the neighbor elements aluminum and phosphorus (thus the most simple being $AlPO_4$) and forming an iso-electronic compound to SiO_2. In comparison with simple alumo-silicates $(MAlO_2)_n(SiO_2)_m$, in the alumophosphates no further metallic counterpart M is needed to balance the valence difference between silicon and aluminum as the phosphate group is formally

*Corresponding author.
E-mail: veith@mx.uni-saarland.de, veith@INM-GMBH.de (M. Veith).

ADVANCES IN ORGANOMETALLIC CHEMISTRY
VOLUME 54 ISSN 0065-3055/DOI 10.1016/S0065-3055(05)54002-4

© 2006 Elsevier Inc.
All rights reserved.

charged −3. The alumo-phosphates show a rich structural chemistry, which ranges from almost molecular to polymeric with classical inorganic solid-state structure principles being predominant.[11–33] Again, like in the alumo-silicates the formation of cages and holes is a typical structural motif.

There have been many attempts to construct molecular counterparts of alumo-phosphates via substitution of oxygen atoms bonded to phosphorus by organic groups. Generally speaking, by exchange of oxygen atoms at the phosphorus atoms through organic substituents (passing from PO_4^{3-} to RPO_3^{2-} or to $R_2PO_2^{-}$), the structures of the compounds become simpler, changing from polymeric (PO_4^{3-})[11–33] to oligomeric and cage-like (RPO_4^{2-})[34–41] or to cyclic $(R_2PO_2^{-})$.[42–44] Of course during this exchange process the molecules also become steadily more soluble in organic solvents.

Astonishingly, the molecular counterparts of alumo-silicates have attracted relatively less attention as may be seen by studying the structural results obtained so far (see below). Again, as with the alumo-phosphates, the most well-established cage structures are those with silicon bearing one organic ligand (or amino-ligand) at the silicon atom but also some scarce reports have been available with two organic groups on the silicon atom. Two surveys of the work especially done in the group of H. W. Roesky are available,[45,46] together with some more general ones.[47,48]

We will restrict this chapter to alumo-poly siloxanes which may be derived from a poly siloxane of the general formula $(R_2SiO)_n$[1,49] by replacing several R_2SiO units by AlOX with X = OH in most cases, and their base derivatives. Before discussing the results obtained in our laboratory, we will highlight some general structural details of compounds of the general formula $(R_xSiO_y)_n(AlOX)_m$. The survey of our own findings may also be considered as a state of the art report, as many of the results have not been published before. Because of the quick expansion of the field, we will restrict ourselves to the hydroxy compounds and their base adducts, and neglect most of our results in the area of molecular alumo-silicates in which the protons are substituted by metallic elements.[50–52]

II
STRUCTURES OF COMPOUNDS OF THE GENERAL COMPOSITION $(RSiO_{1.5})_n(AlOX)_m$

We first will try to assemble the structural principles of molecular alumo-silicon oxides and will begin with compounds which have one organic or other non-bridging ligand on the silicon atom leading to the general formula $(RSiO_{1.5})_n(AlOX)_m$. Such compounds may be viewed as cut out of a three-dimensional oxidic network and as every silicon atom still displays bonds to three other oxygen atoms, such compounds should be polycyclic and cage-like.

The first crystal structure evidence of a molecular cage entity within an alumo-silicon oxide was reported in 1987 with tetrakis(tertamethylammonium)alumino-silicate.[53] Although there is no organic ligand on the silicon atom, the compound contains a separated silsesquioxane-like $Si_4Al_4O_{12}$ unit (half of the silicon atoms of the silsesquioxane being substituted by aluminum atoms) with all silicon and

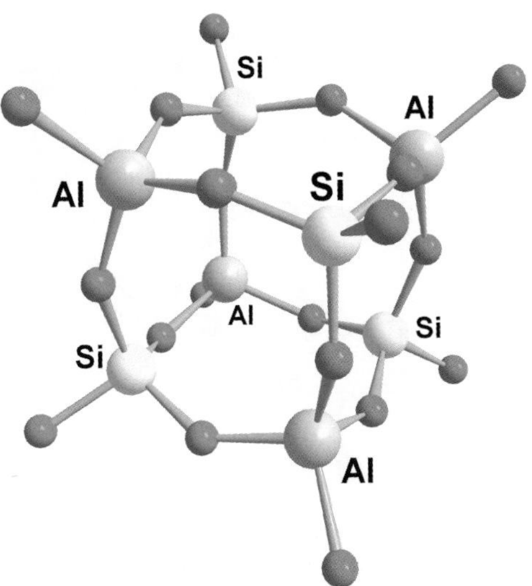

FIG. 1. Cube-like structure in tetrakis(tetramethylammonium)aluminosilicate; only the anion is shown with small balls representing oxygen atoms and the bigger ones aluminum and silicon.[53]

aluminum atoms being tetra-valent and having an OH-group as terminal ligand. In Fig. 1, the inner $Si_4Al_4O_{12}$ cage unit is drawn, which is not only common to the tetrakis(tetramethylammonium)aluminosilicate discussed here but also to a number of other compounds $Si_4Al_4O_{12}X_4Y_4$ that have carbon–cobalt clusters or amino-groups on the silicon atoms (X) and cyclic ether molecules as donors on the aluminum atoms (Y).[54–57] The compounds are all molecular and in contrast to the $Si_4Al_4O_{12}(OH)_8^{4-}$ anion they are uncharged.

Another frequently encountered structural motif is $(R\text{-}Si)_7(AlX)O_{12}$ which can be derived from the silsesquioxane $(R\text{-}Si)_8O_{12}$ by replacement of one silicon-organo corner by aluminum bonded generally to a further ligand X (neutral or charged). A simple example of this structure-type is $(cy\text{-}Si)_7\,[Al\text{–}(OPPh_3)]O_{12}$ (cy = cyclohexyl), where the triphenylphosphine-oxide is completing the electron deficit at aluminum through a donor bond with oxygen as bridging atom.[58] When $OPPh_3$ in this compound is substituted by the negatively charged entity $(R\text{-}Si)_7(SiO^-)O_{12}$ (tetramethylstibonium serving as positively charged counter-ion), two of these silsesquioxane-like cages can be connected at the aluminum corner with oxygen serving as bridge between the aluminum center and the silicon corner of the second cage: $\{[(cy\text{-}Si)_7O_{12}]Al\text{–}O\text{–}Si[O_{12}(Si\text{-}cy)_7]\}^-$.[59] Another possibility is to link two $(cy\text{-}Si)_7O_{12}(Al)$ entities by O^{2-} forming again two corner linked cages: $\{[(cy\text{-}Si)_7O_{12}]Al\text{–}O\text{–}Al[O_{12}(Si\text{-}cy)_7]\}^{2-}$, a di-anion which is isoelectronic with the monoanion $\{[(cl\text{-}Si)_7O_{12}]Al\text{–}O\text{–}Si[O_{12}(Si\text{-}cy)_7]\}^-$ just discussed before.[60]

While all these connected entities still have closed cages of the cube-shaped silsesquioxane type, there exist a number of connected entities that consist of opened silsesquioxane structures with aluminum-oxy species as linkers. One structure type can be expressed by the general formula $\{[(RSi)_7O_{11}(O\text{–}SiR'_3)]Al[(R'_3Si\text{–}O)O_{11}$

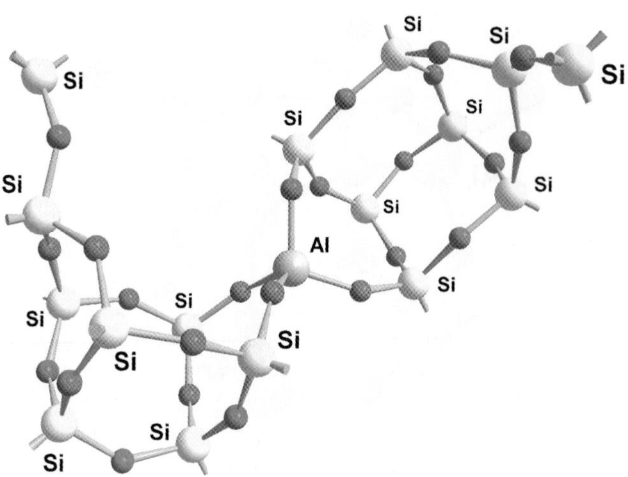

FIG. 2. The structure of an anion of the general formula $\{[(RSi)_7O_{11}(O\text{–}SiR'_3)]Al[(R'_3Si\text{–}O)O_{11}(SiR)_7]\}^-$ with organic groups omitted for clarity.[61–63]

$(SiR)_7]\}^-$ which has a negative charge and which has an alkylammonium ion as positive counter-part.[61–63] The crystal structure of one of the anions is depicted in Fig. 2, in which the tetra-coordinate aluminum atom may be recognized. Instead of using a three-valent aluminum atom as linker between $[(RSi)_7O_{11}(O\text{–}SiR'_3)]^{2-}$ units, two $[Al\text{–}CH_3]^{2+}$ or four $[Al(CH_3)_2]^+$ cations can also be employed.[64] A formal Al_2O_3 entity may also connect together three different $(RSi)_8O_{12}$ units, resulting in an opening of the silsesquioxane cages.[63]

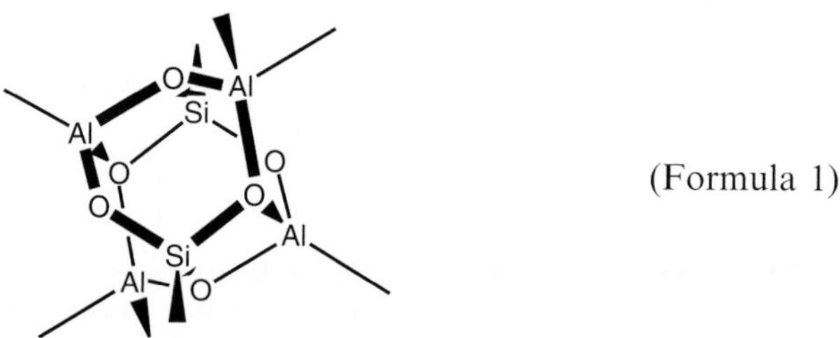

(Formula 1)

There are at least two cage compounds known, in which the Al/Si ratio is not unity, but twice as many aluminum atoms are present as silicons.[60,65] Both compounds have organic substituents on the aluminum atoms. They may be derived from a drum-shaped Si_6O_6 cage with alternating silicon and oxygen atoms, four of the silicon atoms being replaced by aluminum with one or two organic ligands: $(R'Si)_2(AlR)_2(AlR_2)_2O_6$ (see Formula 1). H. W. Roesky and coworkers could also demonstrate that OH-groups bonded to silicon under certain circumstances may be protected from condensation, even if aluminum is present in the compound. It seems that bulky amino-groups on silicon like $N(SiMe_3)(C_6H_3Ph_2)$ have to be used for this protection, which besides their steric bulk also have an electronic effect on silicon.[66,67]

III
STRUCTURES OF COMPOUNDS OF THE GENERAL COMPOSITION $(R_2SiO)_n(AlOX)_m$

Substituting two of the silicon oxygen bonds in the SiO_2 entity of alumo-silicon oxides, $(SiO_2)_n(AlOX)_m$, instead of one bond (see before) results in the general formula $(R_2SiO)_n(AlOX)_m$. As now each silicon atom only displays two bonds to oxygen, the cage-like structures found typically with compounds of the general formula $(RSiO_{1.5})_n(AlOX)_m$, because of fewer connections, should change to cyclic in $(R_2SiO)_n(AlOX)_m$, neglecting in a first approach the bonding of aluminum. Indeed, all of the compounds characterized so far have cyclic structures in common. These compounds may also be considered as alumo-derivatives of poly siloxanes.

The first structural report of an alumo-poly siloxane appeared as early as 1966: the molecule $(Me_2SiO)_4(OAlBr_2)_2(AlBr)$ has an $Si_4Al_2O_6$ 12-membered cycle which is spirocyclically bridged by an Al–Br unit (see Fig. 3).[68] Only in 2000 was the corresponding structure of the chloro-derivative (which is iso-typic to the bromo-compond) published.[69] In both compounds, all silicon and aluminum atoms are tetra-coordinate with the exception of the spiro-cyclic central bromo(chloro)-aluminum –moiety, which is penta-coordinate with four oxygen bonds.

In 1997, we synthesized and characterized the first example of a compound which strictly corresponds to the general formula $(R_2SiO)_n(AlOX)_m$: in $(Ph_2SiO)_8[AlO(OH)]_4$ the molecular structure can be derived from a 24-membered $Si_8Al_4O_{12}$ cycle which is polycyclic because the aluminum atoms behave as intrinsic Lewis-acids toward the OH-Lewis-bases.[70] This results in four further Al–O bonds creating an inner eight-membered Al_4O_4 cycle; in an alternative description the structure of the molecule can also be described to consist of four $Si_2Al_2O_4$ rings which share aluminum corners with the neighboring rings. In Fig. 4 a view of the polycyclic structure is given. As the hydrogen atoms of the hydroxy groups in $(Ph_2SiO)_8[AlO(OH)]_4$ are very acidic the

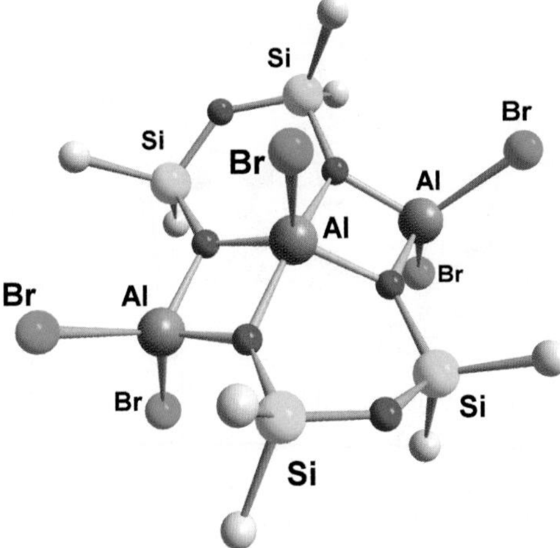

FIG. 3. Molecular structure of $(Me_2SiO)_4(OAlBr_2)_2(AlBr)$ without hydrogen atoms.[68]

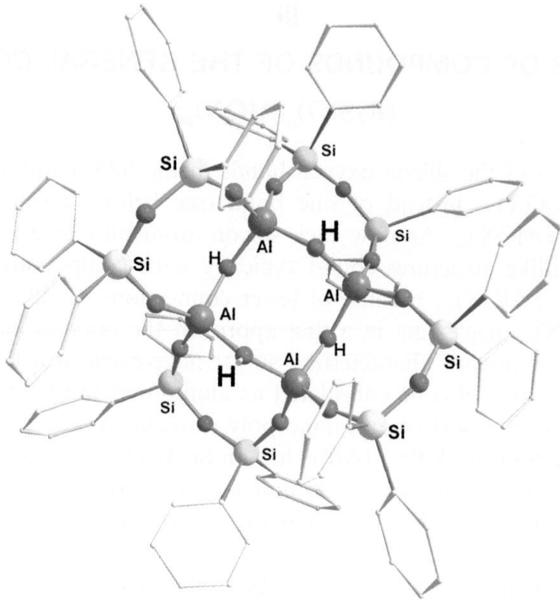

FIG. 4. The naked $(Ph_2SiO)_8[AlO(OH)]_4$ molecule from X-ray diffraction.[70]

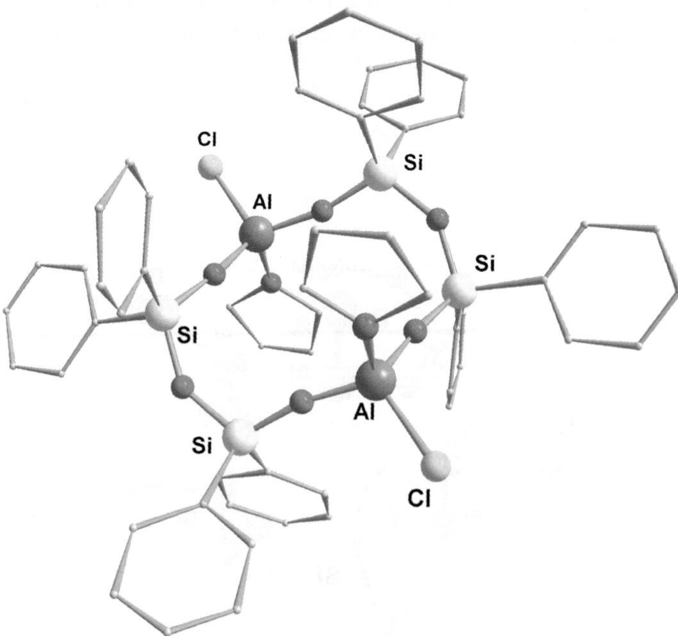

FIG. 5. The molecular structure of $(Ph_2SiO)_4(AlCl \cdot THF)_2$ as deduced from X-ray diffraction.[71]

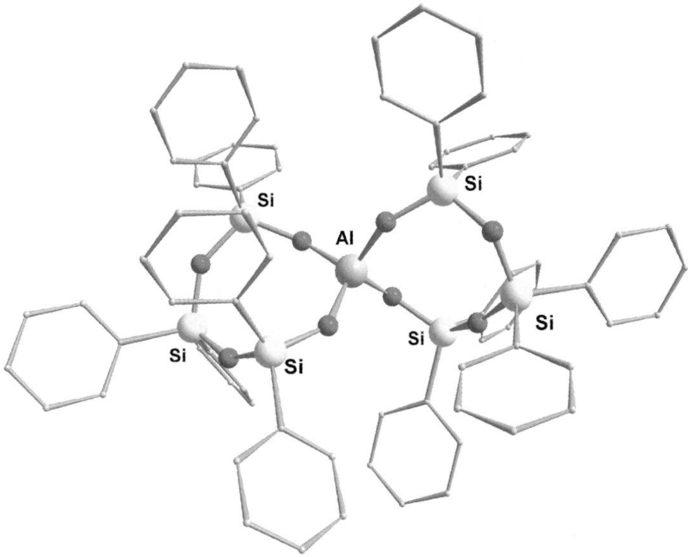

FIG. 6. Part of the molecular structure of the $[(Ph_2SiO)_6(AlO_2)]^-$ anion.[71]

molecule always crystallizes with bases (for example diethyl ether) coordinated to the hydrogen atoms.[70]

In 2001, two other examples of cyclic alumo-siloxanes have been described by V. C. Kessler and coworkers.[71] In $(Ph_2SiO)_4(AlCl \cdot THF)_2$ a 12-membered $Si_4Al_2O_6$ ring forms the molecular backbone of the molecule with no further bonding within the cycle because the aluminum atoms are electronically saturated by THF bonding (see Fig. 5). In the $[(Ph_2SiO)_6(AlO_2)]^-$ anion, with its charge compensated by a pyridinium cation, the aluminum atom is in the center of an oxygen tetrahedron and spirocyclicly connects two eight-membered Si_3AlO_4 cycles as may be deduced from Fig. 6.[71] A cyclic alumino phosphate with an O–Si-substituent $[(RO)_2PO_2]_2[Al(OSiR'_3)_2][Al(R)(OSiR'_3)]$, and thus combining in some respect phosphates with silicates has been reported recently.[72] In the following section, we will describe some typical reactions and structures of compounds which can be derived from the molecule $(Ph_2SiO)_8[AlO(OH)]_4$.

IV
GENERAL REACTIONS AND STRUCTURES OF COMPOUNDS DERIVED FROM $(Ph_2SiO)_8[AlO(OH)]_4$

A. *Synthesis of $(Ph_2SiO)_8[AlO(OH)]_4$ and reactions with water and ammonia*

The synthesis of $(Ph_2SiO)_8[AlO(OH)]_4$ is performed by reaction of diphenylsilandiol $Ph_2Si(OH)_2$ in diethyl ether with *tert*-butoxy alane $(tBuO\text{-}AlH_2)_2$.[73] As may be seen from Eq. (1) the products of this reaction are, besides $(Ph_2SiO)_8[AlO(OH)]_4$, dihydrogen and *tert*-butanol.[70]

$$8\ Ph_2Si(OH)_2 + 2\ (tBuO\text{-}AlH_2)_2 \rightarrow (Ph_2SiO)_8[AlO(OH)]_4 + 4\ tBuOH + 8\ H_2 \quad (1)$$

The products clearly show that all ligands on the aluminum atoms in $(tBuO\text{-}AlH_2)_2$ are completely exchanged by siloxides or hydroxides, liberating the respective ligands saturated by protons. At the same time two disilanols must have been condensed to form in an intermediate step $(HO)Si(Ph)_2O(Ph)_2Si(OH)$ and water. While the water molecule ends up as a hydroxide group on the aluminum atoms, the condensation of $Ph_2Si(OH)_2$ to $(HO)Si(Ph)_2O(Ph)_2Si(OH)$ seems to be catalyzed by the aluminum species present in the reaction mixture, which may act as a catalyst in these condensation reactions.[49] The polycyle $(Ph_2SiO)_8[AlO(OH)]_4$ is isolated from the organic solution by crystallization of the etherate adduct $(Ph_2SiO)_8[AlO(OH)]_4 \cdot 4\ OEt_2$ in 50–65% yield. We have found recently that $(Ph_2SiO)_8[AlO(OH)]_4$ may also be prepared by an alternative route exchanging $(tBuO\text{–}AlH_2)_2$ by $[(tBuO)_3Al]_2$ [Eq. (2)].[74]

$$8\ Ph_2Si(OH)_2 + 2\ [(tBuO)_3Al]_2 \rightarrow (Ph_2SiO)_8[AlO(OH)]_4 + 12\ tBuOH \quad (2)$$

Instead of dihydrogen only the free alcohol tert-butanol is formed, but the yield of the polycycle $(Ph_2SiO)_8[AlO(OH)]_4$ is lower than in the hydrido case (around 30%).[74] The polycycle is isolated as the tert-butanol adduct $(Ph_2SiO)_8[AlO(OH)]_4 \cdot 4\ tBu\text{–}OH$. In the next chapter, the structure of adducts of $(Ph_2SiO)_8[AlO(OH)]_4$ with different organic bases functioning as electron donors versus the protons of the hydroxide groups and ranging from monobasic ethers and amines to di-basic diamines are assembled.

The polycyle $(Ph_2SiO)_8[AlO(OH)]_4$ is stable under exclusion of oxygen and moisture and dissolves in most organic solvents. With water it reacts under destruction of the molecule unless water is added in small stoichiometric amounts in an organic solvent. In the latter case, the water is formally trapped by one of the [AlO(OH)] units and by rearrangement of the whole molecule $(Ph_2SiO)_{12}[AlO(OH)]_6[Al(OH)_3] \cdot 3\ OEt_2$ is formed [see also Eq. (3)].[75]

$$3/2\ (Ph_2SiO)_8[AlO(OH)]_4 + H_2O + \{AlO(OH)\} + 3\ OEt_2 \rightarrow (Ph_2SiO)_{12}[AlO(OH)]_6[Al(OH)_3] \cdot 3\ OEt_2 \quad (3)$$

We have not looked at the other presumably oligo-diphenyl-siloxane which forms as a further side-product. The polycyclic $(Ph_2SiO)_{12}[AlO(OH)]_6[Al(OH)_3] \cdot 3\ OEt_2$ is obtained as a colorless solid and is easily separated by crystallization from the other products. As may be deduced from Fig. 7 the compound may be better described as a polycycle which contains an inner $Al(OH)_6^{3-}$ octahedron which is linked to three double tetrahedral $[Al_2O_2(OH)]^+$ cations connected by six $(\text{–}O\text{–}Si(Ph)_2O(Ph)_2Si\text{–}O\text{–})$ bridges. All hydroxyl groups can be easily distinguished by their participation in $O\text{–}H \cdots O$ hydrogen bonds, six of them being found in the inner cage and three of them at the periphery to the oxygen atoms of the three diethyl ether molecules (see Fig. 7).

Similar rearrangements of the mother polycycle $(Ph_2SiO)_8[AlO(OH)]_4$ may be found when in Eq. (3) the water molecule is replaced by ammonia. Besides other compounds we were able to isolate and to structurally characterize the anion $\{(Ph_2SiO)_{12}[Al_5O_7(OH)][Al(OH)_2(NH_3)_2] \cdot OEt_2\}^-$.[75,76] Again there is an octahedrally coordinated aluminum atom (two nitrogen atoms and four oxygen atoms forming the coordination polyhedron), which is incorporated in a polycyclic arrangement of AlO_4 tetrahedra as may be seen by inspection of Fig. 8.

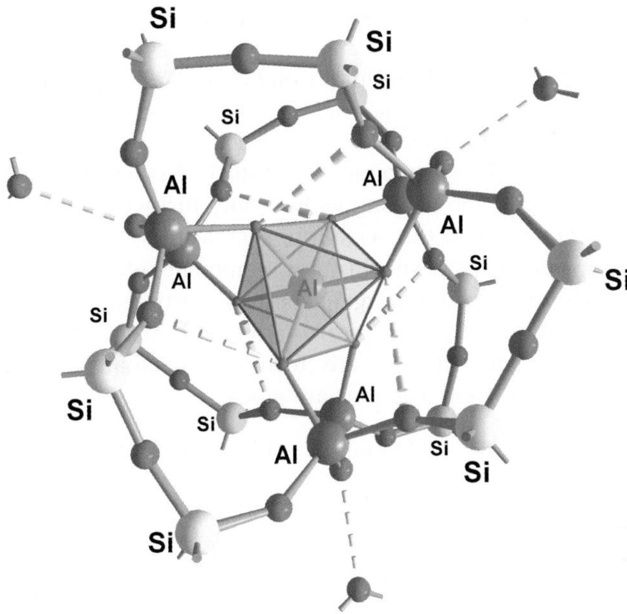

FIG. 7. The molecular structure of the inner skeleton of $(Ph_2SiO)_{12}[AlO(OH)]_6[Al(OH)_3] \cdot 3\ OEt_2$. The organic groups are omitted for clarity. The hydrogen bridges are represented as dotted lines. In the middle of the molecule the formal $Al(OH)_6^{3-}$ octahedron is highlighted.[75]

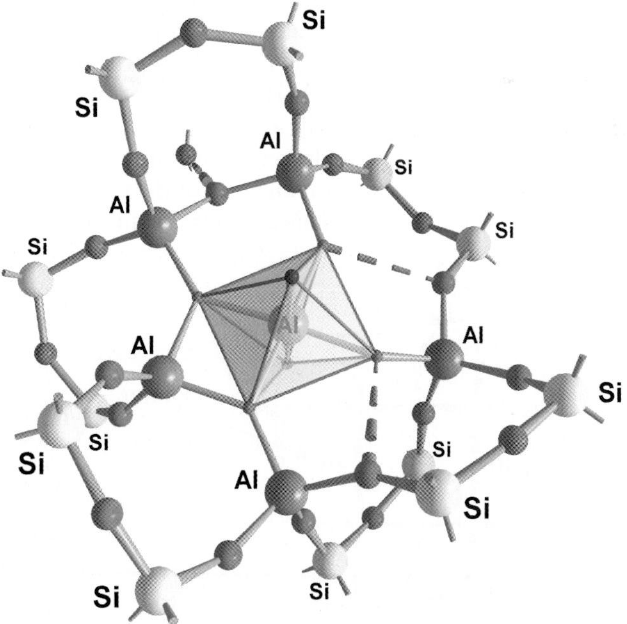

FIG. 8. The inner skeleton of the anion $\{(Ph_2SiO)_{12}[Al_5O_7(OH)][Al(OH)_2(NH_3)_2] \cdot OEt_2\}^-$. The inner octahedron has four equatorial oxygen atoms and two axial nitrogen atoms.[75,76]

B. *Simple Base-Adducts of $(Ph_2SiO)_8[AlO(OH)]_4$*

The polycycle $(Ph_2SiO)_8[AlO(OH)]_4$ is isolated as the diethylether adduct.[70] Inspecting the crystal and molecular structure of $(Ph_2SiO)_8[AlO(OH)]_4 \cdot 4\ OEt_2$ in more detail reveals that three diethyl ether molecules are coordinated to the hydrogen atoms of the hydroxyl groups. The fourth diethyl ether molecule in the crystal occupies a void in the packing without any further contact apart from the van der Waals interactions to the neighboring molecules. The hydroxyl groups in the molecule are acidic (as it is known from alumo-siloxides[1–3]) and it is therefore not astonishing that organic and inorganic bases are attracted by these OH-groups. Indeed, we have found that a great variety of bases may be added to $(Ph_2SiO)_8[AlO(OH)]_4$ and adducts are formed immediately. In Eq. (4) and Scheme 1 a summary of these reactions are given.

$$(Ph_2SiO)_8[AlO(OH)]_4 + n\ \text{Donor} \rightarrow (Ph_2SiO)_8[AlO(OH)]_4 \cdot n\ \text{Donor} \quad (4)$$

As can be concluded from Scheme 1 amines as well as ethers and alcohols may act as donor molecules and serve as acceptors for the protons of the polycycle $(Ph_2SiO)_8[AlO(OH)]_4$. There is a dependence of the steric requirement on the donor guest molecule and the number of guests being accepted in the Lewis acid–base complexes. In a simple picture the molecular structure of $(Ph_2SiO)_8[AlO(OH)]_4$ resembles two baskets which are fused together by a common $Al_4(OH)_4$ bottom (see Fig. 4); two of the hydroxyl groups are pointing toward the upper basket part whereas the other two are pointing in opposite directions within the second basket. While triethylamine can only interact with two out of the four hydroxyl groups, diethyl ether can interact with three, and the flattened pyridine as well as dimethyl amine, the cyclic tetra-hydro furane and the alcohol *iso*-propanol with all four (see Scheme 1).[70,77,78] Interestingly, the inner $Al_4(OH)_4$ ring is not altered in these reactions nor are the siloxy ligands chemically attacked. As the starting compound is polycyclic and all four $Si_2Al_2O_4$ rings are linked to the inner $Al_4(OH)_4$ ring, adopting approximately S_4 symmetry, an opening on one face of the molecule (basket) leads to a closure of the opposite face (basket).[77] This explains the different numbers of guests

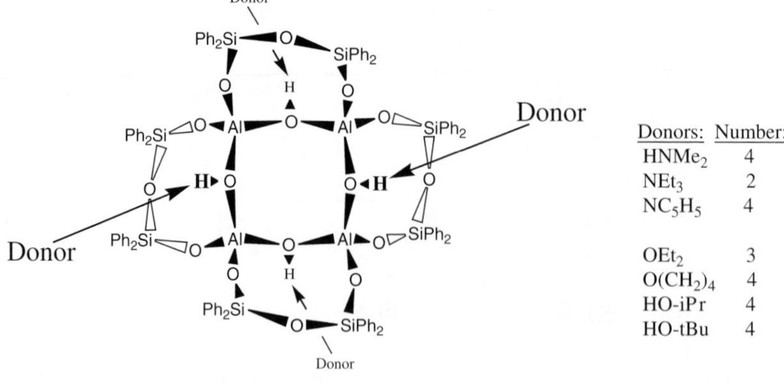

SCHEME 1.

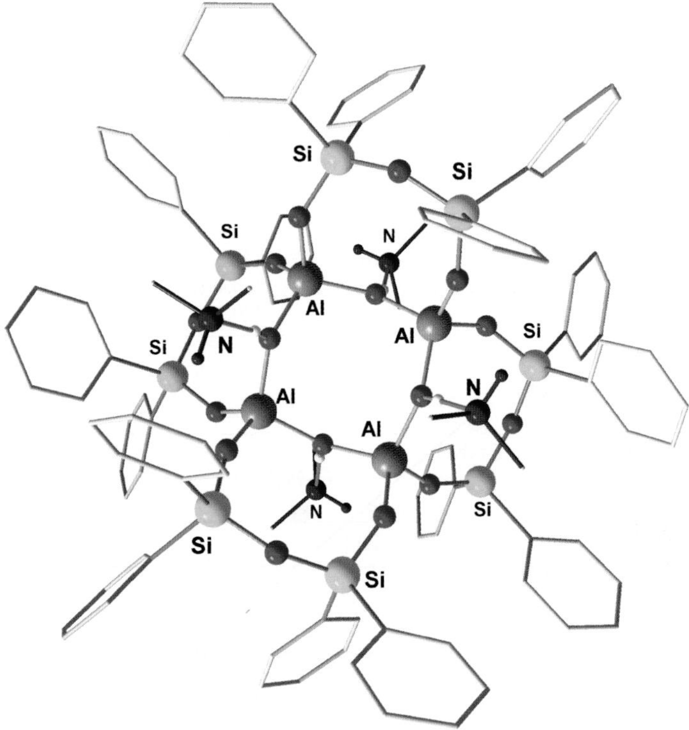

FIG. 9. Molecular structure of $(Ph_2SiO)_8[AlO(OH)]_4 \cdot 4\ HNMe_2$ from X-ray diffraction. Only the hydrogen atoms of the hydroxyl and the amines are drawn.[78]

which can be accepted by the polycycle $(Ph_2SiO)_8[AlO(OH)]_4$. In Fig. 9, the structural representation of the adduct of four dimethyl amine molecules with $(Ph_2SiO)_8[Al$-$O(OH)]_4$ (recently characterized by X-ray diffraction) is shown for illustration.

It is quite difficult to tell whether in these Lewis acid–base adducts the hydrogen atom of the hydroxyl groups still is situated on the oxygen atom or has migrated to the basic center of the guest molecule or is somewhat in between, as exemplified in Eq. (5) with an amine.

$$= Al-O-H \cdots NR_3 \leftrightarrow\ = Al-O^- \ HNR_3{}^+ \tag{5}$$

The IR and NMR spectra of the adducts cannot be used to answer this open question as the intensities of the OH groups compared with the other C–H bonds or the other groups are small and may be hidden by other absorptions. A better insight into the bonding can be obtained by closer inspection of characteristic bond lengths of O···H···O or O···H···N hydrogen bridges, of Al–O bond lengths which are adjacent to the O–H groups and of Al–O–Al angles. In Table I these parameters are given for some adducts characterized in our laboratories.

As can be clearly seen, there is no dependence of the O···O or O···N distance on the Al–O distance of the adjacent bonds. This hydrogen bond seems to be influenced very much by the steric requirements of the bases the order of which follows

TABLE I

SIMPLE BASE ADDUCTS OF $(Ph_2SiO)_8[AlO(OH)]_4$ (MEAN VALUES)

Hydrogen bridge N···N(O) (Å)		Al–O(H) (Å)	Al–O–Al (°)	Number of guest molecules
N···H–O	2.548(5)	1.769(3)	132.4(1)	$4NC_5H_5$
N···H–O	2.543(4)	1.752(3)	135.8(1)	$4HNMe_2$
N···H–O	2.718(6)	1.729(4)	130.4(1)	$2Net_3$
O···H–O	2.628(5)	1.806(3)	129.4(1)	$2THF/1H_2N(CH_2)_4NH_2$
O···H–O	2.760(9)	1.801(6)	130.2(2)	4THF
O···H–O	2.603(9)	1.794(6)	130.4(2)	4iPr–O–H
O···H–O	2.627(6)	1.796(4)	130.5(1)	$3OEt_2$

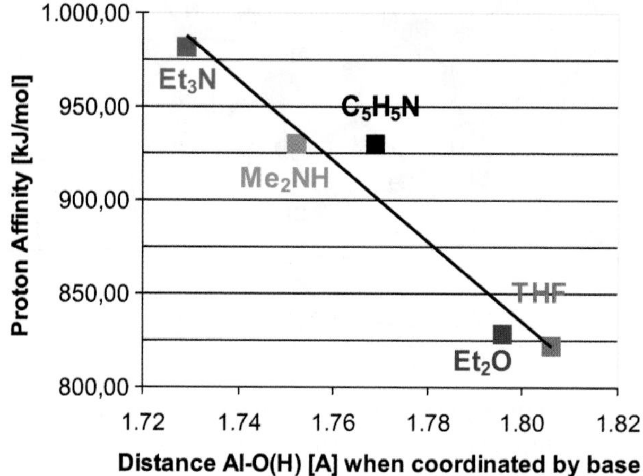

FIG. 10. The dependence of the Al–O(H) distance in the inner ring of $(Ph_2SiO)_8[AlO(OH)]_4$ from the proton affinity of the coordinated bases.

approximately $HNMe_2 \approx NC_5H_5 < iPr–OH \approx OEt_2 \approx THF < NEt_3 < tBuOH$. This series is taken from simple steric calculations using a standard program.[79] Indeed the experimentally determined O···O or O···N distances reflect exactly these model calculations (compare Table I).

Contrarily to the distances of the bases which are very much influenced by the steric requirements, the Al–O bond lengths in the adducts within the $Al_4(OH)_4$ ring of $(Ph_2SiO)_8[AlO(OH)]_4$ are more revealing. Considering Eq. (5) a shift of the proton in direction of the base should lead to a negative charge on the oxygen atom bonded the two aluminum atoms and should therefore lead to a shortening of the Al–O bond lengths. Indeed, it follows from Table I that the Al–O distance ranges from 1.72 to 1.81 Å. We have plotted the Al–O distances versus standard proton affinities[80] of the bases measured in the gas phase (Fig. 10) and find a neat dependence of the Lewis acidity (respectively basicity) with the bond lengths. Instead of the proton affinities also pK_a values or ionization energies of the bases could have been plotted versus the Al–O bond-lengths, but as it has been shown the pK_a values are not always measured under the same conditions and they change comparatively when the values in solution (water) are compared with those of the gas phase.[81,82]

It therefore can be assumed that with strong bases like triethylamine the proton is more likely to be situated on the nitrogen atom.[77] The Al–O distances found within compounds in which the protons have been completely substituted by lithium cations like $(Ph_2SiO)_8[AlO_2]_4Li_4 \cdot 4\ OEt_2$[77] and which have a considerable negative charge at the oxygen atoms have mean Al–O distances of 1.729(9) Å which corresponds exactly to the lower end of the distances found in the adducts (Table I). It can also be deduced from the plot given in Fig. 10 that the pyridine adduct does not fit so well on the curvature which is easily understood by the fact that the pyridine molecules stick in between two parallelly oriented phenyl groups bonded to the silicon atoms. This aromatic π-stacking thus creates a component of attraction which is opposite to the N···H–O interaction.

The mean Al–O–Al angle within the $Al_4(OH)_4$ ring of $(Ph_2SiO)_8[AlO(OH)]_4$ has also be collected in Table I. The more the charge is situated on the oxygen atoms, the more the angle should grow and become more obtuse (see Refs. 1–3). This is generally the case, with the exception of $(Ph_2SiO)_8[AlO(OH)]_4 \cdot 2NEt_3$, in which the angle is small. This feature might be due to intramolecular adjustments owing to the opening and closing of the basket cavities in this case.

C. *Base-Adducts of $(Ph_2SiO)_8[AlO(OH)]_4$ with Poly-Methylene-Diamines $H_2N(CH_2)_nNH_2$*

Instead of coordinating mono-hapto bases to the protons of $(Ph_2SiO)_8[AlO(OH)]_4$, we have explored the field of di-hapto bases like poly methylene diamines as possible donors. As the molecule $(Ph_2SiO)_8[AlO(OH)]_4$ has pairs of hydroxyl groups directed in the same direction a coordination of $H_2N(CH_2)_nNH_2$ could either lead to an intramolecular loop or to intermolecular aggregations of molecules. This field of chemistry is actually very much expanding and we give in this review only a small out-look, essentially with the diamines 1,3-diaminopropane, 1,4-diaminobutane and 1,5-diaminopentane.

In Eq. (6) and Scheme 2 a general reaction is given, illustrating that three sorts of compounds may be obtained when $(Ph_2SiO)_8[AlO(OH)]_4$ is allowed to react with different molarities of diamines $H_2N(CH_2)_nNH_2$.

$$(Ph_2SiO)_8[AlO(OH)]_4 + m\ H_2N(CH_2)_nNH_2 \\ \rightarrow \{(Ph_2SiO)_8[AlO(OH)]_4 \cdot m\ H_2N(CH_2)_nNH_2\} \quad \text{with } m = 1,2,3 \qquad (6)$$

Looking at these reactions in more detail reveals that with $n = 3,4,5$ the 1:2 adduct $\{(Ph_2SiO)_8[AlO(OH)]_4 \cdot 2\ H_2N(CH_2)_nNH_2\}$ is generally formed if the molar ratio of the reactants is in between 1 and 2, and that even a large excess (>3) of the diamine only leads to the 1:3 adduct. The 1:1 adduct until now has been obtained in one case (1,4-diaminobutane) using THF as solvent and has been isolated as $\{(Ph_2SiO)_8[AlO(OH)]_4 \cdot H_2N(CH_2)_4NH_2 \cdot 2\ THF\}$. From NMR studies using different concentrations, it can be concluded that in most reactions the equilibrium is shifted to the right side of the equation. Using other chain lengths in the diamines

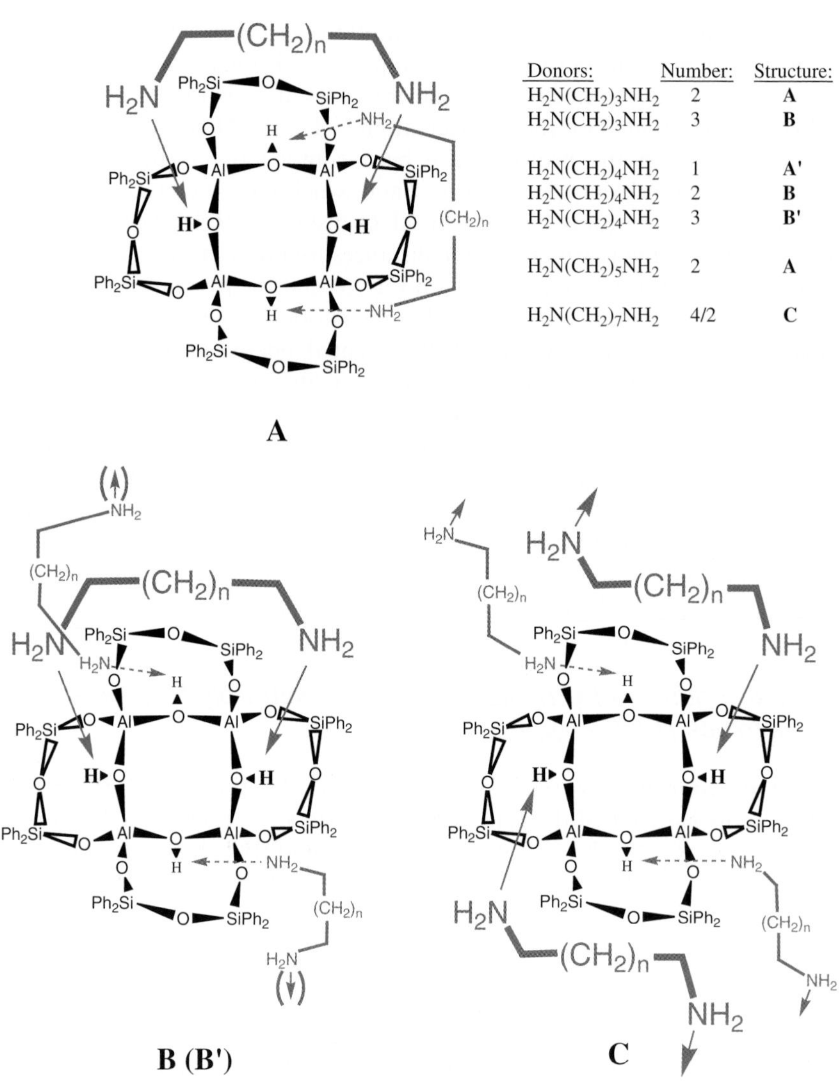

Donors:	Number:	Structure:
$H_2N(CH_2)_3NH_2$	2	**A**
$H_2N(CH_2)_3NH_2$	3	**B**
$H_2N(CH_2)_4NH_2$	1	**A'**
$H_2N(CH_2)_4NH_2$	2	**B**
$H_2N(CH_2)_4NH_2$	3	**B'**
$H_2N(CH_2)_5NH_2$	2	**A**
$H_2N(CH_2)_7NH_2$	4/2	**C**

SCHEME 2.

has an important influence on the structure of the products. With ethylenediamine the starting polycycle $(Ph_2SiO)_8[AlO(OH)]_4$ is destroyed leading to a rearrangement of the molecule.[78] With 1,7-diaminoheptane $(Ph_2SiO)_8[AlO(OH)]_4$ reacts only intermolecularly forming a three-dimensional network: the composition of the compound is $(Ph_2SiO)_8[AlO(OH)]_4 \cdot 2\ H_2N(CH_2)_7NH_2$, each 1,7-diaminoheptane always connecting two molecules (see also below).[78]

The simplest form of diamino adduct of $\{(Ph_2SiO)_8[AlO(OH)]_4\}$ is found in the compound $\{(Ph_2SiO)_8[AlO(OH)]_4 \cdot H_2N(CH_2)_4NH_2 \cdot 2\ THF\}$. As may be seen also by inspection of Fig. 11 the diamine is coordinated to two hydroxyl groups situated

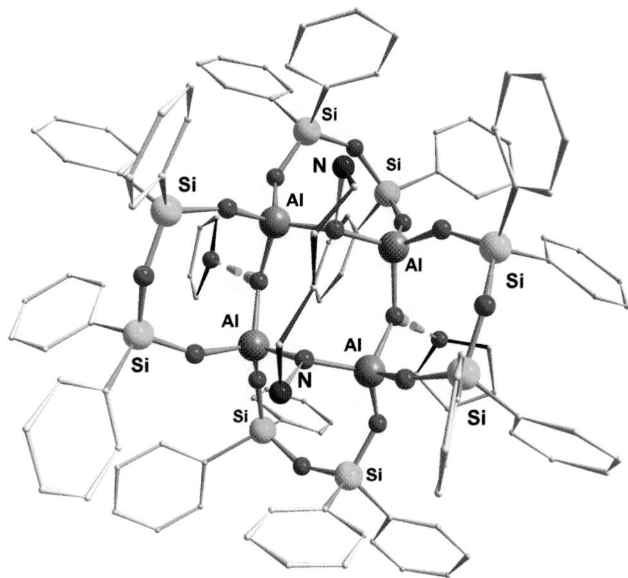

FIG. 11. View of the molecular structure of $\{(Ph_2SiO)_8[AlO(OH)]_4 \cdot H_2N(CH_2)_4NH_2 \cdot 2\ THF\}$ with hydrogen atoms omitted for clarity.[78]

in one of the "baskets" of $(Ph_2SiO)_8[AlO(OH)]_4$, while the other hydroxyl groups are coordinated by the oxgen atoms of the THF molecules (the formula A′ in Scheme 2 denotes that the second diamine of formula A is replaced by two THF molecules). All hydroxyl groups are participating in hydrogen bridges (either O···H···O or O···H···N) and the diamine is "chelating" the two hydroxyls forming a 10-membered $O–H–N(H_2)–C(H_2)–C(H_2)–C(H_2)–C(H_2)–N(H_2)–H–O$ loop.

In Fig. 12 the crystal structure of $\{(Ph_2SiO)_8[AlO(OH)]_4 \cdot 2\ H_2N(CH_2)_3NH_2\}$ is depicted. Here, the two 1,3-diaminopropane molecules are bonded through O–H···N hydrogen bridges on each side of the molecule in such a way that both nitrogen atoms are involved in bonding and a loop is created connecting the two hydroxyl groups situated on the same face of the molecule (compare also formula A of Scheme 2). The two loops are pointing in opposite directions and the point symmetry of the molecule is not far from S_4. The same sort of compound can also be obtained with 1,4-diaminobutane in the place of the diaminopropane (although until now we were not able to isolate the compound in crystalline form; see also below) or with 1,5-diaminopentane (see below). In Table II some pertinent bond lengths and angles are given for diamino adducts of $(Ph_2SiO)_8[AlO(OH)]_4$ having three, four or five CH_2 groups between the amino ends. Even with 1,3-dimethylaminopropane a 1:2 adduct is formed, a structural representation of which can be found in Fig. 13.

The most striking structural element in these 1:2 adducts of $(Ph_2SiO)_8[AlO(OH)]_4$ with poly methylene diamines is the asymmetry in the bonding of the two diamines. In Table II always two pairs of O···N and Al–O distances are given: the first pair is taken from one side of the molecule and the second from the opposite side. It is evident that although the two diamines are chemically and structurally identical and

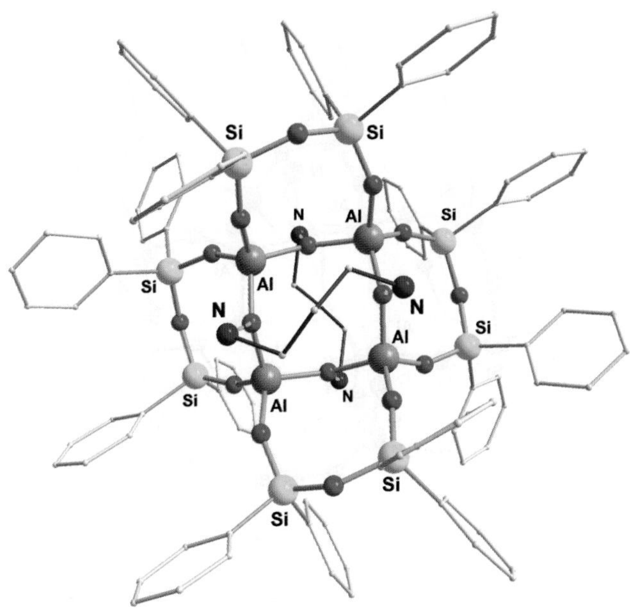

FIG. 12. The structure of the adduct $\{(Ph_2SiO)_8[AlO(OH)]_4 \cdot 2\ H_2N(CH_2)_3NH_2\}$.[78]

TABLE II

SOME ADDUCTS OF $(Ph_2SiO)_8[AlO(OH)]_4$ WITH POLY-METHYLENE DIAMINES

Hydrogen bridge N···O (Å)		Al–O(H) (Å)	Al–O–Al (°)	Number of guest molecules
N···H–O	2.571(6)	1.738(4)	134.7(1)	$1H_2N(CH_2)_4NH_2$/ 2 THF
N···H–O mean	2.69(2)	1.78(2)	129.3(5)	$2Me_2N(CH_2)_3NMe_2$
Diamine 1	2.713(8)	1.813(5)	128.7(2)	
Diamine 2	2.663(8)	1.742(5)	129.9(2)	
N···H–O mean	2.63(3)	1.77(3)	131.0(3)	$2H_2N(CH_2)_3NH_2$
Diamine 1	2.656(7)	1.798(5)	129.7(1)	
Diamine 2	2.606(7)	1.743(5)	132.2(1)	
N···H–O mean	2.57(3)	1.78(2)	132.3(4)	$2H_2N(CH_2)_5NH_2$
Diamine 1	2.603(7)	1.803(5)	131.9(1)	
Diamine 2	2.539(7)	1.747(5)	132.6(1)	

although the two basket-like faces of the molecule $(Ph_2SiO)_8[AlO(OH)]_4$ are equal, the bonding interaction of one diamine is different from the other. It seems as if the opening of one face of the molecule is influencing the other one, accommodating the two diamino molecules in different ways. That this interpretation is valid comes from a closer inspection of the structural features of the compounds: the opening of the two faces of $(Ph_2SiO)_8[AlO(OH)]_4$ is indeed different on the two sides of the molecule as may be seen by comparing similar intra-molecular distances on both sides of the diamino complexes.

The asymmetric bonding of the two diamines in complexes of the type $\{(Ph_2SiO)_8[AlO(OH)]_4 \cdot 2\ H_2N(CH_2)_nNH_2\}$ is found in the crystal structures of

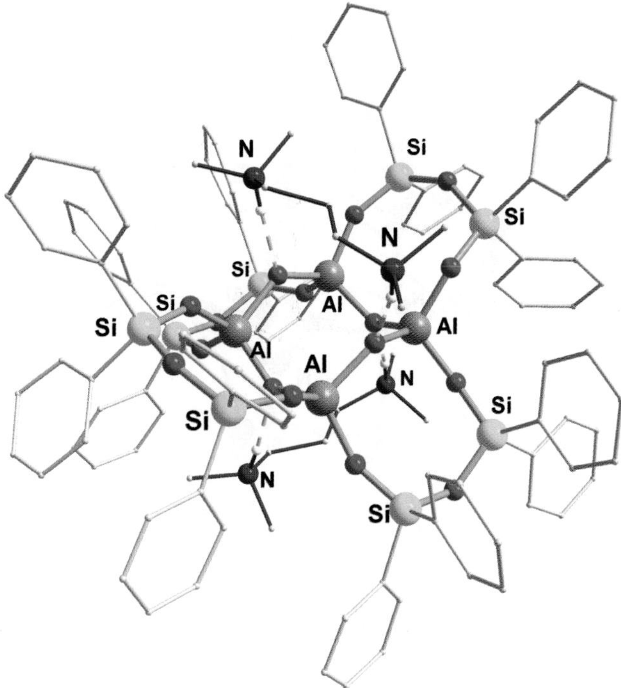

FIG. 13. The double 1,3-dimethylaminopropane adduct of $(Ph_2SiO)_8[AlO(OH)]_4$.[78]

the compounds, but could also be a major property in solution. Interestingly, a further diamine can easily interact with the 1:2 adducts of $(Ph_2SiO)_8[AlO(OH)]_4$ forming a 1:3 adduct, although from an entropic point of view this reaction should not be favored. That this is nevertheless occurring could be explained by the fact that one of the two diamines in $\{(Ph_2SiO)_8[AlO(OH)]_4 \cdot 2\ H_2N(CH_2)_nNH_2\}$ is more loosely bonded than the other one (compare Table II).

In Fig. 14 a structural representation of the 1:3 adduct of $(Ph_2SiO)_8[AlO(OH)]_4$ with three molecules of $H_2N(CH_2)_3NH_2$ is shown (compare also formula B of Scheme 2; in the crystal lattice there is a fourth 1,3-diaminopropane molecule present in the unit cell which is loosely coordinated to one of the end-on coordinated diamines). Comparing Fig. 12 with Fig. 14 (or the formulas A and B of Scheme 2) reveals that one of the diamino-loops has been opened to give place to the accommodation of a second diamine (on the same side of the host molecule). For this reason one of the diamines is still bonding in a chelating mode whereas the two others only have one end of the diamine hydrogen bridged to the hydroxyl groups. The free ends of the 1,3-diaminopropanes have no further contacts apart from van der Waals, and the sequence N–C–C–C–N is too short to connect to other molecules in the crystal.

Before discussing the reaction of $\{(Ph_2SiO)_8[AlO(OH)]_4 \cdot 2\ H_2N(CH_2)_4NH_2\}$ with a further molecule of 1,4-diaminobutane we have to report on the behavior of diethyl ether solutions of $\{(Ph_2SiO)_8[AlO(OH)]_4 \cdot 2\ H_2N(CH_2)_4NH_2\}$ on standing: indeed a

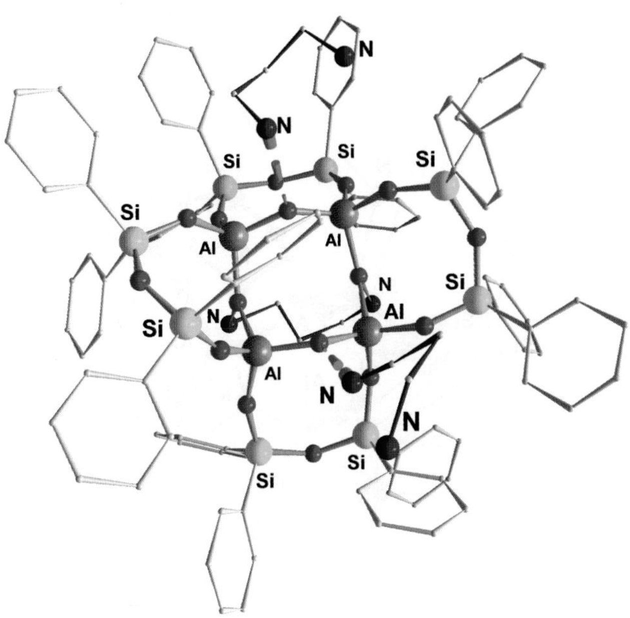

FIG. 14. The molecular structure of $\{(Ph_2SiO)_8[AlO(OH)]_4 \cdot 3\ H_2N(CH_2)_3NH_2\}$.[78]

crystalline precipitation takes place and the crystals formed in such a way are difficult to dissolve again. As revealed by an X-ray structure determination the compound formed on standing is a polymer, the structure of which is shown in Fig. 15. Molecular entities of the formula $\{(Ph_2SiO)_8[AlO(OH)]_4 \cdot H_2N(CH_2)_4NH_2\}$ and having one diamino loop are inter-molecularly connected to a one-dimensional sinusoidal array by a second 1,4-diaminobutane with the two hydroxyl groups not occupied by the intra-molecular loop serving as acidic connecting centers. In Scheme 2 the structure is denoted as B′ indicating that the bonding at $(Ph_2SiO)_8[AlO(OH)]_4$ is similar to B with the difference that the ending amino groups are serving as ligand to a neighboring $(Ph_2SiO)_8[AlO(OH)]_4$ molecule. Although the stoichiometric ratio of 1,4-diaminobutane to $(Ph_2SiO)_8[AlO(OH)]_4$ is maintained at 2:1 (as every connecting diamine belongs always to two $(Ph_2SiO)_8[AlO(OH)]_4$ units, the stoichimetry of amine to alumo-siloxane is 2:1), the crystal structure of $\{(Ph_2SiO)_8[AlO(OH)]_4 \cdot 2\ H_2N(CH_2)_4NH_2\}$ is completely different from that of $\{(Ph_2SiO)_8[AlO(OH)]_4 \cdot 2\ H_2N(CH_2)_3NH_2\}$ which consists of separated molecular entities which only show van der Waals contacts to neighboring molecules. Contrarily, in $\{(Ph_2SiO)_8[AlO(OH)]_4 \cdot 2\ H_2N(CH_2)_4NH_2\}$ the molecules are connected through O···H···N hydrogen bonds from the amino-hydroxy partners and the 1,4-diaminobutane occupying inversion centers in the crystal lattice. Clearly the length of the methylene chain in the diamines plays an important part, with three CH_2-groups being too short for one-dimensional aggregation and with four CH_2-groups just long enough to connect the $(Ph_2SiO)_8[AlO(OH)]_4$ molecules.

It is clear that the balance between molecular van der Waals lattices of structure type A and those of polymers B′ with chain lengths of the diamines $H_2N(CH_2)_nNH_2$

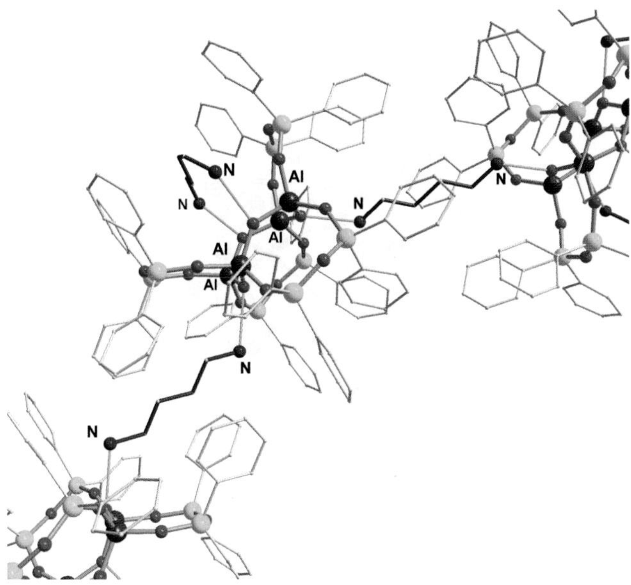

FIG. 15. A section of the one-dimensional structure of $\{(Ph_2SiO)_8[AlO(OH)]_4 \cdot (\eta^2 -H_2N(CH_2)_4NH_2)$ $(H_2N(CH_2)_4NH_2)\}$.[78]

$n>3$ may be influenced by the concentration of the compounds $\{(Ph_2SiO)_8[AlO(OH)]_4 \cdot 2\ H_2N(CH_2)_nNH_2\}$ in the organic solvents. Indeed, we have succeeded using this simple approach to crystallize $\{(Ph_2SiO)_8[AlO(OH)]_4 \cdot 2\ H_2N(CH_2)_5NH_2\}$ in a packing where no intermolecular bonding is observed (Fig. 16). This compound again belongs to structure-type of Scheme 2 like the corresponding adducts with $H_2N(CH_2)_3NH_2$ and $Me_2N(CH_2)_3NMe_2$ (Figs. 12 and 13).

The bite angle of the diamine $H_2N(CH_2)_nNH_2$ with respect to the open baskets and OH-groups of $(Ph_2SiO)_8[AlO(OH)]_4$ also seems to have some influence on the base adducts and their structures. We have noticed that taking 1,7-diaminoheptane in the place of the diaminobutanes or -pentanes also has an important impact on the structure of the adduct. Of course a longer CH_2-chain length should increase the flexibility of the diamines, but should on the other hand hinder the formation of the loops because the energy gain by the chelating effect becomes (owing to the larger rings) less important. In Fig. 17 a section of the lattice of $\{(Ph_2SiO)_8[AlO(OH)]_4 \cdot 2\ H_2N(CH_2)_7NH_2\}$ is shown revealing that all diamines are strictly bonded in the intermolecular connecting fashion (see also formula C of Scheme 2). In the crystal a three-dimensional polymer is found with the centers of the $(Ph_2SiO)_8[AlO(OH)]_4$ units, almost forming a diamond lattice.[78]

An interesting phenomenon was recently discovered when $\{(Ph_2SiO)_8[AlO(OH)]_4 \cdot 2\ H_2N(CH_2)_4NH_2\}$ is reacted with a further equivalent of 1,4-diaminobutane. A two-dimensional polymer is obtained of the composition $\{(Ph_2SiO)_8[AlO(OH)]_4 \cdot 3\ H_2N(CH_2)_4NH_2\}$, the structure of which is sketched in Fig. 18. The structure can be derived from that of $\{(Ph_2SiO)_8[AlO(OH)]_4 \cdot 2\ H_2N(CH_2)_4NH_2\}$ by

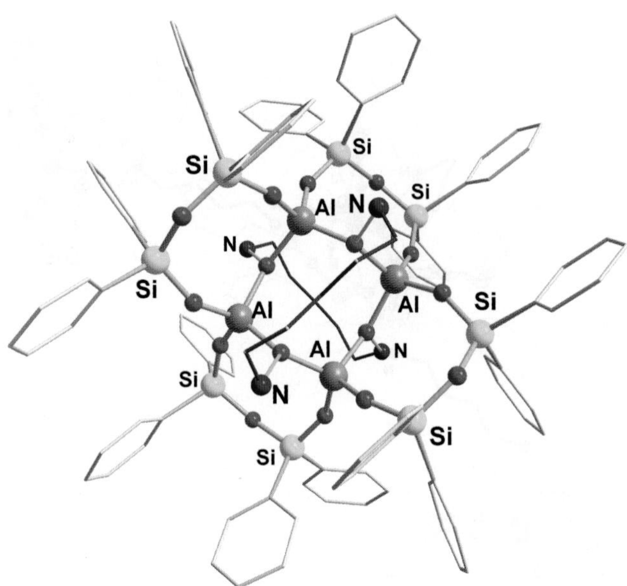

FIG. 16. Molecular structure of $\{(Ph_2SiO)_8[AlO(OH)]_4 \cdot 2H_2N(CH_2)_5NH_2\}$.[78]

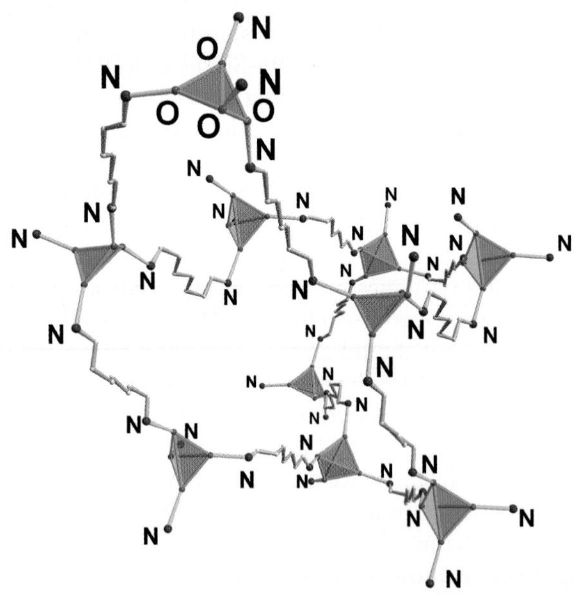

FIG. 17. A simplified representation of the crystal structure of $\{(Ph_2SiO)_8[AlO(OH)]_4 \cdot 2\ H_2N(CH_2)_7NH_2\}$. The $(Ph_2SiO)_8[AlO(OH)]_4$ molecules are sketched as flattened tetrahedra with the OH groups as corners. Most of the organic groups as well as the positions of all the aluminum, silicon and oxygen atoms are omitted for clarity with exception of the heptane chain.[78]

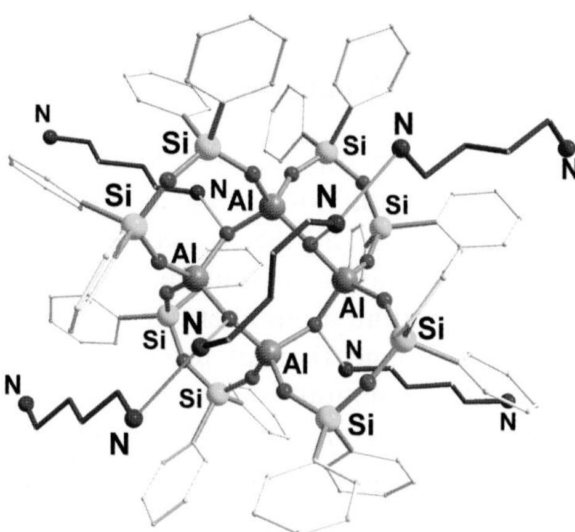

FIG. 18. A section of the crystal structure of $\{(Ph_2SiO)_8[AlO(OH)]_4 \cdot 3\ H_2N(CH_2)_4NH_2\}$. See also Scheme 3 for comparison. One of the alumo-siloxane with all 1,4-diaminobutane molecules coordinated to it is shown. All end amino groups shown in the picture are bonded to further alumo-siloxane entities. Two sorts of hydrogen bridges are present: O–H···N and N–H···N.[78]

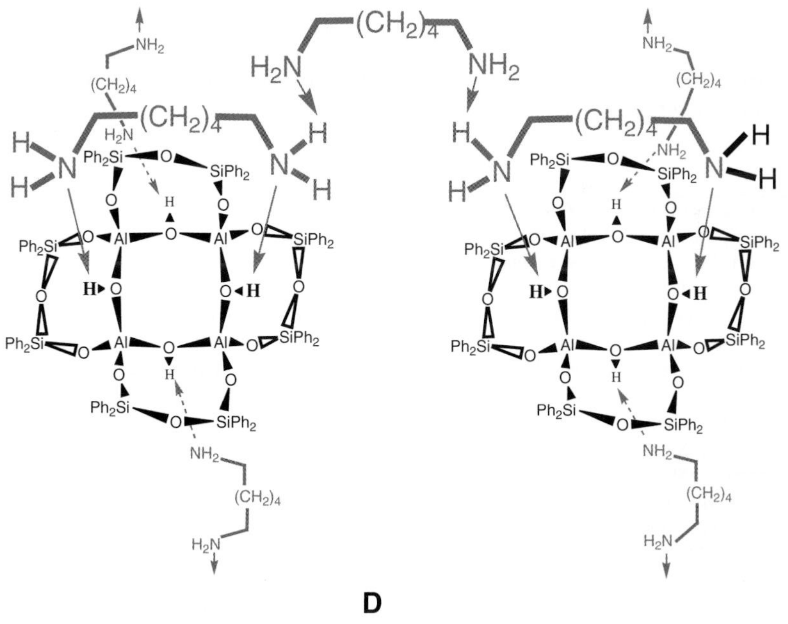

D

SCHEME 3.

forming one-dimensional chains. The extra $H_2N(CH_2)_4NH_2$ is connecting the one-dimensional chains by N–H···N bonding at the nitrogen atoms involved in the loops (see also structure D in Scheme 3).

V
CONCLUSIONS

The adducts of the polycyclic compound $(Ph_2SiO)_8[AlO(OH)]_4$ with different bases show that this molecular amino-siloxane can be easily used for supramolecular assemblies.[83–85] Of course, all the structures shown in this review are in the crystalline state and the equilibria in solution, until now, have not been addressed to in detail to get quantitative numbers in form of the equilibrium constants. Qualitatively, we have found that all reactions are clearly shifted to the adduct sides and that the transformation from molecular to polymeric adducts can be stimulated by changing the concentrations of the solutions. Apparently the OH-groups in the polycyclic alumo-siloxane $(Ph_2SiO)_8[AlO(OH)]_4$ behave as ideal protic centers. The basket shapes of the organo-siloxy ligands nicely accompany the accommodation of the diamines by adopting different geometries. We are actually studying some applications in surface chemistry.

Acknowledgments

I would like to thank all my coworkers for their valuable contributions to this field of chemistry, especially Maria Jarczyk as the person of the very first moments, Hinka Hreleva for her enthusiasm, Volker Huch for the help in the crystal structure work and Andreas Rammo for his interest and discussions. I also thank Julia Biegler, Nicole Kern and Andreas Leinenbach for their contributions. The Fonds of the Chemische Industrie as well as the Deutsche Forschungsgemeinschaft (GRK 532, SFB 277) are acknowledged for their financial help.

References

(1) Cundy, C. S.; Cox, P. A. *Chem. Rev.* **2003**, *103*, 663.
(2) Jones, C. W. *Science* **2003**, 439.
(3) Turro, J. N. *Acc. Chem. Res.* **2000**, *33*, 637.
(4) Thomas, J. M. *Progr. Inorg. Chem.* **1987**, *35*, 1.
(5) Ullmann Encyclopedia, "Zeolites", 5th ed., Vol. A28, Urban Schwarzenberg Verlag, München/Berlin, Germany. 1995, p. 1.
(6) Ihlein, G.; Junges, B.; Junges, U.; Laeri, F.; Schüth, F.; Vietze, U. *Appl. Organomet. Chem.* **1998**, *12*, 305.
(7) Schüth, F.; Schmidt, W. *Adv. Engen. Mater.* **2002**, *4*, 269.
(8) Schüth, F.; Sing, K. S. W.; Weitkamp, J. *Handbook of Porous Solids*, Wiley-VCH Verlag, Weinheim, 2003.
(9) Budd, P. M.; Makhseed, S. M.; Ghanem, B. S.; Msayib, K. J.; Tattershall, C. E.; McKeown, N. B. *Mater. Today,* **2004**, *April*, 40.
(10) Anderson, M. W.; Ohsuna, T.; Sakamoto, Y.; Liu, Z.; Carlsson, A.; Terasaki, O. *Chem. Commun.* **2004**, 907.
(11) Cassidy, J. E.; Jarvis, J. A. J.; Rothon, R. N. *J. Chem. Soc. Dalton Trans.* **1975**, 1497.
(12) Loiseau, T.; Mellot-Draznieks, C.; Sassoye, C.; Girard, S.; Guillou, N.; Huguenard, C.; Taulelle, F.; Ferey, G. *J. Am. Chem. Soc.* **2001**, *123*, 9642.
(13) Sugiyama, K.; Hiraga, K.; Yu, J.; Zheng, S.; Qiu, S.; Xu, R.; Terasaki, O. *Acta Crystallogr. Sect. C: Cryst. Struct. Commun.* **1999**, *55*, 1615.
(14) Vidal, L.; Gramlich, V.; Patarin, J.; Gabelica, Z. *Eur. J. Solid State Inorg. Chem.* **1998**, *35*, 545.
(15) Pluth, J. J.; Smith, J. V. *Acta Crystallogr. Sect. C: Cryst. Struct. Commun.* **1987**, *43*, 866.
(16) Riou, D.; Loiseau, Th.; Ferey, G. *J. Solid State Chem.* **1992**, *99*, 414.

(17) Xu, Y.-H.; Zhang, B.-G.; Chen, X.-F.; Liu, S.-H.; Duan, C.-Y.; You, X.-Z. *J. Solid State Chem.* **1999**, *145*, 220.
(18) Yu, J.; Sugiyama, K.; Hiraga, K.; Togashi, N.; Terasaki, O.; Tanaka, Y.; Nakata, S.; Qiu, S.; Xu, R. *Chem. Mater.* **1998**, *10*, 3636.
(19) Jones, R. H.; Thomas, J. M.; Xu, R.; Huo, Q.; Cheetham, A. K.; Powell, A. V. *Chem. Commun.* **1991**, 1266.
(20) Simon, N.; Loiseau, T.; Ferey, G. *J. Chem. Soc. Dalton Trans.* **1999**, 1147.
(21) Jones, R. H.; Chippindale, A. M.; Natarajan, S.; Thomas, J. M. *Chem. Commun.* **1994**, 565.
(22) Chippindale, A. M.; Natarajan, S.; Thomas, J. M.; Jones, R. H. *J. Solid State Chem.* **1994**, *111*, 18.
(23) Chippindale, A. M.; Powell, A. V.; Jones, R. H.; Thomas, J. M.; Cheetham, A. K.; Huo, Q.; Xu, R. *Acta Crystallogr. Sect. C: Cryst. Struct. Commun.* **1994**, *50*, 1537.
(24) Wright, P. A.; Maple, M. J.; Slawin, A. M. Z.; Patinec, V.; Aitken, R. A.; Welsh, S.; Cox, P. A. *J. Chem. Soc. Dalton Trans.* **2000**, 1243.
(25) Chippindale, A. M.; Walton, R. I. *J. Solid State Chem.* **1999**, *145*, 731.
(26) Gray, M. J.; Jasper, J. D.; Wilkinson, A. P.; Hanson, J. C. *Chem. Mater.* **1997**, *9*, 976.
(27) Feng, P.; Bu, X.; Stucky, G. D. *Inorg. Chem.* **2000**, *39*, 2.
(28) Kedarnath, K.; Choudhury, A.; Natarajan, S. *J. Solid State Chem.* **2000**, *150*, 324.
(29) Muncaster, G.; Sanker, G.; Catlow, C. R. A.; Thomas, J. M.; Bell, R. G.; Wright, P. A.; Coles, S.; Teat, S. J.; Clegg, W.; Reeve, W. *Chem. Mater.* **1999**, *11*, 158.
(30) Leech, M. A.; Cowley, A. R.; Prout, K.; Chippindale, A. M. *Chem. Mater.* **1998**, *10*, 451.
(31) Oliver, S.; Kuperman, A.; Lough, A.; Ozin, G. A. *Inorg. Chem.* **1996**, *35*, 6373.
(32) Xu, Y.-H.; Yu, Z.; Chen, X.-F.; Liu, S.-H.; You, X.-Z. *J. Solid State Chem.* **1999**, *146*, 157.
(33) Kongshaug, K. O.; Fjellvag, H.; Lillerud, K. P. *Microporous Mesoporous Mater* **2000**, *40*, 313.
(34) Hix, G. B.; Carter, V. J.; Wragg, D. S.; Morris, R. E.; Wright, P. A. *J. Mater. Chem.* **1999**, *9*, 179.
(35) Cabeza, A.; Aranda, M. A. G.; Bruque, S.; Poojary, D. M.; Clearfield, A.; Sanz, J. *Inorg. Chem.* **1998**, *37*, 4168.
(36) Yang, Y.; Walawalkar, M. G.; Pinkas, J.; Roesky, H. W.; Schmidt, H.-G. *Angew. Chem. Int. Ed. Engl.* **1998**, *37*, 96.
(37) Wulff-Molder, D.; Meisel, M. Z. *Kristallogr. New Cryst. Struct.* **1998**, *213*, 353.
(38) Mason, M. R.; Perkins, A. M.; Ponomarova, V. V.; Vij, A. *Organometallics* **2001**, *20*, 4833.
(39) Yang, Y.; Schmidt, H.-G.; Noltemeyer, M.; Pinkas, J.; Roesky, H. W. *J. Chem. Soc. Dalton Trans.* **1996**, 3609.
(40) Azais, T.; Bonhomme-Coury, L.; Vaissermann, J.; Maquet, J.; Bonhomme, C. *Eur. J. Inorg. Chem.* **2002**, 2838.
(41) Maeda, K.; Akimoto, J.; Kiyozumi, Y.; Mizukami, F. *Chem. Commun.* **1995**, 1033.
(42) Sangokoya, S. A.; Pennington, W. T.; Robinson, G. H.; Hrncir, D. C. *J. Organomet. Chem.* **1990**, *385*, 23.
(43) Chakraborty, D.; Horchler, S.; Kratzner, R.; Varkey, S. P.; Pinkas, J.; Roesky, H. W.; Uson, I.; Noltemeyer, M.; Schmidt, H.-G. *Inorg. Chem.* **2001**, *40*, 2620.
(44) Corker, J. M.; Browning, D. J.; Webster, M. *Acta Crystallogr. Sect. C: Cryst. Struct. Commun.* **1996**, *52*, 583.
(45) Murugavel, R.; Voigt, A.; Walawalker, M. G.; Roesky, H. W. *Chem. Rev.* **1996**, *96*, 2205.
(46) Murugavel, R.; Walawalker, M. G.; Dari, M.; Roesky, H. W.; Rao, C. N. R. *Acc. Chem. Res.* **2004**, *37*, 763.
(47) King, L.; Sullivan, A. C. *Coord. Chem. Rev.* **1999**, *6*, 1.
(48) Lickiss, P. D. *Adv. Inorg. Chem.* **1995**, *42*, 147.
(49) Lickiss, P. D., Rappaport, Z., Apeloig, Y., Eds., *The Chemistry of Organic Silicon Compounds*, Vol. 3, Wiley, New York, USA, p. 695 2001.
(50) Veith, M.; Jarczyk, M.; Huch, V.Organic-Inorganic Hybrids, ISBN 0953691308, **2000**, Paper 19.
(51) Veith, M., Rammo, A., Jutzi, P., Schubert, U., Eds., *Silicon Chemistry-From the Atom to Extended Systems*, Wiley-VCh, Weinheim, Germany, p. 360 2003.
(52) Veith, M.; Hreleva, H.; Biegler, J.; Huch, V.; Rammo, A. *Phosphor. Sulf. Silicon* **2004**, *179*, 651.
(53) Smolin, Yu. I.; Shepelev, Yu. F.; Ershov, A. S.; Khobbel, D. *Dokl. Akad. Nauk SSSR (Russ.) (Proc. Nat. Acad. Sci. USSR)* **1987**, *297*, 1377.

(54) Klemp, A.; Hatop, H.; Roesky, H. W.; Schmidt, H.-G.; Noltemeyer, M. *Inorg. Chem.* **1999**, *38*, 5832.
(55) Montero, M. L.; Voigt, A.; Teichert, M.; Uson, I.; Roesky, H. W. *Angew. Chem. Int. Ed. Engl.* **1995**, *34*, 2504.
(56) Ritter, U.; Winkhofer, N.; Murugavel, R.; Voigt, A.; Stalke, D.; Roesky, H. W. *J. Am. Chem. Soc.* **1996**, *118*, 8580.
(57) Chandrasekhar, V.; Murugavel, R.; Voigt, A.; Roesky, H. W.; Schmidt, H.-G.; Noltemeyer, M. *Organometallics* **1996**, *15*, 918.
(58) Feher, F. J.; Budzichowski, T. A.; Weller, K. J. *J. Am. Chem. Soc.* **1989**, *111*, 7288.
(59) Feher, F. J.; Weller, K. *J. Organometallics* **1990**, *9*, 2638.
(60) Feher, F. J.; Weller, K. J.; Ziller, J. W. *J. Am. Chem. Soc.* **1992**, *114*, 9686.
(61) Edelmann, F. T.; Gun'ko, Y. K.; Giessmann, S.; Olbrich, F.; Jacob, K. *Inorg.Chem.* **1999**, *38*, 210.
(62) Duchateau, R.; Harmsen, R. J.; Abbenhuis, H. C. L.; van Santen, R. A.; Meetsma, A.; Thiele, S. K.-H.; Kraneburg, M. *Chem. Eur. J.* **1999**, *5*, 3130.
(63) Skowronska-Ptasinska, M. D.; Duchateau, R.; van Santen, R. A.; Yap, G. P. A. *Eur. J. Inorg. Chem.* **2001**, 133.
(64) Skowronska-Ptasinska, M. D.; Duchateau, R.; van Santen, R. A.; Yap, G. P. A. *Organometallics* **2001**, *20*, 3519.
(65) Montero, M. L.; Uson, I.; Roesky, H. W. *Angew. Chem. Int. Ed. Engl.* **1994**, *33*, 2103.
(66) Klemp, A.; Hatop, H.; Roesky, H. W.; Schmidt, H.-G.; Noltemeyer, M. *Inorg. Chem.* **1999**, *38*, 5832.
(67) Chandrasekhar, V.; Murugavel, R.; Voigt, A.; Roesky, H. W.; Schmidt, H.-G.; Noltemeyer, M. *Organometallics* **1996**, *15*, 918.
(68) Bonamico, M.; Dessy, G. *J. Chem. Soc. A* **1968**, 291.
(69) McMahon, C. N.; Obrey, S. J.; Keys, A.; Bott, S. G.; Barron, A. R. *J. Chem. Soc. Dalton Trans.* **2000**, 2151.
(70) Veith, M.; Jarczyk, M.; Huch, V. *Angew. Chem. Int. Ed. Engl.* **1997**, *36*, 117.
(71) Gun'ko, Y. K.; Reilly, R.; Kessler, V. G. *New J. Chem. (Nouv. J. Chim.)* **2001**, *25*, 528.
(72) Fujdala, K. L.; Tilley, T. D. *J. Am. Chem. Soc.* **2001**, *123*, 10133.
(73) Veith, M.; Faber, S.; Wolfanger, H.; Huch, V. *Chem. Ber.* **1996**, *129*, 381.
(74) Veith, M.; Rammo, A.; Leinenbach, A., results to be published.
(75) Veith, M.; Jarczyk, M.; Huch, V. *Phosphor., Sulf. Silicon* **1997**, *124 & 125*, 213.
(76) Veith, M.; Rammo, A.; Jarczyk, M.; Huch, V. *Monatshefte für Chemie* **1999**, *130*, 15.
(77) Veith, M.; Jarczyk, M.; Huch, V. *Angew. Chem. Int. Ed.* **1998**, *37*, 105.
(78) Veith, M.; Hreleva, H.; Biegler, J.; Huch, V. *Phosphor. Sulf. Silicon* **2004**, *179*, 651; further results to be published.
(79) Program 'HyperChem', release 7.5, Hypercube Inc., USA 2002.
(80) Hunter, E. P.; Lias, S. G. *J. Phys. Chem. Ref. Data* **1998**, *27*, 413.
(81) Arnett, E. M.; Jones, F. M.; Taagepara, M.; Henderson, W. G.; Beauchamp, J. L.; Holtz, D.; Taft, R. W. *J. Am. Chem. Soc.* **1972**, *94*, 4724.
(82) Aue, D. H.; Webb, H. M.; Bowers, M. T. *J. Am. Chem. Soc.* **1972**, *94*, 4726.
(83) Vögtle, F. *Supramolecular Chemistry: An Introduction*, Wiley, New York, 1993.
(84) Lehn, J. M. *Supramolecular Chemistry-Concepts and Perspectives*, VCH, Weinheim, Germany, 1995.
(85) Vilar, R. *Angew. Chem. Int. Ed.* **2003**, *42*, 1460.

Progress in the Chemistry of Stable Disilenes

MITSUO KIRA[a,*] and TAKEAKI IWAMOTO[a,b]

[a]*Department of Chemistry, Graduate School of Science, Tohoku University, Aoba-ku, Sendai 980-8578, Japan*

[b]*PRESTO, Japan Science and Technology Agency, 4-1-8 Honcho Kawaguchi, Saitama 332-0012, Japan*

I
INTRODUCTION

It is not surprising that silicon chemists in the early days tried to synthesize compounds having silicon-based multiple bonds like Si=Si, Si=C, Si=O, etc., since multiply bonded compounds of carbon are common.[1] In the earliest efforts to synthesize silicon–silicon doubly bonded compounds (disilenes), Kipping *et al.*

*Corresponding author.
E-mail: mkira@mail.tains.tohoku.ac.jp (M. Kira).

ADVANCES IN ORGANOMETALLIC CHEMISTRY
VOLUME 54 ISSN 0065-3055/DOI 10.1016/S0065-3055(05)54003-6

© 2006 Elsevier Inc.
All rights reserved.

reported the reduction of diphenyldichlorosilane with sodium in 1921.[2] Among a number of products they isolated, a compound with the composition of $[Ph_4Si_2]$ was finally determined to be octaphenylcyclotetrasilane. The elusive nature of disilenes and other doubly bonded compounds of heavier group-14 elements was recognized soon but it took several decades to obtain convincing evidence for the existence of a transient disilene. Peddle and Roark showed that tetramethyldisilene **A** was generated during the thermolysis of disilene-bridged naphthalene **B** at 350 °C; **A** was transferred to anthracene in its presence to give **C** [Eq. (1)].[3] Using (*E*)- and (*Z*)-isomers of 1,2-diphenyl-1,2-dimethyldisilene-bridged naphthalenes, the stereochemistry around the Si=Si double bond was found to be maintained during the disilene transfer, showing a significant energy barrier for the *E*,*Z*-isomerization.[4]

$$\mathbf{B} \xrightarrow{350\ ^\circ C} \text{naphthalene} + [Me_2Si{=}SiMe_2]\ (\mathbf{A}) \xrightarrow{\text{anthracene}} \mathbf{C} \quad (1)$$

In 1981, West *et al.* synthesized the first stable disilene **1** *via* the dimerization of the corresponding silylene generated by the photolysis of a trisilane and characterized the structure by conventional spectroscopies [Eq. (2)].[5] Availability of **1** and other stable disilenes has stimulated theoretical and experimental studies of various aspects of disilenes such as their bonding and structure, spectroscopic properties, reactivities, applications to the synthesis of novel types of organosilicon compounds, etc.

$$Mes_2Si(SiMe_3)_2 \xrightarrow[-Me_3SiSiMe_3]{h\nu} [Mes_2Si:] \xrightarrow{\times 2} Mes_2Si{=}SiMes_2\ (\mathbf{1}) \quad (2)$$

Mes = 2,4,6-trimethylphenyl

Among a number of review articles on stable disilenes published until now,[6–9] the most comprehensive would be those written by Okazaki and West in 1996, which is hereafter abbreviated to "review OW",[7] and by Weidenbruch in 2001.[9] At the time when review OW was written, the number of known stable disilenes was 27 and the type was limited only to acyclic disilenes with rather simple substitution modes. During less than 10 years since then, the number of stable disilenes increased dramatically up to more than 70, and the types include not only acyclic disilenes but also endo- and exocyclic disilenes, bicyclic disilenes, and tetrasiladienes. Very recently, the syntheses of the first stable trisilaallene and disilyne were achieved. All these recent experimental achievements have stimulated renewed theoretical studies, and the interplay between theory and experiments has contributed to the innovation of the understanding of the bonding, structure, and reactivity of disilenes and to the development of the potentials of silicon chemistry. Since review OW appeared in a volume of *Adv. Organomet. Chem.* series, the present article is intended to be a supplemental review covering the studies performed during 1996–2004.

II

SYNTHESIS

In review OW,[7] only four types of synthetic methods are introduced for stable disilenes as shown in Scheme 1: (1) photolysis of linear trisilanes (method A), (2) photolysis of cyclic trisilanes (method B), (3) reductive dehalogenation of dihalosilanes (method C), and (4) reductive dehalogenation of 1,2-dihalodisilanes (method D).

Using these methods, 27 acyclic disilenes with the substitution types of $A_2Si{=}SiA_2$, $A_2Si{=}SiB_2$, (*E*)- and (*Z*)-ABSi=SiAB, and $A_2Si{=}SiAB$ as shown in Chart 1 had been synthesized, at the time when review OW was written. In this article, the numbers for labeling these disilenes are kept to be identical with those used in review OW for convenience. The properties of 47 stable disilenes and two disilynes newly synthesized since then are summarized in Tables I–IV.

In this section, discussion is focused on recent advances in novel synthetic methods for various types of stable disilenes.

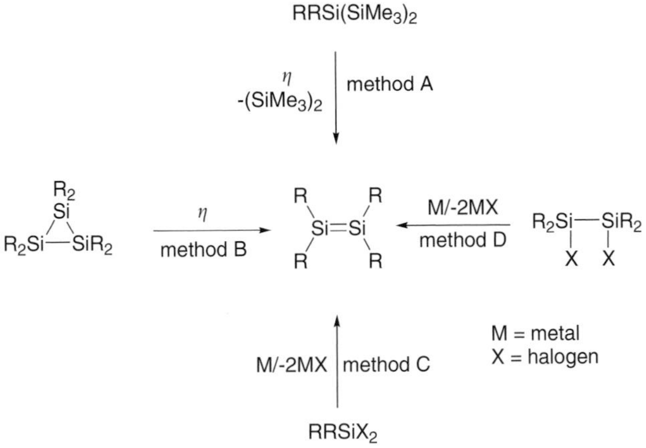

$RRSi(SiMe_3)_2$ —[$h\nu$, $-(SiMe_3)_2$; method A]→ $R_2Si{=}SiR_2$

cyclo-$(R_2Si)_3$ —[$h\nu$; method B]→ $R_2Si{=}SiR_2$

$R_2SiX{-}SiXR_2$ —[M/-2MX; method D]→ $R_2Si{=}SiR_2$

$RRSiX_2$ —[M/-2MX; method C]→ $R_2Si{=}SiR_2$

M = metal
X = halogen

SCHEME 1.

$A_2Si{=}SiA_2$

1, A = Mes
2, A = Tip
11, A = Xyl
12, A = Dmt
19,A = Dep
21, A = *t*-Bu
22, R = *t*-$BuMe_2Si$
23, A = i-Pr_2MeSi
24, A = *i*-Pr_3Si
25, A = $CH(SiMe_3)_2$

ABSi=SiAB

3, A = Mes, B = *t*-Bu
4, A = Mes, B = 1-Ad
5, A = Mes, B = Xyl
6, A = Mes, B = Dmt
7, A = Mes, B = $N(SiMe_3)_2$
8, A = Tip, B = Mes
9, A = Tip, B = $SiMe_3$
10, A = Tip, B = t-Bu
20, A = Mes, B = Dep
26, A = Mes, B = Dip
27, A = Mes, B = Tbt

$A_2Si{=}SiB_2$

13, A = Mes, B = Xyl
14, A = Mes, B = Dmt
15, A = Mes, B = Tip

$A_2Si{=}SiAB$

16, A = Mes, B = Xyl
17, A = Xyl, B = Dmt
18, A = Dmt, B = Xyl

1-Ad = 1-Adamantyl
Dep = 2,6-Diethylphenyl
Dip = 2,6-Diisopropylphenyl
Dmt = 2,6-dimethyl-4-tert-butylphenyl
Mes = 2,4,6-Trimethylphenyl
Tbt = 2,4,6-Tris[bis(trimethylsilyl)methyl]phenyl
Tip = 2,4,6-Triisopropylphenyl
Xyl = 2,6-Dimethylphenyl

CHART 1.

TABLE I
PROPERTIES OF STABLE ACYCLIC DISILENES

$R^1R^2Si^a{=}Si^bR^3R^4$ (R^1 and R^2 on Si^a; R^3 and R^4 on Si^b)

R^1	R^2	R^3	R^4		^{29}Si NMR[a] δ (Si)	Geometric parameters[b] d (Å)	β (deg)	τ (deg)	UV–vis[c] λ_{max}[nm (ε)] [solvent]	Habit (mp) (°C)	Ref. (CCDC)[d]
i-Pr_3Si	i-Pr_3Si	Mes	Mes	(**28**)	−0.8 (Si^a) +152.3 (Si^b)	—	—	—	418 (3100) 285 (5900)	—	13
t-Bu_2MeSi	t-Bu_2MeSi	Mes	Mes	(**29**)	+8.2 (Si^a) +148.6 (Si^b)	—	—	—	437 (3200) 289 (6700)	Red brown oil	13
i-Pr_3Si	i-Pr_3Si	Tip	Tip	(**30**)	+14.0 (Si^a) +137.2 (Si^b)	—	—	—	492 (1900) 299 (4500)	—	13
t-Bu_2MeSi	t-Bu_2MeSi	Tip	Tip	(**31**)	+14.9 (Si^a) +142.0 (Si^b)	—	—	—	517 (3600) 281 (7500)	—	13
(t-$Bu_2MeSi)_2MeSi$	t-Bu_2MeSi	t-Bu_2MeSi	Me	(**32**)	+103.8 (Si^a) +158.9 (Si^b)	2.1984(5)	4.35 (Si^a), 6.04 (Si^b)	29.0	429 (7100) 318 (1700) 234 (25 700)	Red cubic crystal (150–152)	16 (CCDC-226552)
t-Bu_2MeSi	t-Bu_2MeSi	t-Bu_2MeSi	t-Bu_2MeSi	(**33**)	+155.5 [toluene-d_8, 200 K]	2.2598(18)	5.29 6.44	54.5	612 (1300) 375 (2000) 290 (8200) 220 (26500)	Blue crystals (189–191)	64 (CCDC-248084)
t-$BuMe_2Si$	i-Pr_2MeSi	t-$BuMe_2Si$	i-Pr_2MeSi	(E)-(**34**)	+141.8	2.196(3)	0.65	0.0	411 (–) 360 (–) [3-methylpentane]	Orange crystals	63
t-$BuMe_2Si$	i-Pr_2MeSi	i-Pr_2MeSi	t-$BuMe_2Si$	(Z)-(**34**)	+141.9	—	—	—	—	—	63
t-$BuMe_2Si$	t-$BuMe_2Si$	i-Pr_2MeSi	i-Pr_2MeSi	(**35**)	+132.4 (Si^a) +156.6 (Si^b)	2.198(1) (2.1942(8))[e]	0.0 (0.0)[e]	8.97 (11.73)[e]	413 (5000) 359 (2000)	Orange crystals (110, decomp.)	14 (CCDC-207190)
t-$BuMe_2Si$	i-Pr_3Si	t-$BuMe_2Si$	i-Pr_3Si	(E)-(**36**)	145.2 (rt) 142.5 (233 K) [toluene-d_8]	—	—	—	519, 365 [3-methylpentane]	Dark violet oil	157
t-$BuMe_2Si$	i-Pr_3Si	i-Pr_3Si	t-$BuMe_2Si$	(Z)-(**36**)	145.2 (rt) 145.5 (233 K) [toluene-d_8]	—	—	—		Dark violet	157
t-$BuMe_2Si$	t-$BuMe_2Si$	i-Pr_3Si	i-Pr_3Si	(**37**)	+142.0 +152.7	2.2011(9)	0.0	27.95	—	Bloody red Crystals	14 (CCDC-207191)
Tip_2HSi	Tip	Tip	Tip_2ClSi	(Z)-(**38**)	+95.99 +102.71 (nd)	2.2149(9)	0.29 (Si^a) 0.34 (Si^b)	18.33[d]	411 (46820)	Bright orange Crystals (171–173)	18 (CCDC-175689)

Tip_2HSi	Tip	Tip	Tip_2BrSi	(*Z*)- (**39**)	+92.59, +104.12 (nd)	2.2088(10)	0.31(Si[a]) 0.64 (Si[b])	17.17[d]	—	Bright orange Crystals (166–168)	18 (CCDC-175690)
Tip	Tip	Tip	$SiMe_3$	(**40**)	+97.7 (Si[a]) +50.9 (Si[b])	—	—	—	—	Orange oil	20
t-Bu_3Si	Ph	*t*-Bu_3Si	Ph	(*E*)-(**41**)	+128.0	2.182(2)	6.69	0.0[d]	398 (1560)	Bright yellow crystals (172)	62 (CCDC-140134)
(*t*-$Bu_3)_2MeSi$	Cl	(*t*-$Bu_3)_2MeSi$	Cl	(*E*)-(**42**)	—	2.163(4)	0.0	0.0	—	Orange red Crystals	22 (CCDC-171748)
t-Bu_3Si	H	*t*-Bu_3Si (or H)	H (or *t*-Bu_3Si)	(*E*)-(**43**)	+141.32 (^1J(Si–H) = 149.8 Hz, ^2J(Si–H) = 0.9 Hz)	—	—	—	—	Yellow	26
$(Me_3Si)_2N$	C_5Me_5	$(Me_3Si)_2N$	C_5Me_5	(*E*)-(**44**)	—	2.1683(5)	9.0 (Si[a]) 6.9 (Si[b])	4.9	—	Yellow crystals	19 (CCDC-242267)
Tep = (2,4,6-$Et_3C_6H_2$)	Tep	Tep	Tep	(**45**)	—	2.143(1)	6.8, 13.6	10.9	—	—	9
Tip_2OHSi	Tip	Tip	Tip_2Si–O–C_6H_4–OH	(**46**)	—	2.220(1)	2.2, 2.7	17.7	—	—	9
Tip_2OHSi	Tip	Tip	Tip_2Si–O–C_6H_2(OH)(*t*-Bu)(*t*-Bu)	(**47**)	—	2.220(1)	0.9	15.1	—	—	9

[a]In benzene-d_8 at rt unless otherwise noted.
[b]Values determined by X-ray analysis; *d*: Si=Si bond distance, β: angle between R^1–Si^a–R^2 (or R^3–Si^b–R^4) plane and Si^a-Si^b axis, γ: angle between R^1–Si^a–R^2 and R^3–Si^b–R^4 planes.
[c]In hexane unless otherwise noted.
[d]Cambridge Crystallographic Data Center.
[e]Crystallographically independent isomer.

TABLE II
PROPERTIES OF STABLE CYCLIC DISILENES

Compound	^{29}Si NMR[a]	Geometric parameters[b]			UV–vis[c]	Habit (mp) (°C)	Ref. (CCDC)[d]
	δ (Si)	d (Å)	β (deg)	R–Si=Si–R (deg)	λ_{max} [nm (ε)] [solvent]		
t-BuMe₂Si, SiMe₂Bu-t, Si, Si^a=Si^b, t-BuMe₂Si, Si(SiMe₂(t-Bu))₃ **48**	+81.9 (Si^a) +99.8 (Si^b)	2.132(2)	—	21.4(2)	482 (2640) 401 (1340) 315 (sh, 7690) 245 (sh, 3670) 217 (5210) [3-methylpentane]	Dark red crystals (150 decomp.)	24,69 (CCDC-215072)
t-Bu₂MeSi, SiMe(t-Bu)₂, Si, Si=Si, t-Bu₂MeSi, SiMe(t-Bu)₂ **49**	+97.7	2.138(2)	—	31.9(2)	466 (440) 297 (sh, 1490) 259 (sh, 3610) 223 (7490)	Red-orange crystals (207–209)	29 (CCDC-118470)
t-Bu₂MeSi, SiMe(t-Bu)₂, Ge, Si=Si, t-Bu₂MeSi, SiMe(t-Bu)₂ **50**	+92.2	2.146(1)	—	37.0(2)	469 (1890) 308 (6300) 259 (18220) 230 (33440)	Hexagonal ruby crystals (205–207)	30
t-BuMe₂Si, SiMe₂Bu-t, t-BuMe₂Si–Si–Si–SiMe₂Bu-t, Si=Si, t-BuMe₂Si, SiMe₂Bu-t **51**	+160.4	2.174(4)	—	2.8	465 (6810) 359 (1060) 308 (sh, 4070) 271 (sh, 10600)	Orange crystals (263–265)	23 Fold angle 37.1 37.0
I, Si(t-Bu)₃, (t-Bu)₃Si–Si–Si–I, Si=Si, (t-Bu)₃Si, Si(t-Bu)₃ **52**	+164.4	2.257(2)	—	44.9	—	Red platelet (128–130)	25 (CCDC-101877)
H, Si(t-Bu)₃, (t-Bu)₃Si–Si–Si–H, Si=Si, (t-Bu)₃Si, Si(t-Bu)₃ **53**		2.360(2)	—	—	—	Red	26 (CCDC-156923)

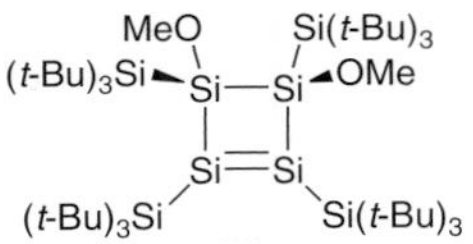 **54**	—	2.258	—	—	—	Orange platelet crystals (160–161)	25
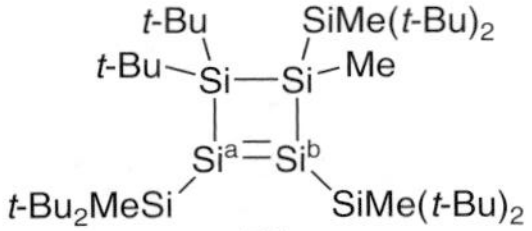 **55**	+158.1 (Si^a) +182.7 (Si^b)	—	—	—	—	Yellow crystals 174–177	27
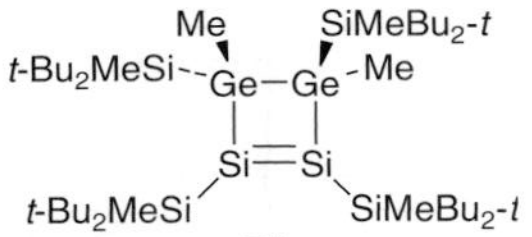 **56**	+167.6	—	—	—	441 (2890) 388 (1160) 337 (820)	Orange crystals (175–177)	28
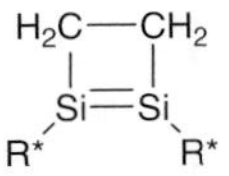 **57** ($R^* = SiMe(Si(t\text{-}Bu)_3)_2$)	+151.39 (toluene-d_8,60 °C)	2.175(1)	—	1.36(17)	—	Orange solid (225, decomp.)	32 (CCDC-238730)
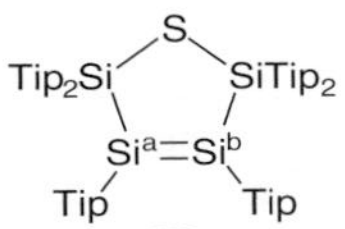 **58**	+86.0	2.173(1)	12.48 (Si^a) 5.65 (Si^b)	14.95	423 (16240)	Yellow needles (>300, decomp.)	31 (CCDC-116227)
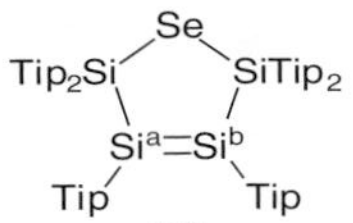 **59**	+90.3	2.1812(12)	12.86 (Si^a) 6.09 (Si^b)	15.06	431 (10140)	Yellow crystals (279–281)	31 (CCDC-116228)

TABLE II (Continued)

Compound	^{29}Si NMR[a] δ (Si)	Geometric parameters[b] d (Å)	β (deg)	R–Si=Si–R (deg)	UV–vis[c] λ_{max} [nm (ε)] [solvent]	Habit (mp) (°C)	Ref. (CCDC)[d]
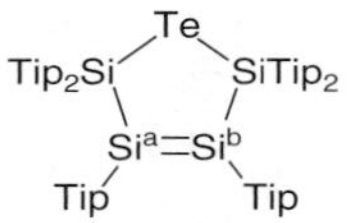 **60** (Te, Tip_2Si, $SiTip_2$, Si^a=Si^b, Tip, Tip)	+97.9	2.1984(17)	8.45 (Si^a) 12.32 (Si^b)	15.65	459 (11440)	Orange crystals (228–229, decomp.)	31 (CCDC-116229)
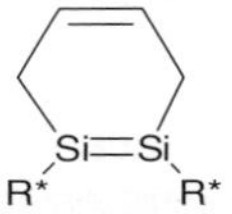 **61** (Si=Si, R*, R*; R* = SiMe(Si(t-Bu)$_3$)$_2$)	+120.94	—	—	—	—	Orange solid (100, decomp.)	32
62 (NR, RN, Si=Si, RN–Si, Si–NR, RN, NR, RN, NR; R = t-Bu)	+119.5	2.2890(14)	32.3, 33.8	25.1	360 (–) 476 (–)	Red crystals	33
63 (R, Si, Si=Si, Si; R = t-Bu)	+102.5	2.2621(15)	0.48, 0.49	12.08(15)	493 (11600) 419 (3800) 359 (6500)	Red-orange crystals (257.5–260)	35 (CCDC-232428)

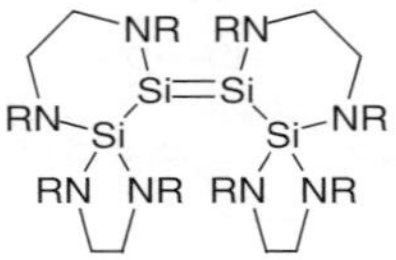

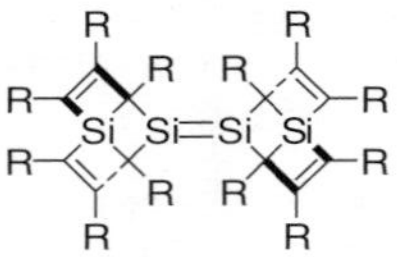

[a]In benzene-d_8 at rt unless otherwise noted.
[b]Values determined by X-ray analysis; d: Si=Si bond distance, β: Angle between R^1–Si^a–R^2 (or R^3–Si^b–R^4) plane and Si^a–Si^b axis.
[c]In hexane unless otherwise noted.
[d]Cambridge Crystallographic Data Center.

TABLE III
PROPERTIES OF STABLE CONJUGATED DISILENES

Compound	^{29}Si NMR[a]	Geometric parameters[b]		UV–vis[c]	Habit (mp) (°C)	Ref. (CCDC)[d]
	δ (Si)	d (Å)	R–Si=Si–R (deg)	λ_{max} [nm(ε)[c]] [solvent]		
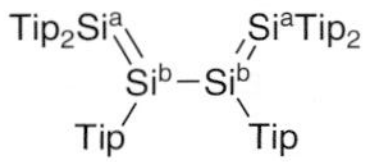 $Tip_2Si^a{=}Si^b(Tip){-}Si^b(Tip){=}Si^aTip_2$ **64**	+52.3 (Si^a) +89.5 (Si^b)	2.175(2)	51[e]	518 (25700) 392 (30200)[THF]	Reddish brown crystals (237–238)	17 (CSD-59437) Si–Si 2.321(1) Å
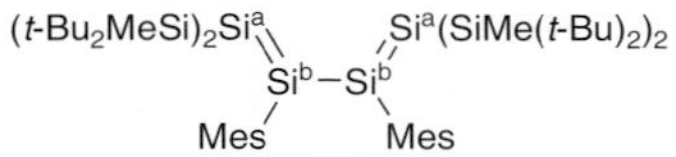 $(t\text{-}Bu_2MeSi)_2Si^a{=}Si^b(Mes){-}Si^b(Mes){=}Si^a(SiMe(t\text{-}Bu)_2)_2$ **65**	+71.5 (Si^a)	2.2003(12)	14.6	531 (3900), 413 (2300)	Deep purple crystals (131)	21 (CCDC-243357) Si–Si 2.3376(11) Å
	+150.8 (Si^b)	2.1983(12)	16.6 72[e]	357 (3000), 322 (2400) 250 (sh, 10600) 233 (sh, 18800)		
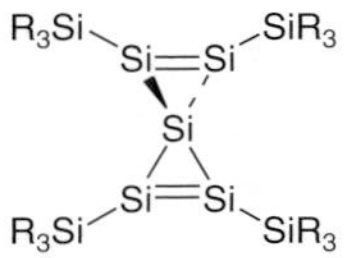 R_3Si, SiR_3 substituted Si=Si–Si–Si=Si (spiro) R = t-BuMe$_2$Si **66**	+154.2	2.186(3)	30.0(5)	560 (2530), 500 (3640) 428 (11700), 383 (18100) 304 (87000), 265 (sh, 66100) 250 (sh, 86300) 235 (sh, 77800) [3-methylpentane]	Dark red crystals (216–218)	36 (CCDC-115175)
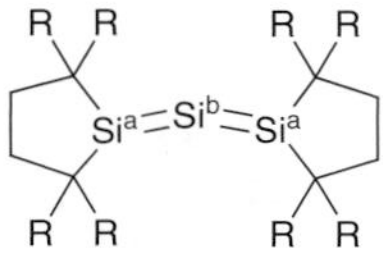 $Si^a{=}Si^b{=}Si^a$ (cyclic, R R / R R) R = SiMe$_3$ **67**	+196.9 (Si^a) +157.0 (Si^b)	2.177(1) 2.188(1)	136.49(6)[f]	584 (700) 390 (21300)	Dark green Crystals (198–200)	40

[a]In benzene-d_8 at rt unless otherwise noted.
[b]Values determined by X-ray analysis; d: Si=Si bond distance.
[c]In hexane unless otherwise noted.
[d]Cambridge Crystallographic Data Center.
[e]Si=Si–Si=Si dihedral angle.
[f]Si=Si=Si bond angle.

TABLE IV

PROPERTIES OF RADICALS AND IONS INCLUDING DISILENES AND DISILYNES

Compound	^{29}Si NMR[a]	Geometric parameters[b]			UV–vis[c]	Habit (mp) (°C)	Ref. (CCDC)[d]
	δ (Si)	d (Å)	β (deg)	R–Si=Si–R (deg)	λ_{max} [nm(ε)] [solvent]		
68 ($Tip_2Si^a{=}Si^b(Tip)Li(dme)_2$)	+94.5 (Si^a) +100.5 (Si^b) [toluene-d_8]	2.192(1)	5.65 (Si^a) 12.57 (Si^b)	18.25	417.2 (760)	Orange crystals (120–121)	20 (CCDC-223197)
69 ($(t\text{-}Bu_2MeSi)_2Si^a{=}Si^b(Mes)Li(thf)_2$)	+63.4 (Si^a) +277.6 (Si^b)	2.2092(7)	4.90 (Si^a) 5.23 (Si^b)	15.2	—	Red crystals	21 (CCDC-243358)
70 (cyclotetrasilenylium; t-Bu, Bu-t, t-Bu$_2$MeSi, SiMe(t-Bu)$_2$; $B(C_6F_5)_4^-$)	+77.3 (Si^a) +315.7 (Si^b) [CD_2Cl_2]	2.240(2) 2.244(2)	—	—	—	Yellow crystals	42
71 (cyclotetrasilenide; t-Bu, Bu-t, t-Bu$_2$MeSi, SiMe(t-Bu)$_2$; [Li(thf)]$^+$)	-24.7 (Si^a, Si^c) +224.5 (Si^b) [THF-d_8]	2.2245(7) (Si^a–Si^b) 2.3153(7) (Si^b–Si^c)	—	23.75 (R–$Si^a{=}Si^b$–R)[d] 60.82 (R–$Si^b{=}Si^c$–R)[d]	—	Green crystals	43 (CCDC-174532)
72 (cyclotetrasilenyl radical; t-Bu, Bu-t, t-Bu$_2$MeSi, SiMe(t-Bu)$_2$)	—	2.226(1) 2.263(1)	—	—	541 (9400) 483 (2100) 365 (sh, 4700) 331 (sh, 6300) 302 (9600) 241 (sh, 23500)	Red purple crystals	27

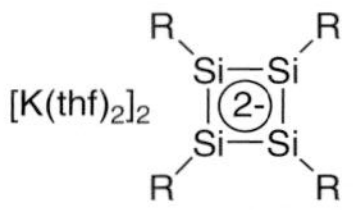 $[K(thf)_2]_2$ [cyclo-R_4Si_4 dianion (2−)], R = t-Bu_2MeSi **73**	+17.0	2.2989(8)–2.3576(8)	—	—	—	Dark green crystals	28
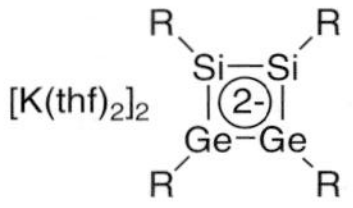 $[K(thf)_2]_2$ [cyclo-$R_4Si_2Ge_2$ dianion (2−)], R = t-Bu_2MeSi **74**	+113.7	—	—	—	—	Dark green crystals	28
R—Si≡Si—R, R = SiMe(Si(t-Bu)$_3$)$_2$ **75**	+91.5	—	—	—	—	Orange red	22
R—Si≡Si—R, R = Si(i-Pr)[CH(SiMe$_3$)$_2$]$_2$ **76**	+89.9	2.0622(9)	—	180.0	690 (14) 483 (120) 328 (5800) 259 (10300)	Emerald green Crystals (127–128, decomp.)	41 (CCDC 245523)

[a]In benzene-d_8 at rt unless otherwise noted.
[b]Values determined by X-ray analysis; d: Si=Si bond distance, β: angle between R^1–Si^a–R^2 (or R^3–Si^b–R^4) plane and Si^a–Si^b axis.
[c]In hexane unless otherwise noted.
[d]Cambridge Crystallographic Data Center.

A. *Acyclic Disilenes*

In the last decade, several new synthetic routes have been developed to allow the synthesis of various stable acyclic disilenes including (*E*)- and (*Z*)-ABSi$=$SiAC-type disilenes, but disilenes with the types of $A_2Si{=}SiBC$ and ABSi$=$SiCD are still unknown.

Much effort has been made for the synthesis of $A_2Si{=}SiB_2$-type disilenes. As stated in review OW, cophotolysis of $A_2Si(SiMe_3)_2$ and $B_2Si(SiMe_3)_2$ gave a statistical mixture of three disilenes, $A_2Si{=}SiA_2$, $B_2Si{=}SiB_2$, and $A_2Si{=}SiB_2$, but the isolation of pure $A_2Si{=}SiB_2$-type disilenes was unsuccessful; A = Mes, B = Xyl (**13**) and A = Mes, B = Dmt (**14**).[10] Reductive dehalogenation of $Tip_2SiCl{-}SiClMes_2$ with lithium naphthalenide gave successfully the corresponding $A_2Si{=}SiB_2$-type disilene **15**.[11]

Applying a unique method for the synthesis of 1,1-dilithiosilanes **78** from the corresponding silacyclopropenes **77**,[12] a convenient synthetic route to $A_2Si{=}SiB_2$-type disilenes **28**–**31** was developed by Sekiguchi and coworkers:[13]

(3)

Two $A_2Si{=}SiB_2$-type tetrakis(triakylsilyl)disilenes **36** and **37** were synthesized using Method D [Eq. (4)].[14] In the synthetic route, the preparation of hydridosilyllithiums **80** by the silametallation of the corresponding dihydridosilanes **79** is involved.[15]

(4)

An ABSi$=$SiAC-type disilene is synthesized by the reduction of dibromosilane **81** with KC_8.[16] The facile disilanylsilylene–silyldisilene rearrangement of silylene **82** formed by the reduction is proposed to give disilene **32**:

$$(t\text{-}Bu_2MeSi)_2MeSi(Br)_2Si\text{-}Si(t\text{-}Bu_2MeSi)_2Me \; \mathbf{81} \xrightarrow[-78\,C\text{ to rt}]{KC_8/THF} [\,(t\text{-}Bu_2MeSi)_2MeSi\text{-}Si{:}\text{-}SiMe(t\text{-}Bu_2MeSi)(SiMe(t\text{-}Bu)_2)\,] \; \mathbf{82} \longrightarrow Me(t\text{-}Bu_2MeSi)Si{=}Si(SiMe(SiMe(t\text{-}Bu)_2)_2)(SiMe(t\text{-}Bu)_2) \; \mathbf{32},\ 71\% \tag{5}$$

Several ABSi═SiAC-type disilenes are synthesized by the 1,4-addition of stable 1,3-tetrasiladiene **64**. The reactions of tetrasiladiene **64**[17] with $HSiCl_3$ and LiBr/ CF_3COOH give disilenes (*Z*)-**38** and (*Z*)-**39**, respectively, by the formal 1,4-addition of the hydrogen halides to **64** [Eq. (6)].[18] The reactions of gaseous hydrogen halides with **64** in solution afford however the corresponding 1,4-dihalotetrasilanes by the twofold 1,2-additions to the double bonds of **64**.

$$Tip_2Si{=}Si(Tip)\text{-}Si(Tip){=}SiTip_2 \; \mathbf{64} \xrightarrow{\text{reagent}} HTip_2Si(Tip)Si{=}Si(Tip)SiTip_2X \tag{6}$$

(*Z*)-**38**, X = Cl, reagent = $HSiCl_3$
(*Z*)-**39**, X = Br, reagent = $LiBr/CF_3CO_2H$

Jutzi *et al.* synthesized an (*E*)-ABSi═SiAB-type disilene by the reaction of stable silyliumylidene cation $Me_5C_5Si{:}^+$ **83** with lithium bis(trimethylsilyl)amide **84**.[19] Formation of disilene **44** is explained by the dimerization of the initially formed $(Me_5C_5)[(Me_3Si)_2N]Si$:

$$Me_5C_5Si{:}^{\oplus}\,B(C_6F_5)_4^{\ominus} \; \mathbf{83} + (Me_3Si)_2NLi \; \mathbf{84} \longrightarrow (Me_5C_5)[(Me_3Si)_2N]Si{=}Si[N(SiMe_3)_2](C_5Me_5) \; \mathbf{44} \tag{7}$$

Several lithiodisilene derivatives are synthesized as shown in Eqs. (8)–(10). Disilenyllithium **68** was proposed by Weidenbruch *et al.* as a possible intermediate in the reduction of $Tip_2Si{=}SiTip_2$ (**2**) with excess lithium [Eq. (8)].[17] Scheschkewitz obtained the same disilenyllithium **68** by the reduction of Tip_2SiCl_2 with lithium in dimethoxyethane (DME).[20] Disilenyllithium **68** complexed with DME is isolated as orange crystals by recrystallization from hexane. The reaction of disilenyllithium **68** with Me_3SiCl gives an $A_2Si{=}SiAB$-type disilene **40** [Eq. (9)].

$$Tip_2Si{=}SiTip_2 \; \mathbf{2} \xrightarrow[-78\,C°\sim rt]{Li/THF} [\,Tip_2Si{=}Si(Tip)Li\,] \; \mathbf{68'} \tag{8}$$

$$\mathrm{Tip_2SiCl_2} \xrightarrow[25\,^\circ\mathrm{C}]{6\mathrm{Li/DME}} \underset{\mathbf{68},\ 51\%}{\mathrm{Tip_2Si{=}Si(Tip)Li(dme)_2}} \xrightarrow{\mathrm{Me_3SiCl}} \underset{\mathbf{40},\ 95\%}{\mathrm{Tip_2Si{=}Si(Tip)SiMe_3}} \quad (9)$$

The reaction of 2,3-diaryl(tetrasilyl)tetrasila-1,3-diene **65** with *t*-BuLi gives the corresponding 1,1-disilyl(aryl)lithiodisilene **69** in 67% yield as orange crystals [Eq. (10)].[21] The formation of lithiodisilene **69** is explained by a single-electron transfer followed by cleavage of the central Si–Si bond.

$$\underset{\mathbf{65}}{\mathrm{R_2Si{=}Si(Mes){-}Si(Mes){=}SiR_2}} \xrightarrow[-78\ \mathrm{C^\circ{\sim}rt}]{t\text{-}\mathrm{BuLi/THF}} 2\ \underset{\mathbf{69},\ 67\%}{\mathrm{R_2Si{=}Si(Mes)Li(thf)_n}} \qquad \mathrm{R}=t\text{-}\mathrm{Bu_2MeSi} \quad (10)$$

As a functional disilene, Wiberg synthesized a halogen-substituted disilene. Treatment of trihalosilane **85** in pentane with a THF solution of tri(*tert*-butyl) silylsodium gives (*E*)-1,2-dichloro-1,2-disilyldisilene **42** in about 12% yield:[22]

$$\underset{\mathbf{85}}{\mathrm{R_2MeSi{-}SiBr_2Cl}} \xrightarrow[\mathrm{R}=t\text{-}\mathrm{BuMe_2Si}]{t\text{-}\mathrm{Bu_3SiNa}} \underset{\mathbf{42},\ \sim 12\%\ \text{yield}}{(\mathrm{R_2MeSi})\mathrm{ClSi{=}SiCl}(\mathrm{SiR_2Me})} \quad (11)$$

B. *Cyclic Disilenes*

1. Cyclotetrasilenes

The first cyclic disilene, hexakis(*tert*-butyldimethylsilyl)cyclotetrasilene (**51**), is synthesized in 13.6% yield by the reduction of a 1:2 mixture of tetrabromodisilane **86** and dibromosilane **87** with lithium naphthalenide in THF [Eq. (12)].[23] Interestingly, the initial product of the reductive condensation is not cyclotetrasilene **51** but bicyclo[1.1.0]tetrasilane **88**. Thus, reduction of a mixture of **86** and **87** with lithium naphthalenide at –78 °C followed by hydrolysis at 0 °C affords mainly a hydrolysis product of bicyclotetrasilane **88**, **89** [Eq. (13)]. As discussed in Section IV.A.4 in more detail, compound **51** should be formed by the thermal isomerization of **88** during workup at higher temperatures.

$$\underset{\mathbf{86}}{\mathrm{R_2SiBr_2}} + \underset{\mathbf{87}}{\mathrm{RBr_2Si{-}SiBr_2R}} \xrightarrow[-78\,^\circ\mathrm{C\ to\ rt}]{\mathrm{LiNp/THF}} \underset{\mathbf{51},\ 13.6\%}{\text{cyclo-}(\mathrm{R_2Si{-}SiR_2{-}SiR{=}SiR})} \qquad \mathrm{R}=t\text{-}\mathrm{BuMe_2Si} \quad (12)$$

$$\mathbf{86} + \mathbf{87} \xrightarrow[\text{-78 °C to 0 °C}]{\text{LiNp/THF}} [\mathbf{88}] \rightarrow \begin{cases} \xrightarrow[\text{0 °C to rt}]{\Delta} \mathbf{51} \\ \xrightarrow{\text{H}_2\text{O, 0 °C}} \mathbf{89} \end{cases} \quad (13)$$

The yield of cyclotetrasilene **51** is much improved using tris(*tert*-butyldimethylsilyl)silyl-substituted dibromochlorosilane **90** as a precursor for **51**; the reductive dehalogenation of **90** using sodium dispersion in toluene at room temperature (rt) provides **51** in 64% yield:[24]

$$\underset{\mathbf{90}}{R_3Si{-}SiBr_2Cl} \xrightarrow[\text{rt, 64\%}]{\text{Na/toluene}} \mathbf{51} \qquad R = t\text{-BuMe}_2\text{Si} \quad (14)$$

1,2-Diiodocyclotetrasil-3-ene **52** is obtained in 61% yield by the reaction of tetrakis(tri-*tert*-butylsilyl)tetrasilatetrahedrane **91** with an equimolar amount of iodine in heptane at rt [Eq. (15)].[25] The treatment of **91** with I_2 in heptane at 0 °C in the dark leads initially to a heptane-insoluble black solid, which slowly converts into **52** when dissolved in benzene at rt. The black intermediate could be species **92**, which would form **52** by way of the elimination of I^- [Eq. (16)]. The mechanism would explain the reason why the two iodine atoms in **52** are arranged in a *trans* fashion.

$$\mathbf{91} \xrightarrow[\text{61\%}]{\text{I}_2\text{/hentane}} \mathbf{52}\text{, 61\%} \qquad R = t\text{-Bu}_3\text{Si} \quad (15)$$

$$\mathbf{91} \xrightarrow{\text{I}_2} [\mathbf{92}] \longrightarrow \mathbf{52} \quad (16)$$

During the reduction of 1,2-dibromodisilane **93** with t-Bu_3SiNa in THF, cyclotetrasilene **53** is obtained as orange crystals in a low yield together with yellow crystals of marginally stable dihydridodisilene **43** [Eq. (17)].[26] A plausible mechanism proposed for the formation of **53** is the intermediary formation of the corresponding disilyne **94** by the dehydrobromination of **93** with t-Bu_3SiNa, followed by the dyotropic migration of two t-Bu_3Si groups to give tetrasiladiene **95** [Eq. (18)].

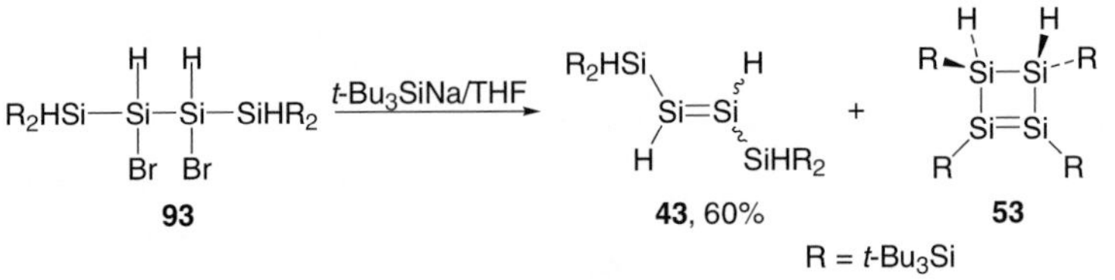

(17)

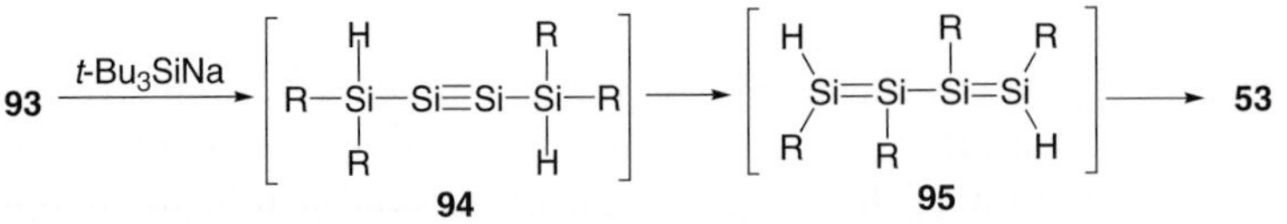

(18)

The reaction of stable cyclotetrasilenylium cation **70** [TPFPB = $(C_6F_5)_4B$] with methyllithium in diethyl ether gives cyclotetrasilene **55** in 97% yield as yellow crystals:[27]

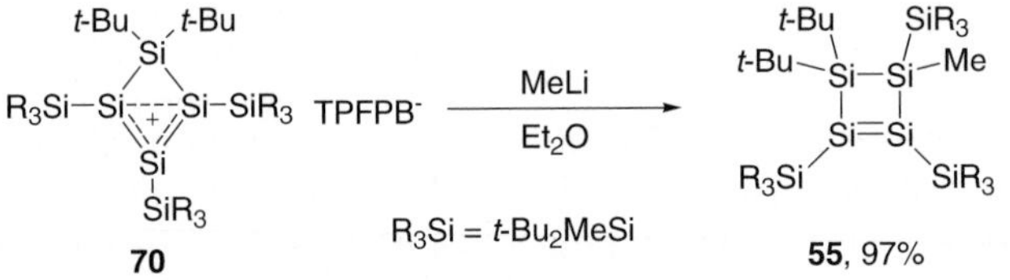

(19)

The four-membered cyclic disilene with mixed group-14 elements **56** is synthesized in 95% yield by the reaction of the dipotassium salt of tetrakis(di-*tert*-butylmethylsilyl)-1,2-disila-3,4-digermacyclobutadiene dianion **74** with Me_2SO_4; the two Me groups in **56** are bound selectively to Ge atoms:[28]

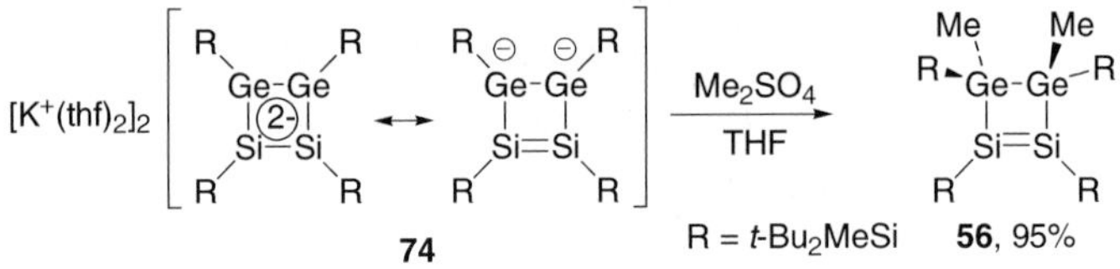

(20)

2. Cyclotrisilenes

The first cyclotrisilene **48** is synthesized by the reduction of dibromochlorosilane **90** with potassium graphite. Compound **48** is isolated as air-sensitive dark red crystals in 65% yield:[24]

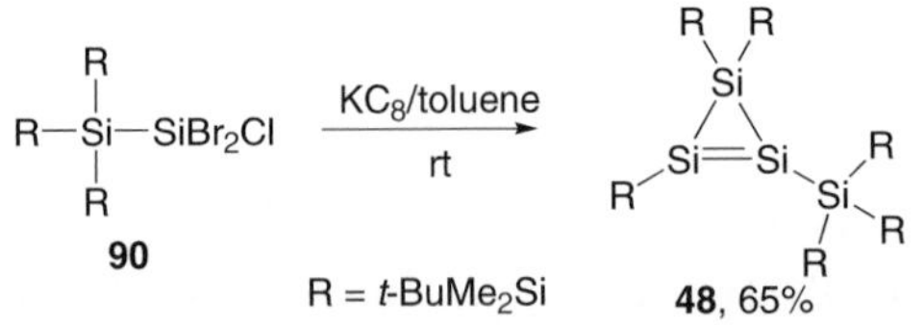

(21)

It is interesting to note that the reduction of **90** with Na in toluene affords the corresponding cyclotetrasilene **51** as shown in Eq. (14). Although the reason why different disilenes, **48** and **51**, are produced preferably from trihalosilane **90** depending on the reduction conditions is still unclear, two possible pathways for the formation of the products may be considered as shown in Eq. (22). The reduction would involve the formation of bromo(trisilylsilyl)silylene **96**, which may dimerize to disilene **97** at low temperature in THF, while **96** may isomerize to disilene **98** at rt in toluene.

90
KC$_8$ or Na
[R_3Si–Si:–Br] 96
THF, -78 °C → [$R_3Si(Br)Si{=}Si(Br)SiR_3$] 97 —KC$_8$→ [$R_3Si{-}Si{\equiv}Si{-}SiR_3$] → [$(R_2Si)(R_3Si)Si{=}Si{:}$, R] → 48
toluene, rt → [$R_2Si{=}Si(R)Br$] 98 —Na→ [$R_2Si{=}Si(R){-}Si(R){=}SiR_2$] → 51

(22)

Reductive coupling of t-Bu_2MeSi-substituted dibromosilane **99** and tetrabromodisilane **101** with sodium metal affords symmetric cyclotrisilene **49** as red orange crystals in 9.1% yield [Eq. (23)].[29] In a similar manner, the co-reduction of dichlorogermane **100** and **101** gives 1-disilagermirene **50** as the first mixed cyclotrimetallene in 29% yield.[30]

R_2EX_2 + $R{-}SiBr_2{-}SiBr_2$ (Na/toluene; R = t-Bu_2MeSi) → cyclo-$ER_2{-}Si(R){=}Si(R)$ (23)

99: E = Si, X = Br
100: E = Ge, X = Cl
101
49: E = Si, 9.1%
50: E = Ge, 29%

3. Other Endocyclic Disilenes

A series of five-membered cyclic disilenes containing a chalcogen atom in a ring **58–60** are obtained in high yields using the reactions of tetrasilabutadiene **64** with chalcogens in toluene upon heating [Eq. (24)].[31] The additions are much slower than those of chalcogens to tetramesityldisilene. For the reactions of **64** with selenium and tellurium, a small amount of triethylphosphine is required as an additive. The addition would proceed *via* the [2 + 1] cycloaddition of a chalcogen atom to one of the Si$=$Si double bonds of **64** followed by the rearrangement into the presumably less strained five-membered ring.

64 (Tip$_2$Si=Si(Tip)–Si(Tip)=SiTip$_2$) —reagent→ cyclo-X–SiTip$_2$–Si(Tip)=Si(Tip)–SiTip$_2$ (24)

reagent = S, Se/PEt$_3$, Te/PEt$_3$
X = S (**58**, 86%)
Se (**59**, 81%)
Te (**60**, 82%)

The reduction of dichlorodisilene **42** with lithium naphthalenide in THF at −78 °C in the presence of ethylene provides orange-red crystals of cyclic disilene **57** as a formal [2 + 2] adduct of disilyne **75** to ethylene:[32]

$$\underset{\mathbf{42}}{(\text{MeR}_2\text{Si})\text{ClSi}{=}\text{SiCl}(\text{SiR}_2\text{Me})} \xrightarrow[\text{-78 °C to rt}\quad R = t\text{-Bu}_3\text{Si}]{\text{LiNp/THF/H}_2\text{C=CH}_2} \underset{\mathbf{57}}{\text{H}_2\text{C–CH}_2\text{ (cyclic) } (\text{MeR}_2\text{Si})\text{Si}{=}\text{Si}(\text{SiR}_2\text{Me})} \tag{25}$$

A similar reduction of **42** in the presence of excess 1,3-butadiene gives yellow-orange crystals of six-membered cyclic disilene **61**, which would be formed by the [2 + 4] cycloadditon of disilyne **75** to 1,3-butadiene:[32]

$$\mathbf{42} \xrightarrow[\text{-78 °C to rt}]{\text{LiNp/THF/1,3-butadiene}} \underset{\mathbf{61},\ R = t\text{-Bu}_3\text{Si}}{(\text{MeR}_2\text{Si})\text{Si}{=}\text{Si}(\text{SiR}_2\text{Me})\text{ (six-membered ring)}} \tag{26}$$

4. Exocyclic Disilenes

A (Z)-isomer of exocyclic ABSi=SiAB-type disilene, disilyldiaminodisilene **62**, is obtained by a tetramerization of stable diaminosilylene **102** [Eq. (27)].[33,34] Quite interestingly, **62** dissociates in solution rapidly into the corresponding silylene **103** and then slowly into the diaminosilylene **102**.

$$4\ \underset{\mathbf{102}}{\text{Si:}} \rightleftharpoons 2\ \underset{\mathbf{103}}{[\ \text{Si:}\ ]} \rightleftharpoons \underset{\mathbf{62}}{\text{Si}{=}\text{Si}} \qquad R = t\text{-Bu} \tag{27}$$

The reduction of tri(*tert*-butyl)cyclopropenyl-substituted tribromosilane **104** with KC_8 in 2-methyltetrahydrofuran gives a quite unique disilene **63** having a lattice framework [Eq. (28)].[35] Disilene **63** constitutes an $A_2Si{=}SiA_2$ type disilene but has C_2-symmetric chirality through the silicon–silicon double bond.

$$\underset{\mathbf{104}}{\text{R}_3\text{C}_3\text{SiBr}_3} \xrightarrow[R = t\text{-Bu}]{3\ \text{KC}_8\text{/2-MeTHF}} \underset{\mathbf{63},\ 47\%}{\text{Si}{=}\text{Si}} \tag{28}$$

C. *Silicon-Based Dienes*

1. Tetrasila-1,3-diene

In 1997, Weidenbruch *et al.* successfully synthesized the first silicon analog of butadiene, hexaaryltetrasila-1,3-diene **64**.[17] Treatment of disilene **2** with lithium in DME at rt followed by the addition of 0.5 equivalent of mesityl bromide at −30 °C gives **64** as reddish brown crystals [Eq. (29)]. In a proposed mechanism, a half of the lithiodisilene **68′**, which is formed by the reductive Si–Ar bond cleavage [Eq. (8)], is converted to the corresponding bromodisilene **105** by the halogen–metal exchange reaction with MesBr, and then the coupling reaction of **105** with unreacted lithiodisilene **68′** provides **64** [Eq. (30)].

$Tip_2Si{=}SiTip_2$ (**2**) → 1) Li/DME/rt; 2) MesBr(0.5 equiv)/-30 °C → **64**, 53% (29)

68' → MesBr (0.5 equiv), - MesLi → [**105**] → + **68'** → **64** (30)

2,3-Dimesityl-1,1,4,4-tetrasilyltetrasila-1,3-diene **65** is synthesized as deep purple crystals in 11% isolated yield using the reaction of 1,1-dilithiosilane **78** with tetrachloro-1,2-dimesitylsilane **106** in THF at rt:[21]

78 + **106** → THF/rt → **65**, 11%; R_3Si = t-Bu_2SiMe (31)

2. Spiropentasiladiene

The first spiroconjugated disilene, spiropentasiladiene **66**, is obtained in 3.5% yield as a by-product of the synthesis of cyclotrisilene **48** using the reaction of dibromochlorosilane **90** with potassium graphite at –78 °C in THF [Eq. (32)].[36] Spiropentasiladiene **66** forms air-sensitive dark red crystals but is thermally very stable with a melting point of 216–218 °C in contrast to the parent carbon spiropentadiene, which survives only at lower than –100 °C.[37] The yield of spiropentasiladiene **66** is much improved by using tetrabromodisilane **107** as a precursor [Eq. (33)].[38]

$R_3Si{-}SiBr_2Cl$ (**90**) → KC_8/THF, -78 °C - rt; R = t-$BuMe_2Si$ → **66**, 3.5% + **48** (32)

$R_3SiSiBr_2{-}SiBr_2SiR_3$ (**107**) → KC_8/THF, -40 °C → **66**, 41% (33)

3. Trisilaallene

Trisilaallene **67** was synthesized as the first stable compound with a formally sp-hybridized silicon atom and also the first cumulative Si=Si doubly bonded compound using the following two-step route [Eq. (34)]. The insertion of isolable dialkylsilylene **108** into a Si–Cl bond of tetrachlorosilane[39] giving **109** followed by the reduction with potassium graphite in THF at –40 °C for 2 days affords dark green crystals of trisilaallene **67** in overall 42% yield.[40]

$$\textbf{108}, R = SiMe_3 \xrightarrow[THF]{SiCl_4} \textbf{109} \xrightarrow[-40\,^\circ C,\ 2\ days]{KC_8\ (4\ equiv)/THF} \textbf{67}, 42\%\ yield\ (overall) \tag{34}$$

D. *Disilynes*

Much effort has been devoted to the synthesis of stable disilynes. Wiberg *et al.* have reported that the reduction of *trans*-1,2-dichlorodisilyldisilene **42** with lithium naphthalenide gives an orange-red solution of disilyne **75** [Eq. (35)].[22] The ^{29}Si NMR spectrum of the solution showed a signal at +91.5 ppm, which is assigned to the unsaturated silicon nuclei of **75** (Section III.F.4). The mass spectrum of the solution supported the existence of disilyne **75**. Although no further spectroscopic evidence for the existence of disilyne **75** is obtained, it is trapped by several reagents to afford novel disilenes as shown in Eqs. (25) and (26). Very recently, Sekiguchi and coworkers have achieved the synthesis of the first isolable disilyne **76** [Eq. (36)].[41] Emerald green crystals of disilyne **76** are obtained by the reductive debromination with KC_8 of tetrabromodisilane **110** with extremely bulky silyl substituents, bis[bis(trimethysilyl)methyl](isopropyl)silyl groups.

$$(R_2MeSi)ClSi{=}SiCl(SiR_2Me)\ \textbf{42} \xrightarrow[R\,=\,t\text{-}Bu_3Si]{LiNp/THF} [MeR_2Si{-}Si{\equiv}Si{-}SiR_2Me]\ \textbf{75} \tag{35}$$

$$R_3Si{-}SiBr_2{-}SiBr_2{-}SiR_3 \xrightarrow{KC_8/THF} R_3Si{-}Si{\equiv}Si{-}SiR_3 \tag{36}$$

110, $R_3Si = [(Me_3Si)_2CH]_2(i\text{-}Pr)Si$ **76**, 73%

E. *Cations, Anions, and Radicals with Si=Si Double Bonds*

A number of ions and radicals with neighboring Si=Si double bonds have recently been synthesized as stable compounds. Treatment of cyclotrisilene **49** with $[Et_3Si(benzene)]^+ \cdot [B(C_6F_5)_4]^-$ gives the cyclotetrasilenylium ion-tetrakis(pentafluorophenyl)borate ion salt **70**, which is isolated as air- and moisture-sensitive yellow crystals in 91% yield [Eq. (37)].[42] Using deuterium-labeled cyclotrisilene **49**-d_6, a mechanism involving silyl cation intermediates **111** and **112** is proposed for the formation of **70** [Eq. (38)].

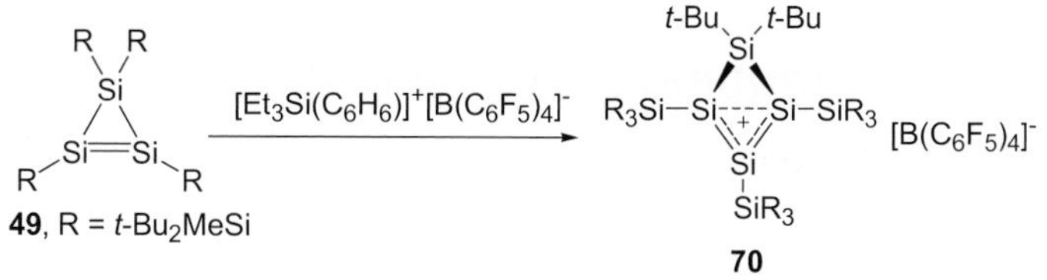

(37)

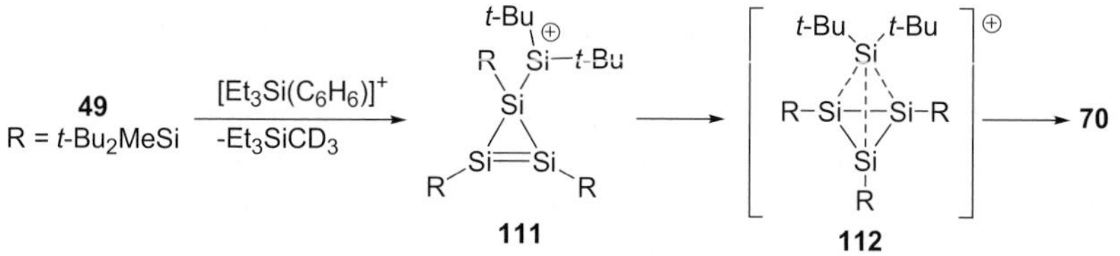

(38)

Cyclotetrasilenylium cation **70** is converted reversibly to the corresponding radical **72**[27] and anion **71**[43] as shown in Eq. (39):

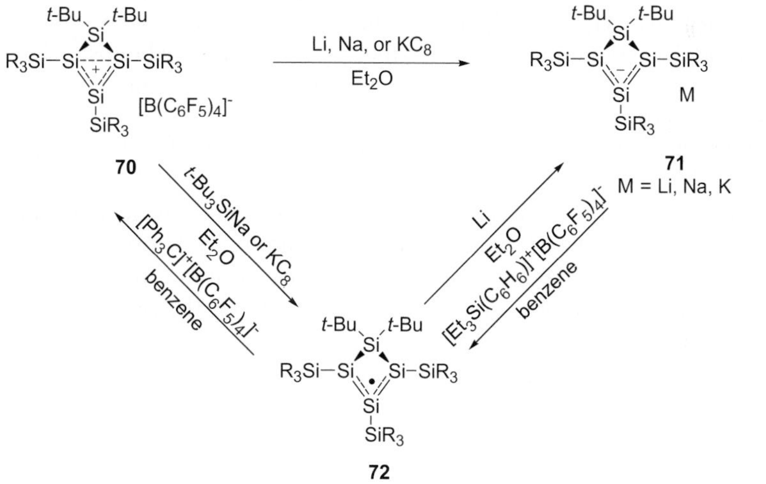

(39)

Silicon and germanium analogs of cyclobutadiene dianions **73** and **74** are obtained by the reduction of the corresponding cyclic precursors **113** and **114** in good yield:[28]

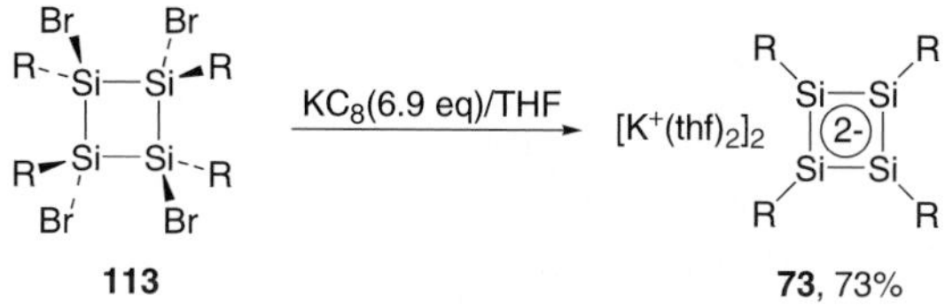

(40)

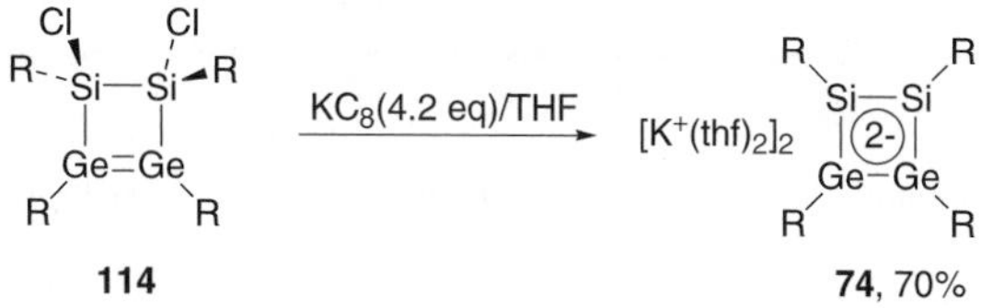

(41)

III
BONDING AND STRUCTURES

A. *An Overview of Si=Si Double Bonds*

For a detailed discussion on the structure of disilenes, it is helpful to understand the intrinsic difference in the bonding between disilene and ethylene. The C=C double bond of ethylene as well as many other alkenes is usually described as being constructed by a σ-bond formed by σ overlap between two sp^2-hybridized orbitals and a π-bond formed by π overlap between the two pure pπ-orbitals. Because the π overlap is effective to make a strong π-bond but very sensitive to the angle between the two pπ orbital axes, the geometry around the C=C bond is planar, the double bond is almost twice as strong as a C–C single bond, and the rotational barrier around the C=C bond is over 60 kcal/mol. Such a bonding scheme is not applicable to double bonds of silicon and heavier group-14 elements. Theoretical calculations have shown that the equilibrium geometry of dimetallenes of heavier group-14 elements ($H_2E{=}EH_2$; E = Si, Ge, and Sn) is not planar but significantly *trans*-bent and the bonding is best described by double dative bonds.[44] The dimer of H_2Pb: does not have direct bonding between two lead atoms any more but a hydrogen-bridged cyclic structure (Chart 2).

The bonding nature of the E=E double bonds is outlined using the CGMT (Carter–Goddard–Malrieu–Trinquier) model.[45–47] In this model, an E=E double bond is formed by the dimerization of its components, divalent metallic species (H_2E:). According to the MO analysis for the bonding by Malrieu and Trinquier,[47] when two EH_2 groups having an *n* and a pπ-type orbital approach each other to form a planar E=E double bond, an overlap between the two *n* orbitals form σ and σ* orbitals and an overlap between the two pπ orbitals forms π and π* orbitals (Fig. 1a). To form the direct E=E bond, the sequence of the orbital energies should be in the order $\pi < \sigma^*$; if $\pi > \sigma^*$, both σ and σ* orbitals should be occupied, and hence, no stabilization by this approach is expected (Fig. 1b). If interaction energies between two *n* orbitals and between two pπ orbitals are F_σ and F_π (both < 0), respectively, and the *n* orbital is lower in energy than the pπ orbital with ΔE_{ST} (singlet–triplet energy difference), the condition of $\pi < \sigma^*$ should be written as $\Delta E_{ST} < -(F_\sigma + F_\pi)$. Another condition to form the direct E=E bond is the orbital

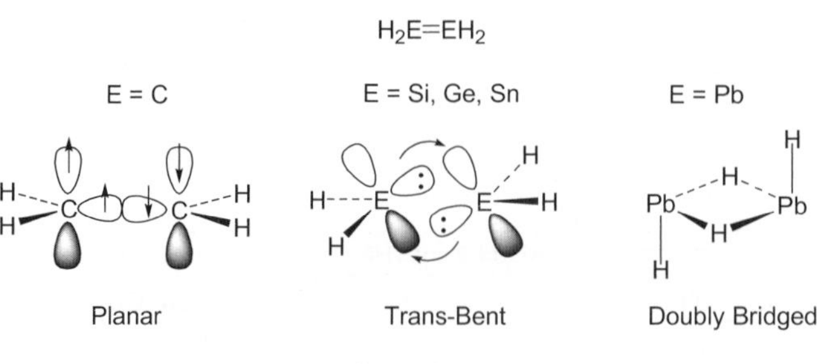

CHART 2.

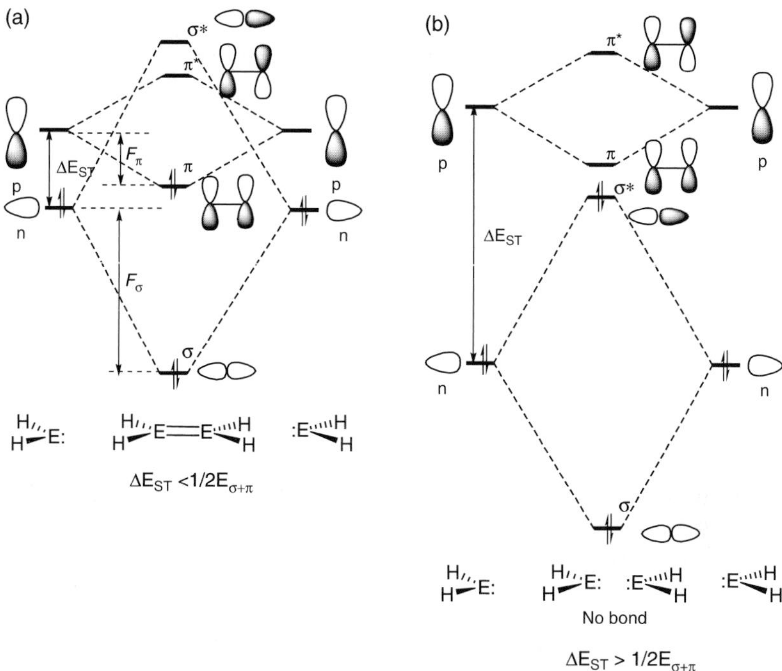

FIG 1. Schematic representation of orbital interactions between two divalent species of group-14 elements (E = Si, Ge, Sn). (a) $\Delta E_{ST} < 1/2E_{\sigma+\pi}$, (b) $\Delta E_{ST} > 1/2E_{\sigma+\pi}$.

sequence $\sigma < \pi^*$, i.e. $\Delta E_{ST} > (F_\sigma + F_\pi)$, but this condition is always maintained if $\Delta E_{ST} > 0$. Therefore, the E=E double bond is formed when $(F_\sigma + F_\pi) < \Delta E_{ST} < -(F_\sigma + F_\pi)$. Because F_σ and F_π are $(-1/2)E_\sigma$ and $(-1/2)E_\pi$, respectively, where E_σ and E_π are the σ and π-bond energies, respectively, the conditions are rewritten as follows: $(-1/2)E_{\sigma+\pi} < \Delta E_{ST} < (1/2)E_{\sigma+\pi}$, where $E_{\sigma+\pi} = E_\sigma + E_\pi$. The bond dissociation energy (BDE) for the E=E bond should be dependent on ΔE_{ST}; BDE $= E_{\sigma+\pi} - 2\Delta E_{ST}$, when $\Delta E_{ST} > 0$.

The reason why the bending from planar geometry occurs is ascribed to the secondary interaction between σ* and π orbitals and/or between σ and π* orbitals; see Ref. 48 for other aspects of the bending. While neither σ*–π nor σ–π* interaction exists in a planar double-bond geometry, *trans*-bending allows the σ*–π and σ–π* interactions, which stabilize π and σ orbitals, respectively, with decreasing *n*–*n* and pπ–pπ overlap. A simplified model in which less important σ–π* interaction is ignored indicates that if $\Delta E_{ST} > (1/4)E_{\sigma+\pi}$, *trans*-bent geometry is more stabilized with the bent angle $\theta = \cos^{-1}[2-4\Delta E_{ST}/E_{\sigma+\pi}]$ than the planar geometry (Chart 3). Since usually, ΔE_{ST} values for $H_2E{=}EH_2$ (M = Si, Ge, Sn) are around 20 kcal mol^{-1} and the $E_{\sigma+\pi}$ values are 60–75 kcal mol^{-1}, the condition, $(1/4)E_{\sigma+\pi} < \Delta E_{ST} < (1/2)E_{\sigma+\pi}$, is satisfied, and hence, *trans*-bent geometry is favored for disilene, digermene, and distannene. On the other hand, ΔE_{ST} of H_2Pb: (41 kcal mol^{-1}) is much larger than its $(1/2)E_{\sigma+\pi}$ (23–30 kcal mol^{-1}), and hence,

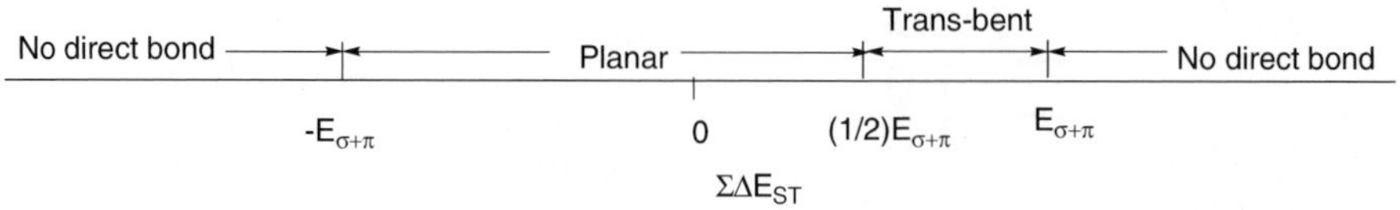

CHART 3.

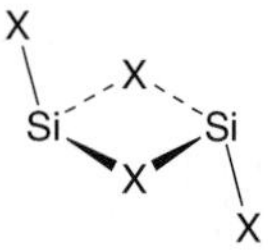

CHART 4.

the Pb=Pb double bond is not formed and a hydrogen-bridged cyclic structure become more stable as a dimer of H_2Pb:. The CGMT model is applicable for unsymmetrically substituted dimetallenes $R_2M{=}M'R_2'$ with different ΔE_{ST} values for R_2M: and $R_2'M'$: by replacing ΔE_{ST} in the above discussion to $(1/2)\Sigma\Delta E_{ST}$.

The geometry around E=E double bonds depends on the substituents on unsaturated silicon atoms because the $\Sigma\Delta E_{ST}$ for the corresponding divalent species are highly substituent dependent.[49] For instance, Allen *et al.*,[50] have shown theoretically that the geometry around the double bonds of tetrasilyldisilene $(H_3Si)_2Si{=}Si(SiH_3)_2$ is planar. The theoretical results are compatible with the CGMT model because ΔE_{ST} for $(H_3Si)_2Si$: is ca. 5–10 kcal mol^{-1} [51,52] and satisfies the condition, $\Delta E_{ST} < (1/4)E_{\sigma+\pi}$. Remarkable dependence of the geometries and bond dissociation energies of disilenes on the substituents have been shown by theoretical calculations. At the MP3/6–31G*//HF/6–31G* level, the bent angle θ and BDE for $H_2Si{=}SiH_2$ ($\Delta E_{ST} = 16.4$ kcal mol^{-1}) are 12.9° and 57.2 kcal mol^{-1} respectively, while θ and BDE for *cis*-FHSi=SiFH ($\Delta E_{ST} = 37.8$ kcal mol^{-1}) are >40° and 29.5 kcal mol^{-1}.[49] Disilenes with the more electronegative substituents have larger bent angles and smaller BDE, because the ΔE_{ST} values for the component silylenes become larger with increasing electron-withdrawing ability of the substituents. In $H_2Si{=}SiF_2$, the bent angle at the H_2Si silicon atom is 73.6°.[49] Dimers of F_2Si: ($\Delta E_{ST} = 74$ kcal mol^{-1}),[49] and $(H_2N)_2Si$: ($\Delta E_{ST} = 80.4$ kcal mol^{-1} MP4SDTQ/6–311G**//6–31G**),[53] are no longer the corresponding disilenes but doubly bridged cyclic dimers **115** (Chart 4). The dimers of silylenes with electropositive substituents (R_2Si:, R = Li, BeH, BH_2, SiH_3) are planar as expected.

The above theoretical analysis for a variety of dimer structures of silylenes requires inevitably a definition of *disilenes* different from that of *alkenes*, molecules with carbon–carbon double bonds. Geometry around a typical C=C double bond is planar and the double bond length (134 pm) is shorter than the corresponding single bond (154 pm). BDE of ethylene to two methylenes is ca. 170 kcal mol^{-1} which is 1.9 times larger than for the C–C single bond (90 kcal mol^{-1} for $H_3C–CH_3$); the BDE of ethylene really almost doubles the BDE for ethane;

exceptionally, small BDE of $F_2C{=}CF_2$ (ca. 70 kcal mol^{-1}) has been ascribed to the large ΔE_{ST} of F_2C: as analyzed by Carter and Goddard.[45] Although these geometric and energetic parameters for ethylene are modified by the steric and electronic effects of substituents, we may take the modification as simply substituent effects; no other dimers of carbenes like bridged dimers are competitive to alkenes, and hence, are taken into account. On the other hand, while usually the Si=Si bond distances of *disilenes* are significantly shorter than those for the Si–Si single bonds, this may not be the indication of the formation of double bonds, because the BDE of *disilenes* are usually smaller than those of the corresponding disilanes; even for parent disilene, $H_2Si{=}SiH_2$, the BDE (57.2 kcal mol^{-1}) is smaller than that of $H_3Si{-}SiH_3$ (>70 kcal mol^{-1}). Through the article, the term *disilene(s)* is used for compounds with direct Si–Si bonding between two component silylenes. Although the BDE of a disilene may be smaller than that of the corresponding disilane, the direct bonding should have both σ- and π-bonds unless the two atomic orbitals forming the π-bond are perpendicular to each other, and, in this sense, *disilenes* are doubly bonded compounds.

The bending potential surface in $H_2Si{=}SiH_2$ is very shallow and the energy difference between optimized *trans*-bent geometry and the planar geometry is only 1–2 kcal mol^{-1}.[49,54] Probably, the potential curves for the other deformations around the Si=Si double bond may also be shallow compared to those around the C=C double bond in ethylene. The Si=Si double bonds in disilenes would be much softer than C=C double bonds, and hence, the geometry around the Si=Si double bonds should be easily deformed. Typically, the geometry around the Si=Si bond of tetraaryldisilenes is expected to be *trans*-bent but that of tetramesityldisilene (**1**) determined by X-ray crystallography is almost planar due to the crystal packing forces. While tetrasilyldisilene is predicted theoretically to be a planar disilene, the geometry of tetrakis(trialkylsilyl)disilenes ($R_2Si{=}SiR_2$; R = *t*-BuMe$_2$Si (**22**), *i*-Pr$_2$MeSi (**23**), *i*-Pr$_3$Si (**24**), *t*-Bu$_2$MeSi (**33**)) determined by X-ray crystallography is not planar but *trans*-bent or twisted depending on the substituents. Geometrical parameters around the Si=Si double bond are easily modified by substituent steric effects, which cause Si=Si bond elongation, twisting around the Si=Si bond, deformation of bond angles around the unsaturated silicon atoms, etc. The Si–Si bond length (d), bend-angle (β), and twist angle (τ) of disilenes determined by the X-ray structural analysis are given in Tables I–IV; herein, the bent angle β is defined as the angle between Si=Si bond and R–Si(sp^2)–R plane and the twist angle τ is defined as the dihedral angle between two R–Si(sp^2)–R planes (Chart 5).

Comparison of energy levels of π- and π*-orbitals between ethylene and disilene is informative. At the B3LYP/6–311++G(2d,p)//B3LYP/6–31G(d) level, the energy levels of the π- and π*-orbitals in disilene are higher and lower than those in ethylene, respectively, and the π–π* splitting in disilene is about a half of that in ethylene (Fig. 2).[55] The relative π and π* energy levels in disilene suggest that reactivity of disilene for both nucleophiles and electrophiles is higher than that of ethylene. The smaller π–π* splitting of disilene is suggestive of a larger singlet biradical nature of disilene than that of ethylene.[56]

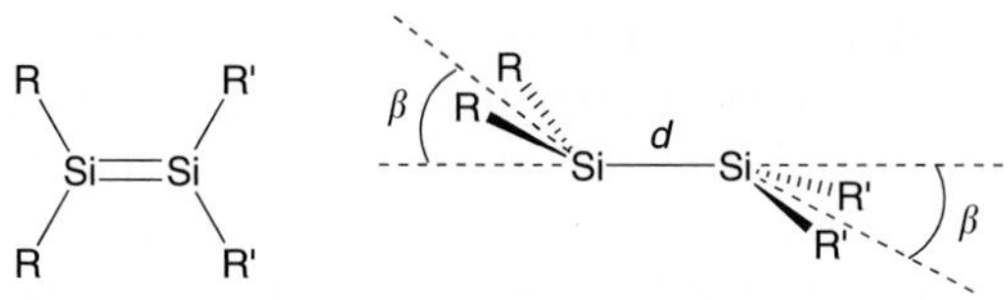

CHART 5.

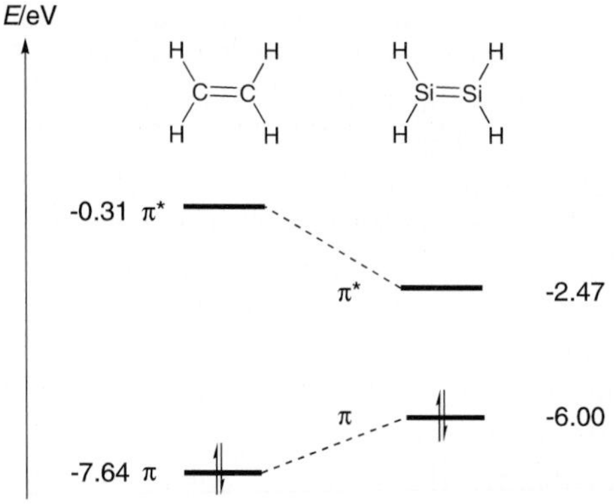

FIG 2. Frontier orbitals and their energy levels (eV) of ethylene and disilene at the B3LYP/6–311++G(2d,p)//B3LYP/6–31G(d) level.

B. *Strain Energies of Cyclic Disilenes*

Ring strain energies (SE) of cyclic disilenes are worth discussing because the ring size effects on the SE values in a series of cyclooligosilanes and cyclooligosilenes are remarkably different from those in the corresponding cycloalkanes and cycloalkenes. While the SEs of cyclic disilenes have not been determined experimentally, they are evaluated theoretically at the B3LYP/6–311++G(3df,2p)//B3LYP/6–31G(d) level as the heats of the following homodesmotic reactions [Eqs. (42)–(44), Table V].[36,57,58]

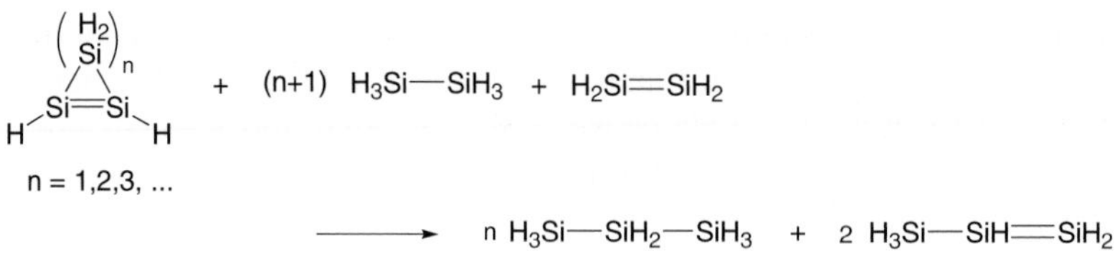

$$\text{cyclo-}(SiH_2)_n(SiH{=}SiH) + (n+1)\ H_3Si{-}SiH_3 + H_2Si{=}SiH_2 \longrightarrow n\ H_3Si{-}SiH_2{-}SiH_3 + 2\ H_3Si{-}SiH{=}SiH_2 \quad (42)$$

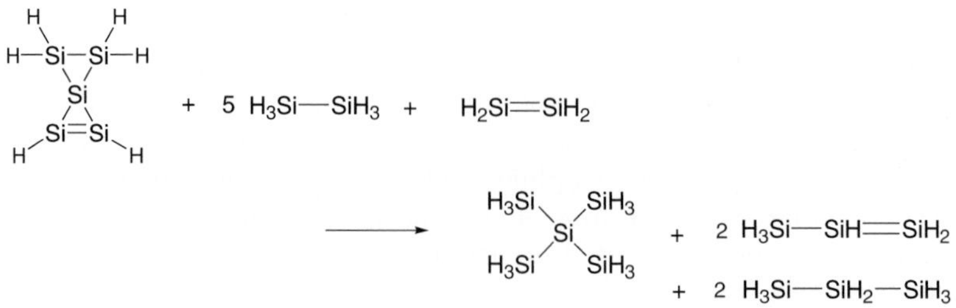

n = 1,2,3, ...

$$\text{spiro-}(H_2Si{-}SiH_2)Si(HSi{=}SiH) + 5\ H_3Si{-}SiH_3 + H_2Si{=}SiH_2 \longrightarrow Si(SiH_3)_4 + 2\ H_3Si{-}SiH{=}SiH_2 + 2\ H_3Si{-}SiH_2{-}SiH_3 \quad (43)$$

$$(\mathrm{HSi{=}SiH})_2\mathrm{Si} + 4\ \mathrm{H_3Si{-}SiH_3} + 2\ \mathrm{H_2Si{=}SiH_2} \longrightarrow \mathrm{Si(SiH_3)_4} + 4\ \mathrm{H_3Si{-}SiH{=}SiH_2} \qquad (44)$$

The fact that SE values of a series of cycloalkanes and cycloalkenes increase with decreasing the ring size is well known and explained by the Baeyer's angle strain and conformational strain. A similar tendency is found in cyclooligosilanes and cyclooligosilenes. Although cyclopropene has the highest SE (55.5 kcal mol^{-1}) among cycloalkanes and cycloalkenes and almost twice of the SE of cyclopropane (25.8 kcal mol^{-1}) as expected by the largest Baeyer's strain, the SE of cyclotrisilene (34.6 kcal mol^{-1}) is comparable to that of cyclotrisilane (35.4 kcal mol^{-1}).[36] The SE of cyclotrisilene is much smaller than that of cyclopropene, while the SE of cyclotrisilane is even larger than that of cyclopropane. While introduction of a C=C double bond into a carbon-based small ring increases the SE value remarkably, the introduction of a Si=Si double bond into a silicon-based small ring causes the reduction of SE. The results are in good accord with the more flexible nature of the Si=Si double bond than C=C double bond; the former bond is less sensitive to the angle strain than the latter. Detailed theoretical studies of ring strains of cycloalkenes and their silicon analogs by Inagaki and coworkers have shown another aspect of the effects of introduction of a double bond on the SE of a cycloalkane or a cyclooligosilane.[58] The ring strain is relaxed by the cyclic orbital interaction between a π-orbital and a σ*-orbital on the saturated atoms in the ring. While the three-, four-, and five-membered silicon rings are considerably stabilized by the cyclic interaction, it is less important for the stabilization of carbon rings.

TABLE V
STRAIN ENERGIES (SE) OF CYCLIC ALKENES AND DISILENES (IN KCAL MOL^{-1})[a]

E	cyclo-$(EH_2)_3$	cyclo-$EH_2(HE{=}EH)$	cyclo-$(EH_2)_4$	cyclo-$(EH_2)_2(HE{=}EH)$	cyclo-$(EH_2)_5$	cyclo-$(EH_2)_3(HE{=}EH)$
C	25.8	55.5	22.76	28.7	4.7	4.5
Si	35.4	34.6	12.9	9.1	3.0	0.9

E	spiro-E$(H_2E{-}EH_2)_2$	spiro-E$(H_2E{-}EH_2)(HE{=}EH)$	spiro-E$(HE{=}EH)_2$
C	55.4	85.7	114.2
Si	63.9	68.3	61.1

[a]Strain energies were calculated at the B3LYP/6-311++G(3df,2p)//B3LYP/6-31G(d) level using homodesmotic reactions shown in Eqs. (42)–(44).

In sharp contrast to spiropentadienes that decomposes even below −100 °C,[37]spiropentasiladiene derivative **66** is very thermally stable with a melting point of 216–218 °C (Section III.C.3).[36] The reason is, of course, in part due to steric protection by four extremely bulky tris(tert-butyldimethylsilyl)silyl groups but the major reason may be ascribed to the small SE of the spiropentasiladiene ring system compared with that of carbon-based spiropentadiene. The SE value calculated for parent spiropentasiladiene Si_5H_4 is almost a half of that for carbon-based spiropentadienes (114.2 kcal mol^{-1}) as a consequence of the large difference in the SE values between cyclotrisilene and cyclopropene.

C. *Molecular Structures*

After review OW, the molecular structures in the solid state of 13 acyclic disilenes, 13 endo- and exocyclic disilenes, 4 silicon-based dienes, 6 disilenes with radical, cation, and anion center, and one disilyne have been determined by X-ray analysis. Although the structural characteristics of a number of these new disilenes are summarized in a recent review by Weidenbruch,[9] the structural parameters of all new disilenes reported after review OW are listed in Tables I–IV.

1. Acyclic Disilenes

As shown in Section III.A, the geometry around the Si=Si double bond in disilenes is expected to be usually *trans*-bent with a small bend angle. Because the Si=Si double bond is rather soft and the potential energies are not sensitive to the expansion of the bond, bending and twisting around the double bonds and so on, the molecular geometry of disilenes in the solid state depends not only on the substituents but on the the crystal packing force. The Si=Si bond distances (d) of acyclic neutral disilenes vary between 2.138(2) Å (for (E)-**4**)[59] and 2.2598(18) Å (for **33**).[21] Silyl substituents tend to elongate the Si=Si double bond compared with carbon-based substituents. While d values of tetraaryl- and tetraalkyldisilene range from 2.138(2) (for (E)-**4**)[59] to 2.228(2) Å (for (E)-**27**),[60,61] those of disilyldiaryldisilenes and tetrasilyldisilenes are from 2.182(2) (for (E)-**41**)[62] to 2.220(1) Å (for **46** and **47**)[9] and from 2.196(3) (for (E)-**34**)[63] to 2.2598(18) Å (for **33**),[64] respectively.

The bent angles (β) range from 0° to 14.6° for disilene (E)-**27**[60,61] and twist angle τ from 0° to 54.5° for **33**[64] (Chart 5).

Interestingly, tetramesityldisilene **1** forms three different single crystals, unsolvated orange colored crystals,[65] yellow crystals with toluene,[66] and yellow crystals with THF.[67] Although no significant interaction exists between a disilene and solvents in the single crystals, the geometrical parameters are different from each other; $d = 2.143(2)$ Å, $\beta = 12$ and 14°, $\delta = 3°$ for unsolvated **1**; $d = 2.160(1)$ Å, $\beta = 18°$, $\tau = 12°$ for **1**·(toluene); and $d = 2.146$ Å, $\beta = 0°$, $\tau = 13°$ for **1**·(THF).

Structural parameters of tetrakis(trialkylsilyl)disilenes depend significantly on the steric bulkiness of the silyl substituents. Disilenes **23** and **24** are slightly *trans*-bent without twisting, while disilenes **22**, **35**, and **37** are twisted with τ of 8.9, 10.4 (avg.), and 28.0°, respectively; no bending was found around the unsaturated silicon atoms for these disilenes.[14,68] Disilene (*E*)-**34** is almost planar.[63] Tetrakis(trialkylsilyl)disilene **33** having the bulkiest *t*-Bu_2MeSi groups shows the largest twist angle δ of 54.5° with β of 5.29° and 6.44° at the unsaturated silicon atoms and the longest *d* value among acyclic disilenes.[64]

2. Cyclic Disilenes

The Si=Si double bond distances in the three-membered ring [2.132(2), 2.138(2), and 2.146(1) Å for **48**, **49**, and **50**, respectively] [24,29,30,69] are much shorter than those of four- and five-membered cyclic disilenes. The short Si=Si distance in the three-membered ring is reproduced by the theoretical calculations at the B3LYP/6–31G(d) level; at the optimized structures, the calculated distances are 2.121, 2.160, and 2.162 Å for parent cyclotrisilene **116a**, cyclotetrasilene **117**, and cyclopentasilene **118**, respectively (Chart 6).[58] The origin for the short distance for cyclotrisilenes would be similar to that for cyclopropene; among cycloalkenes, cyclopropene is known to have the shortest C=C bond distance to form the effective banana-type σ-bond. The geometry around the Si=Si double bond in cyclotrisilenes **48–50** is not planar but significantly twisted with the R–Si=Si–R angle between 21.4(2) and 37.0(2)°. Since cyclotrisilenes **116a**[36,58] and **116b**[36] have theoretically a planar Si=Si double bond at the B3LYP/6–31G(d) level, the twisting found in **48–50** may be ascribed to the steric repulsion between bulky vicinal substituents.

The *d* values of cyclotetrasilenes range from 2.174(4) Å for **51**[23] to unusually long 2.360(2) Å for **53**.[26] Among five-membered cyclic disilenes with a chalcogen atom in the ring **58–60**, the *d* values increase with the increase of the size of the chalcogen atom.[31] The structure of exocyclic disilene **62** is highly distorted with the large *d*, β, and τ values of 2.289 Å, 33.1° (avg.), and 25.1°, respectively.[33,34] The *d* value for another exocyclic disilene **63** [2.2621(15) Å] is the longest among carbon-substituted disilenes, while its β (0.5°) and τ [12.08(15)°] values are not remarkably large.[35]

3. Silicon-Based Dienes

Tetrasila-1,3-dienes **64**[17] and **65**[21] exist in the s-*cis* form in the solid state with the large Si=Si–Si=Si dihedral angle of 51 and 72°, respectively, probably due to the severe steric hindrance between bulky substituents. Despite the large distortion from coplanarity between the two double bonds in **64** and **65**, significant

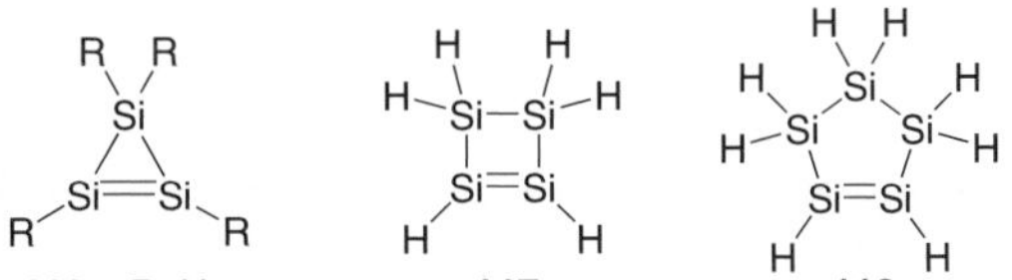

Chart 6.

conjugation between the two double bonds is evidenced by the elongated Si=Si bonds and the shortened central Si–Si σ-bond; the Si=Si double and Si–Si single bond distances in **64** are 2.175(2) and 2.321(2) Å, respectively.

Spiroconjugated disilene **66** has longer Si=Si double bonds (2.184(2) Å) and shorter endocyclic Si–Si single bonds (2.323(2) and 2.320(2) Å) than simple cyclotrisilenes **48**[69] and **49**,[29] whose Si=Si and ring Si–Si bond lengths are 2.132(2)–2.138(2) Å and 2.327(2)–2.364(2) Å, respectively, indicating the effective interaction between the bonding π-orbital of a Si=Si bond in one ring and low-lying Walsh-type Si–Si σ* orbitals in the other ring (σ*-aromaticity).[70] The two three-membered rings in **66** are not perpendicular to each other; the dihedral angle between the two ring planes is 78.3° (Fig. 3).[36] The geometry around each Si=Si bond in **66** is not planar but considerably twisted with the dihedral angle R_3Si–Si=Si–SiR_3 (R = *t*-BuMe_2Si) of 30.0(5)°.

Trisilaallene **67** shows unusual structural features in the solid state.[40] In contrast to carbon allenes having a rigid linear structure, the trisilaallene skeleton is bent and fluxional. The central silicon atom (Si2) in crystals is found at four positions labeled Si2A–Si2D in Fig. 4, indicating that four structurally similar isomers exist, A–D. The populations for A–D are independent of crystals but are remarkably temperature-dependent: the populations are 46, 23, 22, and 10% at 0 °C, but 76, 18, 7, and 0% at −150 °C, respectively. The energy differences between isomers A–D are estimated to be within 1.2 kcal mol^{-1}, suggesting a dynamic disorder mediated by a rotation of the Si2 atom around the Si1–Si3 axis.

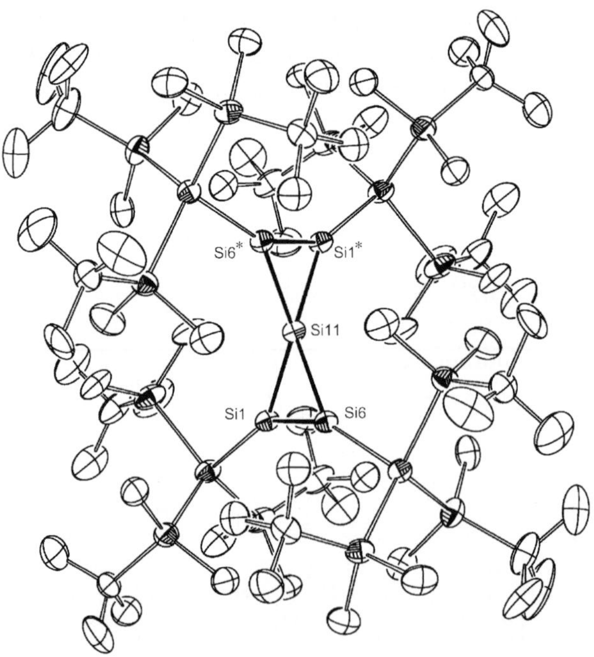

FIG 3. Molecular structure of spiropentasiladiene **66**.

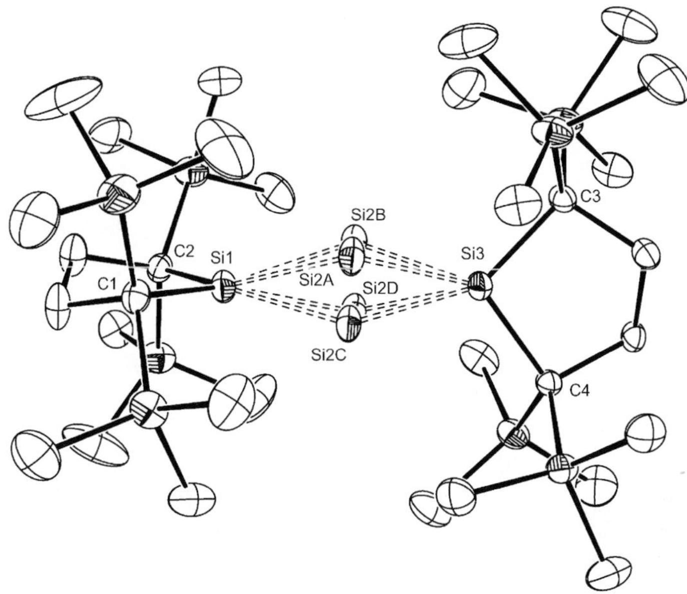

FIG 4. Molecular structure of trisilaallene **67**.

The major isomer A of trisilaallene **67** at $-150\,^{\circ}C$ has a significantly bent Si=Si=Si skeleton with a bond angle of 136.49(6)°, indicating that the bonding at the central silicon atom cannot be described using simple sp-hybridization. Two Si=Si bond lengths in **67** (2.177(1) and 2.188(1) Å) are in the range of those for typical stable disilenes (2.14–2.29 Å). Two five-membered rings are almost perpendicular to each other; the dihedral angle between the C–Si(sp^2)–C planes is 92.58°. Each Si=Si double bond has a *trans*-bent arrangement with the sum of the bond angle around each terminal silicon atom of $354.6 \pm 0.4°$. Since there is a significant interaction between two π orbitals in **67** (vide infra), **67** is characterized as bent-allenic with two cumulative *trans*-bent Si=Si double bonds.

The DFT calculations of model trisilaallene $(H_3C)_2Si{=}Si{=}Si(CH_3)_2$ (**119**) at the B3LYP/6-31 + G(d,p) level (Chart 7) have shown that the most stable structures **119Z** and **119Z′** are neither linear nor bent allenic but zwitterionic with Si–Si–Si angles of 92.4 and 74.7°, respectively; the energy difference between **119Z** and **119Z′** is only 0.3 kcal mol^{-1}.[71,72] The cationic part of the zwitterionic structures is characterized as allyllic and cyclopropenyl cation for **119Z** and **119Z′**, respectively. The structural difference between experimental **67** and theoretical **119Z** and **119Z′** is ascribed to the severe steric hindrance between two bulky silacyclopentane rings caused when a zwitterionic structure is applied to **67**. Actually, large-scale MO calculations of **67** reproduce nearly the molecular structure of **67** in the solid state.[73] Structure **119B**, where the coordinates of four carbon and three silicon atoms of **119** are fixed at those observed for **67**, is only 8.8 kcal mol^{-1} higher in energy than **119Z**. MO analysis has shown that **119B** is best characterized as bent-allenic, because the terminal two pπ-type orbitals of **119B** have significant interaction with both $n\sigma$ and

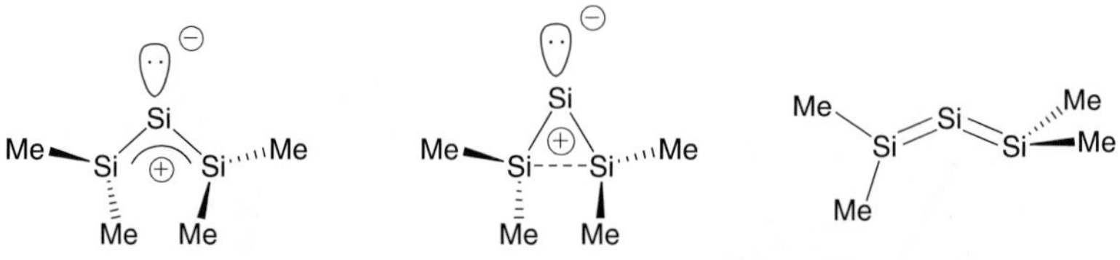

119Z (zwitterionic, allylium) **119Z'** (zwitterionic, cyclopropenylium) **119B** (bent allenic)

CHART 7.

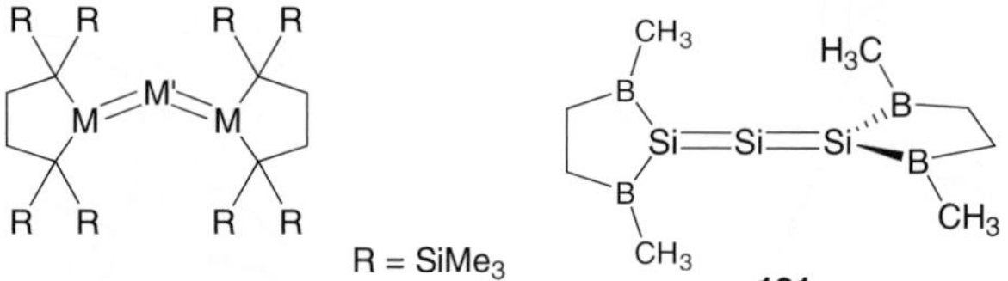

R = SiMe$_3$

120a, M = M' = Ge
120b, M = Ge, M' = Si
120c, M = Si, M' = Ge

121

CHART 8.

pπ orbitals at the Si2 atom. The bulky terminal substituents in **67** not only serve as sterically protecting groups for Si$=$Si bonds but also alter the electronic structure.

Related group-14 element trimetallaallenes **120a**–**120c** have similar bent allenic structure but show significant differences in the dynamic structure from **67**; 1,3-disilagermaallene **120c** shows a similar dynamic disorder with **67** but no such disorder is observed for 1,3-digermametallaallenes **120a** and **120b**, which have strongly bent terminal germanium atoms (Chart 8).[74]

Significant substituent effects on the skeletal geometry of trisilaallene have been revealed theoretically by Apeloig *et al.*;[72] trisilaallene **121** having electron-donating boryl-substituents is designed theoretically as a linear trisilaallene.

4. Disilyne

Isolable disilyne **76** has the shortest Si–Si distances among compounds with Si–Si bonds; the Si$\equiv$Si bond distance (2.0622(9) Å) is 3.8 and 13.8% shorter than typical Si$=$Si double (2.14 Å) and Si–Si single bond distances (2.34 Å), respectively.[41] This shortening is a half as much as that in the acetylenes. The geometry around the unsaturated silicons of **76** is *trans*-bent with a bend angle of 137.44(4)°, in good accord with previous theoretical predictions for silicon–silicon triply bonded compounds.[75]

5. Ionic Disilenes

The Si$=$Si bond distance of lithiodisilene **68** (2.192(1) Å)[20] is significantly longer than the corresponding distance for disilene **2** (2.144 Å),[76] being in qualitative agreement with the theoretical Si$=$Si bond of $H_2Si{=}SiH^-$ (**121**) compared to that of $H_2Si = SiH_2$.[49] The Si–Li bond lengths of **68** (2.853(3) Å) and **69** (2.702(9) Å)[21] are longer and comparable to those of usual silyllithium compounds. Interestingly, the Si–Si(Li)–C angle (107.6(1)°) of **68** is much smaller than the corresponding angles in **2** (120.88° and 121.68°).

Cyclotetrasilenylium ion **70** is free from the counter anion in the solid state.[42] The four-membered ring of **70** is significantly folded; the dihedral angle between the trisilenylium Si–Si–Si plane and the other Si–Si–Si plane is 46.6°. The trisilenylium Si–Si bond distances (2.240(2) and 2.244(2) Å) are intermediate between the Si=Si double bond (2.138 (2) Å) and the Si–Si single bond (2.364 (3) and 2.352 (3) Å) of the structurally similar cyclotrisilene **49**.[29] The interatomic distance between terminal silicon atoms in the cationic part (2.692(2) Å) is only 15% longer than a normal Si–Si single bond, indicating the homoaromaticity of **70**.

In contrast to cyclotetrasilenylium ion **70**, the four-membered ring of cyclotetrasilenyl radical **72** is almost planar with a dihedral angle of 4.7°.[43] The distances between unsaturated silicon atoms of **72** (2.226(1) and 2.263(1) Å) are also intermediate between the Si=Si double bond and the Si–Si single bond in **49**. In contrast to cation **70**, radical **72** has a long interatomic distance between terminal unsaturated silicon atoms [3.225(2) Å] indicating no significant 1,3-orbital interaction in **72**.[27]

The four-membered ring of cyclotetrasilenyl anion **71** is folded with a dihedral angle of 27°.[43] Anion **71** has a σ-allyllithium structure; a lithium atom binds to a terminal unsaturated silicon atom with the Si–Li distance of 2.569(4) Å, while the distances between lithium and the other unsaturated silicon atoms are 2.789(4) and 2.814(4) Å. The distances between neighboring unsaturated silicon atoms are 2.2245(7) and 2.3153(7) Å, indicating Si=Si double and Si–Si single bond characters, respectively.

The structures of tetrasilacyclobutadiene complex $[K^+(thf)_2]_2 \cdot$ **73** and digermadisilacyclobutadiene complex $[K^+(thf)_2]_2 \cdot$ **74** in the solid state are crystallographically isomorphous.[28] In contrast to well-known cyclobutadiene complexes having a planar four-membered ring moiety, four-membered rings of dianions **73** and **74** are significantly folded [the dihedral angle for $(K^+(thf)_2)_2 \cdot$ **73**, 34°] with two η^2-1,3-coordinated potassium cations occupying positions above and below the ring. The Si–Si bond distances in the four-membered ring of **73** are not equal but different with the distances of 2.2989(8) and 2.3576(8) Å. All skeletal Si atoms are significantly pyramidalized in **73**; the sums of the three bond angles around unsaturated silicon atoms are 341 and 326°. These structural features provide evidence for the non-aromatic nature of the tetrametallacyclobutadiene rings in the complexes.

D. *UV–vis Absorption and Fluorescence Spectra*

1. Acyclic Disilenes

The color of stable disilenes in the solid state ranges from pale yellow to blue. The longest absorption maxima of acyclic disilenes in solution are assignable to the $\pi \rightarrow \pi^*$ transitions and are strongly dependent on the electronic and steric effects of substituents; as shown in Tables I–IV. The λ_{max} values are 393–493 nm for tetraalkyldisilenes, 394–452 nm for dialkyldiaryldisilenes, 400–460 nm for tetraaryldisilenes, 468–483 nm for diaminodiaryldisilenes, 398–517 nm for diaryldisilyldisilenes, 429 nm for a alkyltrisilyldisilene, 476 nm for a diaminodisilyldisilene, 411–612 nm for tetrasilyldisilenes, and 417 nm for a triaryllithiodisilene.

The $\pi \rightarrow \pi^*$ absorption band for tetra-*tert*-butyldisilene **21** (λ_{max} 433 nm) is significantly red-shifted from that of tetraalkyldisilene **25** (λ_{max} 399 nm) probably due to the twisting around the Si=Si bond in **21**.[77] The unusually red-shifted absorption band of a tetraalkyldisilene having lattice substituents (**63**, λ_{max} 493 nm) is due to not only the lengthened and twisted Si=Si bond [$d = 2.261(15)$ Å, $\tau = 12.08(15)°$] but also important stereoelectronic interaction between the 3pπ orbital of the Si=Si double bond and the C–C σ- and π-orbitals of lattice-shape alkyl substituents, as supported by DFT calculations.[35]

Absorption bands of tetraaryldisilenes appear at a little longer wavelength than those of tetraalkyldisilenes. Leites *et al.* have found that the absorption and fluorescence spectra of tetramesityldisilene **1** depend on the forms of crystals, suggesting dependence of the conjugation between aryl π and Si=Si π systems on the rotational conformations around C_{Ar}–Si bond.[78] While both a 3-methylpentane solution of **1** and a thin film of **1** show the $\pi \rightarrow \pi^*$ absorption maxima at around 420 nm, crystals of unsolvated **1** exhibit the band at 465 nm. Unsolvated solids and a thin film of **1** show broad fluorescence bands at about 560 and 515 nm, respectively; large Stokes shifts indicate a substantial geometry change upon excitation.

Absorption spectra of tetrakis(trialkylsilyl)disilenes remarkably depend on their molecular structures. Less sterically hindered acyclic tetrasilyldisilenes **22**–**23**, (*E*)- and (*Z*)-**34**, and **35** are yellow to light orange; the $\pi \rightarrow \pi^*$ absorption maxima of these disilenes appear at 410–420 nm both in solution and in KBr pellets. While the absorption spectrum of tetrakis(triisopropylsilyl)disilene **24** is similar to those of disilenes **22** and **23** in KBr pellets, disilene **24** is dark red in a hexane solution and shows a broad band at around 480 nm in addition to the bands at 425, 370, and 295 nm in hexane. As expected from the flexible nature of Si=Si bonds, disilene **24** may be forced to nearly planar geometry in the solid state, while relaxing in solution to take the most stable molecular geometry.[68] The most twisted tetrasilyldisilene **33** ($\tau = 54.5°$) is blue in the solid state and shows the longest absorption band among acyclic disilenes (λ_{max} 612 nm in hexane).[64]

2. Cyclic Disilenes

The $\pi \rightarrow \pi^*$ absorption maxima of cyclic oligosilenes are significantly longer than those of the related acyclic disilenes probably due to the large steric strains of the rings: λ_{max} are 482, 466, and 469 nm for three-membered cyclic disilenes **48**,[24] **49**,[29] and **50**,[30] respectively, and 465 nm for cyclotetrasilene **51**.[23] The $\pi \rightarrow \pi^*$ absorption band of cyclotrisilene **48** [24] is 16 nm red-shifted compared with that of cyclotrisilene **49**,[29] suggesting effective σ–π conjugation in **48** between the disilene π orbital and Si–Si σ orbitals of tris(trialkylksilyl)silyl substituents.

3. Silicon-Based Dienes

Tetrasila-1,3-dienes **64** and **65** show the longest wavelength absorption bands at 518 and 531 nm, which are ca. 100 nm red-shifted from the corresponding disilenes **2**

(λ_{max} 432 nm) and **31** (λ_{max} 437 nm), respectively.[17,21] This indicates that there is significant conjugation between the two Si=Si double bonds in **64** and **65**, although they are twisted around the central Si–Si σ-bond with the dihedral angles of 51 and 72°, respectively (Section III.C.3).

The UV-spectrum of spiropentasiladiene **66**[36] in 3-methylpentane is considerably different from those of cyclotrisilenes **48**[24] and **49**.[29] Spiropentasiladiene **66** shows not only remarkable red-shift of the longest absorption maximum but also four distinct absorption bands in the visible region; λ_{max} (nm) ($\varepsilon/10^4$) = 560 (0.253), 500 (0.364), 428 (1.17), and 383 (1.81). The red-shift is reasonably ascribed to the significant spiroconjugation between the two Si=Si double bonds in **66**. However, the complex spectral pattern suggests that the simple spiroconjugation theory applied to carbon spiropentadiene[79] should be modified. Theoretical calculations at the B3LYP/6–311++G(3df,2p) level have shown that whereas parent spiropentasiladiene **122a** with D_{2d} symmetry, **122a**(D_{2d}), has two split π* orbitals and two degenerate π orbitals similar to carbon spiropentadiene (Fig. 5, left), the degeneracy of the π orbitals is removed when a similar distortion to that found in **66** (Section

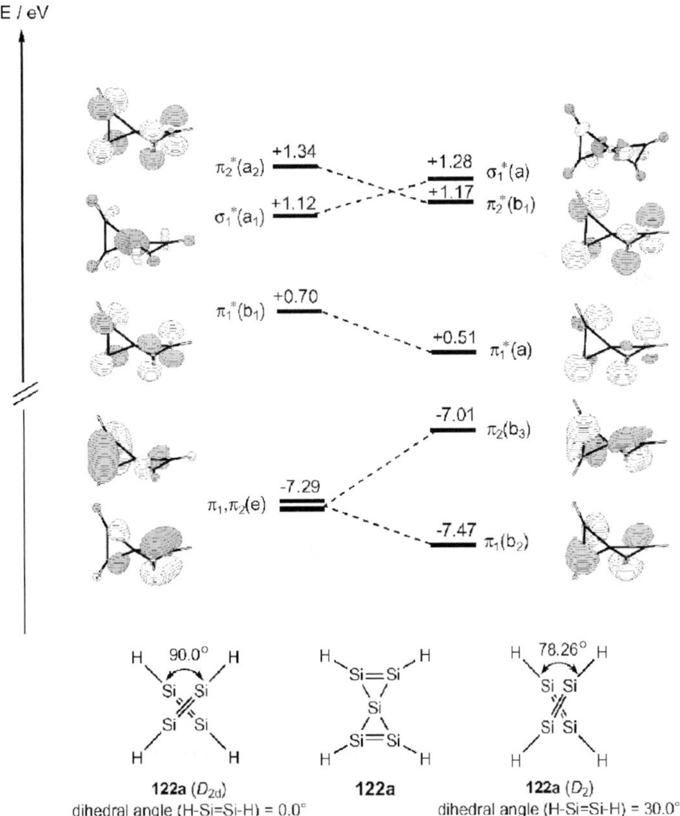

FIG 5. Schematic molecular orbital diagram for parent spiropentasiladiene with D_{2d} and D_2 symmetry, **122a**(D_{2d}) and **122a**(D_2), at the B3LYP/6–311++G(3df,2p)//B3LYP/6–31G(d) level.

III.C.3) is introduced to **122a** to form **122a**(D_2). Orbital splitting of both π* and π orbitals of a spiropentadiene system due to modified spiroconjugation explains the spectral feature of spiropentasiladiene **66**.[36]

Because carbon-based allene has two perpendicular π-bonds in the linear skeleton, no conjugation exists between the two π-bonds and the π→π* transition band appears in the same region as that of ethylene. In contast, the UV–vis absorption maxima of **67** are found at 390 nm (ε 21300) and 584 nm (ε 700) in hexane; the two-splitting-band feature as well as the remarkable red-shift of the longest absorption band of **67** compared with the π–π* transition band for tetramethyldisilene (λ_{max} = 350 nm)[80] and disilene **25** (λ_{max} = 393 nm) [81] is consistent with significant conjugation between the two cumulative Si=Si double bonds. Conjugation between the two Si=Si double bonds in **67** is stronger than the conjugation in spiropentasiladiene **66** and tetrasilabutadienes **64**[17] and **65**.[21]

4. Disilyne

Disilyne **76** is emerald green in the solid state and shows two allowed π→π* absorption bands in hexane at 259 (ε 10300) and 328 (ε 5800) nm and two forbidden π→π* absorption bands at 483 (ε 120) and 690 (ε 14) nm.[41] Theoretical calculations for **76** at the HF/6–311G(d)//B3LYP/6–31G(d) level have shown that HOMO and HOMO-1 are two split π orbitals and LUMO and LUMO+1 are two split lowest π* orbitals. The out-of-plane π orbitals HOMO and LUMO+1 are represented by pure (p_z–p_z)π orbitals, whereas the in-plane π orbitals HOMO-1 and LUMO are represented mainly by (p_y–p_y)π orbitals with a weak contribution of the σ*(Si–Si) orbital of the central bond. The presence of the nondegenerate π and π* molecular orbitals in **76** is responsible for the UV–vis absorption band feature of **76**.

5. Ionic Disilenes

Although radicals and ions including disilenes **68**–**74** have intense color, the UV absorption bands are reported only for **68**[20] and **72**[27] (Tables I–IV).

E. *Cyclic Voltammetry*

Oxidation and reduction potentials of tetraaryldisilenes **1** and **2**,[82] tetrasilyldisilenes **23** and **24**,[83] cyclotrisilene **48**,[84] cyclotetrasilene **51**,[83] and spiropentasiladiene **66**[84] were measured by cyclic voltammetry. Since both reduction and oxidation of these disilenes are irreversible, the peak potentials are shown summarized in Table VI. Interestingly, the oxidation and reduction potentials of tetrasilyldisilenes **23** and **24** are significantly higher and lower, respectively, than those of tetraaryldisilenes **1** and **2**, indicating that both HOMO and LUMO of tetrasilyldisilenes are lower in energy than those of tetraarylldisilenes. The reactivity differences observed between tetraaryldisilenes and tetrasilyldisilenes as shown in Section IV may be ascribed to this intrinsic difference of the frontier orbital levels, in addition to the steric reasons. Cyclotetrasilene **51** has the oxidation and reduction potentials similar to those of acyclic tetrasilyldisilenes, while cyclotrisilene **48** has

TABLE VI
PEAK POTENTIALS OF DISILENES DETERMINED BY CYCLIC VOLTAMMETRY[a,b]

Disilene	E_{peak}(oxidation)	E_{peak}(reduction)
$Mes_2Si{=}SiMes_2$ (**1**)	+0.44 (+0.38)[c]	−2.12 (−2.12)[c]
$Tip_2Si{=}SiTip_2$ (**2**)	+0.56, +1.32	−2.66
$(i\text{-}Pr_2MeSi)_2Si{=}Si(SiMe(i\text{-}Pr)_2)_2$ (**23**)	+0.88	−1.82, −2.41
$(i\text{-}Pr_3Si)_2Si{=}Si(Si(i\text{-}Pr)_3)_2$ (**24**)	+0.70	−1.70, −2.56
Cyclotrisilene **48**	+0.54[d]	−2.20, −2.98[d]
Cyclotetrasilene **51**	+0.72	−1.94, −2.49
Spiropentasiladiene **66**	+0.64, +0.80[d]	−2.05, −2.80[d]

[a]Potentials are measured versus Ag/Ag^+ at rt in THF containing 0.1 M $[Bu_4N][PF_6]$. The ferrocene:ferricinium ion couple is measured at −0.61 V in this system.
[b]Unless otherwise noted, data are taken from Ref. 83.
[c]Data in parentheses are taken from Ref. 82.
[d]Data are taken from Ref. 84.

lower oxidation and higher reduction potentials. The lower reduction potential observed for spiropentasiladiene **66** than those for cyclotrisilene **48** may be ascribed to the spiroconjugation. However, there are significant discrepancies between the observed potentials and those expected from the orbital energy levels calculated theoretically, suggesting the significant geometry changes during the oxidation and reduction processes of some disilenes.

F. *^{29}Si NMR Spectra*

1. Acyclic Disilenes

Acyclic disilenes show the relatively low-field shifted resonances of unsaturated ^{29}Si nuclei depending on the substituents. The ^{29}Si resonances for $A_2Si{=}SiA_2$-type disilenes range from 53–66 ppm for tetraaryldisilenes, to 90–103 ppm for tetraalkyldisilenes, and 131–156 ppm for tetrasilyldisilenes.

West and co-workers determined the ^{29}Si NMR chemical shift tensors of various disilenes by solid-state NMR spectroscopy and discussed the origin on the basis of the theoretical calculations.[85] The experimental tensors are summarized in Table VII. Various disilenes show significant deshielding along one principal axis, which is identified by theoretical calculations to be the axis in the molecular plane and perpendicular to the Si–Si bond (δ_{11} in Fig. 6). The deshielding is mainly attributed to a paramagnetic circular shielding current arising from angular momenum along this axis, which mixes the Si–Si σ orbital with the Si=Si π^* orbital. Deformations of the Si=Si double bond like twisting or *trans*-bending by 15–20° or the aryl-Si=Si conjugation show little effect on the shielding tensors in aryl- or arylalkyl disilenes. In contrast, elongation of the Si=Si bond leads to remarkable deshielding of δ_{11}. The extremely large deshielding of δ_{11} found for tetrasilyldisilenes is ascribed to the low $\sigma(Si{=}Si)\rightarrow\pi^*(Si{=}Si)$ transition energy that is responsible for the shielding tensor δ_{11}. The $\sigma(Si{=}Si)$ orbital level would be raised by the inductive effect of electropositive trialkylsilyl groups and the $\pi^*(Si{=}Si)$ level would be

TABLE VII

CHEMICAL SHIFT TENSORS FOR DISILENES DETERMINED BY SOLID STATE NMR SPECTROSCOPY[a]

Disilene	δ (*iso*) Solution	δ (*iso*) Solid	δ_{11}	δ_{22}	δ_{33}	$\Delta\delta\ (=\delta_{11}-\delta_{33})$
1	63.3	63.2	181	31	−22	203
1·(toluene)	63.3	65.0	185	34	−22	207
1·(THF)	63.0	59.6	165	40	−25	190
2	53.4	50.8	155	30	−31	186
		53.2				
3	90.3	86.1	178	77	3	175
25	90.4	86.1	182	55	21	161
		87.4	199	54	9	190
9	94.4	94.5	296	46	−59	355
23	144.5	143	414	114	−100	514
24	154.5	164	412	149	−69	481

[a]Data were taken from Ref. 85.

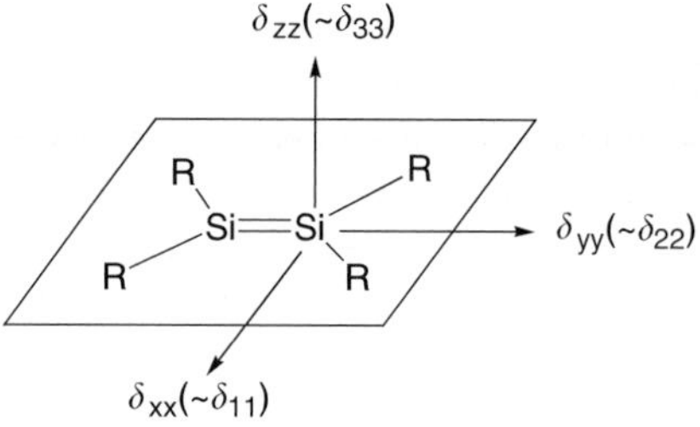

FIG 6. Approximate orientation of the principal shielding tensor components in a disilene.

lowered by the π accepting effects of the silyl substituents; the latter is in good accord with the lower reduction potentials of tetrasilyldisilenes (Section III.E).

The ^{29}Si resonances for $A_2Si{=}SiB_2$-type disilenes appear at 49–62 ppm for diaminodiaryldisilenes and 87–97 ppm for dialkyldiaryldisilenes. However, large differences in the ^{29}Si resonances are observed between aryl-substituted and silyl-substituted silicon nuclei of 1,1-diaryl-2,2-disilyldisilenes **28**–**31**; the ^{29}Si resonances appear at 139–153 ppm for the former and at −1 – 15 ppm for the latter.[13] The former resonances in **28**–**31** are greatly downfield shifted relative to those of tetra-aryldisilenes, while the latter resonances are remarkably upfield shifted from those of tetrakis(trialkylsilyl)disilenes. The rather unusual ^{29}Si resonances for 1,1-diaryl-2,2-disilyldisilenes are interpreted on the basis of DFT calculations.[86] While the substituent effects in $A_2Si{=}SiA_2$ type disilenes can be understood mainly by the energy difference between σ(Si═Si) and π*(Si═Si) orbitals as stated above, the resonances for the two ^{29}Si nuclei of $A_2Si{=}SiB_2$-type disilenes are affected also by the unsymmetric spatial extent and orientation of the occupied and unoccupied molecular orbitals resulting in different nuclear magnetic moments on the respective silicon nuclei.

2. Cyclic Disilenes

The resonances for the unsaturated ^{29}Si nuclei of cyclotrisilenes **48** and **49** (82–100 ppm) are remarkably higher than those of the corresponding acyclic disilenes, while the ^{29}Si resonances are of cyclotetrasilenes **51** and **55** (158–183 ppm) are much lower. This tendency for the ^{29}Si resonances is parallel to that for the resonances of ring unsaturated ^{13}C nuclei among cyclopropene (108.7 ppm), ethylene (123.5 ppm), and cyclobutene (137.2 ppm).[87] The high-field shift of ^{29}Si resonances of cyclotrisilenes and ^{13}C resonances of cyclopropene would have the same origin. These ^{29}Si resonances are well reproduced by GIAO calculations (GIAO/B3LYP/6–311+G(2df,p)//HF/6-31G(d)) for the model compounds.[8]

3. Silicon-Based Dienes

Tetrasila-1,3-dienes show two different resonances due to unsaturated ^{29}Si nuclei. The ^{29}Si resonances depend significantly on the substituents; they appear at 52.3 and 89.5 ppm for **64** and at 71.5 and 150.8 ppm for **65**.[17,21]

The resonance of the unsaturated ^{29}Si nuclei of spiropentasiladiene **66** (154.0 ppm) shifts to a lower field by +60 ppm compared with those of cyclotrisilenes **48** and **49**. The ^{29}Si resonances are well reproduced by DFT calculations for the model compounds; +178.3, +105.3, and +110.1 ppm for H_3Si-substituted spiropentasiladiene **122b** (D_{2d}, Chart 9), spiropentasilene **123** (C_{2v}, Chart 9), and cyclotrisilene **116b** (C_{2v}, Chart 6), respectively, at the GIAO/B3LYP/6-311+G(2df,p)//B3LYP/6-31G(d) level. The remarkable low-field shift for **66** may be ascribed to the lowering of π^*(Si=Si) level by the through-space interaction between π^* orbitals (spiroconjugation). Although the resonance due to the central ^{29}Si nucleus in **66** is not experimentally observed, the corresponding resonance for **122b** (D_{2d}) was evaluated to be −45.9 ppm by the GIAO calculations.

The NMR resonances for unsaturated ^{29}Si nuclei of trisilaallene **67** are found at 157 and 197 ppm in benzene-d_6, and are assigned to the central and terminal silicon nuclei, respectively.[40] The ^{29}Si resonances for a model compound Me_4Si_3 **119** calculated at the GIAO/6–311+G(2df,p)//B3LYP/6–31+G(d,p) level are found to be quite sensitive to the molecular structure; the resonances for central and terminal ^{29}Si nuclei for bent allenic structure **119B** are 144 and 221(212) ppm, respectively, while they are 39 and 355 ppm for zwitterionic **119Z** (Chart 7) and 210 and 98 ppm for linear $Me_2Si{=}Si{=}SiMe_2$ (**119L**). The experimental resonances are in good accord with the theoretical values for **119B**, indicating that the bent-allenic structure of **67** observed in the solid state is maintained in solution.

122b **123**

CHART 9.

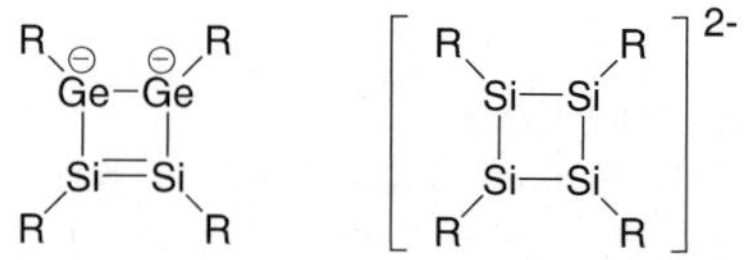

124 125

CHART 10.

4. Disilyne

The resonance of the unsaturated ^{29}Si nuclei of disilyne **76** appears at +89.9 ppm,[41] which is upfield-shifted more than 50 ppm from those of tetrasilyldisilenes 22–24 (142–155 ppm).[68] A parallel tendency is found in the resonances of unsaturated ^{13}C nuclei of silyl-substituted alkenes (188–197 ppm) and alkynes (112–114 ppm).[88]

5. Ionic Disilenes

The resonances for unsaturated ^{29}Si nuclei of 1-lithio-1-aryl-2,2-disilyldisilene **69** appear at 63.1 and 277.6 ppm, and are assigned to the silyl-substituted and the lithium-substituted silicon nuclei, respectively, on the basis of the calculations for $(H_3Si)_2Si{=}Si(Ph)Li$; the resonances are 139.1 and 282.5 ppm for $(H_3Si)_2Si$ and PhLiSi, respectively, at the GIAO/HF/6–311++G(2df,p)//B3LYP/6–31G(d) level.[21] The ^{29}Si NMR resonances for triaryllithiodisilene **68** appear at 100.5 and 94.5 ppm.[20] These resonances are downfield-shifted significantly compared to that of tetraaryldisilene **2** (53.4 ppm).[76] Although the signal at 100.5 ppm of **68** is broadened, no $^{1}J(^{29}Si–^{7}Li)$ is observed even at –80 °C in toluene.

The resonances due to unsaturated ^{29}Si nuclei of cyclotetrasilenylium cation **70** appear at +77.3 (terminal Si) and +315.7 ppm (central Si). The chemical shifts are independent of the solvent such as dichloromethane, benzene, and toluene, implying no covalent interaction between the cation part and solvent molecules. Rather unusually, the central silicon in the cation part is more deshielded than the terminal silicons.[42] The terminal and central ^{29}Si resonances of anion **71** are –31.5 and +273.0 ppm in toluene-d_8, and –24.7 and +224.5 ppm in THF-d_8.[43] The terminal silicon atoms are equivalent, indicating that the lithium cation is fluxional and **71** adopts an allylic anion-type structure in solution. Similarly to allyllithium, the central nucleus of **71** is deshielded, while the terminal nuclei are highly shielded.

Although tetrametallacyclobutadiene dianions **73** and **74** have very similar geometries in the solid state, their ^{29}Si NMR spectra in solution are quite different.[28] The ^{29}Si resonances of ring silicons of **73** and **74** appear at 17.0 and 113.7 ppm, respectively. It is suggested that the negative charge of **74** localizes on more electronegative Ge atoms like **124** in Chart 10, while the negative charge of **73** is considerably delocalized to the four-ring silicon atoms like **125**.

G. *Vibrational Spectra*

Leites and coworkers investigated the Raman spectra of disilenes.[78,89] Disilenes with carbon-substituents (aryl and/or alkyl) **1**, (*E*)-**3**, and **25** exhibit a Raman line at

540, 522, and 454 cm^{-1} for **1**, (*E*)-**3**, and **25**, respectively. This vibration is not a normal mode well-localized in the Si=Si bond but an in-phase combination of the Si=Si and Si–C stretching modes and the vibrations of aromatic rings for aryl-substituted disilenes if aromatic groups are present. The second normal mode with significant Si=Si participation is an out-of-phase combination of the Si=Si and Si–C stretching modes occurring at 681, 687, and 733 cm^{-1} for **1**, (*E*)-**3**, and **25**, respectively.[78] While the frequencies of the Raman lines in 500–550 cm^{-1} of four different solids of **1**, **1**(unsolvated orange crystals), **1**(crystals solvated by toluene), **1**(crystals solvated by THF), and **1**(unsolvated yellow powder) are close to each other, remarkable differences are observed in the relative intensities of the lines, being indicative of the different geometrical arrangements of the aromatic rings among the crystals. Solid **62**, (*E*)-**41**, and (*E*)-**42** exhibit intense Raman lines at 533,[33] 592,[62] and 589 cm^{-1},[22] respectively.

IV
REACTIONS AND MECHANISMS

Most of the fundamental reactions of disilenes have been studied and discussed extensively in previous review articles. During the last decade, many new types of reactions have been found for novel types of disilenes such as cyclic and bicyclic disilenes, and conjugated tetrasiladienes. Mechanistic studies have recently been performed on several fundamental reactions of disilenes both theoretically and experimentally, and have greatly deepened our understanding of their reaction pathways, their potential energy surfaces, the factors determining the rates and stereochemistry, and so on.

A. *Unimolecular Reaction*

1. *E,Z*-Isomerization

As a typical unimolecular reaction of disilenes, the *E,Z*-isomerization is discussed first. In contrast to the isomerization of an alkene that occurs *via* the rotation around the C=C double bond with an activation energy of ca. 60 kcal mol^{-1} the *E,Z*-isomerization of disilenes is known to occur more easily. As shown in review OW, the *E,Z*-isomerization in aryl-substituted disilenes **3**, **4**, **20**, **26**, and **27** proceeds under mild conditions to allow the kinetic studies at 40–80 °C by NMR spectroscopy. Recently, the *E,Z*-isomerization between tetrakis(trialkylsilyl)disilenes (*E*)- and (*Z*)-**33** was found to occur more rapidly with the rates of the NMR time scale at 30 °C:[63]

$$(Si^A)(Si^B)Si{=}Si(Si^B)(Si^A) \rightleftharpoons (Si^A)(Si^B)Si{=}Si(Si^A)(Si^B) \quad (45)$$

(*E*)-**33** (*Z*)-**33**

33, Si^A = *t*-$BuMe_2Si$, Si^B = *i*-Pr_2MeSi

Due to the unique characteristics of Si=Si double bonds as discussed in Section III.A, at least the following three types of mechanisms emerge for the *E*,*Z*-isomerization between (*E*)- and (*Z*)-ABSi=SiAB (**126**), as shown in Eqs. (46)–(48): (1) rotation around the Si–Si bond, (2) dissociation to the corresponding silylenes, and (3) 1,2-migration of a substituent to form the corresponding silylsilylene.

(*E*)-**126** ⇌ [transition state]‡ ⇌ (*Z*)-**126** (46)

(*E*)-**126** ⇌ [ABSi: + :SiBA] ⇌ (*Z*)-**126** (47)

(*E*)-**126** ⇌ [AB_2Si–$\ddot{Si}$A] ⇌ (*Z*)-**126** (48)

Because the activation energy for a simple rotation around Si–Si bond should be roughly the π-bond energy E_π, pathway (1) may be feasible, if $E_\pi < \mathrm{BDE}$. Direct dissociation of a disilene to two silylenes [pathway (2)] would occur with the activation energy comparable to BDE for the disilene, if $E_\pi > \mathrm{BDE}$. If the substituents have a high migratory aptitude, pathway (3) may compete with pathways (1) and (2). At the B3LYP/6-311G(3df,3dp) level, the BDE of $H_2Si{=}SiH_2$ into a pair of singlet SiH_2 is calculated to be 61 kcal mol^{-1}.[90] On the other hand, the E_π value for $H_2Si{=}SiH_2$ is estimated to be 29 kcal mol^{-1} because $E_\sigma + E_\pi = \mathrm{BDE} + 2\Delta E_{ST}$ and ΔE_{ST} for SiH_2 and E_σ for H_3Si–SiH_3 are 21 and 74 kcal mol^{-1}, respectively, at the same theoretical level. Therefore, the direct dissociation of $H_2Si{=}SiH_2$ to two H_2Si: should not be competitive to the rotation near rt. The *E*,*Z*-isomerization of $H_2Si{=}SiH_2$ is expected to occur *via* the pathway (1) with the activation energy of ca. 30 kcal mol^{-1}. The value is in good accord with the activation enthalpies reported for the isomerization of disilenes **3**, **4**, **20**, **26**, and **27** (24.7–30.6 kcal mol^{-1}).[7] In contrast, the activation free energy for the *E*,*Z*-isomerization of tetrasilyldisilene **33** is estimated to be ca. 15 kcal mol^{-1},[63] which is significantly smaller than the E_π of 22 kcal mol^{-1} for $(H_3Si)_2Si{=}Si(SiH_3)_2$; the E_π value is estimated using the BDEs of $(H_3Si)_2Si{=}Si(SiH_3)_2$ and $(H_3Si)_2SiH$-$SiH(SiH_3)$ of 68.3 and 64.3 kcal mol^{-1}[55] respectively, and ΔE_{ST} of :$Si(SiH_3)_2$ of 9.0 kcal mol^{-1} calculated at the B3LYP/6–311++(3d,2p) level.[52] The rather low activation energy for the *E*,*Z*-isomerization of **33** may be ascribed to effective stabilization of the transition state due to the σ(Si–Si)-π conjugation (Chart 11).

1,2-Migration of a substituent on a disilene giving the corresponding silylene [pathway (3)] should be considered as a pathway for the *E*,*Z*-isomerization but occurs usually with much higher activation energies than pathways (1) and/or (2). The pathway (3) and related dyotropic rearrangement are discussed in detail in Section IV.A.3.

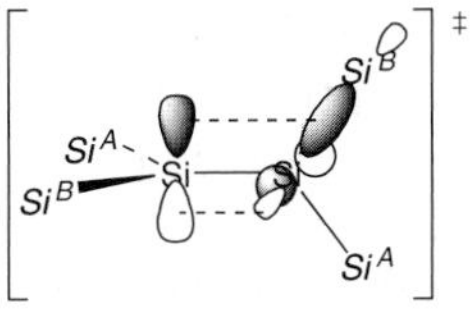

CHART 11.

2. Dissociation into Two Silylenes

As discussed in the previous section, thermal dissociation of disilenes into the corresponding silylenes may occur if the BDE of the disilenes is small. As shown in review OW, a facile thermal dissociation of disilene **27** into silylene **127** occurs at 50 °C [Eq. (49)],[61,91] The formation of silylene **127** is evidenced by its trapping by methanol, triethylsilane, and 2,3-dimethyl-1,3-butadiene. The activation enthalpy and entropy for the dissociation of (*Z*)-**27** to **127** are 25.5 kcal mol^{-1} and 7.8 cal mol^{-1} K^{-1} respectively.[91] The activation free energy for the dissociation at 323 K (22.9 kcal mol^{-1}) is much smaller than that for the *Z*-to-*E* isomerization of **26** (27.8 kcal mol^{-1}), indicating that the *E*,*Z*-isomerization of **27** should occur *via* the pathway (2) in Eq. (47) rather than pathway (1) in Eq. (48).

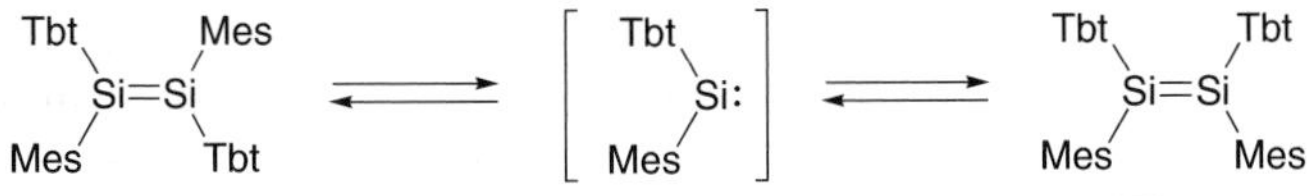

(49)

Tetraalkyldisilene **63** with a lattice framework in *dl*-form dissociates into the corresponding silylene **128** [Eq. (50)],[92] The treatment of (4S*,6S*,4′S*,6′S*)-**63** (*dl*-**63**) with methanol for 6 days at rt affords a racemic mixture of methanol adducts of silylene **128**. Silylene **128** is also trapped by bis(trimethylsilyl)acetylene.

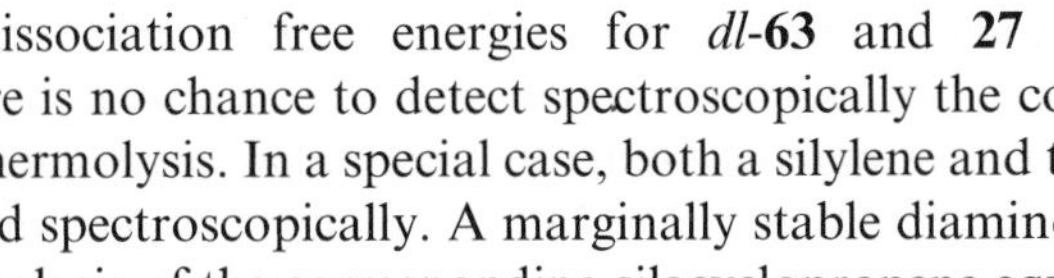

(50)

Because the dissociation free energies for *dl*-**63** and **27** are higher than 20 kcal mol^{-1} there is no chance to detect spectroscopically the corresponding silylenes during the thermolysis. In a special case, both a silylene and the corresponding dimer are observed spectroscopically. A marginally stable diaminosilylene **129** generated by the photolysis of the corresponding silacyclopropene equilibrates with the corresponding disilene **130** in solution in the temperature range of 292 and 77 K [Eq. (51)].[93] At 292 K, silylene **129** is observed both by ^{1}H NMR and UV–vis spectroscopies; the absorption maximum at 335 nm is reasonably assigned to a bis(dialkylamino)silylene.[94] When the solution is cooled to 77 K, a new band emerges at 439 nm, whose band intensity increases with lowering temperatures. The spectral change is reversible between 77 and 292 K. The new band is assigned to disilene **130** on the basis of the substituent effects on the absorption band maximum of disilene, instead of a bridged dimer predicted for parent diaminosilylene;[53] detailed theoretical calculations have shown that the Si–Si bond of **130** is unusually long (2.472 Å), even longer than the normal Si–Si single bond.[95] Because the BDE for **130** is very small (3–8 kcal mol^{-1}), entropically favored silylene **129** can exist in a reasonable concentration at higher temperatures.

2 i-Pr2N(i-Pr2N)Si: ⇄ (K, 3-MP) (i-Pr2N)2Si=Si(N(i-Pr)2)2 (51)

129 (λ_{max} = 335 nm at rt)

130 (λ_{max} = 439 nm)

Stable diaminosilylene **132** (R = *t*-Bu) does not dimerize to the corresponding tetraaminodisilene but undergoes an insertion reaction giving **131**, which further dimerizes to diaminodisilyldisilene **62**. Disilene **62** is stable in the solid state but equilibrates with stable silylene **132** (R = *t*-Bu) *via* **131** in solution [Eq. (52)].[33,34] Intermediacy of **131** is evidenced by the reaction of crystals of **62** with methanol giving a methanol adduct of **131**, **133**.

62, R = *t*-Bu ⇄ [**131**] ⇄ **132**, R = *t*-Bu

131 —MeOH→ **133** (52)

The above unusual equilibrium [Eq. (52)] was investigated in detail using DFT calculations.[33,53] Tetraaminodisilene **134** (R = H), a simple dimer of silylene **132** (R = H) does not correspond to a minimum on the potential energy surface because of large ΔE_{st} values of **132** (R = H) (Chart 12). Instead, two molecules of silylene **132** (R = H) are combined into bridged dimer **135** (R = H), which is 7.8 kcal mol^{-1} more stable than two separate **132** (R = H). However, the approach of two molecules of **132** (R = *t*-Bu) forms neither **134** (R = *t*-Bu) nor **135** (R = *t*-Bu) due to the steric hindrance of the bulky *t*-Bu groups. Thus, the third reaction pathway, the insertion of a silylene into the Si–N bond of another silylene giving aminosilylsilylene **131**, becomes feasible. Although silylene **131** (R = Me) is calculated to

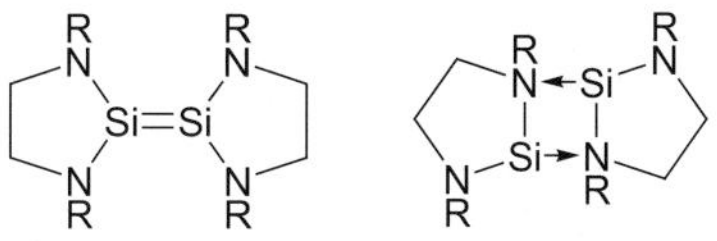

134 135

CHART 12.

be 0.5 kcal mol^{-1} less stable than two separate silylenes **132** (R = Me) at the B3LYP/6–311+G(d,p)//B3LYP/6–31G(d)+ZPE level, the dimerization of **131** (R = Me) into **62** (R = Me) is highly exothermic (27.3 kcal mol^{-1}) at the B3LYP/6–31G(d) level. Dissociation of **62** into **132** is an entropically favored process.

3. 1,2-Migration of Substituents and Related Isomerizations

West *et al.* first demonstrated interesting rearrangements among a set of isomeric disilenes, (*E*)- and (*Z*)-ABSi=SiAB and $A_2Si{=}SiB_2$, indicating the 1,2-migration of substituents on unsaturated silicon atoms in addition to the *E*,*Z*-isomerization discussed in Section IV.A.1.[96] Rearrangements between $A_2Si{=}SiB_2$ and (*Z*)- and/or (*E*)-ABSi=SiAB may occur either *via* a concerted dyotropic rearrangement or *via* a disilene–silylsilylene rearrangement:

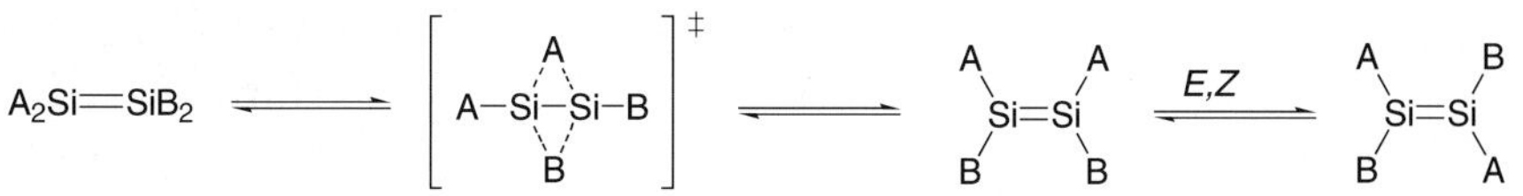

$$A_2Si{=}SiB_2 \rightleftharpoons [\text{A–Si(A)(B)–Si–B}]^{\ddagger} \rightleftharpoons (A)(B)Si{=}Si(A)(B) \overset{E,Z}{\rightleftharpoons} (A)(B)Si{=}Si(B)(A) \quad (53)$$

$$A_2Si{=}SiB_2 \rightleftharpoons A\ddot{S}i{-}SiAB_2 \rightleftharpoons ABSi{=}SiAB \quad (54)$$

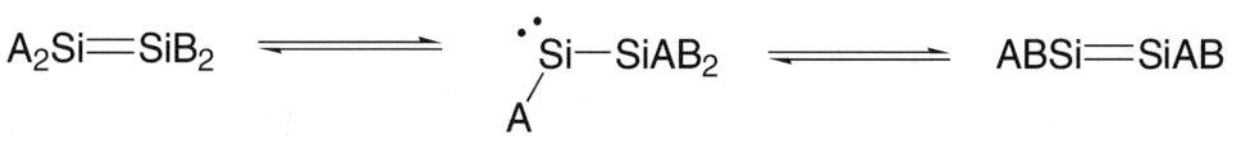

On the basis of a kinetic study for a set of disilenes (A = Mesityl, B = 2,6-Xylyl), the dyotropic rearrangement in Eq. (55) was proposed as a favorable pathway for the intramolecular rearrangement rather than a disilene–silylsilylene rearrangement. An equilibrium among three isomeric disilenes is attained after 70 days at 298 K; $\Delta G^{\ddagger} = 26$ kcal mol^{-1} $\Delta H^{\ddagger} = 15 \pm 2$ kcal mol^{-1}, and $\Delta S^{\ddagger} = -36$ cal mol^{-1} K^{-1} for the isomerization of **5** to **13** at 298 K. The large negative $\Delta S^{\ddagger}$ is compatible with the mechanism involving highly ordered transition state such as that in the dyotropic rearrangement.[96]

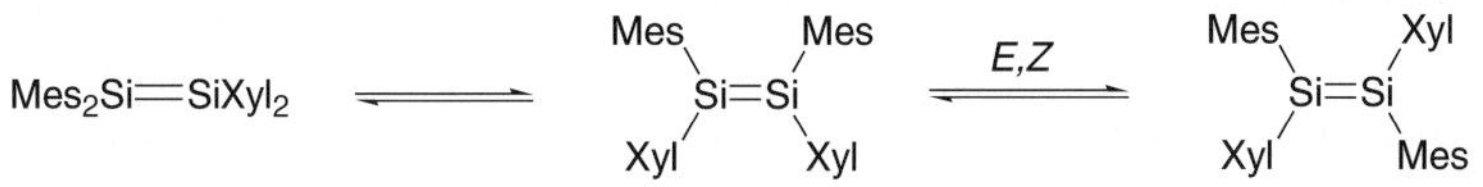

$$\underset{\mathbf{13}}{Mes_2Si{=}SiXyl_2} \rightleftharpoons \underset{(Z)\text{-}\mathbf{5}}{(Mes)(Xyl)Si{=}Si(Mes)(Xyl)} \overset{E,Z}{\rightleftharpoons} \underset{(E)\text{-}\mathbf{5}}{(Mes)(Xyl)Si{=}Si(Xyl)(Mes)} \quad (55)$$

A similar rearrangement is observed between tetrasilyldisilenes **34** and **35** [Eq. (56)].[14] The activation free energy for the rearrangement from **35** to (*Z*)-**34** (or (*E*)-**34**) is 17.4 kcal mol^{-1} at 283 K, which is ca. 1.7 kcal mol^{-1} larger than that for the *E*,*Z*-isomerization and 7.7 kcal mol^{-1} smaller than those for the dyotropic type rearrangement of tetraaryldisilenes.[96]

$$\mathbf{35} \underset{k_{-1}}{\overset{k_1}{\rightleftharpoons}} (Z)\text{-}\mathbf{34}\ (\text{or }(E)\text{-}\mathbf{34}) \underset{k_{-2}}{\overset{k_2}{\rightleftharpoons}} (E)\text{-}\mathbf{34}\ (\text{or }(Z)\text{-}\mathbf{34}) \tag{56}$$

It is interesting to note that the activation free energies for both *E*,*Z*-isomerization and formal dyotropic rearrangement for the tetrasilyldisilene system are remarkably lower than those for the tetraaryldisilenes. A disilene–silylsilylene rearrangement may not be excluded for the rearrangement between **35** and (*Z*)-**34** (or (*E*)-**34**), because of the high 1,2-migratory aptitude of silyl groups.[97] While no silylene insertion product is detected during the rearrangement in the presence of excess triethylsilane as a trapping reagent, this fact may not be evidence for the dyotropic rearrangement. At higher theoretical levels, disilene ($H_2Si{=}SiH_2$) is calculated to be more stable than silylsilylene ($H(H_3Si)Si:$) with rather high activation energy for the rearrangement of disilene to silylsilylene;[54,98] the relative energy is 7.9 kcal mol^{-1} at the G1 method[99] and the activation energy is 17.3 kcal mol^{-1} at the MP3/6–31G(d,p)//HF/6–31G(d) level.[100] These values depend significantly on the theoretical levels. If disilenes **35**, (*E*)-**34**, and (*Z*)-**34** are more stable than the corresponding silylsilylenes and the activation energy for the rearrangement of **35** to the corresponding silylsilylene is significantly lower than the theoretical value, the disilene–silylsilylene rearrangement will be feasible and the insertion of the silylsilylene into a Si–H bond may not compete with the rapid backward rearrangement to the disilenes. Further studies are required for the elucidation of the detailed mechanism for the rearrangement between disilenes $A_2Si{=}SiB_2$ and AB-Si$=$SiAB.

Unusual intramolecular C–H insertion giving **136** occurs upon heating of 1,2-diphenyldisilene **41**:[62]

(57)

4. Skeletal Isomerization of Cyclic Disilenes

A silicon version of the well-studied electrocyclic interconversion of C_4H_6 (cyclobutene, bicyclobutane, butadiene, etc.) has been investigated both experimentally[23,101] and theoretically.[102,103] Irradiation of **51** in 3-methylpentane ($\lambda > 420$ nm) gives a 1:9 mixture of **51** and bicyclo[1.1.0]tetrasilane **88** at the photostationary state as determined by UV–vis spectroscopy.[23] When the mixture is left for 12 h in the dark at rt, **51** is produced quantitatively [Eq. (58)]. The photochemical conversion of **51** to **88** and the thermal reversion can be repeated more than 10 times without any appreciable side reactions.

(58)

The first-order rate constant (k) and activation parameters for the thermal isomerization of **88** to **51** are: $k = 5.67 \times 10^5\,\mathrm{s}^{-1}$ $\Delta H^\ddagger = 16.5\,\mathrm{kcal\,mol}^{-1}$, and $\Delta S^\ddagger = -20.8\,\mathrm{cal\,mol}^{-1}\,\mathrm{K}^{-1}$ at 288 K. The large negative $\Delta S^\ddagger$ value suggests that the transition-state structure of the thermal isomerization would be significantly restricted. There are several possible pathways for the isomerization of **88** to **51** including a 1,2-silyl migration accompanied by cleavage of the central bridge Si–Si bond of **88** [path A, Eq. (59)] and a concerted skeletal isomerization [path B, Eq. (59)]. By labeling the two R groups (R = tert-BuMe$_2$Si) on bridgehead silicon atoms of **88** by two R* groups (R* = tert-Bu(CD$_3$)$_2$Si), the thermal isomerization from **88** to **51** was confirmed to proceed *via* 1,2-silyl migration (path A).[101] By similar labeling experiments, the photochemical isomerization of **51** to **88** was also found to proceed *via* 1,2-silyl migration.

(59)

Irradiation of a C_6D_6 solution of 1-disilagermirene **50** ($\lambda > 300$ nm) gives 2-disilagermirene **137** almost quantitatively *via* silyl migration [Eq. (60)].[30,104] The thermal reaction of **50** in mesitylene at 120 °C for 1 day provides an equilibrium mixture of **50** (~2%) and **137** (~98%), suggesting that **137** is more stable than **50** by ~3 kcal mol^{-1} Theoretical calculations have shown that the model H_3Si-substituted 2-disilagermirene **137′** is more stable than 1-disilagermirene **50′** by 3.9 (MP2/DZd) and 2.3 (B3LYP/DZd) kcal mol^{-1} (Chart 13), being in good agreement with the experimental results.

(60)

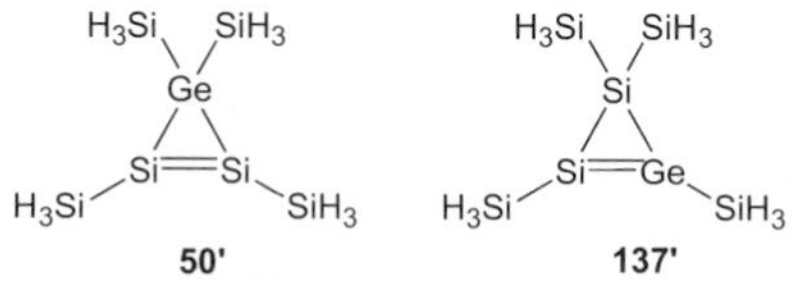

CHART 13.

Irradiation of a dark red toluene-d_8 solution of cyclotrisilene **48** ($\lambda = 254$ nm) at 243 K gives bicyclo[1.1.0]tetrasilane **88** (ca. 90% yield) and a small amount of tris(tert-butyldimethylsilyl)silane (**138**, ca. 10% yield):[69]

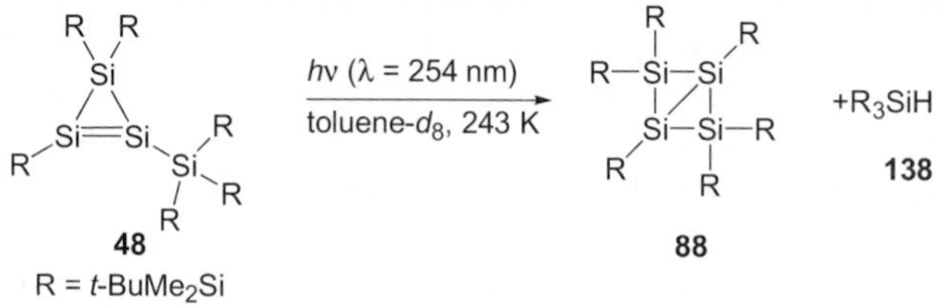

$$\mathbf{48} \xrightarrow[\text{toluene-}d_8\text{, 243 K}]{h\nu\ (\lambda = 254\ \text{nm})} \mathbf{88} + R_3SiH\ (\mathbf{138}) \quad R = t\text{-BuMe}_2\text{Si} \tag{61}$$

As illustrated in Eq. (62), the photochemical isomerization of cyclotrisilene **48** to **88** is rationalized also by a mechanism including 1,2-silyl migration. Whereas three 1,2-silyl migration pathways are possible in this system (paths a–c), only path a leads to **88** *via* biradical **139**; path b is an identity reaction and path c may lead to a minor product **138**.

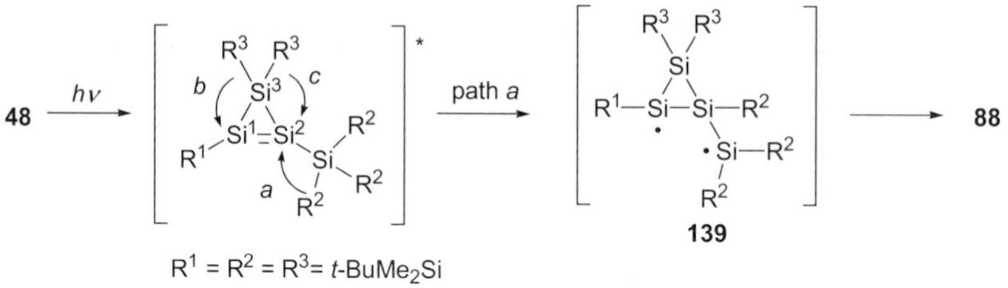

$$\mathbf{48} \xrightarrow{h\nu} [\,\mathbf{48}\,]^* \xrightarrow{\text{path } a} [\,\mathbf{139}\,] \longrightarrow \mathbf{88} \quad R^1 = R^2 = R^3 = t\text{-BuMe}_2\text{Si} \tag{62}$$

Müller has investigated the Si_4H_6 potential energy surface in detail and located several structures as energy minima; *s-trans* and *s-cis* 1,3-tetrasiladienes, 1- and 3-silylcyclotrisilenes, cyclotetrasilene, and *trans*- and *cis*-bicyclo[1.1.0]tetrasilanes using DFT calculations.[103] The relative energies calculated at the B3LYP/6–311+G** level are shown in Fig. 7. In contrast to the C_4H_6 family, bicyclotetrasilane **140** is the most stable isomer in the Si_4H_6 family. Cyclotetrasilene **117** is only 3.0 kcal mol^{-1} higher in energy than **140**. The other isomers **141** to **144** are >20 kcal mol^{-1} less stable than **140**. Tetrasilabutadiene **144** is the most unstable isomer in this series. The relative stability between cyclotetrasilene and bicyclotetrasilane depends significantly on the substituents; persilylcyclotetrasilene is only 0.3 kcal mol^{-1} less stable than the corresponding bicyclotetrasilane at the B3LYP/6–31+G** level.[55]

DFT calculations for the isomerization of **140** to **117** have shown that the 1,2-hydrogen migration from parent bicyclo[1.1.0]tetrasilane to cyclotetrasilene (E_a, 39.8 kcal mol^{-1}) is a more favorable process than the multi-step mechanism *via* the

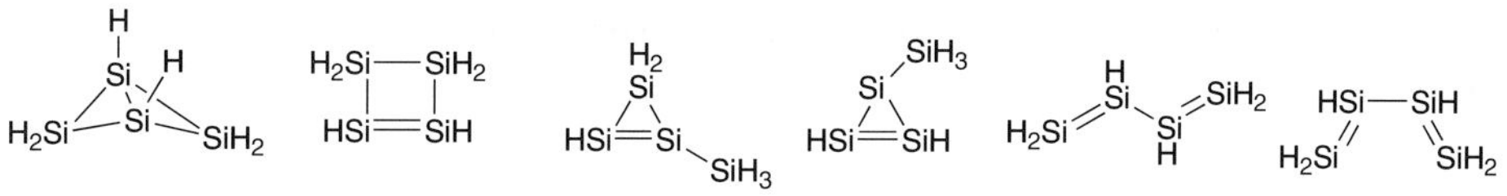

140 (0) **117** (3.0) **141** (21.6) **142** (25.7) **143** (33.3) **144** (34.2)

FIG 7. Relative energies of Si_4H_6 isomers at the B3LYP/6–311+G**//B3LYP/6–311+G**+ZPVE.

route of **140**→**143**→**144**→**117** (E_a, 50.3 kcal mol^{-1}).[103] While the experimental activation enthalpy of 16.5 kcal mol^{-1} for the isomerization of **88** to **51**[23,101] is much smaller than the barrier for the isomerization of **140** to **117** *via* the 1,2-hydrogen shift, the difference is acceptable because the migratory aptitude of silyl group is much higher than hydrogen.[97]

B. *Oxidation and Reduction*

Stable disilenes are usually highly reactive toward oxidation and reduction because the π and π* levels of disilene are much higher and lower than those of ethylene, as shown theoretically (Section III.A) and by the electrochemical oxidation and reduction potentials (Section III.E). Several stable disilenes are however stable in air for a long period. Disilene **27** undergoes very slow decomposition (half-life ca. 84 h) in wet THF solution at rt,[60,61] and air oxidation of disilene **63** is completed in benzene at rt for a week.[35] Disilene **27** survives for 4 months in the solid state. Since the oxidations and reductions of disilenes and their mechanisms have been extensively discussed in review OW, we mostly show herein the results of recent studies.

1. Oxidation

The general profile of the dioxygen oxidation of tetraaryldisilenes is shown in Eq. (63).[7] Both in the solid state and in solution, the initial oxidation product of a tetraaryldisilene is the corresponding 1,2-disiladioxetane **145**, whose intramolecular isomerization gives the thermodynamically more stable 1,3-disiladioxetane **146**. All the steps of the oxidation occur intramolecularly and with the retention of stereochemistry around the Si–Si bond. While a small amount of disilaoxirane **147** is produced in the oxidation in low-temperature solution, **147** is converted to **146** smoothly in the presence of excess oxygen.

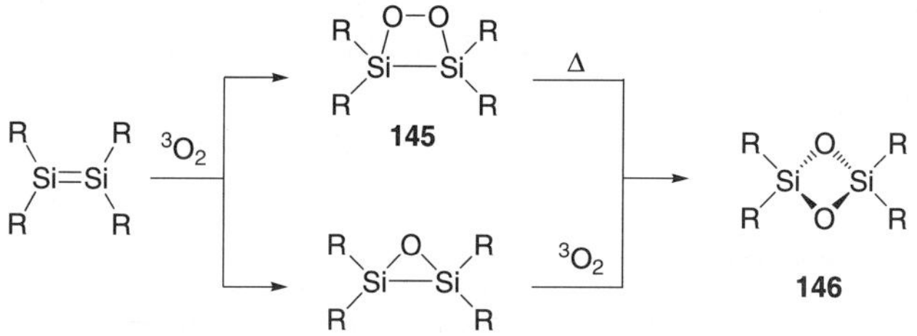

$$R_2Si{=}SiR_2 \xrightarrow{^3O_2} \mathbf{145} \xrightarrow{\Delta} \mathbf{146};\quad R_2Si{=}SiR_2 \xrightarrow{^3O_2} \mathbf{147} \xrightarrow{^3O_2} \mathbf{146} \tag{63}$$

The oxidation of disilene **2** affords unusual oxidation product **149** in addition to **150** *via* **148**:[76,105]

(64)

Dioxygen oxidation of tetrasila-1,3-butadiene **64** provides unusual oxidation product **151**, in which even the sterically well-protected central Si–Si single bond is oxidized together with the two Si=Si double bonds [Eq. (65)].[106] In contrast, when *m*-chloroperoxybenzoic acid (*m*CPBA) is used as a milder oxidant for the reaction, a mixture of **152** and **153**, whose central Si–Si bonds remain intact, is obtained.

(65)

The electrochemical oxidation of disilene **1** gives various monosilanes [Eq. (66)].[107] The ratio among these products is quite dependent on the oxidation conditions such as solvent, electrolytes, and anode materials.

$$\mathrm{Mes_2Si{=}SiMes_2}\ (\mathbf{1}) \xrightarrow[\text{electrolyte, solvent}]{-e} \mathrm{Mes_2SiF_2 + Mes_2SiFH + Mes_2SiFOH + Mes_2Si(OH)_2 + Mes_2SiH(OH) + Mes_2SiH_2} \quad (66)$$

electrolyte = R_4NBF_4, R_4NPF_6, R_4NClO_4,

solvent = CH_3CN, THF

As shown in Section II.D, the oxidation of cyclotrisilene **49** using $[Et_3Si(\text{benzene})]^+ \cdot [B(C_6F_5)_4]^-$ gives cyclotetrasilenylium ion salt $\mathbf{70} \cdot [B(C_6F_5)_4]^-$.[42]

2. Reduction

Because of their low reduction potentials, disilenes undergo the facile reduction giving the corresponding radical anions. Weidenbruch *et al.* first reported ESR spectra of tetramesityldisilene and tetra-tert-butyldisilene anion radicals **154** and **155**, which are generated during the reduction of the corresponding 1,2-dichlorodisilanes with lithium but not by the direct reduction of the corresponding disilenes [Eq. (67)].[108,109] Radical anions **156**–**158** are generated by the reductions of stable tetrakis(trialkylsilyl)disilenes **22**–**24** with potassium metal in DME at rt [Eq. (68)] and investigated by ESR spectroscopy.[8]

$$R_2ClSi{-}SiClR_2 \xrightarrow{\text{Li/THF}} [R_2Si{=}SiR_2]^{\cdot-} \quad (67)$$

1 (R = Mes) → **154** (R = Mes)
21 (R = *t*-Bu) → **155** (R = *t*-Bu)

$$(R_3Si)_2Si{=}Si(SiR_3)_2 \xrightarrow{\text{K/DME}} [(R_3Si)_2Si{=}Si(SiR_3)_2]^{\cdot-} \quad (68)$$

22 (R_3Si = *t*-$BuMe_2Si$) → **156** (R_3Si = *t*-$BuMe_2Si$)
23 (R_3Si = *i*-Pr_2MeSi) → **157** (R_3Si = *i*-Pr_2MeSi)
24 (R_3Si = *i*-Pr_3Si) → **158** (R_3Si = *i*-Pr_3Si)

Diaryldisilyldisilene **41** is reduced using lithium in THF to the corresponding anion radical **159**:[62]

$$(t\text{-Bu})_3Si(Ph)Si{=}Si(Ph)Si(t\text{-Bu})_3 \; (\mathbf{41}) \xrightarrow{\text{Li/THF}} [(t\text{-Bu})_3Si(Ph)Si{=}Si(Ph)Si(t\text{-Bu})_3]^{\cdot-} \; (\mathbf{159}) \quad (69)$$

The lithium salt of anion radical **160** is synthesized by the reduction of disilene **33** with *t*-BuLi in THF and isolated as red crystals in 58% yield [Eq. (70)].[64] X-ray structural analysis reveals interesting structural features of **160**. The central Si1–Si2 bond length is 2.341(5) Å, 3.6% longer than that of the neutral disilene **33** (2.2598(18) Å). The central Si–Si bond is highly twisted with the twist angle of 88°. The geometries around the two unsaturated silicon atoms are quite different; one unsaturated silicon is nearly planar, whereas around the other silicon atom the geometry is significantly pyramidalized (the sum of the bond angles at the Si atom = 352.7°), indicating that one silicon atom has radical character, whereas the other silicon atom has silyl anion character.

$$(R_3Si)_2Si{=}Si(SiR_3)_2 \xrightarrow[\text{-78 °C to rt}]{t\text{-BuLi (1.2 eq)/THF}} [Li(thf)_4]^+[(R_3Si)_2Si{=}Si(SiR_3)_2]^{\cdot-} \quad (70)$$

33, R_3Si = *t*-Bu_2MeSi → **160**, R_3Si = *t*-Bu_2MeSi, 58%

The ESR parameters of disilene anion radicals **154**–**160** are summarized in Table VIII. The hyperfine coupling constants (hfscs) for the unsaturated silicons [$a(Si_\alpha)$]

TABLE VIII

ESR PARAMETERS OF DISILENE ANION RADICALS AND RELATED RADICALS

Radical	$a(Si_\alpha)$ (mT)	$a(Si_\beta)$ (mT)	g	Reference
$[Mes_2Si{=}SiMes_2]^{\bullet -}$ **154**	4.96	—	2.0031	11, 108
$[(t\text{-}Bu)_2Si{=}Si(t\text{-}Bu)_2]^{\bullet -}$ **155**	3.36	—	2.0035	11, 108
$[(t\text{-}BuMe_2Si)_2Si{=}Si(SiMe_2(t\text{-}Bu))_2]^{\bullet -}$ **156**	2.92	—	2.0058	8
$[(i\text{-}Pr_2MeSi)_2Si{=}Si(SiMe(Pr\text{-}i)_2)_2]^{\bullet -}$ **157**	3.18	0.25	2.0063	8
$[(i\text{-}Pr_3Si)_2Si{=}Si(Si(Pr\text{-}i)_3)_2]^{\bullet -}$ **158**	2.45	—	2.0062	8
$[(t\text{-}Bu_3Si)_2PhSi{=}SiPh(Si(t\text{-}Bu)_3)_2]^{\bullet -}$ **159**	2.76	0.98	2.0072	62
$[(t\text{-}Bu_2MeSi)_2Si{=}Si(SiMe(t\text{-}Bu)_2)_2]^{\bullet -}$ **160**	2.45 (298 K)			
	4.50 (120 K)	—	2.0061 (298 K)	64
Me_3Si	18.3	—	2.0031	110
$(Me_3Si)_3Si$	6.5	0.62	2.0050	110
$(Et_3Si)_3Si$	5.72	0.79	2.0063	111
$(t\text{-}BuMe_2Si)_3Si$	5.71	0.81	2.0055	113
$(i\text{-}Pr_3Si)_3Si$	5.56	0.81	2.0061	112
$(t\text{-}Bu_2MeSi)_3Si$	5.80	0.79	2.0056	114

are 2.45–4.96 mT, much smaller than those of trimethylsilyl radical ($a(Si_\alpha) =$ 18.3 mT)[110] and tris(trialkyl)silyl radicals ($a(Si_\alpha) = 5.56$–6.5 mT),[110–114] suggesting that the unpaired electron is delocalized over two unsaturated silicon atoms. The $a(Si_\alpha)$ of **160** is remarkably temperature dependent with $a(Si_\alpha) = 2.45$ and 4.50 mT at 298 and 120 K in 2-methyltetrahydrofuran, indicating frozen spin exchange between two unsaturated silicon atoms at low temperatures:

$$[(R_3Si)_2\dot{S}i-\overset{\ominus}{S}i(SiR_3)_2] \rightleftharpoons [(R_3Si)_2\overset{\ominus}{S}i-\dot{S}i(SiR_3)_2] \qquad (71)$$

$R_3Si = t\text{-}Bu_2MeSi$

Tetrakis(trialkylsilyl)disilenes **22** and **23** react with excess lithium in THF giving the corresponding 2,3-dilithiotetrasilanes **161** and **162** as thermally stable off-white powders [Eq. (72)].[115] Interestingly, under the same conditions, the reduction of tetrakis(triisopropylsilyl)disilene **24** provides quantitatively bis(triisopropylsilyl) dilithiosilane **78** [Eq. (73)];[115] dilithiosilane **78** is known to be obtained also by the lithium reduction of 3,3-bis(triisopropylsilyl)-1,2-bis(trimethyislyl)silacyclopropene.[12]

$$(R_3Si)_2Si{=}Si(SiR_3)_2 \xrightarrow{Li/THF} (R_3Si)_2Si(Li){-}Si(Li)(SiR_3)_2 \quad (72)$$

22 (R_3Si = t-BuMe_2Si) **161** (R_3Si = t-BuMe_2Si)
23 (R_3Si = i-Pr_2MeSi) **162** (R_3Si = i-Pr_2MeSi)

$$(R_3Si)_2Si{=}Si(SiR_3)_2 \xrightarrow{Li/THF} (R_3Si)_2SiLi_2 \quad (73)$$

24 (R_3Si = i-Pr_3Si) **78** (R_3Si = i-Pr_3Si)

As shown in Section II.A, the reductions of $Tip_2Si{=}SiTip_2$ (**2**) and 1,2-diaryl-tetrasilyltetrasilabutadiene **65** using lithium and *t*-BuLi, respectively, undergo the cleavage of the saturated bonds to give the corresponding disilenyllithums [Eqs. (8)–(10)].

C. *1,2-Addition of Water, Alcohols, and Ammonia*

In review OW are introduced a number of 1,2-additions to stable disilenes using various reagents such as halogens, hydrogen halides, alcohols, water, tributyltin hydride, and lithium aluminum hydride. Similarly to alkenes, the disilenes react with halogens and hydrogen halides easily to give the corresponding adducts. Less electrophilic water and alcohols do not add to alkenes without catalysts but add to the disilenes smoothly to give the corresponding 1,2-adducts. Although the addition to alkenes is usually electrophilic, the addition to disilenes may be nucleophilic or electrophilic, because of the relatively high-lying π HOMO and low-lying π* LUMO of disilenes (Fig. 2).

DeYoung, Fink, West, and Michl first observed that the reaction of stereoisomeric disilene (*E*)-1,2-di-tert-butyl-1,2-dimesityldisilene ((*E*)-**3**) with various alcohols gave a mixture of two diastereomeric 1,2-adducts.[116] Sekiguchi, Maruki, and Sakurai found that the diastereoselectivity of the reactions of the transient disilenes (*E*)- and (*Z*)-PhMeSi=SiMePh with alcohols was controlled by the concentration of the alcohols.[80] They concluded that *syn*- and *anti*-adducts, which were preferred at the low and high concentrations of the alcohols, respectively, were formed *via* intra- and intermolecular transfer of an alcoholic proton in a four-membered zwitterionic intermediate. West *et al.* reported the *syn* selective addition of alcohols to (*E*)-**3**, (*E*)-**10**, and (*Z*)-**10** and low stereoselectivity of the ethanol addition to (*E*)-**3** in THF.[117] Apeloig and Nakash have investigated the substituent effects on the rates of the addition reactions of various *p*- and *m*-substituted phenols to tetramesityldisilene and found a concave shape Hammett relationship.[118,119] They have explained the results by a change in mechanism as a function of the substituent on the phenol; at the rate-determining step, nucleophilic or electrophilic attacks are preferred for electron-rich and electron-poor phenols, respectively. The activation energies (E_a) for the addition reactions of p-$CH_3OC_6H_4OH$ and p-$CF_3C_6H_4OH$ are 13.7 and 9.7 kcal mol^{-1} the activation entropies ($\Delta S^{\ddagger}$) are quite negative (−34.9 and −45.3 cal mol^{-1} K^{-1}) for the two reactions, suggesting that the transition states at the rate determining steps are highly congested. The addition of p-$CH_3OC_6H_4OH$ to (E)-(t-Bu)MesSi=SiMes(Bu-t) [(E)-**3**] is *syn* selective (*syn*-**164**:*anti*-**164** = 9:1) in benzene but *anti*-selective (*syn*-**164**:*anti*-**164** = 2:8) in polar THF, even at the low alcohol concentration [Eq. (74)].[119] The formation of an *anti*-adduct (*anti*-**164**) is explained by rotation around the Si–Si bond in the zwitterionic intermediate **163** giving **163′**, followed by fast intramolecular proton transfer. The origin of the remarkable solvent effects has not been elucidated.

(74)

Theoretically, Nagase, Kudo, and Ito first investigated the reaction of parent disilene with water at the HF/6–31G* level and found the *syn*-addition pathway *via* a four-centered cyclic transition state.[120] Apeloig and Nakash have recently investigated in detail the addition reactions of CH_3OH and CF_3OH to $Me_2Si{=}SiMe_2$ at the MP3/6-31G*//HF/6–31G* level.[121] Only pathways leading to the *syn*-adducts were taken into account in these calculations. Recently, Takahashi, Veszpremi, and Kira have examined the detailed reaction pathways for the water addition to disilene using *ab initio* MO calculations and found that two bimolecular addition pathways with different stereochemical outcome are feasible [Eq. (75)].[122] First, two types of weak van der Waals complexes, C_N and C_E, are formed *via* the interactions between disilene π*LUMO and water *n* type HOMO and between disilene πHOMO and water σ*LUMO, respectively (Fig. 8). From C_N, an *anti*-adduct (P_A) is derived *via* the antarafacial approach of the water hydrogen to the pπ lobe of the β-silicon of the disilene (*anti*-pathway), while C_E leads to a *syn*-adduct (Ps) *via* the attack of

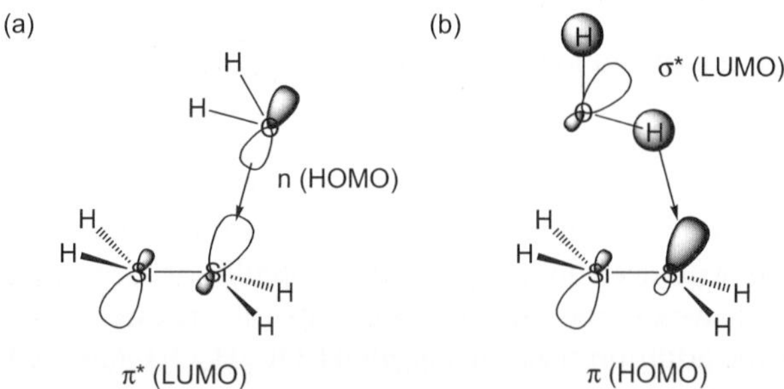

FIG 8. Two initial interaction between disilene and water leading to (a) C_N and (b) C_E.

oxygen lone-pair electrons to disilene π*LUMO (*syn*-pathway). The energy barriers for the rate-determining steps of the *anti*- and *syn*-pathways are low (5.16 and 2.92 kcal mol^{-1} respectively), and hence, the preference of the two pathways will be modified by the substituents on disilene, the nature of the reagents, solvents, etc. Lewis adducts C_L and C_L' formed as an intermediate give the final products *via* TS_L and TS_L', respectively, with almost no barriers. The theoretical view is compatible with the results of Apeloig and Nakash [Eq. (74)]; nevertheless, the theoretical model is remote from the experimental systems. The transition state for the *anti*-pathway (TS_E) is more polar than that for the *syn*-pathway (TS_N), which explains the observed remarkable solvent effects. This type of theoretical approach has been extended to the various types of addition reactions to silicon unsaturated compounds to reveal the significant diversity of the reaction pathways.[123,124] Typically, the addition of CF_3OH to $H_2Si{=}SiH_2$ proceeds theoretically *via* the C_E-TS_L'-P_S pathway without intermediacy of C_L'.[124]

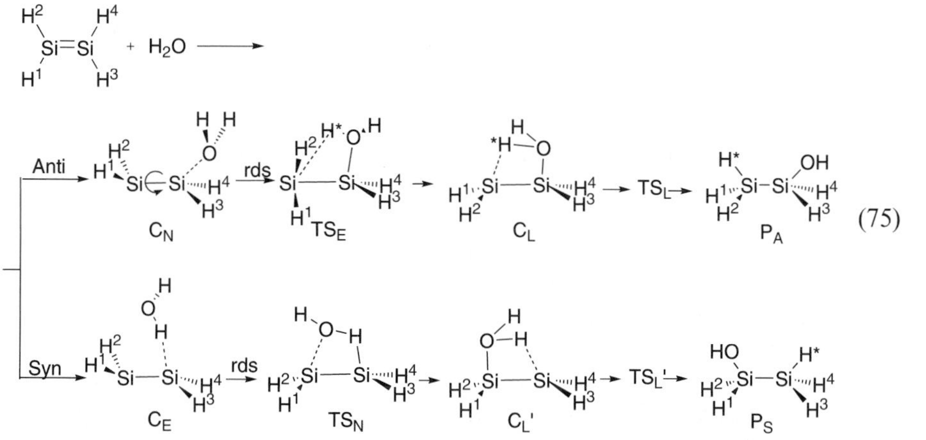

(75)

The reaction of 1,1-diaryl-2,2-disilyldisilene **29** with methanol proceeds regiospecifically to give methoxysilane **165** quantitatively [Eq. (76)].[13] This regiospecificity is explained by the polarity of the Si=Si double bond in **29**, in which the unsaturated silicon atom attached to the two silyl groups is more negative than the other unsaturated silicon atom; the calculated natural atomic charges in the model compound $(Me_3Si)_2Si{=}SiMes_2$ are −0.355 and +0.958 for $(Me_3Si)_2Si$ and Mes_2Si silicon atoms at the B3LYP/6–31G(d) level.

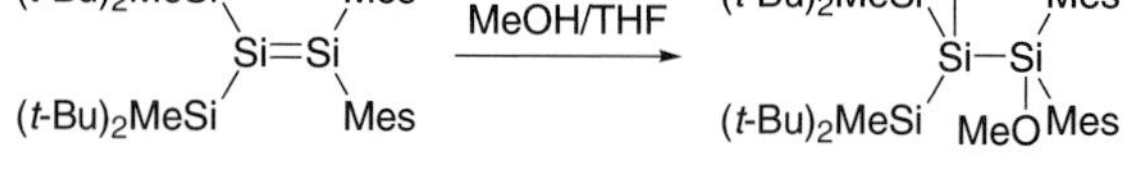

(76)

The addition of water to conjugated disilene **64** occurs stepwise and regioselectively [Eq. (77)]. The initial 1,2-addition of a molecule of water to one of the double bonds gives disilene **166**, which is detected by ^1H NMR of the reaction mixture. When the amount of water is small, intramolecular ring-closing silanol addition of **166** to the residual Si=Si double bond yields **167**. When a large amount of water is available, a second 1,2-addition of water to the remaining double bond provides tetrasilane-1,4-diol **168**.[106]

H_2O/THF

64 **166**

167, 86%

168, 84%

(77)

Trisilaallene **67** with cumulative Si=Si double bonds also undergoes stepwise and regioselective addition of water giving trisilane-1,3-diol **169**:[40]

H_2O/THF

67, R = $SiMe_3$ **169**

(78)

Unsymmetrically substituted cyclotrisilene **48** reacts with water and various alcohols to give *anti*-adduct **170** in a stereospecific manner, where the hydroxyl (alkoxy) group is bound to the sterically less hindered silicon atom [Eq. (79)].[84] Regio and stereoselectivity of the addition is independent of the substituents of alcohols. A similar *anti*-addition of benzyl alcohol to 1-germadisilirene **50** was reported.[104]

R'OH/THF

48
R = *t*-$BuMe_2Si$

170, R = *t*-$BuMe_2Si$
R' = H, Me, Et, C_6H_5, *p*-$MeOC_6H_4$, *p*-$CF_3C_6H_4$

(79)

The reactions of disilenes with ammonia are hardly known. Whereas $Tip_2Si{=}SiTip_2$ (**2**) does not react with ammonia, the addition of ammonia to the conjugated Si=Si double bonds of tetrasiladiene **64** proceeds within several minutes to give

1,4-diaminotetrasilane **171** in 79% yield:[125]

$$\mathrm{Tip_2Si{=}SiTip{-}SiTip{=}SiTip_2}\ (\mathbf{64}) \xrightarrow[\text{toluene}]{2\mathrm{NH_3}} \mathrm{Tip_2Si(HNH{\cdots}NH_2)SiTip_2},\ \mathrm{TipSi(H){-}SiTip(H)}\ (\mathbf{171},\ 79\%) \tag{80}$$

D. *Addition of Halogens and Haloalkanes*

In contrast to alkenes, disilene derivatives react very smoothly with various haloalkanes as shown in Eqs. (81) and (82). Tetramesityldisilene **1** reacts with tert-butyl chloride to give hydrogen chloride adduct **172** together with 2-methylpropene (**173**), while treatment of **1** with benzyl chloride affords 1-benzyl-2-chlorodisilane **174**.[126] The reactions of tetrasilyldisilene **22** with haloalkanes proceed in a similar way to give 1-halo-2-(haloalkyl)disilanes **175**, whereas the reaction with carbon tetrahalides affords the corresponding 1,2-dihalodisilanes **176**.[127]

$$\mathrm{Mes_2Si{=}SiMes_2}\ (\mathbf{1}) \xrightarrow{t\text{-BuCl}} \mathrm{Mes_2Si(Cl){-}Si(H)Mes_2}\ (\mathbf{172}) + \mathrm{H_3C{-}C(=CH_2){-}CH_3}\ (\mathbf{173})$$
$$\mathrm{Mes_2Si{=}SiMes_2}\ (\mathbf{1}) \xrightarrow{\mathrm{PhCH_2Cl}} \mathrm{Mes_2Si(Cl){-}Si(CH_2Ph)Mes_2}\ (\mathbf{174}) \tag{81}$$

$$\mathrm{R_2Si{=}SiR_2}\ (\mathbf{22},\ \mathrm{R} = t\text{-BuMe_2Si}) \xrightarrow{\mathrm{R'{-}X}} \mathrm{R_2Si(R'){-}Si(X)R_2}\ (\mathbf{175})$$
$$(\mathrm{R'{-}X} = \mathrm{CHCl_2{-}Cl,\ CH_2Cl{-}Cl,\ CH_3{-}Br})$$
$$\mathrm{R_2Si{=}SiR_2}\ (\mathbf{22}) \xrightarrow{\mathrm{CX_4}} \mathrm{R_2Si(X){-}Si(X)R_2}\ (\mathbf{176}\ (\mathrm{X = Cl,\ or\ Br})) \tag{82}$$

The formation of these products is apparently suggestive of the radical nature of the reactions; a pair of the corresponding halodisilanyl radical and alkyl radical (**178**) is formed in the first step and in the second step, various types of products will be formed *via* recombination and disproportionation of the radical pair in the cage and abstraction of the second halogen by the halodisilanyl radical out of the cage [Eq. (83)]. The radical mechanism has been confirmed by the detailed product studies, ESR detection of intermediate silyl radical, kinetic studies including substituent effects on the rates, etc. Substituent effects on the rates for the reactions of disilenes are parallel to those for the reactions of silyl radicals suggesting that the pertinent transition state **177** is similar to that for the corresponding halogen abstraction of silyl radicals, while the absolute rate constants for disilenes are 10^8–10^{14} times slower than those for silyl radicals.[83]

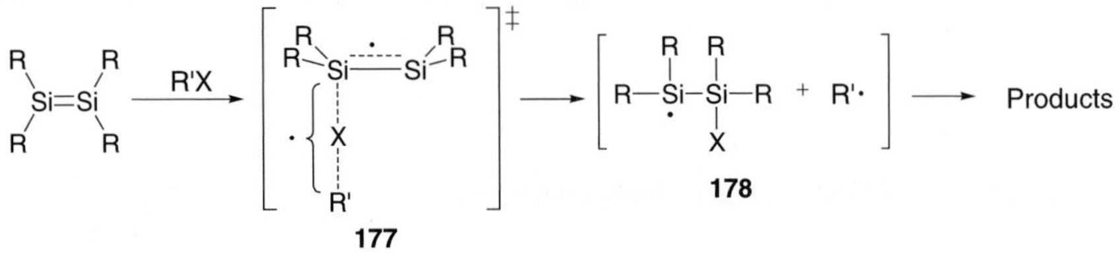

(83)

Although reactions between two neutral closed shell molecules forming a neutral radical pair are rather exceptional, the reactions may be understood by the singlet biradical character of disilenes. A small π-bond energy and a small π–π* energy gap in disilenes are suggestive of the significant extent of the biradical nature of disilenes. Teramae has shown theoretically that parent disilene has weak but significant singlet biradical character of 13% as calculated at the CASSCF/6–31G** level.[56]

The transition structures for the reactions of model disilenes with haloalkanes have been located using *ab initio* MO calculations.[128]

The reactions of cyclic disilenes with haloalkanes proceed in a similar way. Cyclotrisilene **48** reacts with carbon tetrachloride without cleavage of endocyclic Si–Si bonds to give *trans*-1,2-dichlorocyclotrisilane **179** stereospecifically [Eq. (84)].[24] The *anti*-stereochemistry is in good accord with the radical mechanism of the reactions as stated above. Similarly, cyclic disilenes **49** and **50** react with carbon tetrachloride and 1,2-dibromoethane giving the corresponding *trans*-1,2-dihalodisilanes **180** [Eq. (85)].[129] Methyl iodide adds to cyclic disilene **50** in an *anti*-addition manner to give **181**.[104]

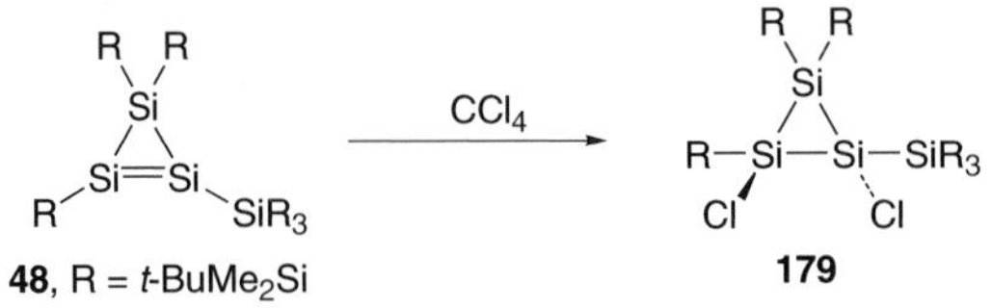

(84)

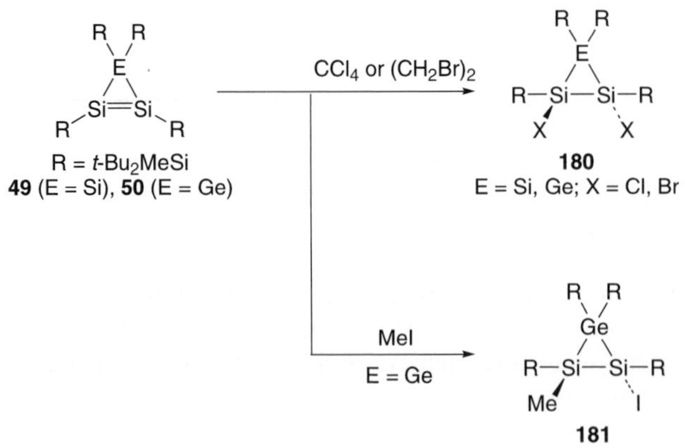

(85)

The reaction of tetrasilabutadiene **64** with Cl_2 gives 1,2,3,4-tetrachlorotetrasilane **182** [Eq. (86)].[125] As chlorine gas is added, the solution of **64** turns from dark red to blue-black to orange, and finally to colorless. On this basis, the addition is proposed to proceed stepwise *via* the initial formation of a charge-transfer complex between **64** and Cl_2 to give **183**, which is converted to disilene **184** and then to **182** by the addition of a second molecule of Cl_2.

64 —Cl_2→ [$Tip_2Si=SiTip–SiTip=SiTip_2$ · Cl_2] **183** → $Tip_2SiCl–SiTip(Cl)–SiTip=SiTip_2$ **184** —Cl_2→ $Tip_2SiCl–SiTip(Cl)–SiTip(Cl)–SiTip_2Cl$ **182**, 84%

(86)

E. *Carbometallation and Hydrometallation*

As shown in review OW, tetramesityldisilene **1** undergoes hydrostannation by tributyltin and hydrogenation by lithium aluminum hydride that may involve a hydrometallation process.[116] Tetrasilyldisilenes undergo unique carbolithiation and hydrolithiation without catalysts. The reaction of tetrakis(*t*-butyldimethylsilyl)disilene **22** with excess methyllithium at 0 °C gives a solution of 1-methyl-2-lithiodisilane **185** and the hydrolysis of the solution affords the corresponding 2-methyltetrasilane **186** [Eq. (87)].[127] Because no similar reactions have been reported for tetramesityldisilene (**1**) and tetra-tert-butydisilene (**21**), the results suggest that disilene **22** is more electrophilic than disilenes **1** and **21**, in good accord with the lower reduction potential of **22** (Section III.E). A similar carbolithiation of unsymmetrically substituted disilene **32** with methyllithium occurs regioselectively to give the corresponding silyllithium **187**; the structure of **187** was determined by X-ray crystallography.[16] The observed regioselectivity is ascribed to the anion-stabilizing effects of trialkylsilyl substituents [Eq. (88)].

$R_2Si=SiR_2$ (**22**, R = *t*-BuMe$_2$Si) —MeLi/THF, 0 °C→ [$R_2MeSi–SiR_2Li$] **185** —H_2O→ $R_2MeSi–SiR_2H$ **186**, 48% (87)

$R(Me)Si=SiR(SiR_2Me)$ (**32**, R = *t*-Bu$_2$MeSi) —MeLi/THF, -78 °C to rt→ $[Li(thf)_4]^+$[$Me–SiR(Me)–Si^{\ominus}R–SiR_2Me$] **187**, 89% (88)

Cyclic disilene **50** reacts with *t*-BuLi to afford rather unusually cyclic germyllithium **188** [Eq. (89)], suggesting the migration of a silyl substituent during the reaction.[130] The formation of **188** is confirmed by NMR spectroscopy and by a quenching experiment using methyl iodide, which forms quantitatively the corresponding methylgermane **189**. Germyllithium **188** is produced also by the reaction of germasilene **50′** with *t*-BuLi.

R_3Si = *t*-Bu_2MeSi
50, E = Si, E' = Ge
50', E = Ge, E' = Si

t-BuLi (1.2 eq)/THF, -78 °C to rt, -$Me_2C{=}CH_2$ → **188**, 52%; MeI → **189**, quant. (89)

F. *Ene Reaction*

Although the thermolysis of tetramesityldisilene **1** giving **191** through a possible intermediate **190** [131]constitutes a formal intramolecular ene reaction [Eq. (90)], no intermolecular ene reactions have been reported in review OW.

$Mes_2Si{=}SiMes_2$ (**1**) → [**190**] → **191** (90)

The reactions of tetrasilyldisilene **22** with 1-butene and 1-hexyne give the corresponding adducts **192** and **193**, respectively [Eq. (91)].[127] There are two possibilities for the mechanism of the ene reaction [Eq. (92)]: (1) a concerted ene addition mechanism *via* a six-membered cyclic transition state (path A) and (2) a step-wise radical mechanism including the addition of a disilene to an alkene giving the corresponding 1,3-diradical followed by an intramolecular hydrogen abstraction of the silyl radical moiety (path B). Path B is not eliminated because disilene has a diradical nature (Section IV.D). However, the reaction of disilene **22** with 1,6-heptadiene gives only ene-adduct **195** probably through **194** in high yield without the formation of **198**; if the reaction proceeds *via* path B, an intramolecular coupling of diradical intermediate **196** might compete with the radical cyclization [132] to give diradical **197** and then final adduct **198** [Eq. (93)].

193 (20%) ← HC≡C−CH_2Pr — **22**, R = *t*-$BuMe_2Si$ — 1-butene → **192** (86%) (91)

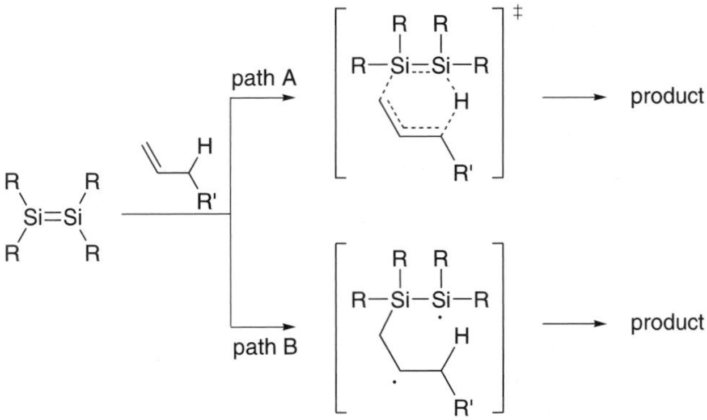

(92)

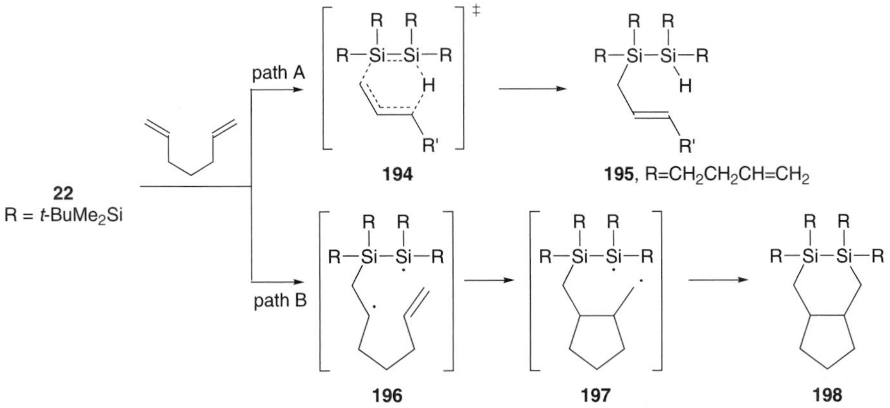

(93)

The ene reaction of cyclic disilene **50** with enolizable ketones such as biacetyl and acetophenone gives initially *cis*-adduct **199**, which gradually isomerizes to the corresponding *trans*-isomer **200**:[104,133]

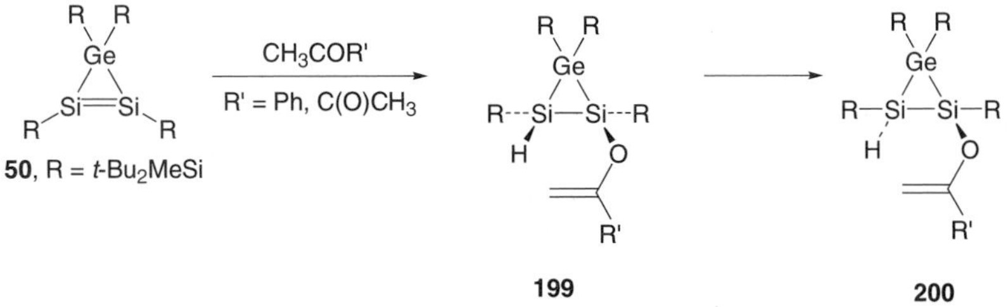

(94)

G. *Cycloaddition*

1. [2+1] and [2+2] Cycloaddition

A number of formal [2 + 1] and [2 + 2] cycloadditions of disilenes were investigated before 1996 and discussed extensively in review OW as useful methods for the synthesis of various three- and four-membered ring compounds with two silicon atoms in the ring.

Styrene and substituted styrenes react with tetramesityldisilene **1**, tetra-tert-butyldisilene **21**, and tetrakis(tert-butyldimethylsilyl)disilene **22** to afford the corresponding disilacyclobutane derivatives.[127,134] Similarly, [2 + 2] additions occur between the disilenes with a C=C double bond in an aromatic ring[135] and acrylonitrile.[136] Bains *et al.* have found that the reaction of disilene **1** with *trans*-styrene-d_1 provides a 7:3 diastereomeric mixture of [2 + 2] adducts, **201** and **202** [Eq. (95)]; the ratio is changed, when *cis*-styrene-d_1 is used.[137] The formation of the two diastereomeric cyclic adducts is taken as the evidence for a stepwise mechanism *via* a diradical or dipolar intermediate for the addition, similar to the [2 + 2] cycloaddition of phenylacetylene to disilene (*E*)-**3**, which gives a 1:1 mixture of stereoisomeric products.[116,137]

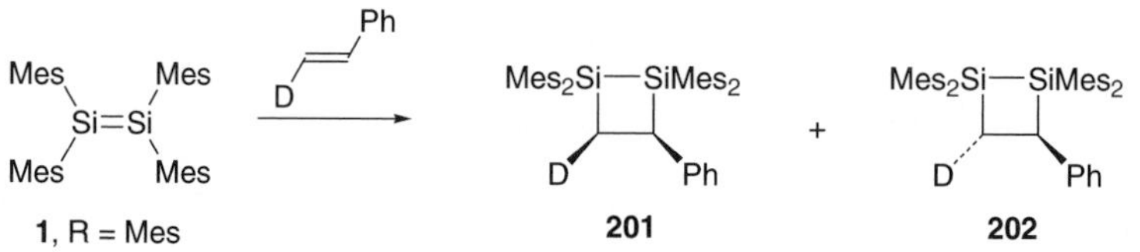

1, R = Mes → **201** + **202** (95)

Cyclotrisilene **49** reacts with 2 mol of phenylacetylene to give unique bicyclic product **203** in good yield [Eq. (96)].[138] On the basis of the results of deuterium labeling experiments, compound **203** is proposed to form *via* two consecutive [2 + 2] cycloadditions of **49** to phenylacetylene, i.e., the initial [2 + 2] cycloaddition giving **204** followed by the isomerization to disilene **206** *via* **205**, and then the second [2 + 2] cycloaddition of **206** [Eq. (97)]. The reaction of 1-disilagermirene **50** with phenylacetylene proceeds in a similar way.[139]

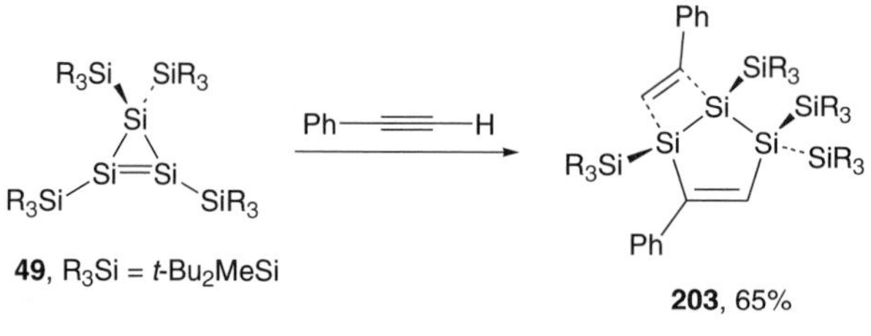

49, R_3Si = *t*-Bu_2MeSi → **203**, 65% (96)

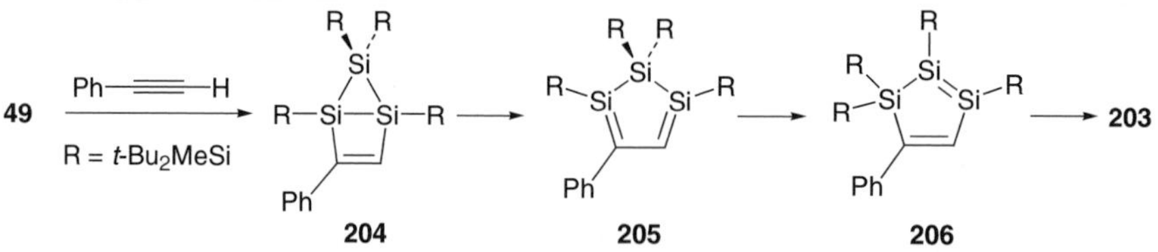

49 → (R = *t*-Bu_2MeSi) **204** → **205** → **206** → **203** (97)

Nonenolizable ketones and aldehydes add to disilenes affording the corresponding [2 + 2] adducts **207** [Eq. (98)] as shown in review OW.[7]

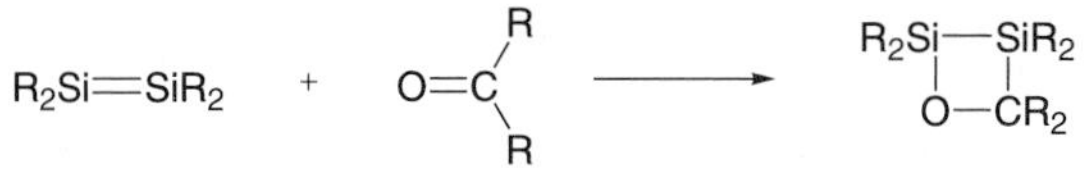

(98)

Stepwise radical mechanisms have been proposed for the apparent [2 + 2] cycloaddition of disilenes with ketones by Baines *et al.*[140,141] They have found that the reactions of tetramesityldisilene **1** with *trans*-2-phenylcyclopropane carbaldehyde (**208a**) and *trans,trans*-2-methoxy-3-phenylcyclopropane carbaldehyde (**208b**), a mechanistic probe developed by Newcomb *et al.*,[142] undergo characteristic cyclopropane ring-opening as shown in Eq. (99).

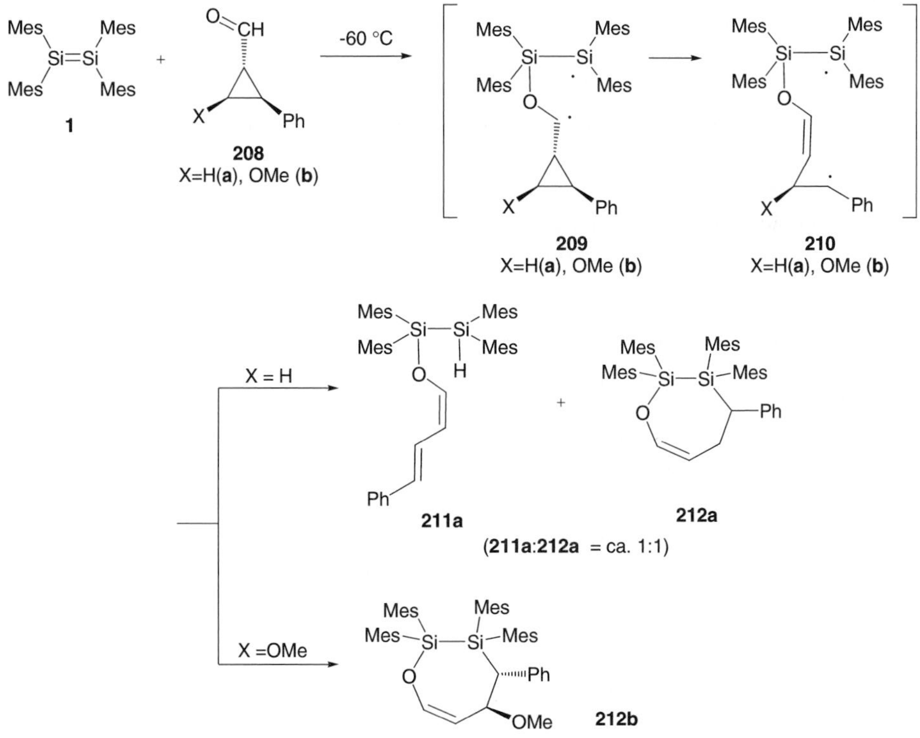

(99)

The regiospecific ring-opening to give **212a** and **212b** indicates that the reactions involve facile cyclopropylcarbinyl-to-homoallyl radical rearrangements (**209a** → **210a** and **209b** → **210b**). A possible mechanism involving zwitterionic intermediate **209b**′ is excluded, because the ring-opening of **209b**′ should preferably afford methoxymethyl-type cation **210b**′ leading to a regioisomer of **212b**, **212b**′ [Eq. (100)]; the cationic ring-opening of **213** is known to produce regioselectively **214**, while the corresponding radical ring-opening gives **215** [Eq. (101)].[141,142]

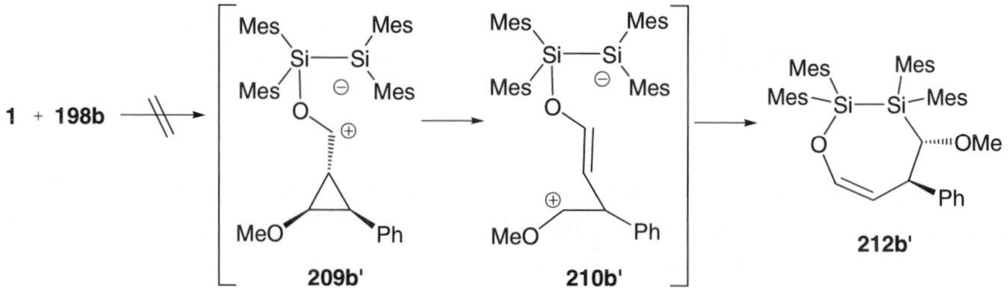

(100)

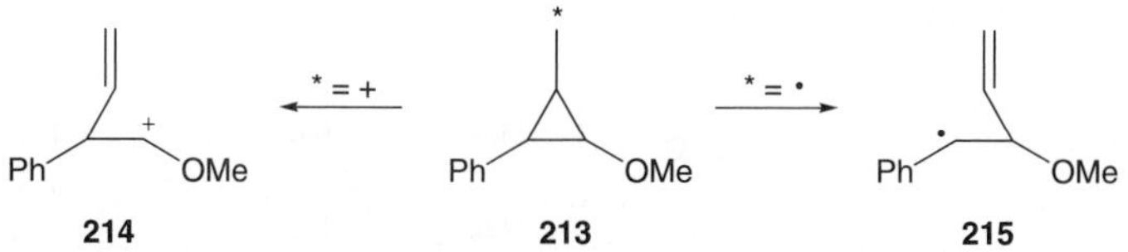

(101)

Tetrasila-1,3-diene **64** reacts with maleic anhydride to give the rather unusual tetracyclic adduct **216** [Eq. (102)]. A [2+2] cycloaddition of one of the Si=Si double bonds of **64** to a C=O group of the acid anhydride followed by the second [2+2] cycloaddition between the remaining Si=Si bond and the C=C double bond would afford the final product **216**.[143]

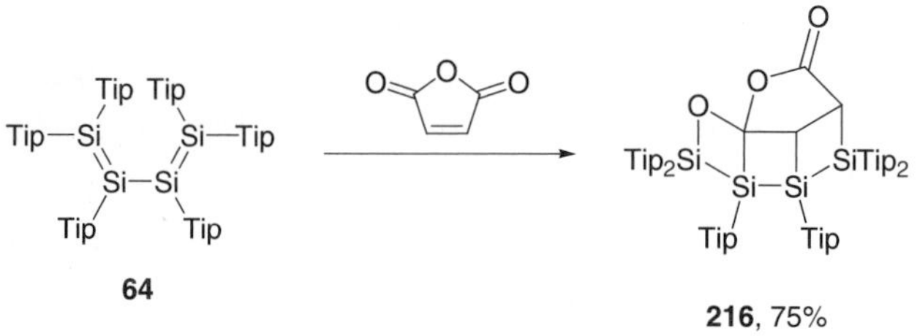

(102)

2. [2+3] and [2+4] Cycloadditions

A number of formal [2+3] and [2+4] cycloadditions of disilenes giving five- and six-membered ring compounds have been discussed in review OW, while their mechanisms have not been elucidated in detail.

Disilene **2** reacts with maleic anhydride to give **217** through formal [2+3] addition accompanied by hydrogen shift:[143]

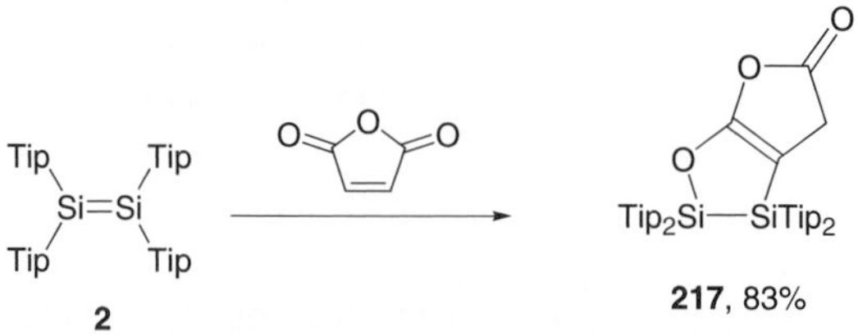

(103)

Stable aryl- and alkyl-substituted disilenes undergo the [2+4] cycloadditions with benzyl, acylamines, and 1,4-diazabutadienes but not with 1,3-butadienes.[77,131] On the other hand, stable tetrasilyldisilene **22** reacts with 2,3-dimethyl-1,3-butadiene at rt giving the corresponding 4,5-disilacyclohexene **218** [Eq. (104)].[127] The reason for the higher reactivity of **22** as a dienophile may be ascribed to the lower-lying π-LUMO of **22** as evidenced by its lower reduction potential (Section III.E).

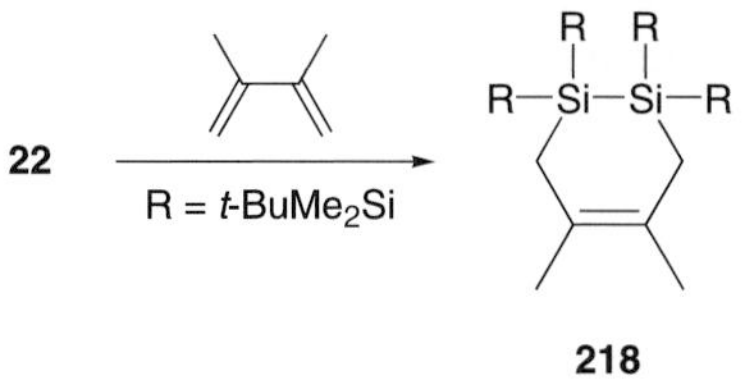

(104)

218

H. *Miscellany*

1-Disilagermirene **50** and its isomer **50′** undergo unique ring expansion during the reaction with $GeCl_2$ and $SnCl_2$ to afford the corresponding disiladigermetene **219** and disilagermastannetene **220**, respectively.[144] The reaction is proposed to involve the addition of $GeCl_2 \cdot$dioxane to the Si=Si double bond of **50** to produce chlorogermylene **221** [Eq. (105)], whose intramolecular insertion of the germylene into the ring Si–Ge bond forms cyclic germasilene **222**. Compound **219** will be formed by the 1,2-silyl migration of **222** to give bicyclo[1.1.0]butane **223** followed by the 1,2-chlorine migration from Ge to Si [Eq. (106)].

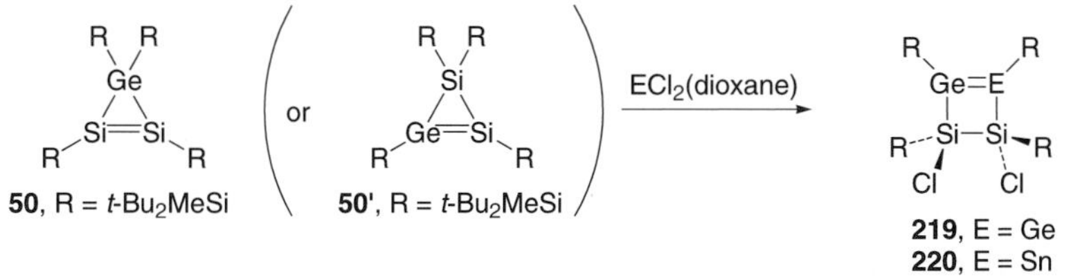

(105)

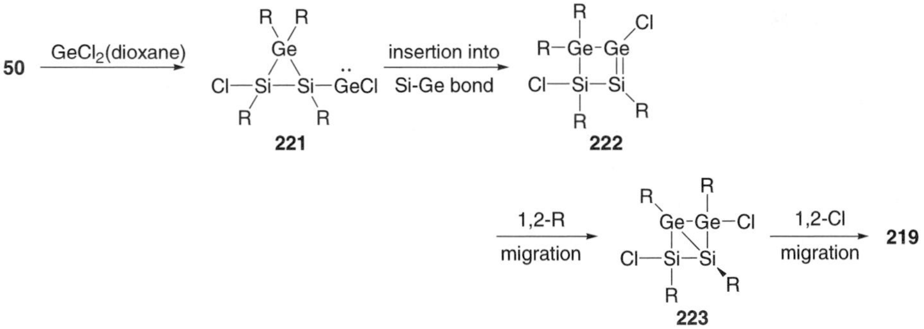

(106)

Neither water nor methanol reacts with the disilene moiety of 3,4-diiodocyclotetrasilene **52** but rather with the silicon-bound iodine atom to form tricyclic compound **224** *via* skeletal rearrangement or 3,4-dimethoxycyclotetrasilene **225** [Eq. (107)].[25] The rapid conversion of **52** into **224** or **225** may be explained by the formation of the ionic intermediate **226** *via* dissociative activation (S_N1 mechanism) that is facilitated by bulky t-Bu_3Si groups. In favor of this mechanism, cyclotetrasilene **52** reacts with 1 equivalent of BI_3 in dichloromethane at rt to give tricyclic cation **226** as a colorless compound, which reacts with H_2O and MeOH in solution to afford **224** and **225**, respectively. Cyclotetrasilene **52** is reduced by tri-*t*-butylsilyl sodium to give tetrasilatetrahedrane **91**.

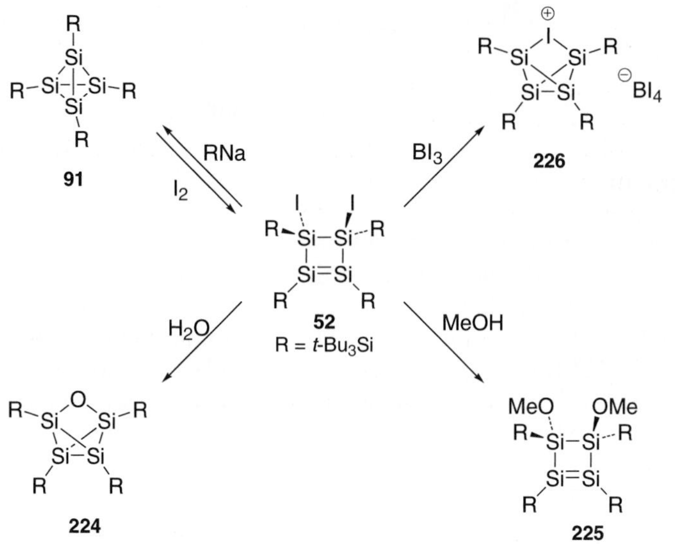

(107)

V
DISILENE-TRANSITION METAL COMPLEXES

A. *Synthesis*

In review OW have been discussed platinum complexes of stable disilene **1** (**227a** and **227b**),[145] which were synthesized by the direct reaction of $(Ph_3P)_2Pt(C_2H_4)$ with disilene **1**. Complexes **228a**–**228c**,[146] **229**,[147,148] and **230**[147] as shown in Chart 14 were synthesized rather indirectly using the reactions of 1,2-dihydrodisilanes with bis(phosphine)metal dichlorides or the reactions of 1,2-dichlorodisilanes with metal-magnesium reagents followed by the reduction with alkali metals. Among

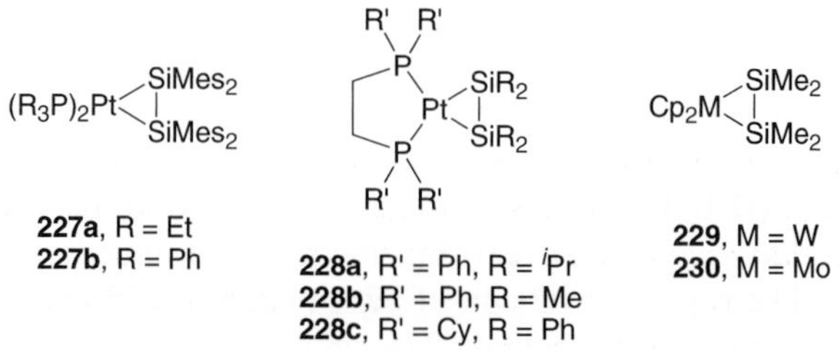

CHART 14.

these disilene–metal complexes, only the molecular structure of **229** has been investigated by X-ray crystallography.

Recently, remarkable progress has been made in the synthesis and structural characteristics of novel disilene complexes. Disilene–platinum and disilene–palladium complexes **231a**–**231c** are synthesized by the reactions of metal dichlorides with the corresponding 1,2-dilithiodisilanes **161** [Eq. (108)].[115,149,150] Complexes **231b** and **231c** are obtained also by the reactions of disilene **22** with bis(phosphine)palladium dichlorides **232b** and **232c**, as shown in Eq. (109). In these reactions, two equivalents of disilene **22** are required for the complete consumption of **232b** and **232c**, suggesting that 1 mol of disilene is consumed in the reduction of **232b** and **232c** to the corresponding bis(phosphine)Pd(0) complexes, which react further with another equivalent of **22** to give **231b** and **231c**, respectively.

$$L_2MCl_2 + (R_3Si)_2Si(Li)-Si(Li)(SiR_3)_2 \xrightarrow{THF} L_2M[Si(SiR_3)_2]_2 \tag{108}$$

232a, M = Pt, L = PMe_3
232b, M = Pd, L = PMe_3
232c, M = Pd, L_2 = dmpe

161
$SiR_3 = SiMe_2(t\text{-Bu})$
dmpe = $Me_2PCH_2CH_2PMe_2$

231a, M = Pt, L = PMe_3 (46% yield)
231b, M = Pd, L = PMe_3 (55% yield)
231c, M = Pd, L_2 = dmpe (43% yield)

$$L_2PdCl_2 + 2\ (R_3Si)_2Si{=}Si(SiR_3)_2 \xrightarrow{THF} \textbf{231b}\text{ (or }\textbf{231c}) + R_3Si-Si(SiR_3)(Cl)-Si(SiR_3)(Cl)-SiR_3 \tag{109}$$

232b (or **232c**) **22**, $SiR_3 = SiMe_2(t\text{-Bu})$ **233**

The first 14-electron disilene–palladium complex is synthesized by the reaction of bis(tricyclohexylphosphine)palladium dichloride (**232d**) with dilithiodisilane **161** [Eq. (110)]. The initial reaction would be the reduction of **232d** with **161** giving zero-valent $(Cy_3P)_2Pd$ (**234**) and free tetrasilyldisilene **22**.[68,151] Because both **234** and **22** are actually observed during the reaction by NMR spectroscopy, coupling between **234** and **22** accompanied by the elimination of the bulky Cy_3P ligand would give the final 14-electron disilene–palladium complex **235**. The reaction of $(Cy_3P)_2PtCl_2$ with **161** forms $(Cy_3P)_2Pt$ and **22** but no further reaction proceeds even at reflux in THF.

$$(Cy_3P)_2PdCl_2 + \textbf{161} \xrightarrow[-Cy_3P]{THF} Cy_3P-Pd\leftarrow[(R_3Si)_2Si{=}Si(SiR_3)_2] \tag{110}$$

232d
$R_3Si = t\text{-}BuMe_2Si$
Cy = Cyclohexyl
235 (85% yield)

Iron complexes of (*E*)- and (*Z*)-1,2-dichlorodisilenes **236E** and **236Z** are synthesized as the first disilene complexes with *E*,*Z*-isomerism by the reaction of the corresponding 2,2,3,3-tetrachlorotetrasilane **237** with an excess of $K_2Fe(CO)_4$ [Eq. (111)].[152] The reaction gives preferably **236Z** at the initial stage but **236Z** isomerizes slowly to its more stable isomer, **236E**. Preferable formation of *Z*-isomer **236Z** at the initial stage is explained by the intermediacy of the corresponding silylsilylene complex **238** [Eq. (112)]. A straightforward mechanism involving the formation of the corresponding 1,2,2-trichlorodisilanyliron anionic complex **239** followed by the nucleophilic attack of the anionic iron to the β-silicon atom cannot explain the observed diastereoselectivity in the synthetic reaction. The isomerization of **236Z** to **236E** is discussed in Section V.C.

$$R_3Si{-}SiCl_2{-}SiCl_2{-}SiR_3\ (\mathbf{237}) \xrightarrow[\text{DME / rt}]{\text{excess } K_2Fe(CO)_4} (CO)_4Fe[Si(Cl)(SiR_3){-}Si(SiR_3)(Cl)]\ (\mathbf{236E},\ 28\%\ \text{yield}) + (CO)_4Fe[Si(Cl)(SiR_3){-}Si(Cl)(SiR_3)]\ (\mathbf{236Z},\ 9\%\ \text{yield}) \quad (111)$$

$R_3Si = t\text{-}Bu_2MeSi$

$$\mathbf{210} \xrightarrow[\text{-KCl}]{K_2Fe(CO)_4} \left[(CO)_4Fe^{-}{-}Si(Cl)(SiR_3)(SiCl_2SiR_3)\ K^{+}\right]\ (\mathbf{239}) \xrightarrow{\text{-KCl}} \left[(CO)_4Fe{=}Si(SiR_3)(SiCl_2SiR_3)\right]\ (\mathbf{238}) \longrightarrow \mathbf{209Z} \xrightarrow{\circlearrowleft} \mathbf{209E} \quad (112)$$

The first disilene-group-4 metal complex, disilene–hafnium complex **240**, is synthesized using a similar reaction to those shown in Eqs. (108) and (109) [Eq. (113)].[153] The reaction of **240** with trimethylphosphine gives the phosphine adduct **241** [Eq. (114)].

$$(Me_3Si)_2Si(K){-}Si(K)(SiMe_3)_2\ (\mathbf{214}) \xrightarrow{\text{1. } MgBr_2,\ Et_2O;\ \text{2. } Cp_2HfCl_2} Cp_2Hf[Si(SiMe_3)_2{-}Si(SiMe_3)_2]\ (\mathbf{240}) \quad (113)$$

$$\mathbf{240} + Me_3P \longrightarrow (Me_3P)Cp_2Hf[Si(SiMe_3)_2{-}Si(SiMe_3)_2]\ (\mathbf{241}) \quad (114)$$

B. *Bonding and Structure*

Structures of complexes **231a–231c**, **235**, **236E** and **236Z**, and **241** are determined by single crystal X-ray analysis. The molecular structures of 16- and 14-electron palladium complexes **231b** and **235** are shown in Fig. 9.

The bonding of an alkene to a transition metal center is usually understood in terms of alkene-to-metal σ-donation and metal-to-alkene π-back donation, according to the Dewar–Chatt–Duncanson model.[154] The structure of an alkene complex is significantly influenced by the relative importance of σ-donation and π-back donation, and hence they are classified into π-complexes having a major contribution of σ-donation and metallacycles having dominant π-back donation. It is an interesting issue whether a similar classification can be applied to disilene–metal complexes. The geometry around the disilene ligand in the π-complex would not be very much different from the free disilene, while that in the metallacycle would be significantly distorted and characterized by an elongated Si^1–Si^2

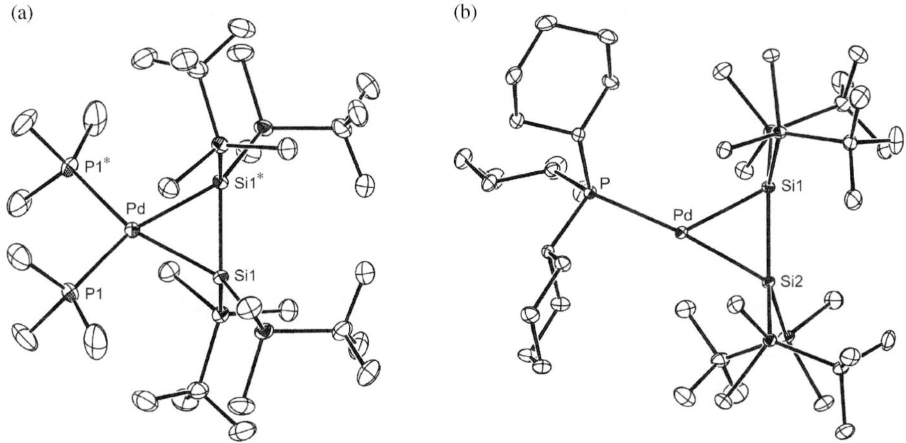

FIG 9. Molecular structures of complexes (a) **231b** and (b) **235**.

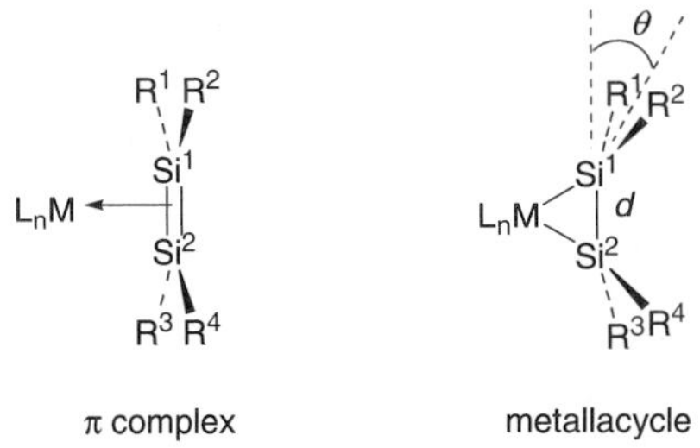

L_nM — $R^1 R^2$ Si^1 Si^2 $R^3 R^4$ — π complex

L_nM — θ — $R^1 R^2$ Si^1 d Si^2 $R^3 R^4$ — metallacycle

CHART 15.

bond length (d) and a large bend-back angle (θ) defined as the angle between the R^1–Si^1–R^2 (or R^3–Si^2–R^4) plane and the Si^1–Si^2 bond (Chart 15).

The bent-back angle θ and bond elongation $\Delta d/d_0$ ($\Delta d = d - d_0$, where d_0 is the Si1–Si2 bond length in the corresponding free disilene) for tungsten complex **229**,[147,148] 16-electron platinum complex **231a**,[149] and 16-electron palladium complex **231b**[150] are estimated to be 30.2° and 3.7%, 29.3° and 5.4%, 27.5° and 4.6%, respectively. In good accordance with the results of theoretical calculations for $(H_3P)_2Pt(Si_2R_4)$ and $(H_3P)_2Pd(Si_2R_4)$ (R = H and SiH_3), complexes **229**, **231a**, and **231b** are characterized as metallacycles. On the other hand, the bent-back angles θ around the central Si atoms of **235** are quite small (4.41° for Si1 and 9.65° for Si2, respectively) and the $\Delta d/d_0$ value is also relatively small (3.2%).[151] On this basis, complex **235** is regarded as the complex having the strongest π-complex character among the known disilene complexes. DFT calculations for a model 14-electron complex [**242**, $(Me_3P)Pd((H_3Si)_2Si{=}Si(SiH_3)_2)$] reveal that the unsymmetrical (T-type) structure observed for **235** is well reproduced by the optimized structure (**242U**), while the P–Pd–Si^1 angle for **242U** is much smaller than that observed for **235** (128.9°) probably due to the steric effects of bulky trialkylsilyl substituents in **235**. Unsymmetrical complex **242U** is only 2.9 kcal mol^{-1} more stable than

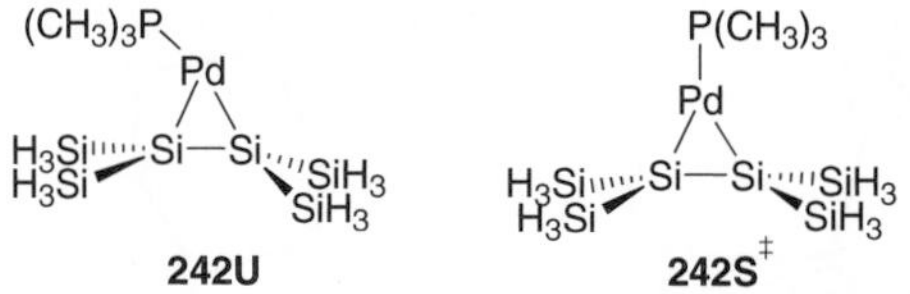

242U **242S‡**

CHART 16.

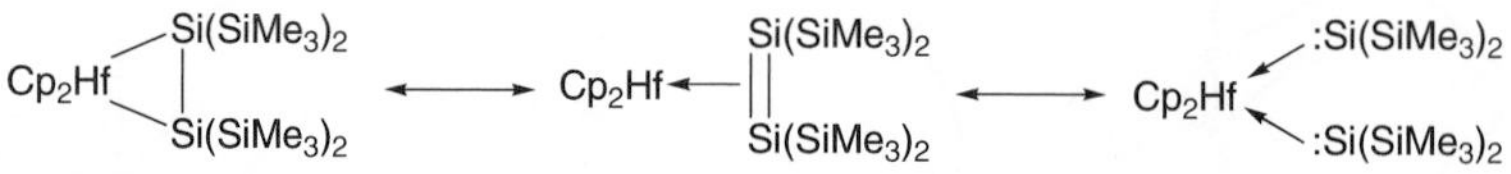

240A **240B** **240C**

CHART 17.

symmetrical (Y-type) complex **242S‡** that is found as a transition structure (Chart 16). Due to the intrinsic electron-deficient nature of the central metal of a 14-electron complex, π-back-bonding is much less effective in complex **235** than complex **231b**.

The NMR resonances for the central ^{29}Si nuclei of tetrakis(trialkylsilyl)disilene complexes in solution vary from −160 ppm for **241** to 127 ppm for **240**, depending on the bonding characteristics of the disilene ligands to the metals; for reference, the ^{29}Si NMR resonances for hexakis(trialkylsilyl)cyclotrisilanes and free disilene **22** are −174 to 149 ppm and +142.1 ppm.[68,155] Among disilene–palladium complexes, the ^{29}Si NMR resonance of **235** observed at +65.3 ppm is considerably downfield-shifted compared to the corresponding resonances of complexes **231b** (−46.5 ppm) and **231c** (−51.9 ppm), indicating the strong π-complex character of **235**. On the basis of its unusually low ^{29}Si NMR shift (127.2 ppm) as well as the DFT calculations for the zirconocene analog of **240**, the hafnium complex **240** was proposed to have partial disilylene complex character (**240C**) in addition to metallacycle (**240A**) and π-complex characters (**240B**) (Chart 17). The NMR resonances of the ^{29}Si nuclei bonded to the hafnium atom for complex **241** appear at −135.7 and −159.7 ppm, in accord with the metallacycle character of **241** as found in its X-ray structure.

C. *Reactions*

Although several reactions of disilene complexes have been reported, the studies are not yet performed systematically. A Si–Si bond of platinum complex **228a** is cleaved by O_2 and H_2 to give complexes **243** and **244** [Eqs. (115) and (116)].[145] The Pt–Si bond of complex **227a** is cleaved by methanol to give complex **245** [Eq. (117)].[145] In the reaction of tungsten complex **229** with $Ph_3P{=}S$, both W–Si and Si–Si bonds are cleaved competitively to give the corresponding sulfur insertion products **246** and **247** [Eq. (118)].[147] The Si–Si bond in complex **229** is cleaved also by CO_2 to give **248** [Eq. (119)].[147] Both Mo–Si and Si–Si bonds are cleaved during the reaction of molybdenum complex **230** with methanol giving products **249** and **250** [Eq. (120)].[156]

228a $\xrightarrow[C_7D_8]{O_2}$ Pt(Ph₂P–CH₂CH₂–PPh₂)(Si(i-Pr)₂–O–Si(i-Pr)₂) **243** (115)

228a $\xrightarrow[C_6D_6]{H_2}$ Pd(Ph₂P–CH₂CH₂–PPh₂)(Si(H)^{i}Pr₂)₂ **244** (116)

227a $\xrightarrow[C_6D_6]{CH_3OH}$ $Et_3P(H)Pt(PEt_3)Si(Mes_2)Si(Mes)_2OCH_3$ **245** (117)

229 $\xrightarrow[-\ PPh_3]{Ph_3P=S}$ Cp_2W(SiMe₂–S–SiMe₂) **246** (75%) + Cp_2W(S–SiMe₂–SiMe₂) **247** (25%) (118)

229 $\xrightarrow[C_6D_6]{CO_2}$ Cp_2W(SiMe₂–O–C(=O)–SiMe₂) **248** (119)

230 $\xrightarrow[C_6D_6]{CH_3OH}$ $Cp_2Mo(SiMe_2(OCH_3))(SiMe_2(H))$ **249** (80%) + $Cp_2Mo(H)(Me_2Si{-}SiMe_2OCH_3)$ **250** (20%) (120)

For the isomerization of **236Z** into **236E**, the first-order rate constant (k_1) in THF-d_8 is $3.34 \times 10^{-4}\,s^{-1}$ at 323 K and the activation enthalpy ($\Delta H^{\neq}$) and activation entropy ($\Delta S^{\neq}$) are 20.2 kcal mol^{-1} and -7.4 cal mol^{-1} K^{-1} respectively [Eq. (121)].[152] The isomerization is very slow in benzene-d_6 with $k_1 = 4.66 \times 10^{-7}\,s^{-1}$ at 323 K.

$$(Z)\text{-}\mathbf{236} \xrightarrow{\Delta,\ k_1} (E)\text{-}\mathbf{236} \quad (121)$$

For the *Z*-to-*E* isomerization, a mechanism *via* silylsilylene complex **238** [Eq. (112)] has been proposed. Mechanisms including the following processes are excluded on the basis of the theoretical calculations: (1) dissociation–association equilibrium between the corresponding disilene and $Fe(CO)_4$, (2) removal of one CO from **236Z** and then Si–Si bond cleavage forming the corresponding bis-silylene complex followed by the reverse Si–Si bond formation, and (3) heterolytic Fe–Si bond cleavage forming the corresponding zwitterionic compound and then recombination.

VI
CONCLUSION

During a quarter of century since the synthesis of the first stable disilene in 1981, remarkable development of the chemistry of Si=Si doubly bonded compounds has been achieved. Known types of disilenes are no longer limited to simple acyclic disilene and have been expanded to cyclooligosilene, spirocyclic disilene, tetrasiladiene, and even trisilaallene and disilyne having formal sp-hybridized silicon atom. Extensive studies on the bonding and structures of these stable disilenes and unsaturated compounds of related heavier group-14 elements have revealed rather unusual characteristics that have never been observed in organic compounds. The beautiful and systematic theory describing the bonding and structure of organic compounds must be modified significantly for compounds of heavier main-group elements. Although the mechanisms for many reactions of disilenes still remain to be elucidated, a few of the well-investigated reactions indicate that a disilene π system serves not only as a nucleophile but also as an electrophile or a biradical.

Remarkable progress of the chemistry of disilenes seems to encourage us to build up the modern general theory of bonding, structure, and reactions applied for the heavier main-group element chemistry in the near future; the general theory should include the theory for organic chemistry as a special case. Because stable disilenes and other unsaturated compounds of heavier group-14 elements show unique spectroscopic properties that cannot be easily obtained from organic compounds, future research may also be directed toward the design and synthesis of organic–inorganic hybrid materials with superior optoelectronic and other functions being inherent in this class of compounds.

Acknowledgment

The authors are very grateful to Prof. Robert West, University of Wisconsin for his encouragement to write this review.

References

(1) Corey, J. Y. (S. Patai, Z. Rappoport, Eds.), *The Chemistry of Organic Silicon Compounds* Volume 1, **1989**, Wiley, New York, p. 1.

(2) Kipping, F. S.; Sands, J. E. *J. Chem. Soc.* **1921**, *394*, 221.

(3) Peddle, G. J. D.; Roark, D. N.; Good, A. M.; McGeachin, S. G. *J. Am. Chem. Soc.* **1969**, *91*, 2807. Roark, D. N.; Peddle, G. J. D. *J. Am. Chem. Soc.* **1972**, *94*, 5837.

(4) Sakurai, H.; Nakadaira, Y.; Kobayashi, T. *J. Am. Chem. Soc.* **1979**, *101*, 487.

(5) West, R.; Fink, M. J.; Michl, J. *Science* **1981**, *214*, 1343.

(6) West, R. *Pure appl. Chem.* **1984**, *56*, 163. Raabe, G.; Michl, J. *Chem. Rev.* **1985**, *85*, 5. West, R. *Angew. Chem., Int. Ed. Engl* **1987**, *26*, 1201. Tsumuraya, T.; Batcheller, S. A.; Masamune, S. *Angew. Chem., Int. Ed. Engl* **1991**, *30*, 902. Grev, R. S. *Adv. Organomet. Chem* **1991**, *33*, 125. Weidenbruch, M. *Coord. Chem. Rev.* **1994**, *130*, 275. Koch, R.; Weidenbruch, M. *Angew. Chem.*,

Int. Ed **2002**, *41*, 1861. West, R. *Polyhedron* **2002**, *21*, 467. Weidenbruch, M. *J. Organomet. Chem.* **2002**, *646*, 39. Weidenbruch, M. *Organometallics* **2003**, *22*, 4348.

(7) Okazaki, R.; West, R. *Adv. Organomet. Chem.* **1996**, *39*, 231.

(8) Kira, M.; Iwamoto, T. *J. Organomet. Chem.* **2000**, *611*, 236.

(9) Weidenbruch, M. (Z. Rappoport, Y. Apeloig, Eds.), *The Chemistry of Organic Silicon Compounds* Volume 3, **2001**, John Wiley & Sons, Chichester, p. 391.

(10) Yokelson, H. B.; Maxka, J.; Siegel, D. A.; West, R. *J. Chem. Soc.* **1986**, *108*, 4239. Yokelson, H. B.; Siegel, D. A.; Millevolte, A. J.; Maxka, J.; West, R. *Organometallics* **1990**, *9*, 1005.

(11) Weidenbruch, M.; Kramer, K.; Peters, K.; Von Schnering, H. G. *Z. Naturforsch., Teil B: Anorg. Chem., Org. Chem.* **1985**, *40B*, 601.

(12) Sekiguchi, A.; Ichinohe, M.; Yamaguchi, S. *J. Am. Chem. Soc.* **1999**, *121*, 10231.

(13) Ichinohe, M.; Arai, Y.; Sekiguchi, A.; Takagi, N.; Nagase, S. *Organometallics* **2001**, *20*, 4141.

(14) Iwamoto, T.; Okita, J.; Kabuto, C.; Kira, M. *J. Organomet. Chem.* **2003**, *686*, 105.

(15) Iwamoto, T.; Okita, J.; Kabuto, C.; Kira, M. *J. Am. Chem. Soc.* **2002**, *124*, 11604.

(16) Ichinohe, M.; Kinjo, R.; Sekiguchi, A. *Organometallics* **2003**, *22*, 4621.

(17) Weidenbruch, M.; Willms, S.; Saak, W.; Henkel, G. *Angew. Chem., Int. Ed. Engl.* **1997**, *1997*, 2503.

(18) Boomgaarden, S.; Saak, W.; Marsmann, H.; Weidenbruch, M. *Z. Anorg. Allg. Chem.* **2002**, *628*, 1745.

(19) Jutzi, P.; Mix, A.; Rummel, B.; Schoeller, W. W.; Neumann, B.; Stammler, H.-G. *Science* **2004**, *305*, 849.

(20) Scheschkewitz, D. *Angew. Chem., Int. Ed.* **2004**, *43*, 2965.

(21) Ichinohe, M.; Sanuki, K.; Inoue, S.; Sekiguchi, A. *Organometallics* **2004**, *23*, 3088.

(22) Wiberg, N.; Niedermayer, W.; Fischer, G.; Nöeth, H.; Suter, M. *Eur. J. Inorg. Chem.* **2002**, 1066.

(23) Kira, M.; Iwamoto, T.; Kabuto, C. *J. Am. Chem. Soc.* **1996**, *118*, 10303.

(24) Iwamoto, T.; Kabuto, C.; Kira, M. *J. Am. Chem. Soc.* **1999**, *121*, 886.

(25) Wiberg, N.; Auer, H.; Nöeth, H.; Knizek, J.; Polborn, K. *Angew. Chem., Int. Ed.* **1998**, *37*, 2869.

(26) Wiberg, N.; Niedermayer, W.; Nöeth, H.; Warchhold, M. *Z. Anorg. Allg. Chem.* **2001**, *627*, 1717.

(27) Sekiguchi, A.; Matsuno, T.; Ichinohe, M. *J. Am. Chem. Soc.* **2001**, *123*, 12436.

(28) Lee, V. Y.; Takanashi, K.; Matsuno, T.; Ichinohe, M.; Sekiguchi, A. *J. Am. Chem. Soc.* **2004**, *126*, 4758.

(29) Ichinohe, M.; Matsuno, T.; Sekiguchi, A. *Angew. Chem., Int. Ed.* **1999**, *38*, 2194.

(30) Lee, V. Y.; Ichinohe, M.; Sekiguchi, A.; Takagi, N.; Nagase, S. *J. Am. Chem. Soc.* **2000**, *122*, 9034.

(31) Grybat, A.; Boomgaarden, S.; Saak, W.; Marsmann, H.; Weidenbruch, M. *Angew. Chem., Int. Ed.* **1999**, *38*, 2010.

(32) Wiberg, N.; Vaisht, S. K.; Fischer, G.; Mayer, P. *Z. Anorg. Allg. Chem.* **2004**, *630*, 1823.

(33) Schmedake, T. A.; Haaf, M.; Apeloig, Y.; Müller, T.; Bukalov, S.; West, R. *J. Am. Chem. Soc.* **1999**, *121*, 9479.

(34) Haaf, M.; Schmedake, T. A.; Paradise, B. J.; West, R. *Can. J. Chem.* **2000**, *78*, 1526.

(35) Matsumoto, S.; Tsutsui, S.; Kwon, E.; Sakamoto, K. *Angew. Chem., Int. Ed.* **2004**, *43*, 4610.

(36) Iwamoto, T.; Tamura, M.; Kabuto, C.; Kira, M. *Science* **2000**, *290*, 504.

(37) Billups, W. E.; Haley, M. M. *J. Am. Chem. Soc.* **1991**, *113*, 5084. Saini, R. K.; Litosh, V. A.; Daniels, A. D.; Billups, W. E. *Tetrahedron Lett* **1999**, *40*, 6157.

(38) Iwamoto, T.; Ohshima, K.; Kira, M. unpublished results.

(39) Ishida, S.; Iwamoto, T.; Kabuto, C.; Kira, M. *Silicon Chem* **2003**, *2*, 137.

(40) Ishida, S.; Iwamoto, T.; Kabuto, C.; Kira, M. *Nature* **2003**, *421*, 725.

(41) Sekiguchi, A.; Kinjo, R.; Ichinohe, M. *Science* **2004**, *305*, 1755.

(42) Sekiguchi, A.; Matsuno, T.; Ichinohe, M. *J. Am. Chem. Soc.* **2000**, *122*, 11250.

(43) Matsuno, T.; Ichinohe, M.; Sekiguchi, A. *Angew. Chem., Int. Ed.* **2002**, *41*, 1575.

(44) Trinquier, G.; Malrieu, J. P.; Riviere, P. *J. Am. Chem. Soc.* **1981**, *104*, 4529.

(45) Carter, E. A.; Goddard, W. A. III *J. Phys. Chem.* **1986**, *90*, 998.

(46) Trinquier, G.; Malrieu, J.-P. *J. Am. Chem. Soc.* **1987**, *109*, 5303. Trinquier, G.; Malrieu, J.-P. (S. Patai, Ed.), *The Chemistry of Functional Group, Suppl. A: The Chemistry of Double-Bonded functional Group, Vol. 2, Part 1*, **1989**, Wiley, New York, . Driess, M.; Grützmacher, H. *Angew. Chem., Int. Ed. Engl.* **1996**, *36*, 828. Trinquier, G. *J. Am. Chem. Soc.* **1990**, *112*, 2130. Trinquier, G.; Malrieu, J. P. *J. Phys. Chem.* **1990**, *94*, 6184.

(47) Malrieu, J.-P.; Trinquier, G. *J. Am. Chem. Soc.* **1989**, *111*, 5916.
(48) Jacobsen, H.; Ziegler, T. *J. Am. Chem. Soc.* **1994**, *116*, 3667.
(49) Karni, M.; Apeloig, Y. *J. Am. Chem. Soc.* **1990**, *112*, 8589.
(50) Liang, C.; Allen, L. C. *J. Am. Chem. Soc.* **1990**, *112*, 1039.
(51) Gordon, M. S.; Bartol, D. *J. Am. Chem. Soc.* **1987**, *109*, 5948. Grev, R. S.; Schaefer III, H. F.; Gaspar, P. P. *J. Am. Chem. Soc.* **1991**, *113*, 5638. Holthausen, M. C.; Koch, W.; Apeloig, Y. *J. Am. Chem. Soc.* **1999**, *121*, 2623.
(52) Yoshida, M.; Tamaoki, N. *Organometallics* **2002**, *21*, 2587.
(53) Apeloig, Y.; Müller, T. *J. Am. Chem. Soc.* **1995**, *117*, 5363.
(54) Apeloig, Y. (S. Patai, Z. Rappoport, Eds.), *The Chemistry of Organic Silicon Compounds*, Wiley, New York, **1989**, p. 57.
(55) Iwamoto, T.; Kira, M. unpublished results.
(56) Teramae, H. *J. Am. Chem. Soc.* **1987**, *109*, 4140.
(57) George, P.; Trachtman, M.; Bock, C. W.; Brett, A. M. *Tetrahedron* **1976**, *32*, 317.
(58) Naruse, Y.; Ma, J.; Inagaki, S. *Tetrahedron Lett* **2001**, *42*, 6553.
(59) Shepherd, B. D.; Powell, D. R.; West, R. *Organometallics* **1989**, *8*, 2664.
(60) Tokitoh, N.; Suzuki, H.; Okazaki, R.; Ogawa, K. *J. Am. Chem. Soc.* **1993**, *115*, 10428.
(61) Suzuki, H.; Tokitoh, N.; Okazaki, R.; Harada, J.; Ogawa, K.; Tomoda, S.; Goto, M. *Organometallics* **1995**, *14*, 1016.
(62) Wiberg, N.; Niedermayer, W.; Polborn, K. *Z. Anorg. Allg. Chem.* **2002**, *628*, 1045.
(63) Kira, M.; Ohya, S.; Iwamoto, T.; Ichinohe, M.; Kabuto, C. *Organometallics* **2000**, *19*, 1817.
(64) Sekiguchi, A.; Inoue, S.; Ichinohe, M. *J. Am. Chem. Soc.* **2004**, *126*, 9626.
(65) Shepherd, B. D.; Campana, C. F.; West, R. *Heteroatom Chem.* **1990**, *1*, 1. Fink, M. J.; Michalczyk, M. J.; Haller, K. J.; West, R.; Michl, J. *Organometallics* **1984**, *3*, 793.
(66) Fink, M. J.; Michalczyk, M. J.; Haller, K. J.; West, R.; Michl, J. *J. Chem. Soc., Chem. Commun.* **1983** 1010.
(67) Wind, M.; Powell, D. R.; West, R. *Organometallics* **1996**, *15*, 5772.
(68) Kira, M.; Maruyama, T.; Kabuto, C.; Ebata, K.; Sakurai, H. *Angew. Chem., Int. Ed. Engl.* **1994**, *34*, 1489.
(69) Iwamoto, T.; Tamura, M.; Kabuto, C.; Kira, M. *Organometallics* **2003**, *22*, 2342.
(70) Goller, A.; Heydt, H.; Clark, T. *J. Org. chem.* **1996**, *61*, 5840. Tsutsui, S.; Sakamoto, K.; Kabuto, C.; Kira, M. *Organometallics* **1998**, *17*, 3819. Nyulászi, L.; Schleyer, P. v. R. *J. Am. Chem. Soc.* **1999**, *121*, 6872.
(71) Iwamoto, T.; Ishida, S.; Kira, M. unpublished results.
(72) Kosa, M.; Karni, M.; Apeloig, Y. *J. Am. Chem. Soc.* **2004**, *126*, 10544.
(73) Nagase, S.; Takagi, N. private communication.
(74) Iwamoto, T.; Masuda, H.; Kabuto, C.; Kira, M. *Organometallics* **2005**, *24*, 197.
(75) Power, P.P. *Chem. Commun.* **2003** 2091; Nagase, S.; Kobayashi, K.; Takagi, N. *J. Organomet. Chem.* **2000**, *611*, 264; Kobayashi, K.; Nagase, S. *Organometallics* **1997**, *16*, 2489; Takagi, N.; Nagase, S. *Eur. J. Inorg. Chem.* **2002**, 2775.
(76) Watanabe, H.; Takeuchi, K.; Nakajima, K.; Nagai, Y.; Goto, M. *Chem. Lett.* **1988**, 1343.
(77) Masamune, S.; Murakami, S.; Tobita, H. *Organometallics* **1983**, *2*, 1464.
(78) Leites, L. A.; Bukalov, S. S.; Mangette, J. E.; Schmedake, T. A.; West, R. *Spectrochimica Acta, Part A* **2003**, *59A*, 1975.
(79) Simmons, H. E.; Fukunaga, T. *J. Am. Chem. Soc.* **1967**, *89*, 5208. Hoffmann, R.; Imamura, A.; Zeiss, G. D. *J. Am. Chem. Soc.* **1967**, *89*, 5215. Duerr, H.; Gleiter, R. *Angew. Chem., Int. Ed. Engl.* **1978**, *17*, 559.
(80) Sekiguchi, A.; Maruki, I.; Sakurai, H. *J. Am. Chem. Soc.* **1993**, *115*, 11460.
(81) Masamune, S.; Eriyama, Y.; Kawase, T. *Angew. Chem., Int. Ed. Engl.* **1987**, *26*, 584.
(82) Shepherd, B. D.; West, R. *Chem. Lett.* **1988**, 183.
(83) Kira, M.; Ishima, T.; Iwamoto, T.; Ichinohe, M. *J. Am. Chem. Soc.* **2001**, *123*, 1676.
(84) Iwamoto, T.; Tamura, M.; Kira, M. unpublished results.
(85) West, R.; Cavalieri, J. D.; Buffy, J. J.; Fry, C.; Zilm, K. W.; Duchamp, J. C.; Kira, M.; Iwamoto, T.; Müller, T.; Apeloig, Y. *J. Am. Chem. Soc.* **1997**, *119*, 4972. Cavalieri, J. D.; West, R.; Duchamp, J. C.; Zilm, K. W. *Phosphorus, Sulfur Silicon Relat. Elem.* **1994**, *93–94*, 213.

(86) Auer, D.; Strohmann, C.; Arbuznikov, A. V.; Kaupp, M. *Organometallics* **2003**, *22*, 2442.
(87) Kalinowski, H.-O.; Berger, S.; Braun, S. *Carbon-13-NMR Spectroscopy*, Wiley, New York, 1988.
(88) Sakurai, H.; Tobita, H.; Nakadaira, Y. *Chem. Lett.* **1982**, 1251.
(89) Leites, L. A.; Bukalov, S. S.; Garbuzova, I. A.; West, R.; Mangette, J.; Spitzner, H. *J. Organomet. Chem.* **1997**, *536/537*, 425. Leites, L. A.; Bukalov, S. S.; West, R.; Mangette, J. E.; Schmedake, T. A. (N. Auner, J. Weis, Eds.), *Organosilicon Chemistry IV*, Wiley-VCH, Weinheim, **2000**, p. 98.
(90) Jursic, B. S. *J. Mol Struct., THEOCHEM* **2000**, *497*, 65.
(91) Suzuki, H.; Tokitoh, N.; Okazaki, R. *Bull. Chem. Soc. Jpn.* **1995**, *68*, 2471.
(92) Tsutsui, S.; Tanaka, H.; Kwon, E.; Matsumoto, S.; Sakamoto, K. *Organometallics* **2004**, *23*, 5659.
(93) Tsutsui, S.; Sakamoto, K.; Kira, M. *J. Am. Chem. Soc.* **1998**, *120*, 9955.
(94) Takahashi, M.; Kira, M.; Sakamoto, K.; Müller, T.; Apeloig, Y. *J. Comput. Chem.* **2001**, *22*, 1536.
(95) Takahashi, M.; Tsutsui, S.; Sakamoto, K.; Kira, M.; Müller, T.; Apeloig, Y. *J. Am. Chem. Soc.* **2001**, *123*, 347.
(96) Yokelson, H. B.; Siegel, D. A.; Millevolte, A. J.; Maxka, J.; West, R. *Organometallics* **1990**, *9*, 1005.
(97) Kira, M.; Iwamoto, T. (Z. Rappoport, Y. Apeloig, Eds.), *The Chemistry of Organic Silicon Compounds* Volume 3, **2001**, Wiley, NewYork, p. 853.
(98) Karni, M.; Kapp, J.; Schleyer, P. v. R.; Apeloig, Y. (Z. Rappoport, Y. Apeloig, Eds.), *The Chemistry of Organic Silicon Compounds* Volume 3, **2001**, Wiley, New York, p. 1.
(99) Boatz, J. A.; Gordon, M. S. *J. Phys. Chem.* **1990**, *94*, 7331. Pople, J. A.; Head-Gordon, M.; Fox, D. J.; Raghavachari, K.; Curtiss, A. L. *J. Chem. Phys.* **1989**, *90*, 5622.
(100) Krogh-Jespersen, K. *Chem. Phys. Lett.* **1982**, *93*, 327.
(101) Iwamoto, T.; Kira, M. *Chem. Lett.* **1998**, 227.
(102) Boatz, J. A.; Gordon, M. S. *J. Phys. Chem.* **1988**, *92*, 3037. Koch, R.; Bruhn, T.; Weidenbruch, M. *J. Mol Struct., THEOCHEM* **2004**, *680*, 91.
(103) Müller, T. (N. Auner, J. Weis, Eds.), *Organosilicon Chemistry IV*, Wiley-VCH, Weinheim, **2000**, p. 110.
(104) Lee, V. Y.; Ichinohe, M.; Sekiguchi, A. *J. Organomet. Chem.* **2003**, *685*, 168.
(105) Millevolte, A. J.; Powell, D. R.; Johnson, S. G.; West, R. *Organometallics* **1992**, *11*, 1091.
(106) Willms, S.; Grybat, A.; Saak, W.; Weidenbruch, M.; Marsmann, H. *Z. Anorg. Allg. Chem.* **2000**, *626*, 1148.
(107) Zhang, Z.-R.; Becker, J. Y.; West, R. *Chem. Commun.* **1998**, 2719.
(108) Weidenbruch, M.; Kramer, K.; Schaefer, A.; Blum, J. K. *Chem. Ber.* **1985**, *118*, 107. Weidenbruch, M.; Thom, K. L. *J. Organomet. Chem.* **1986**, *308*, 177.
(109) Weidenbruch, M.; Pellmann, A.; Pan, Y.; Pohl, S.; Saak, W. *J. Organomet. Chem.* **1993**, *450*, 67.
(110) Cooper, J.; Hudson, A.; Jackson, R. A. *Mol. Phys.* **1972**, *23*, 209.
(111) Kyushin, S.; Sakurai, H.; Betsuyaku, T.; Matsumoto, H. *Organometallics* **1997**, *16*, 5386.
(112) Kyushin, S.; Sakurai, H.; Matsumoto, H. *Chem. Lett.* **1998**, 107.
(113) Kira, M.; Obata, T.; Kon, I.; Hashimoto, H.; Ichinohe, M.; Sakurai, H.; Kyushin, S.; Matsumoto, H. *Chem. Lett.* **1998**, 1097.
(114) Sekiguchi, A.; Fukawa, T.; Nakamoto, M.; Lee, V. Y.; Ichinohe, M. *J. Am. Chem. Soc.* **2002**, *124*, 9865.
(115) Kira, M.; Iwamoto, T.; Yin, D.; Maruyama, T.; Sakurai, H. *Chem. Lett.* **2001**, 910.
(116) DeYoung, D. J.; Fink, M. J.; West, R.; Michl, J. *Main Group Met. Chem.* **1987**, *10*, 19.
(117) Budaraju, J.; Powell, D. R.; West, R. *Main Group Met. Chem.* **1996**, *19*, 531.
(118) Apeloig, Y.; Nakash, M. *J. Am. Chem. Soc.* **1996**, *118*, 9798.
(119) Apeloig, Y.; Nakash, M. *Organometallics* **1998**, *17*, 1260.
(120) Nagase, S.; Kudo, T.; Ito, K. (V. H. Smith Jr., H. F. Schaefer III, K. Morokuma, Eds.) *Appl. Quant. Chem.* Reidel, Dordrecht, The Netherland, **1986**.
(121) Apeloig, Y.; Nakash, M. *Organometallics* **1998**, *17*, 2307.
(122) Takahashi, M.; Veszpremi, T.; Hajgato, B.; Kira, M. *Organometallics* **2000**, *19*, 4660.
(123) Hajgato, B.; Takahashi, M.; Kira, M.; Veszpremi, T. *Chem. Eur. J.* **2002**, *8*, 2126.
(124) Veszpremi, T.; Takahashi, M.; Hajgato, B.; Kira, M. *J. Am. Chem. Soc.* **2001**, *123*, 6629.
(125) Boomgaarden, S.; Saak, W.; Weidenbruch, M.; Marsmann, H. *Z. Anorg. Allg. Chem.* **2001**, *627*, 349.

(126) Fanta, A. D.; Belzner, J.; Powell, D. R.; West, R. *Organometallics* **1993**, *12*, 2177.
(127) Iwamoto, T.; Sakurai, H.; Kira, M. *Bull. Chem. Soc. Jpn.* **1998**, *71*, 2741.
(128) Su, M.-D. *J. Phys. Chem. A* **2004**, *108*, 823.
(129) Lee, V. Y.; Matsuno, T.; Ichinohe, M.; Sekiguchi, A. *Heteroatom Chem* **2001**, *12*, 223.
(130) Lee, V. Y.; Sekiguchi, A. *Chem. Lett.* **2004**, 84.
(131) Fink, M. J.; DeYoung, D. J.; West, R.; Michl, J. *J. Am. Chem. Soc.* **1983**, *105*, 1050.
(132) Schmid, P.; Friller, D.; Ingold, U. K. *Int. J. Chem. Kinet.* **1979**, *11*, 333. Lai, D.; Griller, D.; Husband, S.; Ingold, U. K. *J. Am. Chem. Soc.* **1974**, *96*, 6355.
(133) Lee, V. Y.; Ichinohe, M.; Sekiguchi, A. *Chem. Commun.* **2001**, 2146.
(134) Weidenbruch, M.; Kroke, E.; Marsmann, H.; Pohl, S.; Saak, W. *J. Chem. Soc., Chem. Commun.* **1994**, 1233.Dixon, C. E.; Liu, H. W.; Vander Kant, C. M.; Baines, K. M. *Organometallics* **1996**, *15*, 5701.
(135) Weidenbruch, M.; Flintjer, B.; Pohl, S.; Hasse, D.; Hartens, J. *J. Organomet. Chem.* **1988**, *388*, C1. Weidenbruch, M.; Lesch, A.; Peters, K.; von Schnering, H. G. *J. Organomet. Chem.* **1992**, *423*, 329.
(136) Dixon, C. E.; Cooke, J. A.; Baines, K. M. *Organometallics* **1997**, *16*, 5437.
(137) Dixon, C. E.; Baines, K. M. *Phosphorus, Sulfur Silicon Relat. Elem.* **1997**, *124 & 125*, 123.
(138) Ichinohe, M.; Matsuno, T.; Sekiguchi, A. *Chem. Commun.* **2001**, 183.
(139) Lee, V. Y.; Ichinohe, M.; Sekiguchi, A. *J. Organomet. Chem.* **2001**, *636*, 41.
(140) Mosey, N. J.; Baines, K. M.; Woo, T. K. *J. Am. Chem. Soc.* **2002**, *124*, 13306.
(141) Samuel, M. S.; Jenkins, H. A.; Hughes, D. W.; Baines, K. M. *Organometallics* **2003**, *22*, 1603.
(142) Newcomb, M.; Chestney, D. L. *J. Am. Chem. Soc.* **1994**, *116*, 9753.
(143) Boomgaarden, S.; Saak, W.; Weidenbruch, M. *Organometallics* **2001**, *20*, 2451.
(144) Lee, V. Y.; Takanashi, K.; Ichinohe, M.; Sekiguchi, A. *J. Am. Chem. Soc.* **2003**, *125*, 6012. Lee, V. Y.; Takanashi, K.; Nakamoto, M.; Sekiguchi, A. *Russ. Chem. Bull., Int. Ed.* **2004**, *53*, 1102.
(145) Pham, E. K.; West, R. *Organometallics* **1990**, *9*, 1517.
(146) Pham, E. K.; West, R. *J. Am. Chem. Soc.* **1989**, *111*, 7667. Pham, E. K.; West, R. *J. Am. Chem. Soc.* **1996**, *118*, 7871.
(147) Berry, D. H.; Chey, J. H.; Zipin, H. S.; Carroll, P. J. *J. Am. Chem. Soc.* **1990**, *112*, 452.
(148) Hong, P.; Damrauer, N. H.; Carroll, P. J.; Berry, D. H. *Organometallics* **1993**, *12*, 3698.
(149) Hashimoto, H.; Sekiguchi, Y.; Iwamoto, T.; Kabuto, C.; Kira, M. *Organometallics* **2002**, *21*, 454.
(150) Hashimoto, H.; Sekiguchi, Y.; Sekiguchi, Y.; Iwamoto, T.; Kabuto, C.; Kira, M. *Can. J. Chem.* **2003**, *81*, 1241.
(151) Kira, M.; Sekiguchi, Y.; Iwamoto, T.; Kabuto, C. *J. Am. Chem. Soc.* **2004**, *126*, 12778.
(152) Hashimoto, H.; Suzuki, K.; Setaka, W.; Kabuto, C.; Kira, M. *J. Am. Chem. Soc.* **2004**, *126*, 13628.
(153) Fischer, R.; Zirngast, M.; Flock, M.; Baumgartner, J.; Marschner, C. *J. Am. Chem. Soc.* **2005**, *127*, 70.
(154) Dewer, M. J. S. *Bull. Soc. Chim. Fr.* **1951**, *18*, C71. Chatt, J., Duncanson, L. A., *J. Chem. Soc.* **1953**, 2939.
(155) Kira, M.; Iwamoto, T.; Maruyama, T.; Kuzuguchi, T.; Yin, D.; Kabuto, C.; Sakurai, H. *J. Chem. Soc., Dalton Trans.* **2002**, 1539.
(156) Berry, D. H.; Chey, J.; Zipin, H. S.; Carroll, P. J. *Polyhedron* **1991**, *10*, 1189.
(157) Ohya, S.; Iwamoto, T.; Kira, M. unpublished results.

Metallacycloalkanes – Synthesis, Structure and Reactivity of Medium to Large Ring Compounds

BURGET BLOM,[a] HADLEY CLAYTON,[a,b] MAIRI KILKENNY[a,c] and JOHN R. MOSS[a,*]

[a]Department of Chemistry, University of Cape Town, Rondebosch 7701, South Africa
[b]Department of Chemistry, UNISA, P.O. Box 392, Pretoria 0003, South Africa
[c]Department of Biochemistry, University of Cambridge, 80 Tennis Court Road, Cambridge, CB2 1GA, UK

I
INTRODUCTION

Since many metal fragments are isolobal with CH_2, it should be possible to make a range of metallacycloalkanes. Metallacyclobutanes are well known as a class of compounds and serve as key intermediates in catalytic alkene metathesis.[1] This reaction has gained great importance in recent years through the work of Grubbs[2] and Schrock.[3] Alkene metathesis has many applications in organic chemistry,

*Corresponding author. Tel.: +27 21 650 2535.
E-mail: jrm@science.uct.ac.za (J.R. Moss).

ADVANCES IN ORGANOMETALLIC CHEMISTRY
VOLUME 54 ISSN 0065-3055/DOI 10.1016/S0065-3055(05)54004-8

© 2006 Elsevier Inc.
All rights reserved.

polymer and materials chemistry, as well as organometallic chemistry for ring opening metathesis polymerisation (ROMP), ring closing metathesis (RCM), etc.[4] More recently, metallacycloheptanes have been proposed[5] and shown[6] to be key intermediates in ethylene trimerisation. Other catalytic reactions, where metallacycles have been proposed as intermediates, range from the Fischer Tropsch reaction[7] to ethylene tetramerisation.[8]

Owing to the growing importance of metallacycles, and since the subject has not been reviewed for some time, we embarked on this review. Previous reviews are in 1982 by Puddephatt[9] and Chappel and Cole-Hamilton.[10] This topic has also been highlighted in the book by Collman *et al.*[11]

In this present review, we will discuss metallacyclopentane and larger rings containing only one transition metal. Thus we will cover compounds containing the structural motifs (I-A) and (I-B), shown below; where *n* is 4 or greater. We will not include unsaturated rings or rings containing heteroatoms.

There has been a recent comprehensive review written on Group 10 metallacycles[12] and only the more recent and relevant papers in this group will be covered. In this present review, we will discuss particularly the synthesis, structure, reactivity and applications of metallacycles.

II
METALLACYCLOPENTANES

A. *Group 4 Metallacyclopentanes*

The preparation of saturated five-membered metallacycles containing a group 4 transition metal has been extensively investigated and a number of synthetic routes are available for complexes of this type. Reactions of metallocene dichlorides with 1,4-dilithiobutane or the corresponding di-Grignard reagent represent the most common methods for preparation of these compounds. These methods of synthesis have been used to prepare metallacyclic compounds of Ti, Zr and Hf in moderate to good yields. The second most general route for these metals involves the reduction of tetravalent precursors in the presence of alkenes that first coordinate and then couple. A recurrent feature of this approach is that the cyclisation is often reversible.

1. Titanium

$Cp_2Ti(C_4H_8)$, **1**, prepared by the addition of 1,4-dilithiobutane to Cp_2TiCl_2 in diethylether at $-78\,^{\circ}C$ was isolated as an orange crystalline solid in ca. 20% yield (Scheme 1) ($Cp = \eta{-}C_5H_5$).[13] This compound is reported to be unstable[14] in

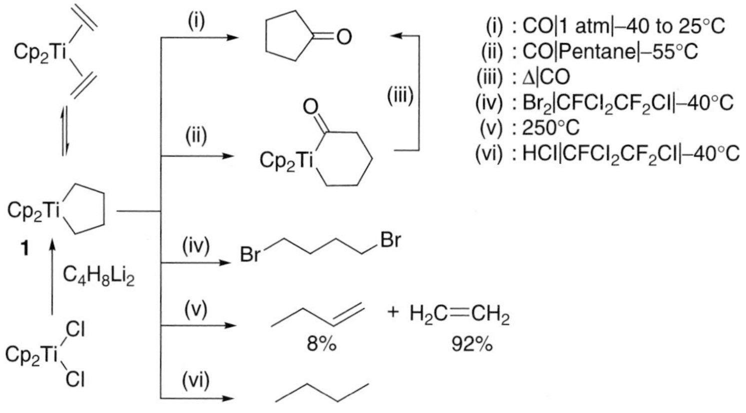

(i) : CO|1 atm|–40 to 25°C
(ii) : CO|Pentane|–55°C
(iii) : Δ|CO
(iv) : Br_2|$CFCl_2CF_2Cl$|–40°C
(v) : 250°C
(vi) : HCl|$CFCl_2CF_2Cl$|–40°C

SCHEME 1.

solution even at −30 °C. The thermal and oxidative instability of **1** precluded analysis and the compound was characterised by its reactions.[13] Studies by Grubbs and Miyashita[15] have shown that decomposition occurs by reversible carbon–carbon bond cleavage to produce an intermediate bis-ethylene complex.

Compound **1** exhibits significantly different reactivity than the acyclic analogue. The metallacycle is more stable than the acyclic complex and whereas $Cp_2Ti^nBu_2$ decomposes *via* the expected β-H elimination pathway to produce butenes and butane, the thermal decomposition products of **1** are ethylene and 1-butene. In addition, the metallacycle is observed to be significantly more reactive towards CO than $Cp_2Ti^nBu_2$ Reaction of **1** with carbon monoxide at −55 °C yields the titanium acyl species, based on infrared data, which then rapidly converts to cyclopentanone at 0 °C (Scheme 1).[13]

Takaya and Mashima[14] have reported the synthesis of stable titanacyclopentanes *via* coupling of two olefin ligands. Reaction of $Cp^*_2Ti(CH_2{=}CH_2)$ with methylenecyclopropane or 2-phenyl-1-methylenecyclopropane results in the formation of a substituted titanacyclopentane ($Cp^* = \eta^5{-}C_5Me_5$). Compounds **2a** (R = H) and **2b** (R = Ph) were isolated as crystalline solids and found to be more stable than the bis(cyclopentadienyl) titanacyclopentane complex, **1**. Compound **2a** melted with decomposition at 156–158 °C while **2b** was found to be slightly less stable, and decomposed at 122–124 °C. The stability of these compounds relative to **1** has been ascribed to a strain and/or steric effect exerted by the cyclopropane ring on the β-carbon of the titanacyclopentane, which makes the decomposition reaction less favourable.

The structure of **2b** was determined by X-ray analysis.[14] The C(α)–Ti–C(α) angle was found to be 80.6° with the titanacyclopentane ring in a puckered form as shown in Fig. 1. The Ti–C(α) bond lengths are reported as 2.214 and 2.189 Å.

Solid **2a** undergoes predominantly β-carbon–carbon bond cleavage when heated rapidly to 200 °C under vacuum to give ethylene (43%), 1,3-butadiene (31%), methylcyclopropane (19%) and the diene $CH_3CH_2C({=}CH_2)CH_2{=}CH_2$ (16%). In comparison, thermal decomposition of **2b** gave mainly the diene CH_3CH_2C

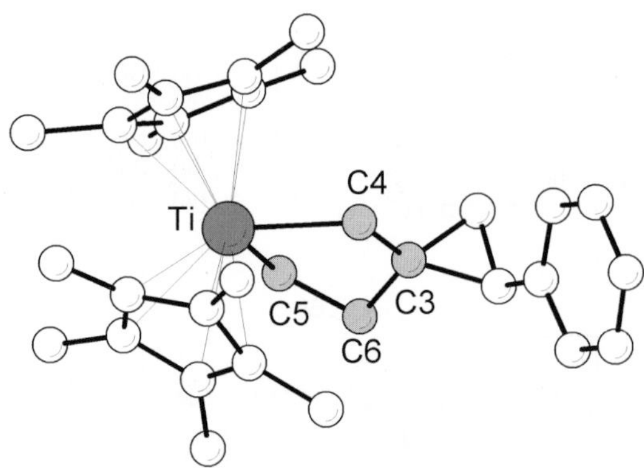

FIG. 1. Molecular structure of **2b** (hydrogen atoms omitted).

($=CH_2$)$CH_2=CHPh$ in 70% yield Eq. (1). It is proposed that the diene is derived from the decomposition of a six-membered metallacycle as shown in Eq. (1).[14]

Cp_2Ti — Rearrangement → Cp_2Ti (1)

2a R = H
2b R = Ph

The cyclisation of alkenes by low valent titanium has also been applied to intramolecular processes. Thus the reduction of titanocenes bearing pendant alkenyl substituents $[TiCl_2(\eta\text{-}C_5Me_4XCH{=}CHR)_2]$ ($X = SiMe_2$, CH_2, CHMe; R = H, Me) provides chiral titanacyclopentanes shown in Scheme 2.[16] The titanacyclopentanes could then be cleaved with HCl to provide new titanocene dichlorides in which the two cyclopentadienyl ligands were linked by a hydrocarbon chain.

An alternative ligand system to the cyclopentadienyl-based compounds is available through the reaction of $[(ArO)_2Ti(C_4Et_4)]$ (ArO = 2,6-diphenylphenoxide) with olefins which affords complexes **3a**–**3e**.[17] Solutions of the titanacyclopentadiene compound, $[(ArO)_2Ti(C_4Et_4)]$, in benzene react rapidly with excess ethene, propene, 1-butene or styrene to produce the titanacyclopentane derivatives **3a**–**3e**, respectively, by coupling of the α-olefins. These titanacyclopentane compounds can be readily isolated pure from the reaction mixtures. The compounds are more stable than the analogous cyclopentadienyl complexes and **3a** has been found to be stable in toluene solution, under 1 atm ethene, at temperatures of up to 80 °C. The reaction of $[(ArO)_2Ti(C_4Et_4)]$ with styrene gave a mixture of two substitutional isomers, identified by NMR spectroscopy as the 2,4- and 2,5-diphenyl derivates **3d** and **3e**, respectively (Scheme 3).

Compounds **3a** and the 2,5-diphenyl isomer **3e** have been structurally characterised by single crystal X-ray diffraction analysis (**3a**, Fig. 2). Bond lengths for Ti–C(α) were measured at 2.084 Å and Ti–O at 1.799 and 1.804 Å, which are within the ranges typical of aryloxide and alkyl groups bound to 4-coordinate Ti(IV).[17]

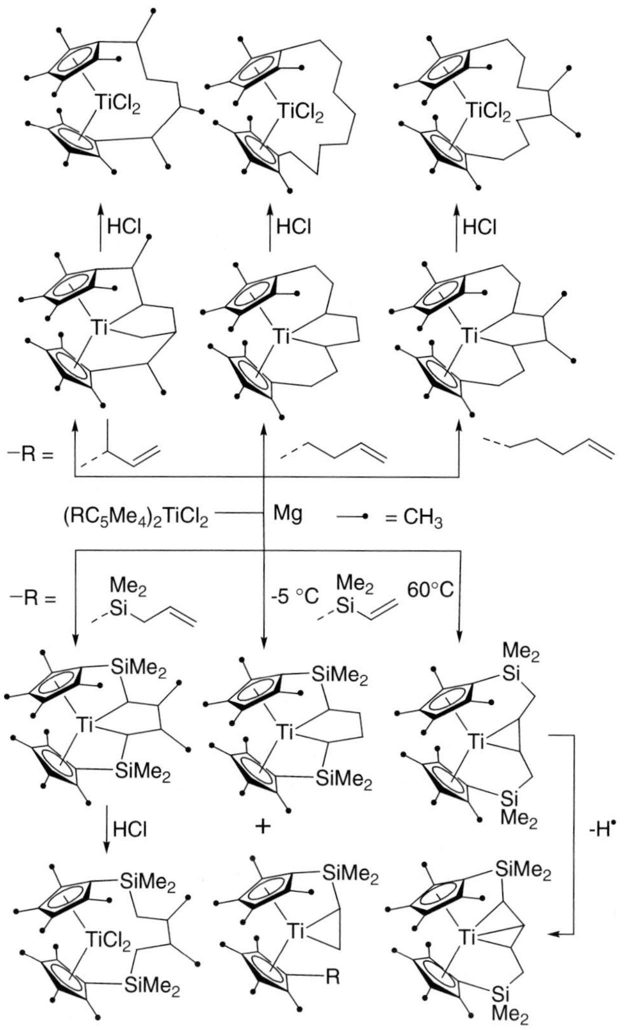

SCHEME 2.

The titanacyclopentane complex **3a** reacts with $Ph_2C{=}O$ in the presence of excess ethylene to produce the oxatitanacycloheptane complex **4**. However, in the absence of excess ethylene, a mixture of oxatitanacyclopentane **5** and dioxatitanacycloheptane **6** is produced (Scheme 4).

Reaction of **3a** with butadiene or isoprene results in the formation of the π-allyl complexes **7a** and **7b**, respectively, which are stable in benzene solution for days at 25 °C.[18] Hydrocarbon solutions of **3a** react with 2,3-dimethylbutadiene to initially form the titanacyclopent-3-ene complex **8**, isolated as a purple crystalline solid, which then subsequently reacts with generated free ethylene to form a mixture of the isomers **9** and **10** (Scheme 5). In an earlier paper, Rothwell and co-workers[19]

(ArO)$_2$TiCl$_2$

--OAr = --O–C$_6$H$_3$(Ph)$_2$ (2,6-diphenylphenoxide)

3a R = H
3b R = Me
3c R = Et
3d
3e

SCHEME 3.

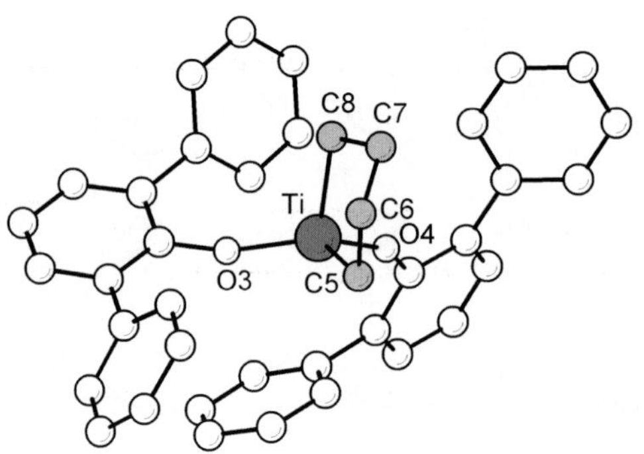

FIG. 2. Molecular structure of **3a** (hydrogen atoms omitted).

3a → (C$_2$H$_4$, Ph$_2$C=O) → 4

3a → (Ph$_2$C=O) → 5 + 6

SCHEME 4.

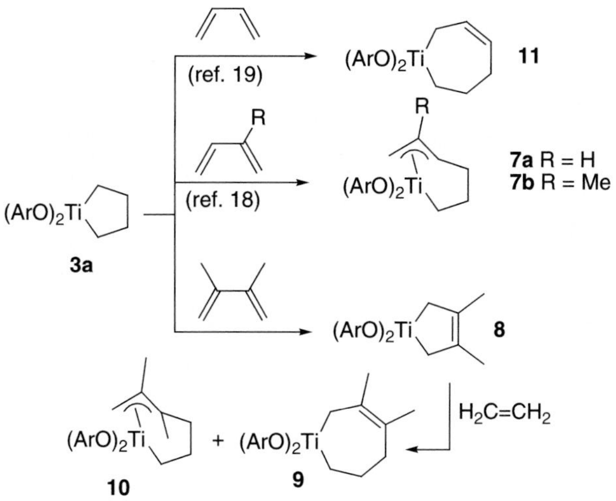

(ArO)$_2$Ti 3a; (ref. 19) → (ArO)$_2$Ti 11; (ref. 18) → (ArO)$_2$Ti 7a R = H, 7b R = Me; → (ArO)$_2$Ti 8; $H_2C{=}CH_2$ → (ArO)$_2$Ti 10 + (ArO)$_2$Ti 9

SCHEME 5.

reported that addition of 1,3 butadiene to **3a** produces a titanocyclohept-3-ene complex **11** in quantitative yield.

The titanacyclopentane complexes **3a**–**3e** react rapidly in C_6D_6 solution with tertiary phosphines to form the analogous titanacyclopropane complex containing a phosphine ligand:

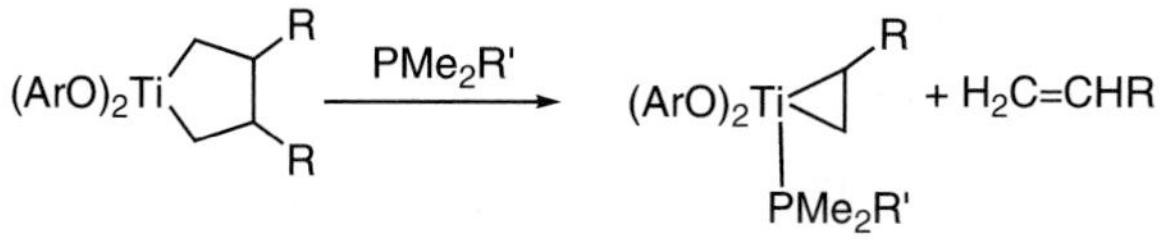

$$(ArO)_2Ti(CH_2CHR)_2 \xrightarrow{PMe_2R'} (ArO)_2Ti(CH_2CHR)(PMe_2R') + H_2C{=}CHR \qquad (2)$$

Compound **3a** was found to react with excess water to produce a mixture of organic products including butane, ethene and ethane.[17] However, when trace amounts of water were introduced into a hexane solution of **3b**, the oxo complex **12** is formed. The molecular structure of **12** shows that it is formed by reaction of 2 equivalents of **3b** and 1 equivalent of water:

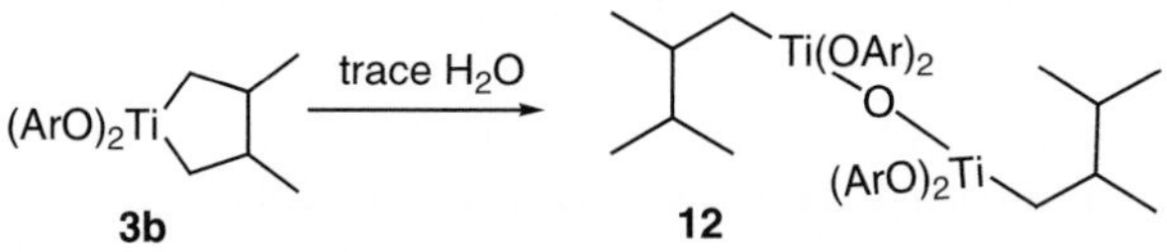

$$\textbf{3b} \xrightarrow{\text{trace } H_2O} \textbf{12} \qquad (3)$$

When a solution of **3d** or **3e** in benzene is heated to 100 °C in the presence of excess styrene, catalytic dimerisation has been found to occur to produce *trans* 1, 3-diphenylbut-1-ene which subsequently isomerises within the reaction mixture to give 1,3-diphenylbut-2-ene:[17]

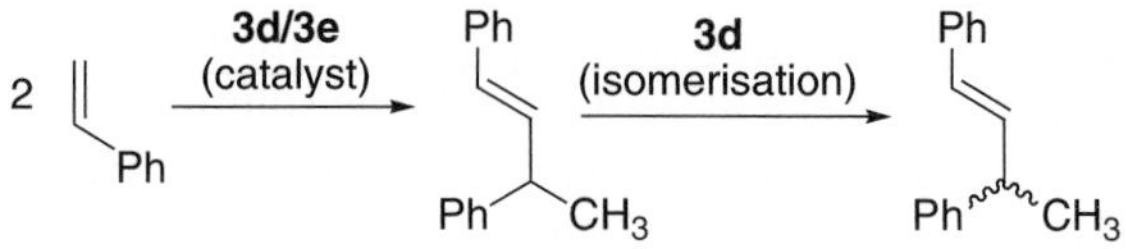

$$2\ \text{CH}_2{=}\text{CHPh} \xrightarrow[\text{(catalyst)}]{\textbf{3d/3e}} \text{PhCH{=}CHCH(Ph)CH}_3 \xrightarrow[\text{(isomerisation)}]{\textbf{3d}} \text{PhCH}_2\text{CH{=}C(Ph)CH}_3 \qquad (4)$$

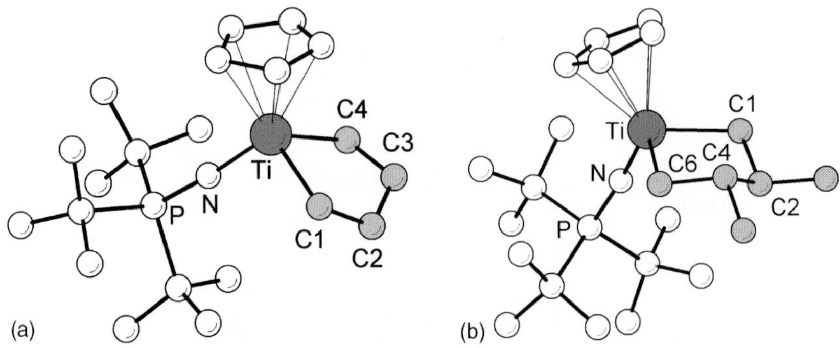

FIG. 3. Molecular structures of (a) **13a** and (b) **13b** (hydrogen atoms omitted).

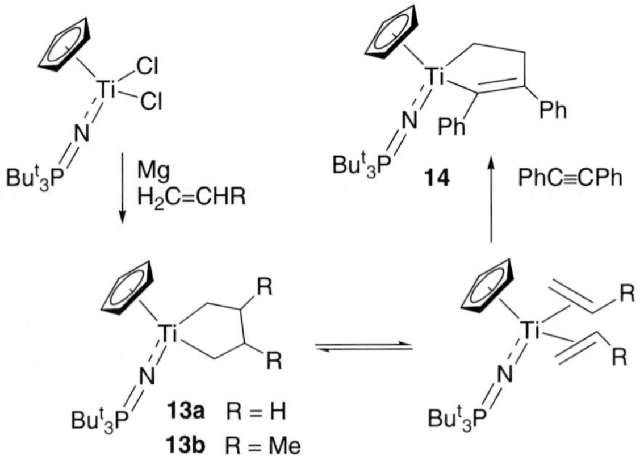

SCHEME 6.

Phosphinamides have also been presented as alternatives to cyclopentadienyl ligands in early transition metal chemistry and are reviewed elsewhere in this volume.[20] Thus, the complex $[TiCl_2(N{=}P^tBu_3)(\eta\text{-}C_5H_5)]$ has been shown to react with magnesium in the presence of ethene or propene to provide titanacyclopentanes $[Ti(C_4H_6R_2)(N{=}P^tBu_3)(\eta\text{-}C_5H_5)]$ (R = H **13a**, Me **13b**, Fig. 3).[20a] Treating this metallacycle with tolane provides the titanocyclopentene $[Ti(C_4H_4Ph_2)(N{=}P^tBu_3)$-$(\eta\text{-}C_5H_5)]$ **14** and this may be taken as evidence for the reversibility of alkene coupling. (Scheme 6).

2. Zirconium

Zirconacyclopentanes are readily formed by the reductive coupling of α-olefins. $[Cp^*_2ZrN_2]_2N_2$ reacts with ethene at 25 °C in toluene solution to afford the zirconacyclopentane **15**, in nearly quantitative yield.[21] In addition, **15** can be obtained by reacting $Cp^*_2ZrH_2$ with ethene at 25 °C. Reaction of $Cp^*_2ZrH_2$ with excess

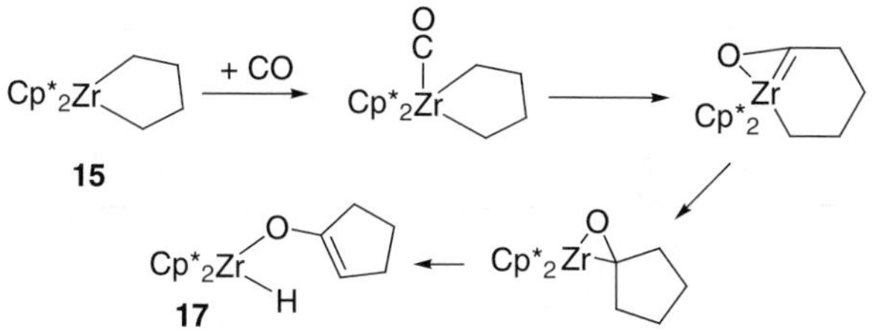

SCHEME 7.

isobutylene, however, does not lead to the formation of the substituted zirconapentacycle, instead the *neo*-butyl hydride complex **16** is formed which was found to be remarkably stable in benzene solutions at 50 °C. However, when treated with ethylene at 25 °C **16** was found to convert to **15** with release of isobutene:[21]

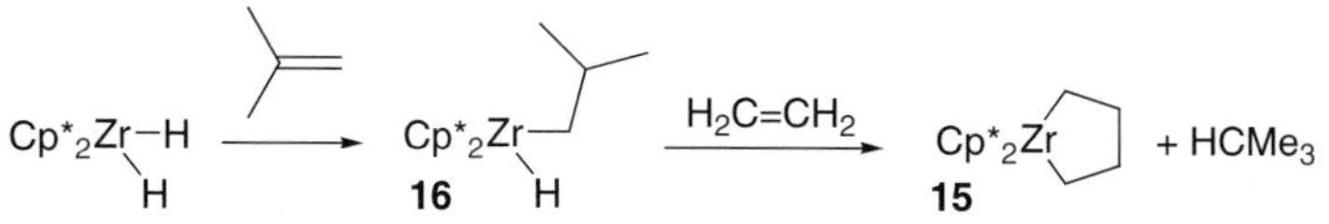

(5)

Compound **15** reacts with both HCl and H_2 to yield predominantly butane.[21] The zirconacycle was found to react instantaneously at 25 °C with CO to form a white crystalline product for which Bercaw *et al.* proposed the structure of the enolate hydride **17** on the basis of spectral evidence. The mechanism suggested by the authors for the reaction is shown in Scheme 7.

Treating the sterically encumbered zirconocene $ZrCl_2(\eta\text{-}C_5H_3^tBu_2)_2$ with ^{n}BuLi in the presence of ethene also provides a structurally characterised zirconocyclopentane $Zr(C_4H_8)(\eta\text{-}C_5H_3^tBu_2)_2$ **18**. The reversibility of metallacycle formation was demonstrated by its reaction with diphenylacetylene to provide the zirconacyclopentene **19** incorporating the alkyne and one equivalent of ethene:[22]

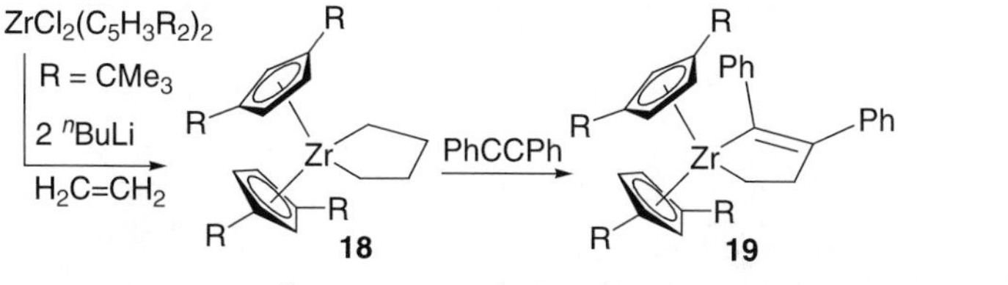

(6)

Reaction of *rac*-(ebthi)Zr(η^2-$Me_3SiC_2SiMe_3$) [ebthi = 1,2 ethylene-1,1′-bis(η^5-tetrahydroindenyl) with excess ethylene at room temperature gives the corresponding zirconacyclopentane **20a** as bright yellow crystals (m.p. 133–136 °C) in 85% yield, while reaction with styrene at 50 °C gave unsymmetrically substituted **20b** as red crystals (m.p. 185–187 °C) in 72% yield.[23] The use of norbornadiene with this reagent allowed the isolation of zirconacyclopentanes bearing alkene functional groups, e.g., **20c**, while the vinylpyridine adduct coupled with norbornadiene (nbd) to provide an α-pyridyl substituted zirconacyclopentane **20d** (Scheme 8, Fig. 4).[24]

A further example of a zirconacyclopentane with an indenyl ligand system is the compound *rac*-(ebi)Zr(C_4H_8) **21** (ebi = ethylenebisindenyl) prepared by the treatment of a tetrahydrofuran (THF) solution of (ebi)$ZrCl_2$ at –78 °C with magnesium

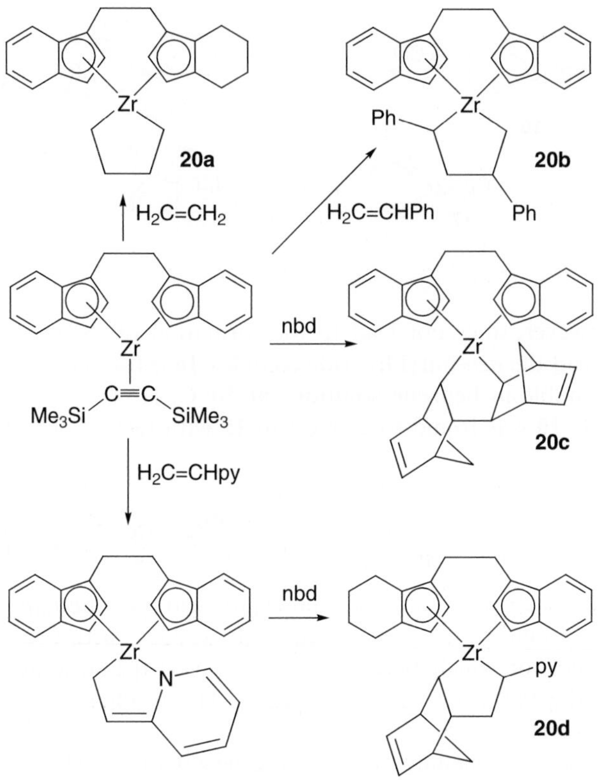

SCHEME 8.

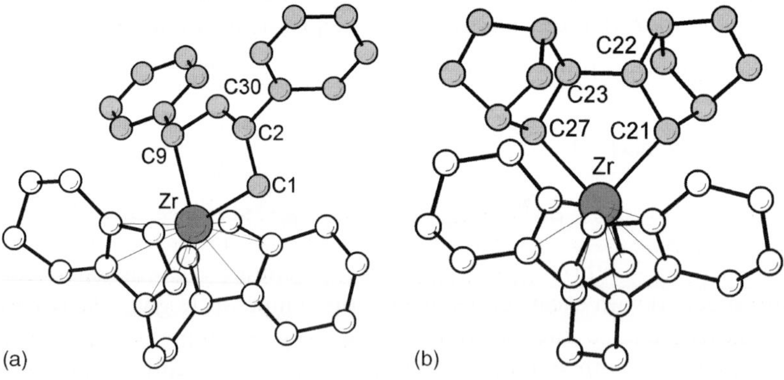

FIG. 4. Molecular structures of (a) **20b** and (b) **20c** (hydrogen atoms omitted).

powder in the presence of ethene and trimethylphosphine. Compound **21** was isolated as a light orange powder in 79% yield.[25] The structure of **21** was determined crystallographically (Fig. 5a).[25] Here, the Zr–C(α) bonds are 2.305 and 2.282 Å with a C(α)–Zr–C(α) angle of 84.78°, which is about 3° larger than normally found.

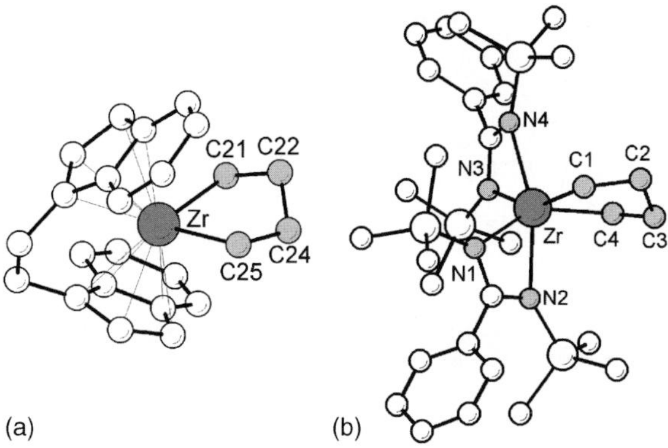

FIG. 5. Molecular structures of (a) **21** and (b) **22**(hydrogen atoms omitted).

Zirconacyclopentanes with *N,N′*-bis(trimethylsilyl)benzamidinato co-ligands have also been synthesised. Reduction of L_2ZrCl_2 ($L = PhC(NSiMe_3)_2$) by 1% Na–Hg in THF under 100 psi ethylene pressure affords the zirconacyclopentane $L_2Zr(C_4H_8)$ **22** as an orange crystalline solid in 67% yield [Eq. (7)].[26] The compound is reported to be remarkably stable, showing no decomposition after 1 year in a dry-box. Single X-ray diffraction studies of **22** (Fig. 5b) show the zirconacyclopentane ring to be twisted, with Zr–C_α bond lengths of 2.252 and 2.224 Å, virtually identical to that measured in the dialkyl complex L_2ZrMe_2 at 2.241 Å.[26]

Ph, SiMe3, N, Me3Si—N, Zr, Cl, Cl, Me3Si—N, N, SiMe3, Ph
Na(Hg)
H2C=CH2
Ph, SiMe3, N, Me3Si—N, Zr, Me3Si—N, N, SiMe3, Ph
22

(7)

Erker and co-workers[27] have shown that alkene groups bound pendantly to the indenyl rings of an *ansa* zirconocene or hafnocene may be coupled intramolecularly to provide metallacyclopentanes such as α,α′-disubstituted zirconacyclopentane **23** and **23c** that are tethered to the indenyl groups (Scheme 9, Fig. 6). In the zirconium case **23a**, the cyclisation is regioselective such that the tethers are bound to the α and α′ carbons of the metallacycle. For hafnium, however, a (ca. 1:1) mixture of the α,α′ **23b** and α,β′ **23c** regioisomers is obtained. Presumably, the regiochemistry observed for zirconium reflects arrival at the thermodynamic product *via* reversible cleavage of the zirconacyclopentane to its constituent alkenes, a supposition supported by the reaction of the metallacycle with but-2-yne to provide a zirconacyclopentane **24** with one liberated pentenyl group. The regioisomers of the hafnocyclopentane example do, however, interconvert at 65 °C to provide a 6:1 equilibrium mixture of **23b** and **23c**.

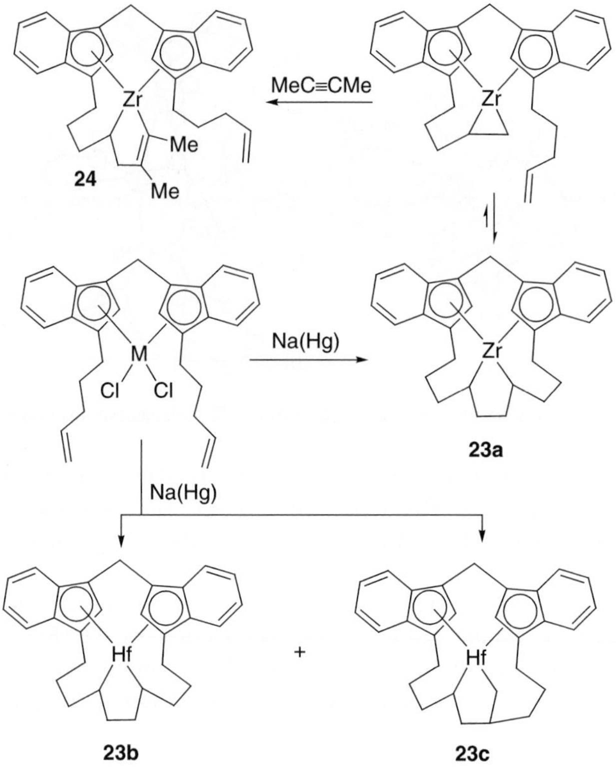

SCHEME 9.

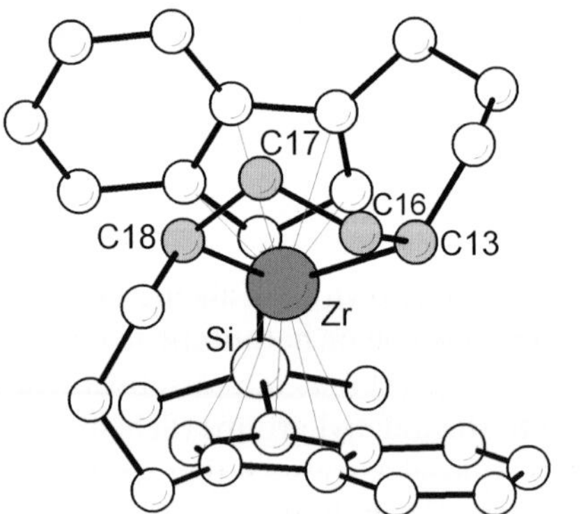

FIG. 6. Molecular structure of **23** (hydrogen atoms omitted).

Nugent has isolated and structurally characterised the zirconocyclopentanes derived from 1,6-heptadiene and either $Cp_2ZrCl_2/_2BuLi$ or $Cp^*ZrCl_3/Na(Hg)$. The former **25** is monomeric while the latter **25a** involved a $Zr_2(\mu\text{-}Cl)_2$ structure. However, of particular note is that while the Negishi reagent provides primarily (97:3)

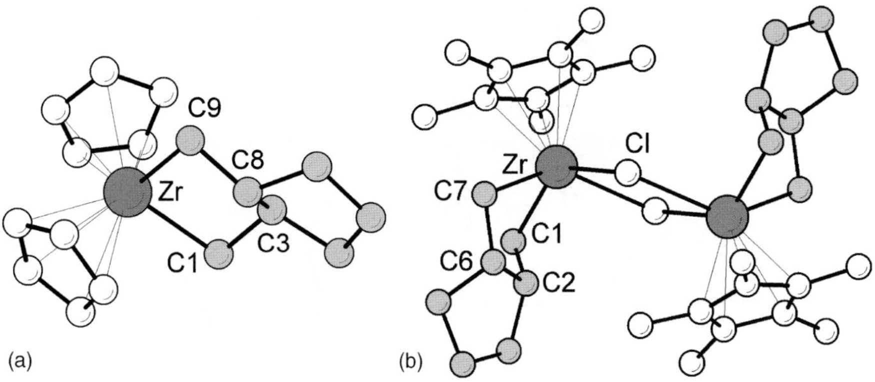

FIG. 7. Molecular structures of (a) **25** and (b) **25a** (hydrogen atoms omitted).

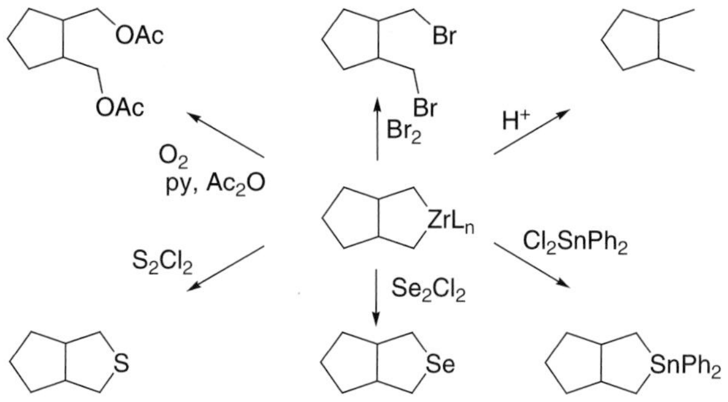

OAc
OAc
Br
Br
Br_2
H^+
O_2
py, Ac_2O
ZrL_n
Cl_2SnPh_2
S_2Cl_2
Se_2Cl_2
S
Se
$SnPh_2$

SCHEME 10.

the *trans*-fused bicycle (Fig. 7a), the half-sandwich variant results essentially (99:1) in the formation of a *cis*-fused metallabicycle (Fig. 7b).[28] These metallacyclopentanes are intermediates en route to a range of organic products obtained by stoichiometric electrophilic cleavage (Scheme 10).

A catalytic application of Nugent's cyclozirconation involves transfer of the metallacycle to magnesium (transmetallation), a reaction achieved through the use of butyl Grignard reagents which effectively regenerate Negishi's dibutylzirconocene reagent. In a study directed at elucidating the mechanism of the zirconium mediated cyclomagnesiation of α,ω-dienes, the reaction of Cp_2ZrCl_2 with ^{n}BuLi and 9, 9-diallylfluorene was shown to provide the structurally characterised zirconacyclopentane **26** (Fig. 8) capable of catalysing (10 mol%) the conversion of 1,7-octadiene and BuMgBr to cyclo-$C_6H_{10}(CH_2MgBr)_2$ (Scheme 11).[29]

A key observation was that the distribution of mono and dimagnesiated products was dependent upon the choice of magnesium alkyl, with dibutylmagnesium proving most effective for dimagnesiation. The stereoselectivity of the cyclozirconation step(s) has been rationalised with recourse to computational studies, the conclusion

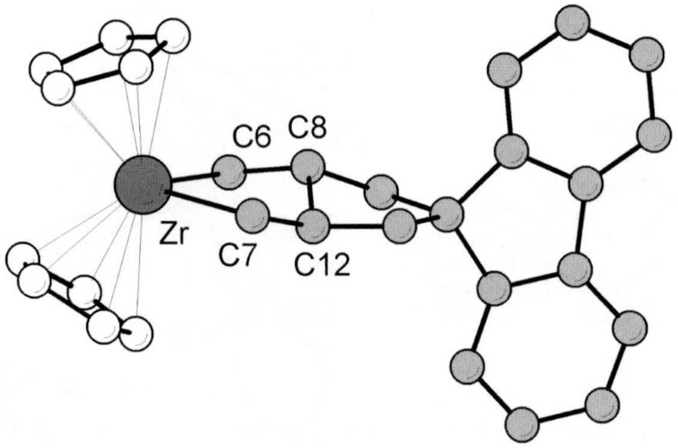

FIG. 8. Molecular structure of **26** (hydrogen atoms omitted).

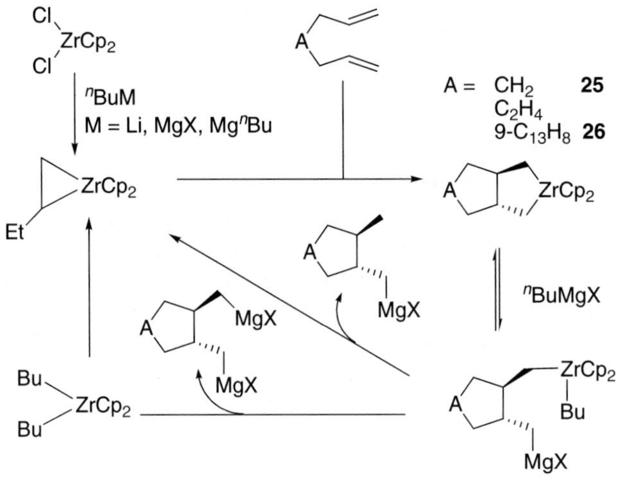

SCHEME 11.

being that the Cp*ZrCl derived metallacycle arises from kinetic control while the zirconocene derivative is more prone to retro-cycloaddition allowing thermodynamic control to operate *via* an easily established equilibrium. A further study of zirconocene mediated carbomagnesiation employing ethylmagnesium reagents revealed different processes depending on the conditions and whether EtMgCl or Et_2Mg dioxane was employed.[30] The former magnesium reagent provided the binuclear complex $[Cp_4Zr_2(Et)_2(\mu\text{-}C_2H_4)]$ while the latter, in THF, provided the mononuclear ethene complex $[Cp_2Zr(\eta\text{-}C_2H_4)(THF)]$, both of which were structurally characterised. In a similar manner, the complexes $[(Me_3SiCp)_4Zr_2(\mu\text{-}C_2H_4)]$, $[\eta^5\text{-}C_9H_7)_2Zr(\eta\text{-}C_2H_4)(THF)$ and $[Cp_2Zr(\eta\text{-}C_2H_4)(py)]$ were also obtained. The mononuclear ethene complexes reacted smoothly with norbornene to provide bicyclic zirconacyclopentanes.

The parent zirconacyclopentane **27a** can be synthesised by either the reaction of Cp_2ZrCl_2 with $BrMg(CH_2)_4MgBr$ as reported by Takahashi and co-workers or by warming dibutylzirconocene (Cp_2ZrBu_2) to room temperature under an ethene atmosphere:[32]

$ZrCl_2(C_5H_5)_2$ —$BrMgC_4H_8MgBr$→ **27a** ←$H_2C{=}CH_2$— Cp_2ZrBu_2 (8)

Negishi and co-workers[33] report that zirconacyclopentanes of type **27** react with benzaldehyde at or below 25 °C to give seven-membered oxazirconacycles **28**, which are then converted into alcohols on hydrolysis with HCl:

27 —O=CHPh→ **28** —HCl→ alcohol (9)

R = H **27a**; R = Et **27b** — R = H **28a**; R = Et **28b** — R = H (46%); R = Et (52%)

Similar results have been reported by Takahashi and co-workers.[34] These authors report that zirconacyclopentanes react with PhCHO under a slightly positive pressure of ethylene to form the oxazirconaheptacycle **28a** as the major product at room temperature, and that the reaction carried out in the absence of additional ethylene produces the five-membered oxazirconacyclopentane **29** instead [Eq. (10)]. Reaction of **27a** with benzaldehyde at higher temperatures of 50 °C was also found to yield the five-membered oxazirconacyclopentane as the major product.

27a —O=CHPh, - C_2H_4→ **29** —HCl→ HO–CH(Ph)Et (10)

Reaction of **27a** with lithiated *cis*-trimethylsilyl-1,2-epoxyoctane followed by aqueous work-up gave an alkenylsilane in 80% yield [Eq. (11)]. The proposed mechanism for this reaction is *via* a zirconacyclohexane intermediate.

27a + Li, $SiMe_3$ epoxide (R = C_6H_{13}) → zirconacyclohexane (H, LiO, R, $SiMe_3$) → H(R)C=C($SiMe_3$)Bu (11)

3. Hafnium

In contrast to the metallacyclopentanes based on titanocenes or zirconocenes, the Cp_2Hf analogues are sufficiently thermally stable to be handled at ambient temperature. $Cp_2Hf(C_4H_8)$ **30** was prepared in 71% yield by Erker *et al.*[35] by reaction of Cp_2HfCl_2 with 1,4-dilithiobutane at −78 °C. The hafnacyclopentane was found to be very stable, with a melting point of 126 °C. Compound **30** has also been prepared by the reaction of Cp_2HfCl_2 with $BrMg(CH_2)_4MgBr$.[36]

Thermolysis of a toluene solution of **30** at 90 °C in the presence of excess butadiene results in the liberation of 1 equivalent of ethene and the formation of the reactive $Cp_2Hf(\eta^2\text{-}CH_2{=}CH_2)$ intermediate, which subsequently reacts with butadiene to form a 2-vinylmetallacyclopentane complex **31**, which is however unstable under the reaction conditions and was isolated as the corresponding isomeric (η^3-allyl)hafnocene complex **32**:[35]

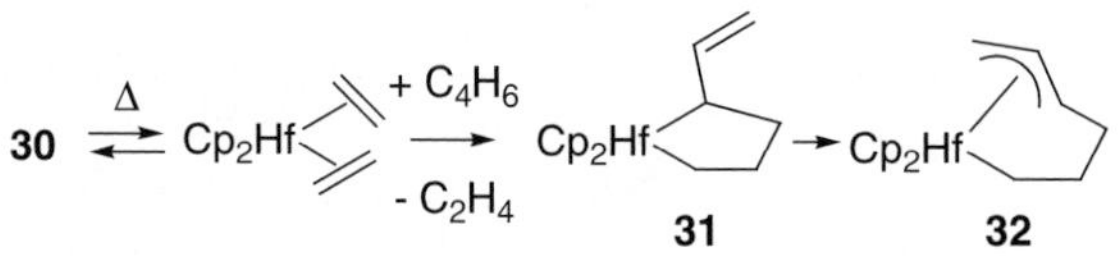

(12)

Bercaw and Moss[37] have reported the reaction of $Cp^*_2HfH_2$ with 1,4-pentadiene in a 1:1 ratio yields the hafnacyclopentane, $Cp^*_2Hf\{CH(CH_3)CH_2CH_2CH_2\}$ **33** as lemon yellow crystals in 57% yield. Compound **33** was also found to be the product of the slow reaction of the 1,7-dihafnaheptane $Cp^*_2(H)Hf(CH_2)_5Hf(H)Cp^*_2$ with excess 1,4-pentadiene at room temperature over a period of 1–4 days (Scheme 12).

As noted above the hafnacyclopentanes **23b** and **23c** have been obtained *via* the intramolecular coupling of pendant alkenyl groups tethered to indenyl ligands, with **23b** being the only structurally characterised hafnacyclopentane (Scheme 9).[29]

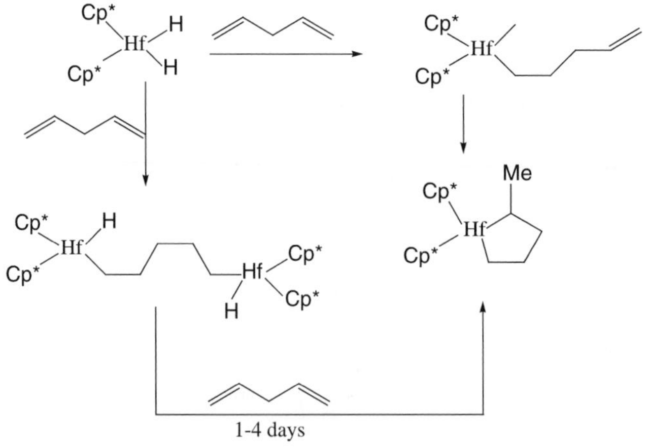

SCHEME 12.

B. *Group 5 Metallacyclopentanes*

Although there are few five-membered metallacycles reported for group 5, there are some examples of metallacyclopentane compounds for niobium and tantalum. The reaction of the labile vanadium naphthalene complex $[CpV(C_{10}H_8)]$ with ethene provided the unusual binuclear complex $[Cp_2V_2(\mu\text{-}C_4H_8)_2]$ (**34** [Eq. (13)], Fig. 9) in which the two vanadium centres are bridged by two butanediyl groups.[38]

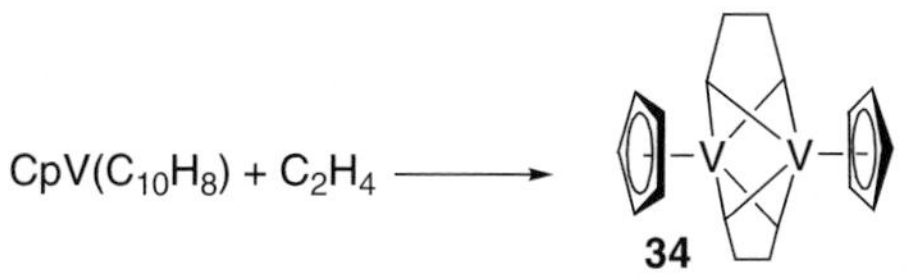

$$CpV(C_{10}H_8) + C_2H_4 \longrightarrow \mathbf{34} \tag{13}$$

A report by Schrock and co-workers[39] describes a route to the preparation of metallacyclopentane compounds of niobium and tantalum, using neopentylidene complexes that react with simple α-alkenes. In this report, $CpTaCl_2({=}CHCMe_3)$, when treated with ethylene, was reported to yield the metallacyclopentane $CpTaCl_2(C_4H_8)$ **35b** as yellow-orange crystals in 95% yield. The tantalacyclopentane was characterised by elemental analysis and ^1H and ^{13}C NMR spectroscopy. The ^{13}C NMR spectrum of **35b** in C_6D_6 shows resonances at 33.5 (triplet), 89.7 (triplet) and 112.8 ppm. In light of the fact that all shifts known for carbon atoms of coordinated ethylene lie between 20 and 60 ppm, downfield from $SiMe_4$, these data support the fact that the compound is, in fact, a metallacyclopentane. Also, the $^1J_{CH}$ coupling constants in this spectrum reflect those of sp^3 carbons rather than coordinated sp^2 carbon atoms, as would be the case in coordinated ethylene.

A second method subsequently reported for the synthesis of complexes of this type involves the use of the complexes $Cp^*Ta(\eta^2\text{-}H_2C{=}CHR)Cl_2$ (R = H, Me, ^{n}Pr) as starting materials [Eq. (14)]. These alkene complexes reacted with a further equivalent of ethene, propene or 1-pentene to yield the corresponding

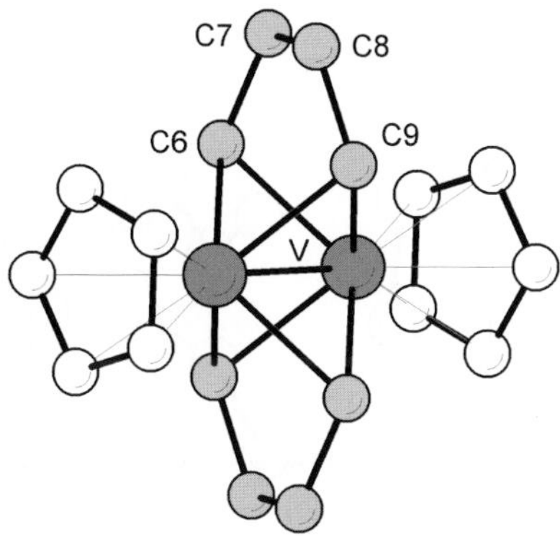

Fig. 9. Molecular structure of **34** (hydrogen atoms omitted).

metallacyclopentanes **36**, as orange crystals in ca. 80% yield, which were found to be in equilibrium with their alkene precursors.[31]

$Cp^*Ta(CH_2{=}CHR)Cl_2 \underset{}{\overset{H_2C{=}CHR}{\rightleftharpoons}}$ tantalacyclopentane (14)

R = H **36a**
R = Me **36b**
R = ^{n}Pr **36c**
R = ^{t}Bu **36d**
R = $(CH_2)_3$ **36e**

The $^{13}C\{^1H\}$ NMR spectrum of **36b** at 7 °C shows only three peaks corresponding to the metallacyclopentane fragment. Since the molecule does not possess a plane of symmetry, it is likely that a fluxional process takes place which converts C_α and $C_{\alpha'}$; C_β and $C_{\beta'}$; C_γ and $C_{\gamma'}$. Deuterium labelling studies described in this paper highlight the preferred mechanism of formation of the metallacyclic ring. For example, reaction of $Cp^*Ta(C_2H_4)Cl_2$ with hex-1-ene-1,1-d$_2$ yields a metallacyclic product, with both deuterium atoms on the C_α [Eq. (15)]. It has also been shown that no pair-wise or single deuterium scrambling occurs with this species, over time, to yield the analogous metallacycle, with the deuterium atoms in the C_β position.[40]

(15)

Churchill *et al.*[41] have reported the crystal structures of both **36a** (Fig. 10a) and the bicyclic tantalacyclopentane derived from 1,6-heptadiene **36e** (Fig. 10b). In this report, open envelop rather than puckered conformations are noted for these metallacyclopentanes.

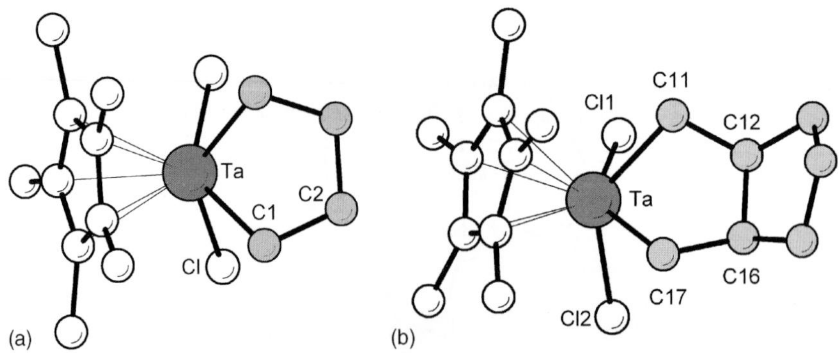

Fig. 10. Molecular structures of (a) **36a** and (b) **36e** (hydrogen atoms omitted).

C. Group 6 Metallacyclopentanes

1. Chromium

Emrich *et al.*[42] have reported the preparation of a chromacyclopentane in high yield (70–90%), using two different methods. The chromacyclopentane **37** can be synthesised by the reaction of the appropriate chromium dichloride with (i) 1,4-dilithiobutane or (ii) active Mg in the presence of ethene [Eq. (16)] (Fig. 11).

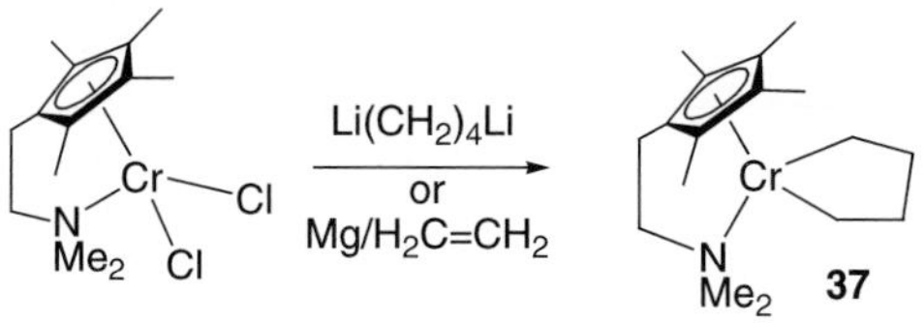

$$\text{Cr complex (Cr–Cl, Cl; N Me}_2\text{)} \xrightarrow[\text{Mg/H}_2\text{C=CH}_2]{\text{Li(CH}_2)_4\text{Li} \\ \text{or}} \text{Cr complex (chromacyclopentane; N Me}_2\text{)}\ \mathbf{37} \qquad (16)$$

The role of metallacycles, as intermediates in chromium-catalysed ethylene trimerisation, is highlighted in this article. It is also shown that when **37** is treated with ethene in a protonolysis reaction, hexane and butane are liberated (1:3 ratio hexane:butane). This observation gives strength to the argument that metallacycles are intermediates in chromium-catalysed ethylene trimerisation. Scheme 13 provides a simplified representation for chromium-based ethylene trimerisation.

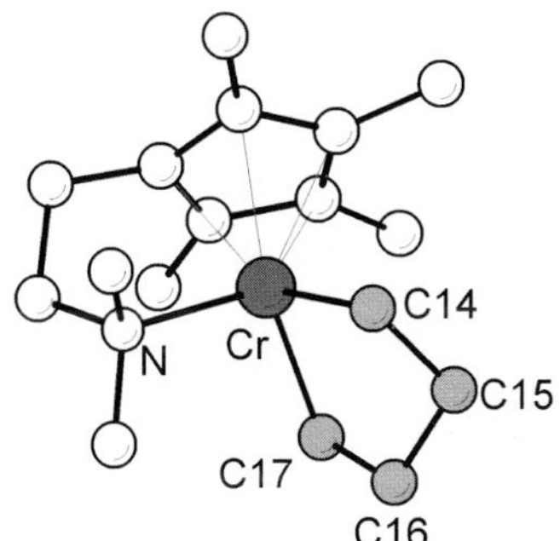

FIG. 11. Molecular structure of **37** (hydrogen atoms omitted).

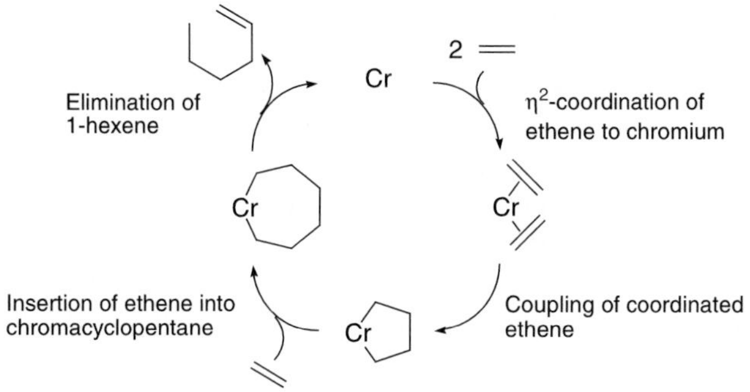

Cr
2 =
η^2-coordination of ethene to chromium
Cr
Coupling of coordinated ethene
Cr
Insertion of ethene into chromacyclopentane
Cr
Elimination of 1-hexene

SCHEME 13.

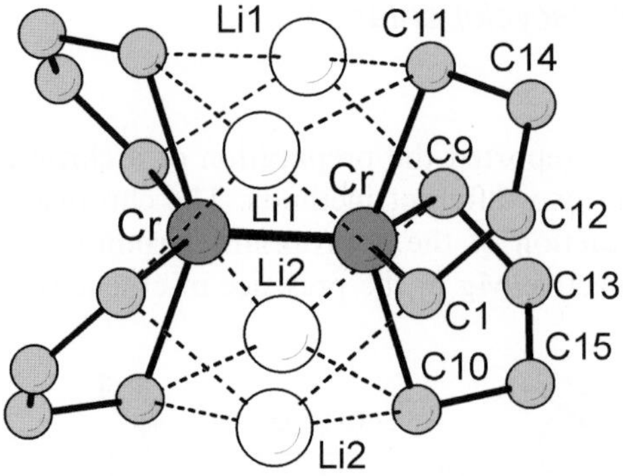

FIG. 12. Molecular structure of **39** (hydrogen atoms omitted).

The unusual chromacyclopentanes **38** arise from the reactions of the bis(allyl) complex [$Cp^*Cr(\eta^3,\eta^3\text{-}C_8H_{12})$] with internal alkynes in which the alkene group of the resulting polycyclic ligand also coordinates to the chromium centre, as was confirmed crystallographically for the diethyl derivative **38b**:

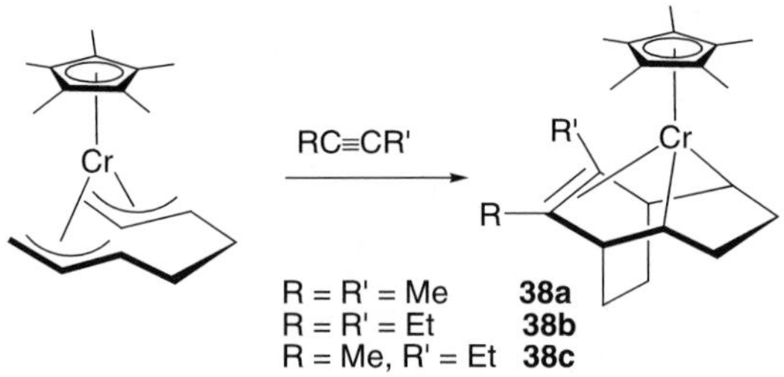

(17)

The single example of a homoleptic chromacyclopentane is provided by the bimetallic complex $Li_4(OEt_2)_4[Cr_2(C_4H_8)_4]$ **39** for which a formal Cr–Cr quadruple bond may be invoked (Fig. 12; Cr–Cr = 1.975(5) Å).[43,44]

2. Molybdenum

Using similar methodology to that for the tantalum metallacyclopentanes, molybdacyclopentane compounds of type **40** have recently been prepared.[45] In this reaction, an alkylidene ligand is lost from an appropriate molybdenum precursor *via* conventional metallacyclobutane formation and alkene metathesis. The resulting ethene complex then coordinates a second ethene to form a bis-ethylene complex that undergoes ring closure to form a metallacyclopentane complex **40** (Scheme 14). The conversion of the monoalkene complex to the metallacyclopentane was also monitored by NMR spectroscopy. It was observed that by increasing the concentration of ethylene, the conversion to the metallacyclopentane was promoted. In the presence of 120 equivalents of ethene, a 61% yield was obtained for the metallacyclopentane complex, compared with 6% upon addition of 8 equivalents of ethene. The reaction was also observed to be fully reversible.

The catalytic use of the mono-alkene molybdenum complex in converting $H_2C{=}CHSnR_3$ to $H_2C{=}CHCH_2SnR_3$ has also been investigated.[43] In this

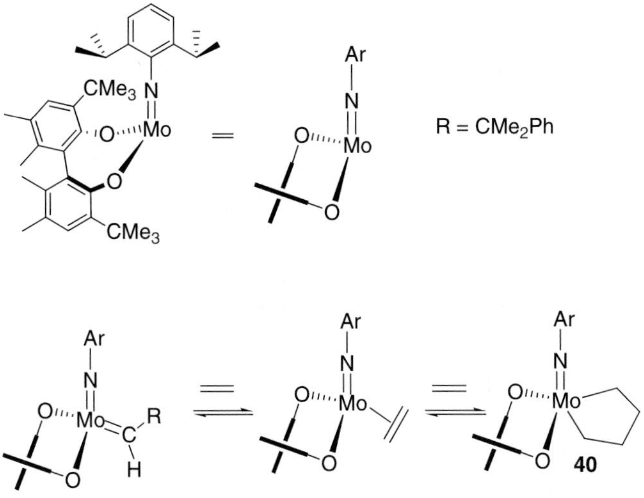

R = CMe₂Ph

SCHEME 14.

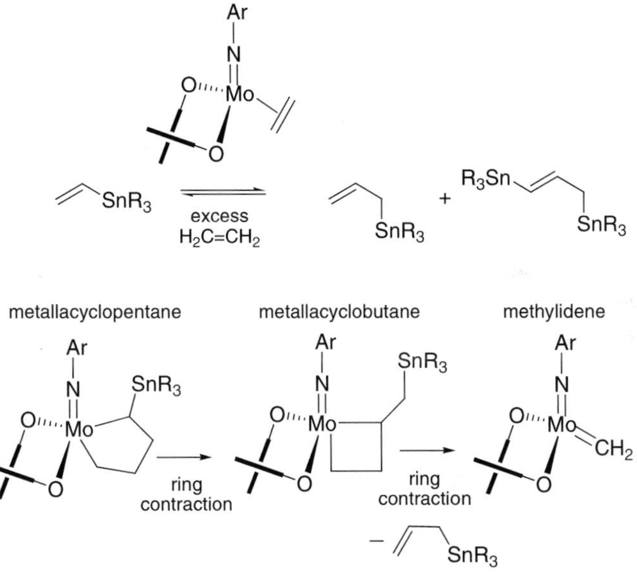

SCHEME 15.

instance, a ring-contraction mechanism is speculated to account for the formation of the allyl-tin product. This argument is supported by a similar occurrence in tantalum metallacycles.[39,40] A schematic representation of this ring-contraction mechanism is shown in Scheme 15, suggesting how the metathesis active methylene complex can form.

In a similar manner to the biphenolate examples **40**, the phenylenediamido derivative **41** could be obtained *via* the reactions of $[MoCl_2(THF)(=NPh)\{C_6H_4$

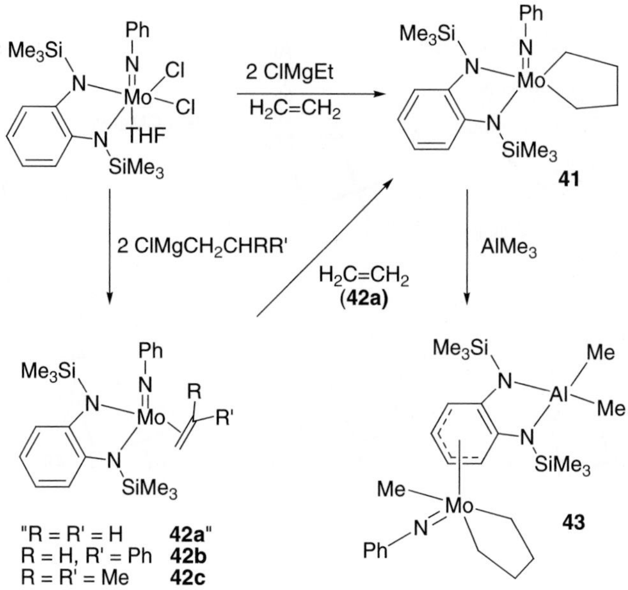

"R = R' = H **42a**"
R = H, R' = Ph **42b**
R = R' = Me **42c**

SCHEME 16.

$(NSiMe_3)_2\}]$ with either $C_4H_8(MgBr)_2$ or two equivalents of EtMgCl in the presence of excess ethene. The use of other Grignard reagents RMgCl (R = ^{n}Bu ^{s}Bu) afforded alkene complexes suggesting that $[Mo(C_2H_4)(=NPh)\{C_6H_4(NSiMe_3)_2\}]$ **42a** is a likely intermediate. Indeed, carrying out the reaction in the presence of PMe_3 results in the formation of the corresponding ethene complex $[Mo(C_2H_4)(=NPh)(PMe_3)_2\{C_6H_4(NSiMe_3)_2\}]$ (Scheme 16).[46]

The 'spectator' phenylenediamido ligand in **41** can be partially transferred to aluminium *via* treatment with $AlMe_3$, resulting in *tetrahapto* coordination of the benzene ring and transfer of one methyl group to molybdenum in the unusual complex **43** (Fig. 13a). The only other structurally characterised molybdenacyclopentane is the metallocene derivative **44a** (Fig. 13b) obtained from $[MoI_2(C_5H_5)_2]$ and $C_8H_8Li_2$ [Eq. (18)]:[47]

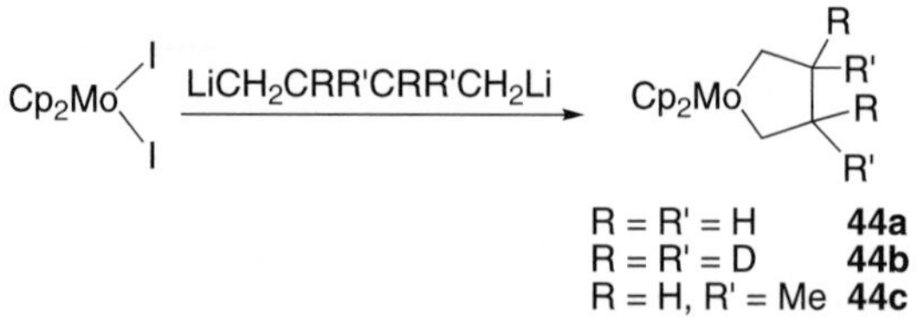

$$Cp_2MoI_2 \xrightarrow{LiCH_2CRR'CRR'CH_2Li} Cp_2Mo(CH_2CRR'CRR'CH_2) \tag{18}$$

R = R' = H **44a**
R = R' = D **44b**
R = H, R' = Me **44c**

3. Tungsten

A 1989 report by Chisholm *et al.*[48] showed that the reaction between $W_2(OR)_6$ (R = ^{i}Pr ^{c}Hex, CH_2^tBu) and ethene (1 atm) yielded the binuclear tungstacyclopentane compounds **45a**–**45c**. This study showed that it is possible to couple ethylene units at

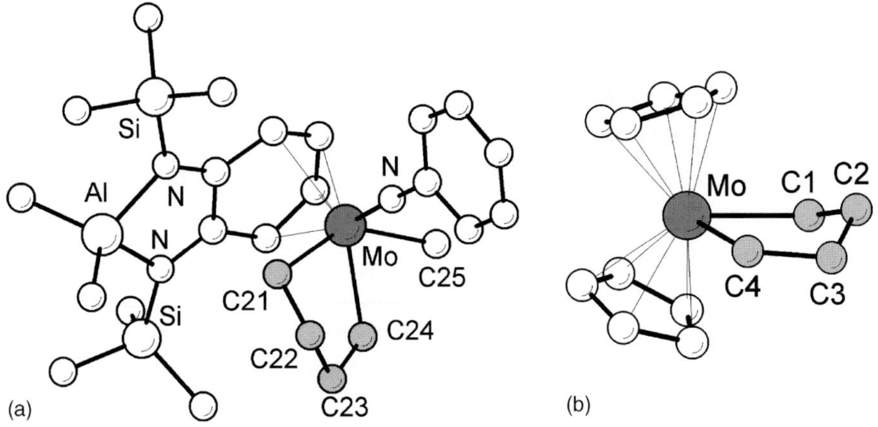

FIG. 13. Molecular structures of (a) **43** and (b) **44a** (hydrogen atoms omitted).

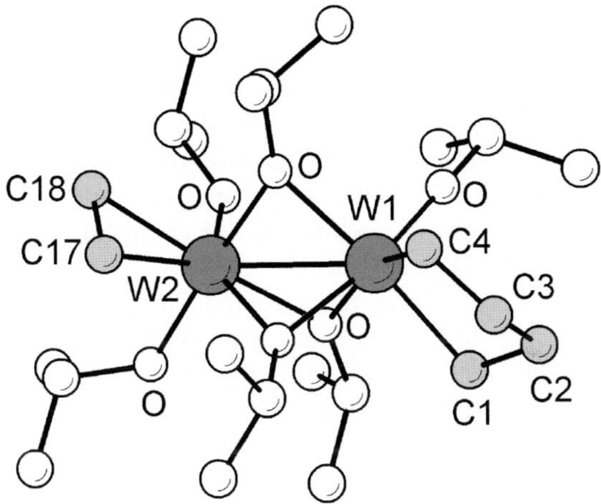

FIG. 14. Molecular structure of **45a** (hydrogen atoms omitted).

a binuclear (or polynuclear) centre. X-ray crystallographic studies show that the compound $W_2(OR)_6(C_4H_8)(\eta\text{-}C_2H_4)$ (**45a**, Fig. 14) to contain a metallacyclic ring, and a coordinated ethene ligand at different tungsten centres:

$$(RO)_3W\equiv W(OR)_3 \xrightarrow[\text{1 atm. / 0°C}]{H_2C=CH_2} (\eta\text{-}C_2H_4)(RO)_2W(\mu\text{-}OR)_2W(OR)(C_4H_8) \qquad (19)$$

$R = {}^iPr$ **45a**

$R = Cy$ **45b**

$R = CH_2{}^tBu$ **45c**

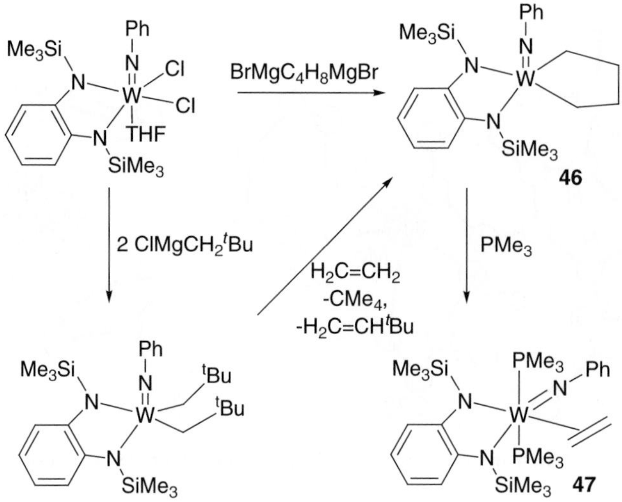

SCHEME 17.

Variable temperature $^{13}C\{H\}$ NMR studies on **45** show large coupling of the α-carbon atoms of the metallacyclopentane ring to the ^{183}W nucleus (70 and 86 Hz), in contrast to the smaller coupling associated with the coordinated ethene (33 and 37 Hz). A two-dimensional ^{13}C NMR shift correlation study confirmed the connectivity observed in the X-ray study, while ^{13}C–^{1}H chemical shift correlation experiments identified the ^{1}H NMR resonances.

Boncella and co-workers[46b] have reported the formation of a metallacyclopentane complex of tungsten **46** and a deactivation pathway for tungsten(IV) alkylidene complexes in alkene metathesis. In this report, an excess of ethene reacts with the precursor $[W(CH_2CMe_3)_2(=NPh)(PMe_3)_2\{C_6H_4(NSiMe_3)_2\}]$ to generate a tungstacyclopentane complex **46**, giving *neo*-pentane and *tert* butylethene as side products. The use of a di-Grignard reagent as the alkylating agent is also reported (Scheme 17). Treating **46** with an excess of PMe_3 results in the formation of the ethene complex **47**, however, this reaction is reversed by excess ethene under mild conditions.

Tungsten(IV) calix[4]arene fragments have been reported to enable ethene rearrangement reactions, and formation of metallacyclopentanes of tungsten **48** by Floriani and co-workers.[49,50] The rationale behind this work was an attempt to mimic processes well known with heterogeneous metal-oxide catalysts in the formation of metal carbon single, double or triple bonds, by using a calix[4]arene fragment as a framework to support the tungsten centres. It is shown in this report that high-valent alkene complexes are formed upon reduction of [*p*-*t*-Bu-calix[4] $(O_4)WCl_2$] in the presence of ethene or propene. These can be deprotonated to generate an alkylidyne complex, which undergoes reprotonation to form the alkylidene complex. The high-valent alkene complexes can undergo coupling with another alkene to form the metallacyclopentane **48**, thought to proceed *via* electron-transfer catalysis (Scheme 18). Photolysis of the **48** results in rearrangement to a

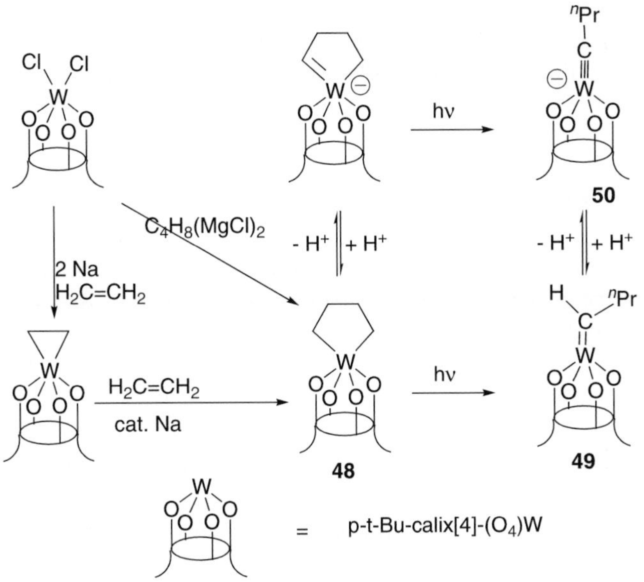

SCHEME 18.

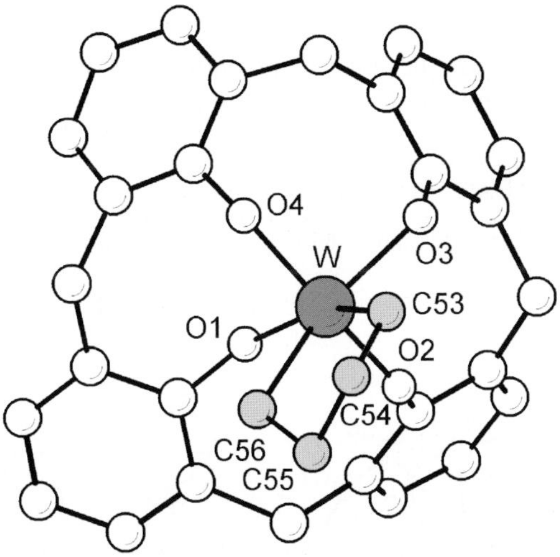

FIG. 15. Molecular structure of **48** (hydrogen atoms and tert Butyl groups omitted).

butylidene complex **49**, deprotonation of which generates the anionic butylidyne derivative **50**. This same butylidyne may be accessed by reversing the sequence, i.e., initial deprotonation followed by photolysis (Scheme 18). The same tungstacyclopentane (Fig. 15) may also be obtained *via* the reaction of [*p*-*t*-Bu-calix[4] $(O_4)WCl_2$] with $C_4H_8(MgBr)_2$.

It is also shown that it is possible to perform deprotonation reactions on the tungstacyclopentane and form an anionic species **51**, in which the unsaturated metallacyclic fragment forms a double bond with the tungsten centre.[49]

D. *Group 7 Metallacyclopentanes*

Reports on metallacyclopentane compounds of the group 7 transition metals are restricted to those of manganese and rhenium. To our knowledge, no metallacyclopentane compounds have been reported for technetium.

1. Manganese

Bis(ethene)manganacyclopentane complexes of the type $[M_A(THF)_x]_2[(C_4H_8)Mn(C_2H_4)_2]$ (M_A = K, Na) **52** have been synthesised by Jonas and co-workers by the reaction of CpMn(biphenyl) with ethylene at temperatures below –25 °C. The compounds were isolated as red crystals and found to be stable at room temperature. The crystal structure of $[Na(PMDETA)_2][(C_4H_8)Mn(C_2H_4)_2]$ (Fig. 16a, PMDETA = pentamethyldiethenetriamine) has been reported.[51] Reduction of manganocene by potassium naphthalenide in ethene saturated THF provides an uncharacterised compound formulated as $K[Mn(C_6H_{12})(C_{10}H_8)_x]$. The bonding of the C_6H_{12} group remains unknown, however, addition of pyridine results in further uptake of ethene to provide the spiro-manganabicyclononane derivative **53** featuring two manganacyclopentane units (Fig. 16b) [Eq. (20)].[51b]

$$\begin{array}{c} Cp_2Mn \\ + \\ 3\,K[C_{10}H_8] \end{array} \xrightarrow[\text{1 atm. / -65°C}]{H_2C{=}CH_2} \text{"}K[C_6H_{12}Mn(C_{10}H_8)_?]\text{"} \xrightarrow{py} \underset{\mathbf{53}}{[K(py)_2][Mn(C_4H_8)_2(py)]} \qquad (20)$$

(a) (b)

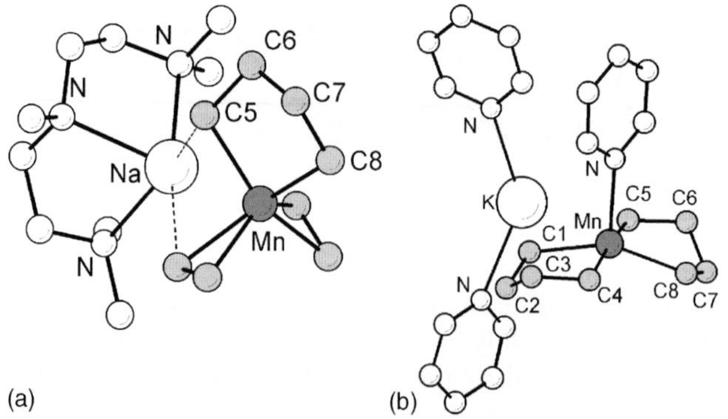

FIG. 16. Molecular structures of (a) **52** and (b) **53** (hydrogen atoms omitted).

2. Rhenium

The earliest report of a rhenacyclopentane was that by Yang and Bergman[52] in 1983. In this report, $CpRe(CO)_2H_2$ was treated with $I(CH_2)_4I$ in THF at room temperature in the presence of the base 1,8-diazabicyclo[5.4.0]undec-7-ene (DBU) to yield the rhenacyclopentane **54** (Fig. 17) as air-stable yellow-crystals in 53% yield. In this synthesis, the low pK_a values of the dihydride of $CpRe(CO)_2H_2$ are exploited to allow attachment of the tetramethylene group to the metal centre. Thermolysis of the rhenacyclopentane at 100 °C resulted in the formation of $CpRe(CO)_3$ (50%), methylcyclopropane (40%) and traces of 1-butene. The authors propose a decomposition pathway *via* loss of CO and/or $\eta^5 \leftrightarrow \eta^3$ isomerisation of the Cp ligand (Scheme 19).

A crystal structure has also been reported for the $CpRe(CO)_2(C_4H_8)$ metallacycle by Yang and Bergamn[53] showing the commonly encountered δ/λ disorder at the β carbon atoms of the metallacyclopentane ring (Fig. 17).

More recently, a bis(triflate) route has been used by Lindner and Wassing[54] to prepare metallacyclopentanes of rhenium. Similar to the report by Yang and Bergman[52] this route exploits the acidity of the hydride $CpRe(CO)_2H_2$, but instead of

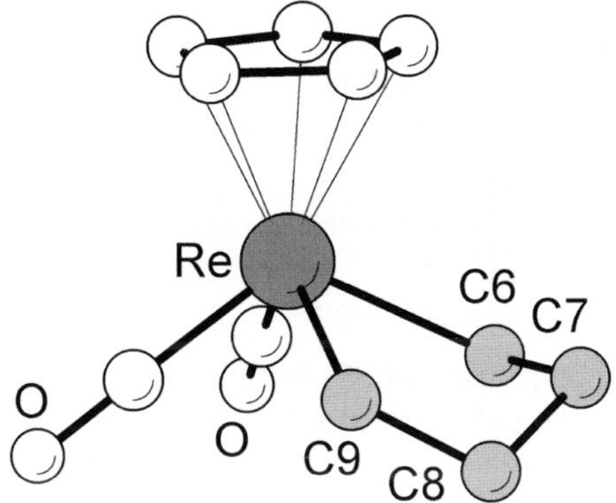

FIG. 17. Molecular structure of **54** (hydrogen atoms omitted).

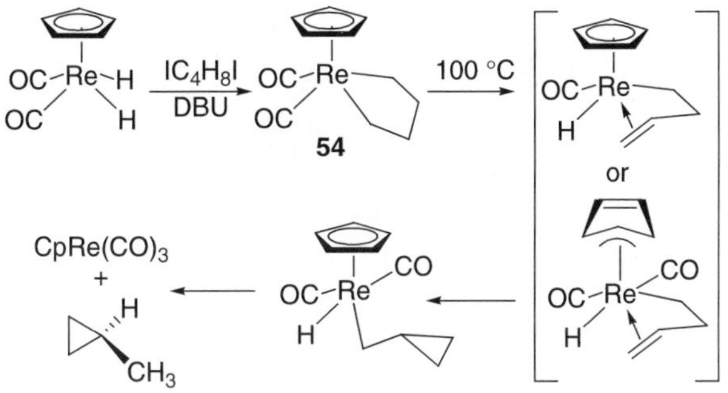

SCHEME 19.

using $I(CH_2)_4I$ as the alkylating agent, $(OTf)(CH_2)_4(OTf)$ is used. The rationale behind using the triflate as a leaving group arose from the desire to make larger metallacycloalkanes of rhenium, which will be discussed later in the review.

E. *Group 8 Metallacyclopentanes*

The dianion route has been used effectively, particularly by Lindner and co-workers[55,56] to prepare metallacyclopentanes for all three metals of the triad. This approach has been employed to provide the iron derivatives $[Fe(C_4H_8)(CO)_3L]$ (L = CO **55a**, PPh_3 (structurally characterised, Fig. 18) **55b**, $P(OMe)_3$ **55c**),[57] from the reaction of the corresponding dianion $[Fe(CO)_3L]^{2-}$. The parent compound **55a** is significant in three respects: firstly, it recalls Stone's archetypal $[Fe(C_4F_8)(CO)_4]$ **56**;[58] secondly, it had previously been discussed from a theoretical standpoint by Hoffmann and co-workers[59] and finally, it is an intermediate in ethene carbonylation mediated by $Fe(CO)_5$. Although **55a** is unreactive towards SO_2, the more electron-rich derivative **55b** inserts SO_2 into the Fe–C bond *trans* to the phosphine to provide a metallacyclic sulfinato-*S* complex:

$$[ML(CO)_3]^{2-} \xrightarrow{C_4H_8(OTf)_2} (OC)_3M(C_4H_8)L \xrightarrow[(55b)]{SO_2} (OC)_3Fe(C_4H_8SO_2)(PPh_3) \tag{21}$$

M = Fe, L = CO **55a**
M = Fe, L = PPh_3 **55b**
M = Fe, L = $P(OMe)_3$ **55c**
M = Ru, L = CO **55d**
M = Os, L = CO **55e**

The metallacyclopentanes **55d** and **55e** were characterised by elemental analysis, mass spectrometry, IR, ^1H and ^{13}C NMR.[55–57] In general, the osmium compounds were found to be much more stable than their ruthenium analogues,[60] both decompose on heating to give *cis*- and *trans*-2-butenes, and in the presence of CO

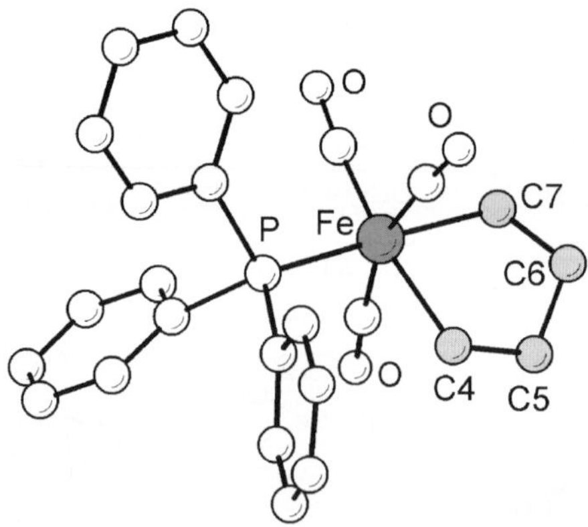

FIG. 18. Molecular structure of **55b** (hydrogen atoms omitted).

provide cyclopentanones. Norton and co-workers showed that, whereas $(CO)_4Os(C_3H_6)$ reacted with CO at 63 °C in CH_2Cl_2, there was no reaction with $(CO)_4Os(C_4H_8)$ under similar conditions,[60] thus demonstrating that chemical reactivity depends on the metallacycloalkane ring size.

Substituted metallacyclopentanes have also been prepared by this route, e.g., for the bicycles **57**[44] and *cis*- or *trans*-**58** [Eq. (22)], the *cis* isomer having been structurally characterised.[61]

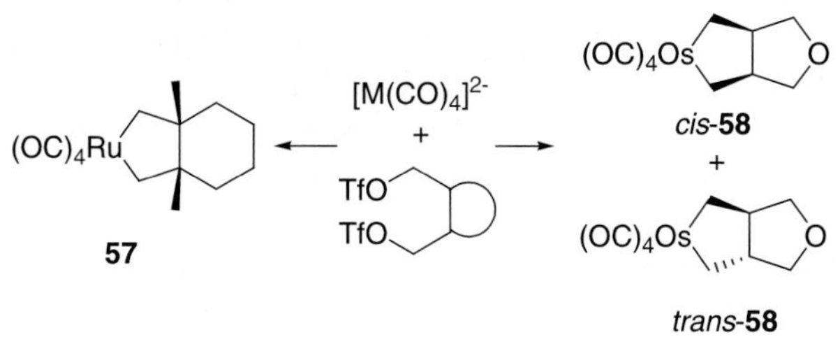

(22)

The reactions of $Ru(CO)_4(C_2H_4)$ with activated alkynes have been shown in some cases to yield metallacyclopentanes as well as metallacyclopentenes and metallacyclopentadienes, depending on the nature of the activating group (CF_3 *vs.* CO_2Et)[62] (Scheme 20).

Ruthenacyclopentanes have also been prepared by using the di-lithio route in combination with $[RuCl_2(PMe_xPh_{3-x})(\eta\text{-}C_6Me_6)]$ ($x = 0,1,2$) [Eq. (23)]. Each of the 'three legged piano stool' type derivatives was structurally characterised; the molecular geometry of **59c** is depicted in Fig. 19. The complexes react with the hydride abstracting reagent $[Ph_3C]BF_4$ to provide η^3-allyl derivatives **60**.[63]

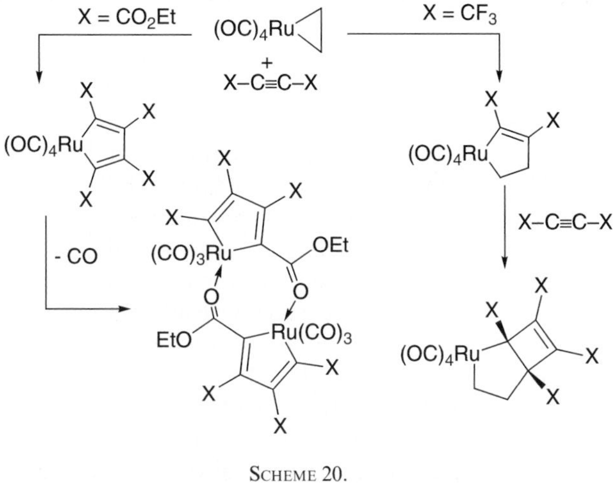

Scheme 20.

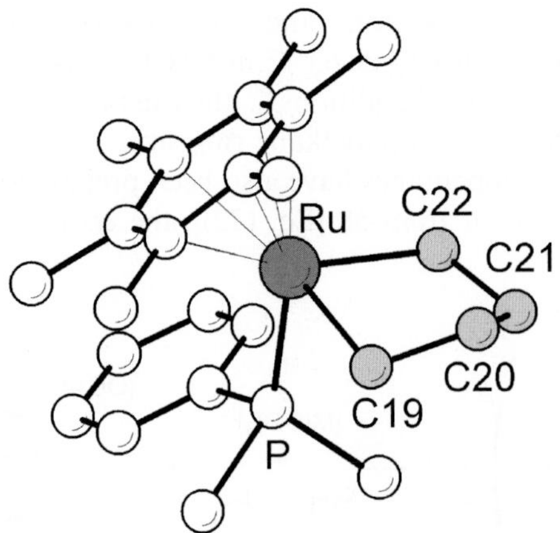

FIG. 19. Molecular structure of **59c** (hydrogen atoms omitted).

$C_4H_8Li_2$ Ph_3C^+

L = PMe_3 **59a**
L = PMe_2Ph **59b**
L = $PMePh_2$ **59c**

60

(23)

Reactions of $CpRuBr_2(\eta^3\text{-}CH_2C(R)CH_2)$ (R = H or Me) with the di-Grignard reagent $BrMg(CH_2)_4MgBr$ were expected to yield the corresponding metallacyclopentanes; however, Ru(II)–alkylbutadiene–neobutyl complexes were ultimately isolated instead. Their formation is, however, consistent with the intermediacy of the expected ruthenacyclopentanes, an interpretation supported by labelling studies employing $BrMgCD_2CH_2CH_2CD_2MgBr$ and $BrMgCH_2CD_2\ CD_2CH_2CD_2MgBr$, which confirmed that the hydrogen atoms were transferred from β-carbons of the putative ruthenacyclopentane (Scheme 21).[64] Computational studies on the 16-electron ruthenacyclopentane $CpRuCl(C_4H_8)$, and related platinum complexes, have been carried out and suggest that β-elimination from the ruthenium compounds is thermodynamically and kinetically feasible. In contrast, β-elimination from the 16-electron platinum species $Pt(PR_3)_2(C_4H_8)$ was found to be very unfavourable.[65]

Cowie and co-workers[66] have isolated a novel example of a binuclear osmacyclopentane **61** (Fig. 20) in which the metallacycle forms a bridge between osmium and rhodium *via* an agostic C–H–Rh interaction. The osmacyclopentane unit is constructed *via* the addition of excess diazomethane to either $[OsRh(CO)_4(dppm)_2]OTf$ or $[OsRh(\mu\text{-}CH_2)(CO)_4(\mu\text{-dppm})_2]OTf$ at low temperature. At ambient temperatures, however, the salt $[OsRh(CH_3)(C_3H_5)(CO)_3(\mu\text{-dppm})_2]OTf$ (**62**) is obtained. Isotopic labelling studies with $^{13}CH_2N_2$ point towards the involvement of a propylene bridged intermediate $[OsRh(\mu{:}\sigma,\sigma'\text{-}CH_2CH_2CH_2)(CO)_4(\mu\text{-dppm})]OTf$, which at room temperature undergoes competitive β-H elimination to generate an allyl ligand (Scheme 22).

SCHEME 21.

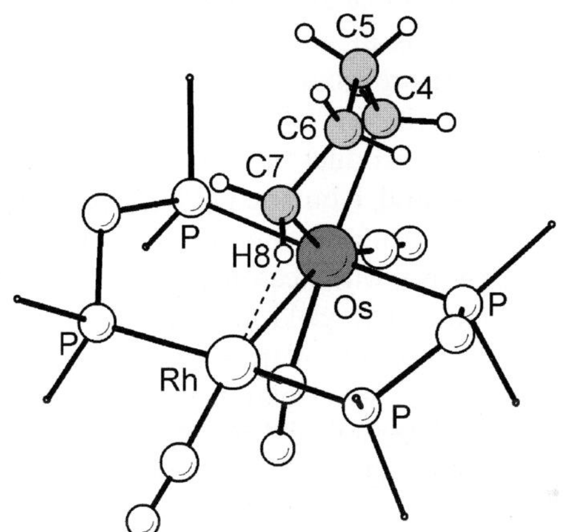

FIG. 20. Molecular structure of **61** (hydrogen atoms omitted).

SCHEME 22.

F. *Group 9 Metallacyclopentanes*

1. Cobalt

The synthesis of the cobaltacyclopentane compound $CpCo(PPh_3)(C_4H_8)$ **63** was first reported by Diversi *et al.* in 1977. Reaction of $CpCo(PPh_3)I_2$ with BrMg $(CH_2)_4MgBr$, or magnesiacyclopentane $Mg(CH_2)_4$, as the alkylating agent in a 1:3 molar ratio in diethyl ether produced **63** as orange-red crystals in ca. 15–20% yield (Scheme 23).[67] The molar ratio of alkylating agent:transition metal was found to be critical. At values lower than 3, only traces of the metallacycle were obtained along with large amounts of unidentified halogenated metal compounds.[67b]

Compound **63** has been structurally characterised by single crystal X-ray diffraction (Fig. 21a).[52,53] The geometry of the five-membered metallacyclic ring is puckered with notably different Co–C(β) distances. The C–C bonds are also observed to be significantly shorter than the standard $(sp^3)C–(sp^3)C$ bond length. Compound **63** was found to be stable at room temperature under an N_2 atmosphere and could be stored for months at $-30\,°C$.[67] However, in benzene solution, **63** decomposes at room temperature giving a mixture of 1-butene (11%), *trans* 2-butene (64%) and *cis* 2-butene (25%).[68] Reaction of **63** with CO at 44 atm pressure afforded cyclopentanone as the product in 33% yield, while reaction with ethylene at 1 atm gave 1-hexene (65%) and 1-butene (11%) as the products.[68]

The 1,4-dicyanocobaltacyclopentane complex **64**, prepared by the reaction of $[CpCo(PPh_3)(\eta^2\text{-}CH_2{=}CHCN)]$ with acetonitrile, was isolated as a mixture of the *cis* and *trans* isomers in 4% and 10% yield, respectively [Eq. (24)]. The *trans*-isomer was recovered as brown-red crystals with a melting point of 164 °C (decomp.) compared to the *cis*-isomer, which was isolated as orange-red crystals with a slightly lower melting point of 153 °C (decomp.).[69] Both isomers are thermally more stable than the unsubstituted analogue **63**.

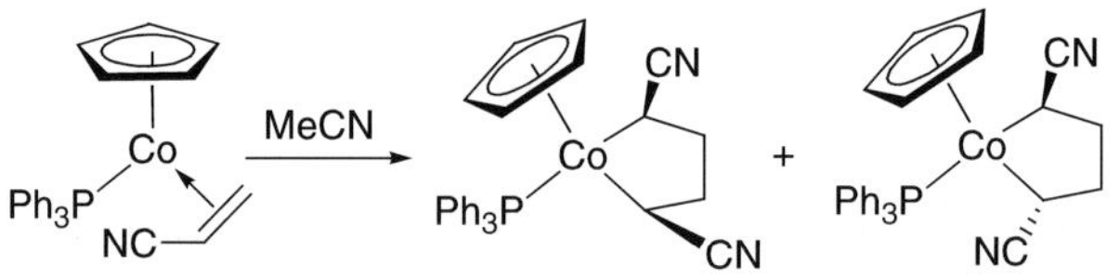

Co Ph3P NC MeCN → cis-64 + trans-64 (24)

Co Ph3P I I → C4H8Mg → 63 → R.T. C6H6 → 64%, 25%, 11%; C2H4 → 65% + 11%; CO → O 33%

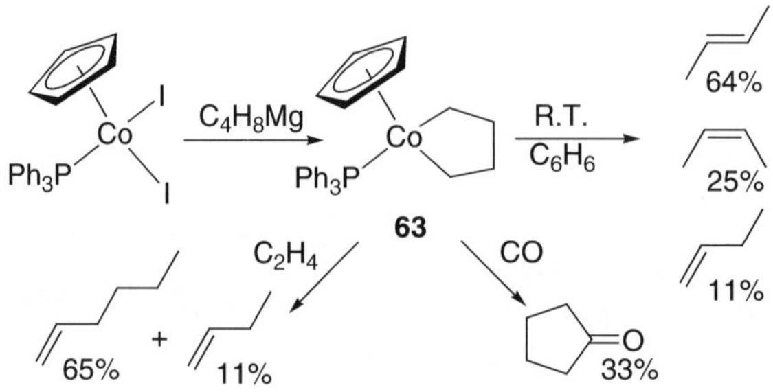

SCHEME 23.

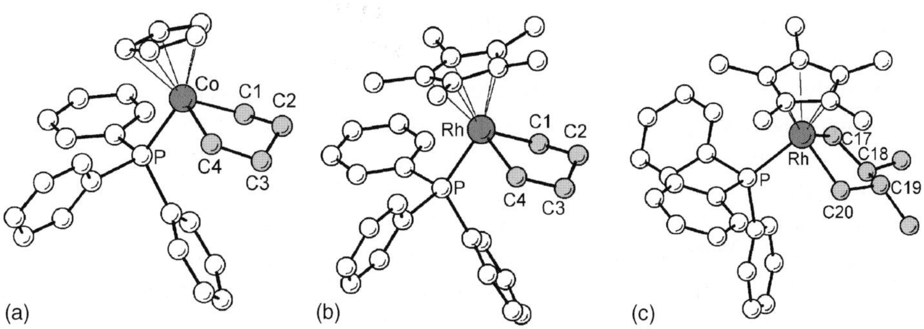

Fig. 21. Molecular structures of (a) **63**, (b) **65a** and (c) **65c** (hydrogen atoms omitted).

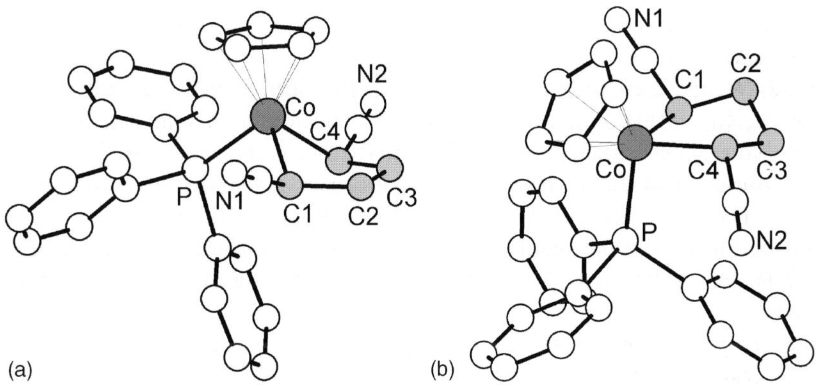

Fig. 22. Molecular structures of (a) *cis*-**64** and (b) *trans*-**64** (hydrogen atoms omitted).

The two stereoisomers of **64** were characterised by single crystal X-ray analysis. The cobaltacyclopentane ring in the *cis*-isomer (Fig. 22a) has an open-envelope conformation while that of the *trans*-isomer (Fig. 22b) is puckered.[69] The Co–C(α) distances are reported to be approximately equal to that found in the unsubstituted analogue **63**, while the Co–P distance is distinctly longer. The open-envelope conformation in the *cis*-isomer results from the steric requirements of the cyclopentadienyl and triphenylphosphine ligands. The cone angles of the Cp and PPh_3 ligands are 136° and 145°, respectively, resulting in an available cone angle at the metal centre of approximately 80° for the metallacyclic ring.

2. Rhodium

The rhodapentacycle, $Cp^*Rh(PPh_3)(C_4H_8)$ **65a** was formed by the reaction of $Cp^*Rh(PPh_3)Cl_2$ with $Li(CH_2)_4Li$ in diethyl ether or alternatively by the reaction of $Cp^*Rh(PPh_3)I_2$ with $BrMg(CH_2)_4MgBr$ in THF.[67] The substituted rhodapentacycles **65b** and **65c** were prepared by the reaction of the alkylating agents

$LiCH_2CHRCHR'CH_2Li$ with $Cp^*Rh(PPh_3)Cl_2$ in diethyl ether [Eq. (25)]. Compounds **65b** and **65c** were isolated as orange-yellow crystals in 15% and 22% yield, respectively.[67] The X-ray crystal structures of the rhodapentacycles **65a** and **65c** have been reported by Diversi *et al.*[70] (Fig. 21).

$Li_2C_4H_6R_2$

R = R' = H **65a**
R = H, R' = Me **65b**
R = R' = Me **65c**
(25)

3. Iridium

The iridapentacycle $Cp^*Ir(PPh_3)(C_4H_8)$ **66a** has been prepared by the reaction of $Cp^*Ir(PPh_3)Cl_2$ with the di-Grignard $BrMg(CH_2)_4MgBr$ or magnesacyclopentane in THF. No reaction takes place when diethyl ether is used as solvent. Compound **66a** was obtained in 25% yield as yellow-green crystals with a melting point of 209 °C.[67b] The substituted iridapentacycles **66b–d** were similarly prepared by the reaction of the alkylating agent $ClMgCH_2CHRCHR'CH_2MgCl$ and $Cp^*Ir(PPh_3)Cl_2$ in THF and isolated as yellow crystals in 20%, 45% and 5% yield, respectively [Eq. (26)].[69]

$(BrMg)_2C_4H_6R_2$

R = R' = H **66a**
R = H, R' = Me **66b**
R = R' = Me **66c**
R = H, R' = ^{i}Pr **66d**
(26)

The X-ray crystal structures of the iridapentacycles **66b** (Fig. 23a) and **66c** (Fig. 23b) were reported by Diversi *et al.*[70] Results show that puckering of the metallacyclopentane ring is dependant on the nature of the substituents on the β-carbon of the ring. The degree of puckering of the ring is higher in the dimethylated complexes **65c** and **66c** than it is in **66b**.

The abstraction of a β-hydrogen from the iridacyclopentane ring was investigated by Diversi *et al.*[70,71] It was found that hydride abstraction occurs regiospecifically to form an intermediate σ-3–butenyl cationic intermediate, which rapidly isomerises to the cationic η^3-allyl complex [Eq. (27)]. Compounds **65a–c** and **66a–c** react under very mild conditions with (i) $BF_3 \cdot Bu^n_2O$ in Bu^n_2O at −78 to 20 °C or (ii) Ph_3CPF_6 or Ph_3CBF_4 in CH_2Cl_2 at room temperature to afford the π-allyl complexes as crystalline solids in high yield:[70]

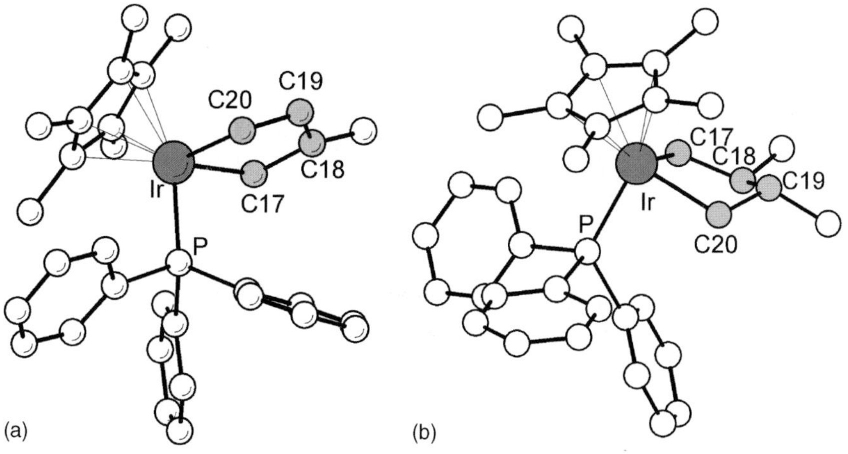
C20
C19
C18
C17
Ir
P
(a)
C17
C18
C19
C20
Ir
P
(b)

FIG. 23. Molecular structures of (a) **66b** and (b) **66c** (hydrogen atoms omitted).

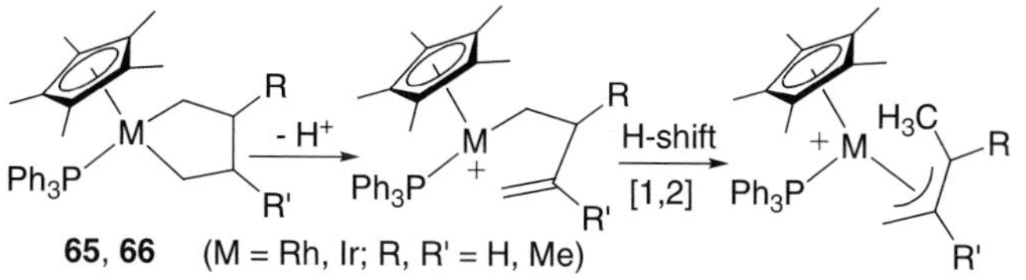
R
M
Ph3P
R'
- H+
R
M
+
Ph3P
R'
H-shift
[1,2]
H3C
+ M
R
Ph3P
R'
65, 66 (M = Rh, Ir; R, R' = H, Me)

(27)

The thermal decomposition of compounds **65a** and **66a** was investigated by differential scanning calorimetry (DSC) and thermogravimetric analysis (TGA).[72] The results show that both compounds decompose on melting, with decomposition occurring in three steps. Decomposition of **65a** is characterised by the release of the C_4H_8 moiety, followed by the C_5Me_5 group and finally the release of the phenyl groups of the PPh_3 ligand. The mass of the residue at 800 °C was found to correspond to the sum of rhodium and phosphorus. A similar result was obtained for the decomposition of **66a**, however, here the third step observed at 830–1000 °C corresponded to the release of PPh_3 and the mass of the residue corresponded to that of metallic iridium.[72]

The dimethylthiophene complex $[Ir(\eta^5\text{-}SC_4H_2Me_2)(\eta\text{-}C_5Me_5)]$ reacts with maleic anhydride to provide both the simple adduct $[Ir(\sigma\text{-}SC_4H_2Me_2)(\eta^2\text{-}C_4H_2O_3)(\eta\text{-}C_5Me_5)]$ and an unusual chelated tricyclic iridacyclopentane **67** (Fig. 24) comprising the thiophene and maleic anhydride units in addition to an oxygen atom, the origin of which remains obscure, although since **67** is also the product of the reaction of the acylallyl complex **68** with maleic anhydride, it is most likely that adventitious oxidation of the precursor to form **68** precedes the anhydride addition (Scheme 24).[73]

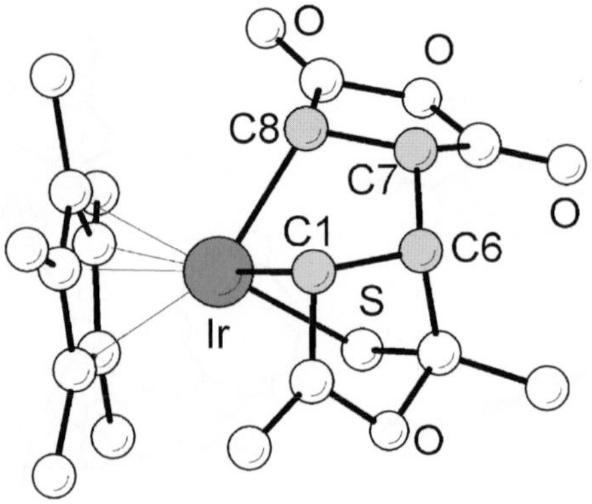

FIG. 24. Molecular structure of **67** (hydrogen atoms omitted).

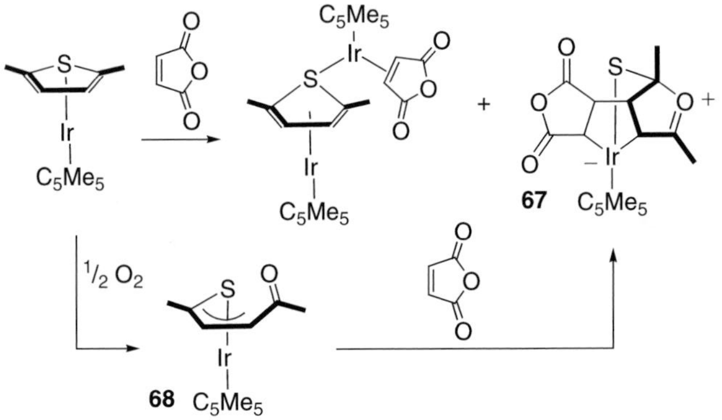

SCHEME 24.

G. *Group 10 Metallacyclopentanes*

The chemistry of group 10 metallacyclopentanes is well established, with metallacyclic complexes of nickel and platinum having been extensively investigated, particularly by the groups of Grubbs[74–79] and Whitesides[80–82] in the 1970s.

1. Nickel

The synthesis and reactivity of a series of nickelacyclopentane compounds with a range of tertiary phosphine ligands was reported by Grubbs *et al.* Compounds of type **69** were prepared by the reaction of 1,4-dilithiobutane with the appropriate dichlorobis(tertiaryphosphine)nickel(II) complex, and isolated as yellow crystals in ca. 40% yield[78] (dppe = bis(diphenyl) (phosphinoethane).

$$L_2NiCl_2 \xrightarrow{Li_2C_4H_8} L_2Ni(C_4H_8) \qquad (28)$$

L = PPh_3 **69a**
L = PCy_3 **69b**
L = P^nBu_3 **69c**
L_2 = dppe **69d**
L = PPh_2Bz **69e**
L = PPh_2Me **69f**

Thermolysis of these complexes at 9 °C produced ethylene, cyclobutane, and butenes. The ratio of the gaseous products was found to be a function of the coordination number of the complex, or intermediate. Thus three coordinate complexes favoured butene formation, while four coordinate complexes favoured reductive elimination to form cyclobutane, and five coordinate complexes produced ethylene as shown in Scheme 25.[83]

The dilithiobutane route was also successfully applied to the synthesis of the homoleptic 'ate' complex $Li_2(solv)_x[Ni(C_4H_8)_2]$ (**70a**, solv = THF, Et_2O, TMEDA) (TMEDA = *N,N,N′,N′*,-tetramethylethylenediamine), which was obtained from the reaction of LiC_4H_8Li with either nickelocene or $K[Ni(NPh_2)_3]$ [Eq. (29)].[84] The permethylated derivative $Li_2(TMEDA)_2[Ni(C_4H_4Me_4)_2]$ **70b** was also obtained and structurally characterised (Fig. 25).[85]

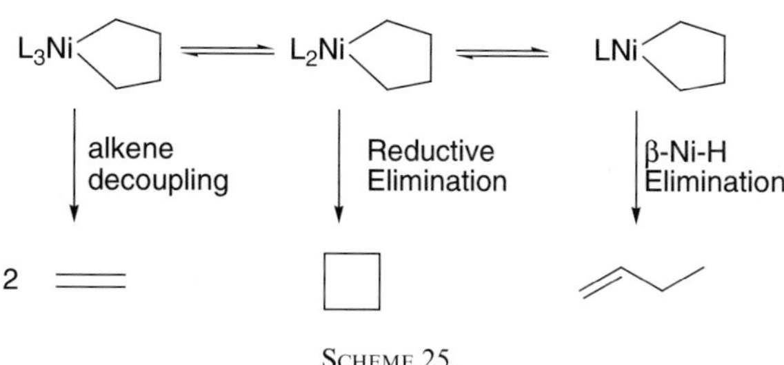

SCHEME 25.

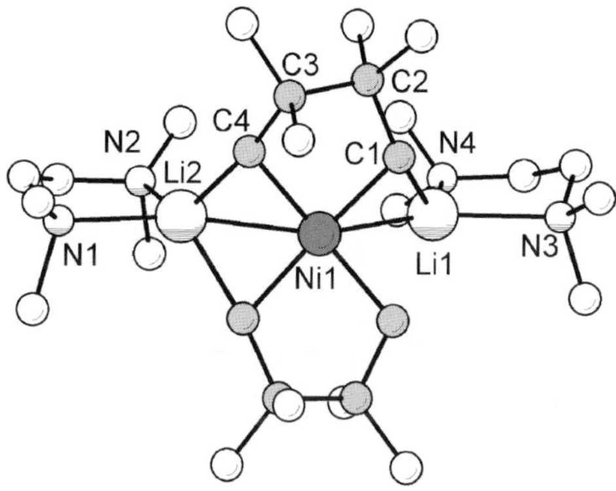

FIG. 25. Molecular structure of **70b** (hydrogen atoms omitted).

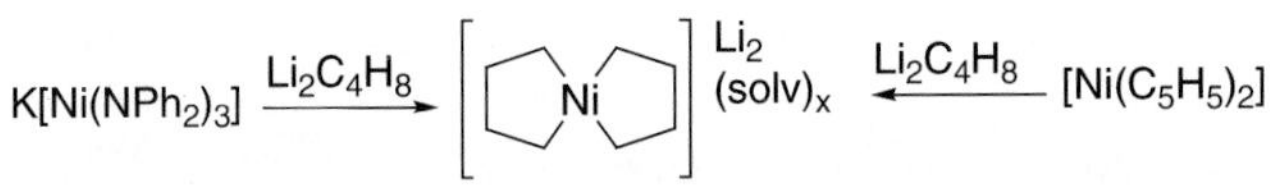

$$K[Ni(NPh_2)_3] \xrightarrow{Li_2C_4H_8} \left[\text{Ni}(C_4H_8)_2 \right] Li_2(solv)_x \xleftarrow{Li_2C_4H_8} [Ni(C_5H_5)_2] \quad (29)$$

Nickelapentacycles have also been prepared by the reaction of 1,4-dihaloalkanes such as 1,4-dibromobutane or 1,4-dibromopentane with Ni(COD)(bipy) (bipy = 2,2′-bipyridyl) to afford complexes of type **71** as dark green crystalline solids:[86]

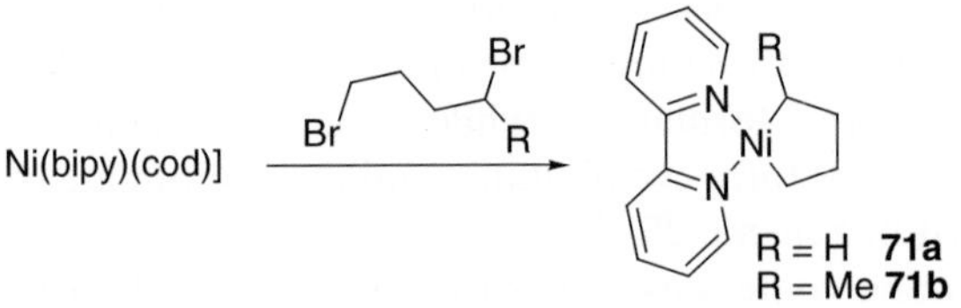

$$Ni(bipy)(cod)] \xrightarrow{BrCH_2CH_2CH_2CH(R)Br} (bipy)Ni(C_4H_7R) \quad (30)$$

R = H **71a**
R = Me **71b**

The derivative **71a** was subsequently structurally characterised[87] and shown to eliminate cyclobutane, 1-2-*E* and 2-*Z* butenes and butane on thermolysis or exposure to $BF_3 \cdot OEt_2$, while addition of maleic anhydride resulted primarily in reductive elimination of cyclobutane. Compounds **71a** and **71b** react with nitrous oxide to give the oxametallacycles **72a** and **72b**, respectively, as shown in Scheme 26. Compound **72a** undergoes cleavage reactions with HCl to give butanol; with I_2 to give tetrahydrofuran and with CO to give δ-valerolactone. Similarly, **72b** undergoes cleavage reactions with HCl to give 2-pentanol; with I_2 to give 2-methyltetrahydrofuran and with CO to give δ-caprolactone.[88]

2. Palladium

The metallacycles of palladium have not been as extensively studied as those of nickel and platinum. Palladapentacycles **73** containing PPh_3, chelating diphosphines (dppe, dcpe) (dcpe = 1,2-*bis*(dicyclohexylphosphino)ethane) or nitrogen donor ligands (TMEDA, bipy) have been prepared by treatment of the corresponding palladium dihalides with 1,4-dilithiobutane.[89,90] Diversi *et al.*[90] have reported the crystal structure of $Pd(C_4H_8)(bipy)$, and Canty has described the corresponding TMEDA derivative. Thermal decomposition of the palladapentacycles gave butenes as the major product with minor amounts of cyclobutane and

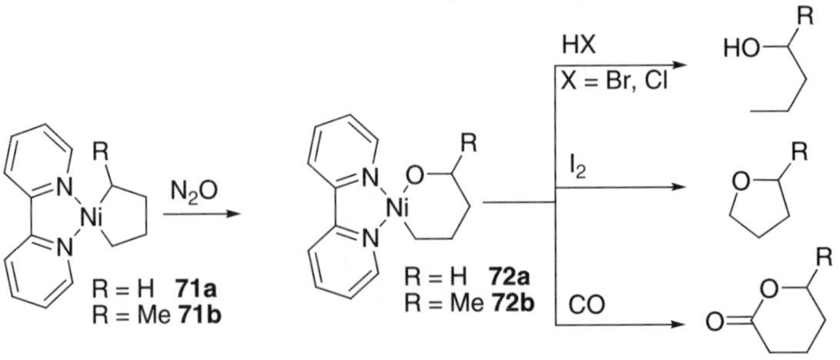

R = H **71a**
R = Me **71b**
N_2O
R = H **72a**
R = Me **72b**
HX
X = Br, Cl
I_2
CO

SCHEME 26.

ethylene being formed. Decomposition on photolysis, however, gave ethylene as the major product with minor amounts of butane and cyclobutane.[92] Pallada-cyclopentanes with nitrogen donor ligands were found to react with CO to give cyclopentanone. However, when the chelating ligand is dppe, no carbonylation products were formed.[89] Palladapentacycles have been found to undergo facile hydride abstraction when treated with the trityl salt $[CPh_3][BF_4]$ to give cationic π-allyl Pd(II) complexes **74**:[71]

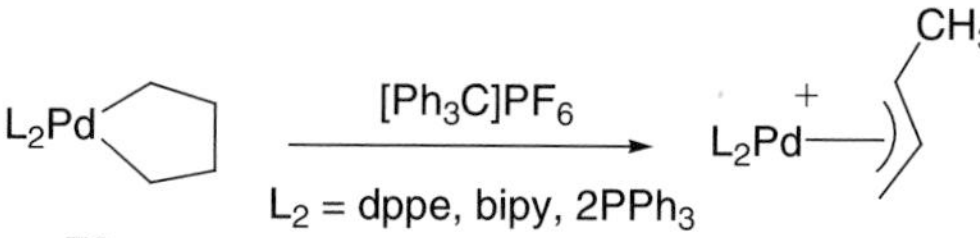

$$L_2Pd(C_4H_8)\ \mathbf{73} \xrightarrow[L_2 = \text{dppe, bipy, } 2PPh_3]{[Ph_3C]PF_6} [L_2Pd(\eta^3\text{-}C_3H_4CH_3)]^+\ \mathbf{74} \tag{31}$$

The remarkable homoleptic palladate $(Et_2O)Li_2[Pd(C_4H_8)_2]$ **75** was obtained from the reaction of LiC_4H_8Li with $[PdCl_2(cod)]$ [Eq. (32)] in which the ether could be replaced with other donors including TMEDA, PMDTA and THF (PMDTA = pentamethyl diethylene triamine), the last of which afforded the crystallographically characterised adduct $(THF)_4Li_2[Pd(C_4H_8)_2]$ (Fig. 26).[93]

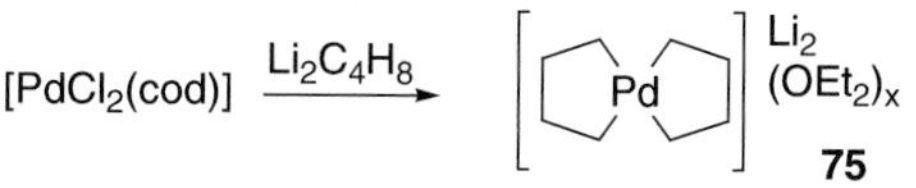

$$[PdCl_2(cod)] \xrightarrow{Li_2C_4H_8} [Pd(C_4H_8)_2]\,Li_2(OEt_2)_x\ \mathbf{75} \tag{32}$$

Nickel-mediated cyclopropene oxidative cyclisation[94] has also been applied to the dimerisation of cyclopropenes by $Pd_2(dba)_3$, including chiral examples that cyclise diastereoselectively (dba = dibenzylideneacetone). The reactions are typically carried out in acetone to provide initially the bis(acetone) adducts (Fig. 27), which may be converted into new derivatives **76** by exchanging the labile acetone ligands with, e.g., acetonitrile, bipy, phosphines, cyclooctadiene or norbornadiene [Eq. (33)]. Cyclopropenes bearing ester functional groups have been widely investigated and in such cases the initial products are generally coordination polymers wherein monomers are presumably linked by intermolecular ester coordination, which is however cleaved by the addition of stronger donor ligands.[95–100]

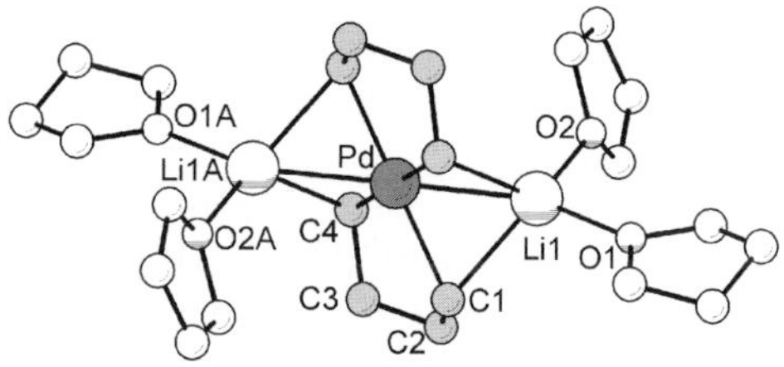

Fig. 26. Molecular structure of **75** (hydrogen atoms omitted).

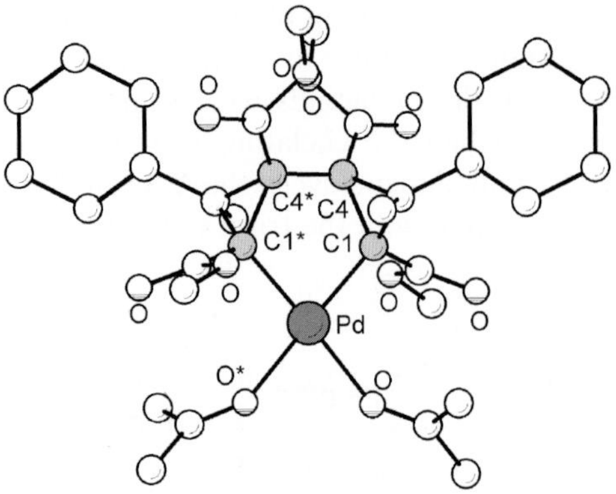

FIG. 27. Molecular structure of **76** (hydrogen atoms omitted, * indicates symmetry generated atoms).

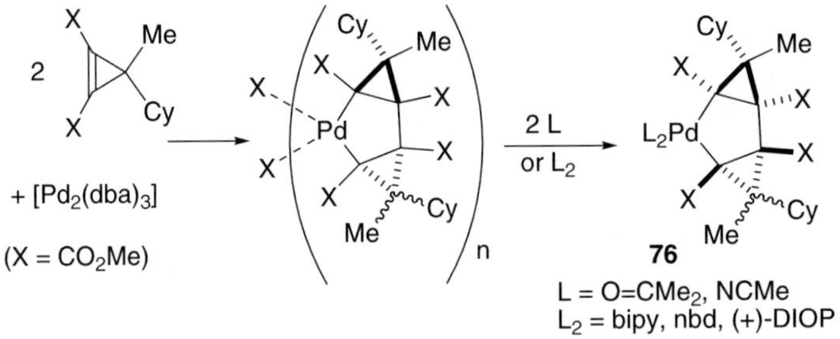

(33)

$L = O{=}CMe_2$, NCMe
L_2 = bipy, nbd, (+)-DIOP

Known palladacyclopentanes are almost exclusively based on divalent palladium. However, Canty has shown that oxidation of $[Pd(C_4H_8)(bipy)]$ by PhSSPh, $(PhCO_2)_2$ or PhSeSePh results in very labile Pd^{IV} derivatives, which decompose at ambient temperatures *via* C–C and C–O, C–S or C–Se bond formation. However, in the case of diphenyldiselenide, it was possible to isolate the oxidative addition product **77**, which was characterised crystallographically confirming the *trans* addition of the diselenide (Fig. 28, Scheme 27).[101] Primary alkyl bromides undergo *trans* oxidative addition to provide palladium(IV) derivatives **78**;[102] however, the oxidative addition of phenyl and triflato groups (delivered as $[IPh_2]OTf$) resulted in both *cis* and *trans* isomers of **79**. The triflate could be metathesised with iodide or chloride, though again each of these derivatives decomposed at room temperature.[103]

In marked contrast, a series of unusually stable Pd^{IV} palladacyclopentanes were derived from the hydrotris(pyrazolyl)borate derivative $K[Pd(C_4H_8)\{HB(pz)_3\}]$ **80**, which is readily accessible *via* reaction of $[Pd(C_4H_8)(TMEDA)]$ **81**[91] with $K[HB(pz)_3]$ on treatment with a wide range of oxidants (Scheme 28).[104–106] Among the various palladium(IV) derivatives $[Pd(C_4H_8)X\{HB(pz)_3\}]$ (X = Me, Et, CH_2Ph, C_3H_5, Cl, Br, I, OH), perhaps the most remarkable is the hydroxy derivative **82** wherein the hydroxide ligand arises from cleavage of water or THF. The hydroxy ligand is particularly prone to intermolecular hydrogen bonding and crystal

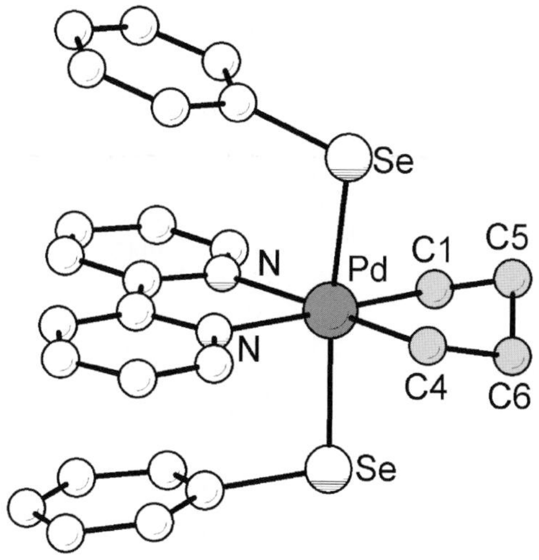

FIG. 28. Molecular structure of **77** (hydrogen atoms omitted).

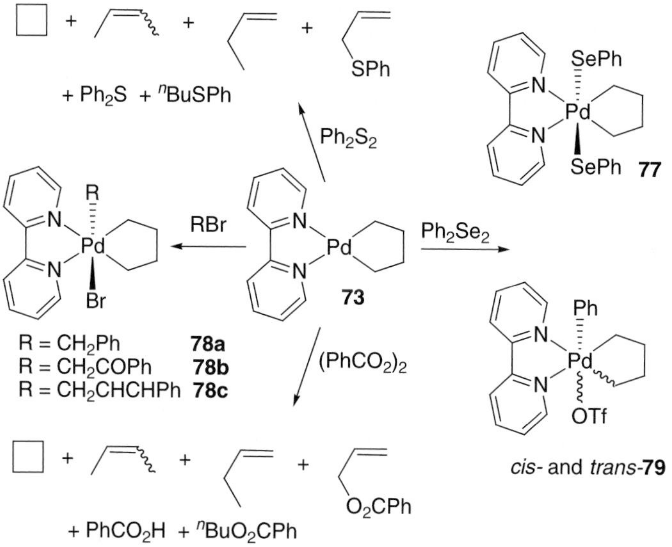

SCHEME 27.

structures have been determined for a range of phenol adducts (viz. C_6F_5OH, PhOH, MeC_4H_4OH).

3. Platinum

In general, two procedures are most often reported for the preparation of platinacyclopentane compounds. The reaction of the di-Grignard reagent BrMg $(CH_2)_4MgBr$ with dichloro(1,5-cyclooctadiene)platinum(II) is followed by the

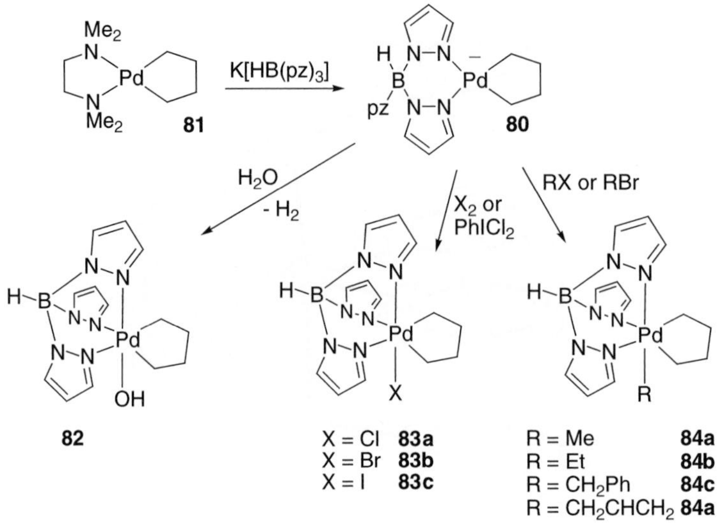

SCHEME 28.

displacement of the labile COD ligand by tertiary phosphines to give the products in about 25% yield. Alternatively, direct reaction of *cis*-dichlorobis-(triphenylphosphine)platinum(II) with 1,4-dilithiobutane gives the product $[Pt(C_4H_8)(PPh_3)_2]$ **85a** in about 60% yield [Eq. (34)], the crystal structure of which was reported by Grubbs *et al.* in 1973.[76]

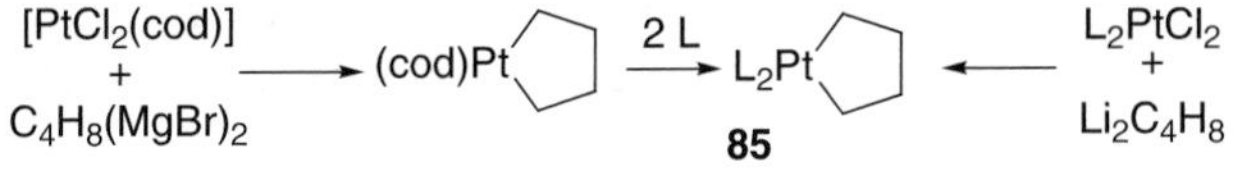

$$[PtCl_2(cod)] + C_4H_8(MgBr)_2 \longrightarrow (cod)Pt\bigcirc \xrightarrow{2\,L} L_2Pt\bigcirc\ (\mathbf{85}) \longleftarrow L_2PtCl_2 + Li_2C_4H_8 \quad (34)$$

Platinacyclopentanes containing methyl substituents at the α or β ring positions **86** have also been made using the appropriate di-Grignard or dilithium alkylating reagents. Direct reaction of $PtCl_2(PPh_3)_2$ with the di-Grignard reagent is reported to be an unsatisfactory method for preparation of these complexes as a mixture of products is obtained containing large quantities of monoalkylated Pt(II) compounds:[81]

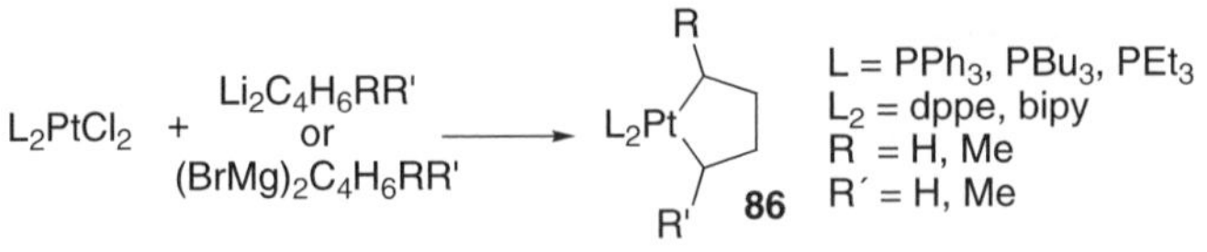

$$L_2PtCl_2 + Li_2C_4H_6RR' \text{ or } (BrMg)_2C_4H_6RR' \longrightarrow L_2Pt(C_4H_6RR')\ (\mathbf{86}) \quad (35)$$

L = PPh_3, PBu_3, PEt_3
L_2 = dppe, bipy
R = H, Me
R′ = H, Me

An alternative method for the synthesis of 2,5-disubstituted platinacyclopentanes is *via* the reaction of $Pt(COD)_2$ with 1,3-butadiene which affords the vinyl-substituted metallacycle **87a** as a white crystalline solid in 76% yield. The COD ligand is labile and may be replaced by either two equivalents of PMe_3 or $CNCMe_3$ (Scheme 29). Each of the compounds **87a**, **87b** and **87c** were structurally characterised, revealing the *trans* arrangement of α-substituents.[107] The α,α′ (head to head) regiochemistry observed for **87** is in contrast to that observed from the reaction of

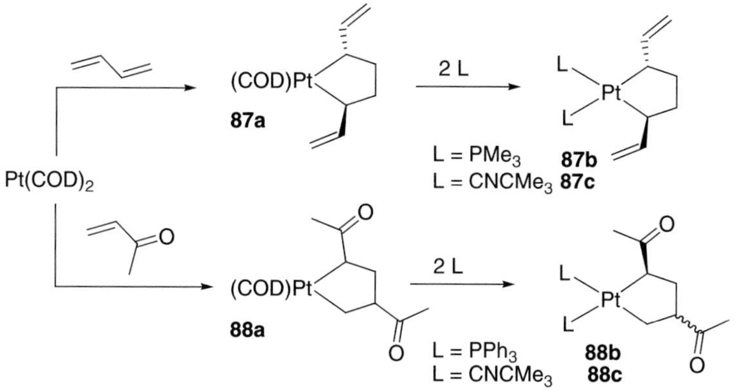

SCHEME 29.

$Pt(COD)_2$ with methylvinylketone in which head to tail dimerisation occurs to provide one isomer of **88a** which was converted into the crystallographically characterised derivative **88b** upon addition of triphenylphosphine. However, the crystallographic analysis revealed that both *cis* and *trans* isomers of the head to tail dimerisation product were present in the crystal.[108]

Puddephatt and co-workers[109] have recently reported the synthesis of the platinum analogue of **73** by reaction of $[PtCl_2(SEt_2)_2]$ with dilithiobutane to give the presumed dimeric platinacycle $[\{Pt(C_4H_8)(\mu\text{-}SEt_2)\}_2]$ intermediate, which when treated with 2,2′-bipyridyl gives the monomeric platinapentacycle $[Pt(C_4H_8)(bipy)]$ **85**. In a similar manner to that described above for palladium, a variety of platinates of the form $L_xLi_2[Pt(C_4H_8)_2]$ **89**[110] and $L_xLi_2[Pt(C_4H_4Me_4)_2]$ **88** ($L_x = Et_2O$, THF, TMEDA, PMDETA)[111,112] have been prepared *via* the dilithiobutane route, including structurally characterised examples.

As for palladium discussed above, the majority of platinacyclopentanes are based on the divalent metal. However, the structurally characterised complex $[Pt(C_4H_8)(Ph)I(bipy)]$ **91** has been obtained by treating $[Pt(C_4H_8)(bipy)]$ **85** with diphenyliodonium triflate followed by triflate-iodide metathesis [Eq. (36)]. As in the palladium example **79** described above, both *cis* and *trans* oxidative addition products are observed.[103] Prior to this, the structurally characterised platinum(IV) derivative $[Pt(C_4H_8)I_2(PMe_2Ph)_2]$ **92** had been obtained *via trans* oxidative addition of iodine to $[Pt(C_4H_8)(PMe_2Ph)_2]$ [Eq. (37)].[113]

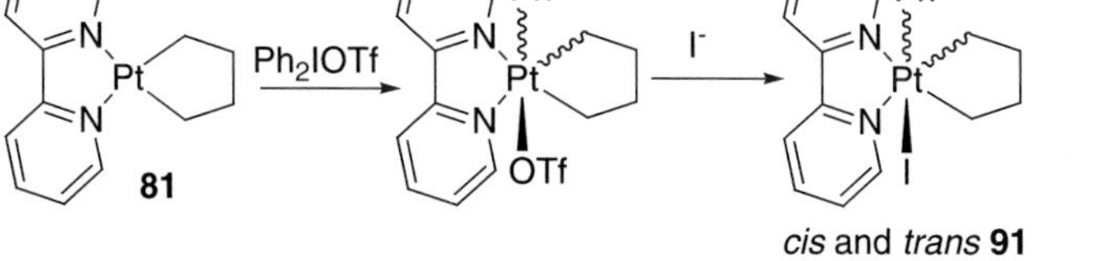

(36)

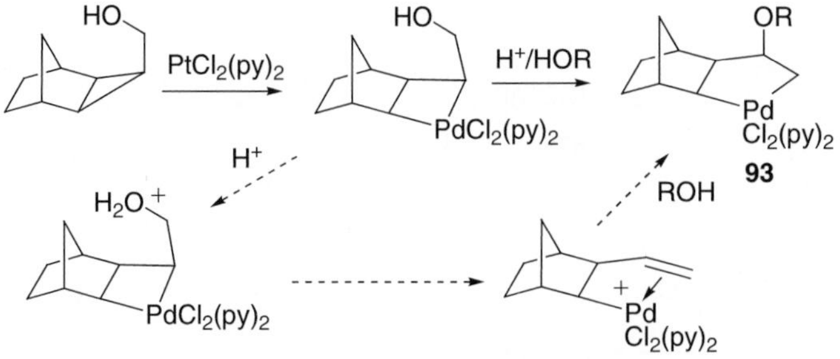

SCHEME 30.

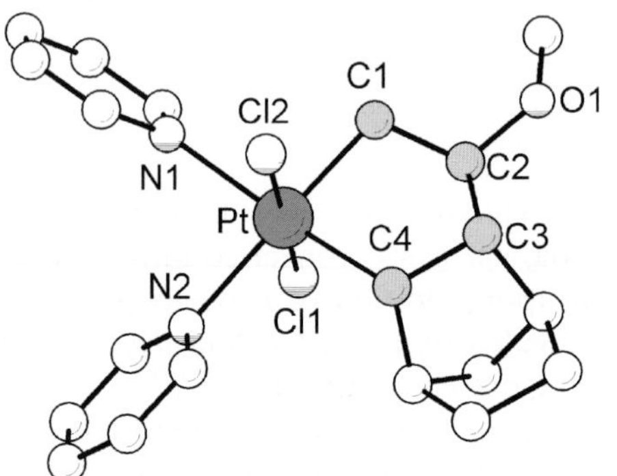

FIG. 29. Molecular structure of **93** (hydrogen atoms omitted).

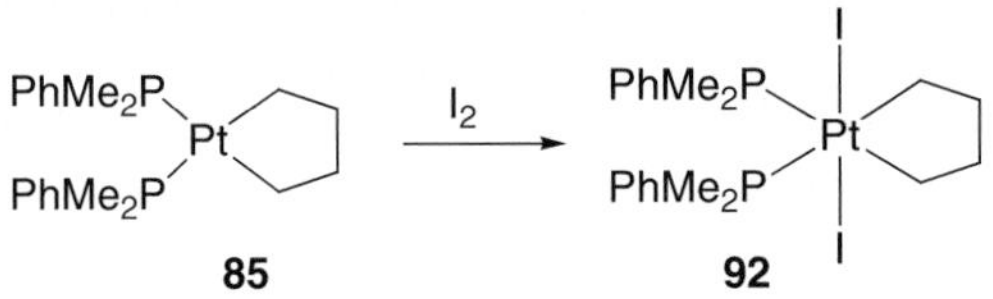

(37)

An alternative approach to platinum(IV) metallacyclopentanes relies on the facile insertion of platinum(II) reagents into hydroxymethylcyclopropanes followed by ring expansion of the resulting platinacyclobutanes as illustrated (Scheme 30, Fig. 29) for the synthesis of the structurally characterised example **93**.[114]

III
METALLACYCLOHEXANES

A. *Group 4 Metallacyclohexanes*

1. Titanium

Compound **94** was prepared by the reaction of 1,5-dilithiopentane with Cp_2TiCl_2 in diethyl ether.[13,74] The product was isolated as air-sensitive orange-red crystals in 27% yield:[74]

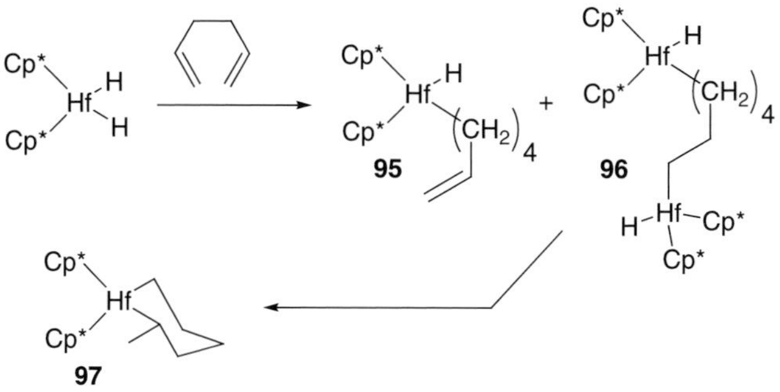

SCHEME 31.

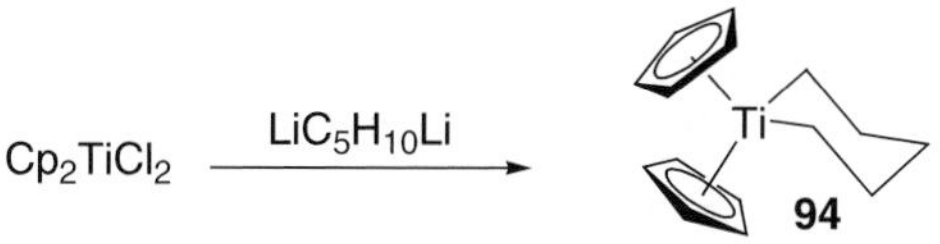

$$Cp_2TiCl_2 \xrightarrow{LiC_5H_{10}Li} \mathbf{94} \quad (38)$$

Compound **94** was found to be less reactive to CO than the analogous titanacyclopentane **1**. Carbonylation of the hexacycle gave less than 1% yield of cyclohexanone.[13] Thermal decomposition of **94** at 250 °C gave the hydrocarbon products 1-pentene (30%), 2-pentene (65%), and cyclopentane (5%).

2. Hafnium

The reaction of $Cp^*_2HfH_2$ with 1,5-hexadiene in a 1:1 ratio produces a mixture of $Cp^*_2HfH(CH_2)_4CHCH_2$ **95** and $Cp^*_2(H)Hf(CH^*_2)_6Hf(H)Cp^*_2$ **96** which on standing at room temperature for 4 days appears to form the hafnacyclohexane **97**, based on NMR data.[37] Similarly, solutions of **96** were found to react slowly over a period of 1–4 days at room temperature with excess 1,5-hexadiene to form **97** (Scheme 31).

B. *Groups 5, 6, 7*

Although there are no reports of metallacyclohexane compounds for groups 5 or 6, there is a report of a group 7 rhenacyclohexane compound. The few reports can possibly be attributed to the fact that most of the routes to the metallacycles of this type make use of ethylene co-ordination, followed by coupling. Using this methodology, it would be impossible to generate a metallacyclohexane complex. In the case of the rhenium complex, a dihydride complex is reacted with an appropriate 1,5-di-triflate to form the rhenacyclohexane $CpRh(CO)_2(C_6H_{12})$ **98**. This compound was fully characterised by NMR, elemental analysis and mass spectrometry.[54] The $^{13}C\{^1H\}$ NMR spectrum showed one singlet at 91.3 ppm corresponding to the five equivalent Cp carbons, and three distinct singlets at 32.1, 29.3 and −4.8 ppm assigned to the methylene ring carbons. Decomposition of the compound is reported to yield mainly 1-pentene and cyclopentane *cf* Scheme 19.

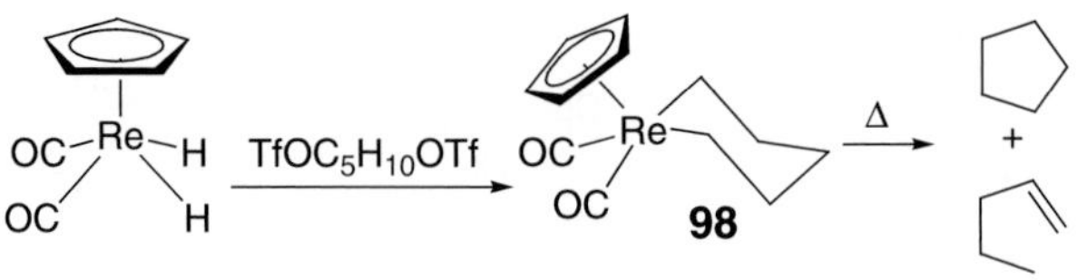

OC–Re(H)(H)(OC) $\xrightarrow{TfOC_5H_{10}OTf}$ OC–Re(OC) **98** $\xrightarrow{\Delta}$ + (39)

C. Group 8 Metallacyclohexanes

There are no known metallacyclohexanes of Fe, although the Ru and Os compounds $(CO)_4M(C_5H_{10})$ (M = Ru **99**, Os **100**) have been prepared and characterised.[56] The synthetic protocol followed that developed for the corresponding metallacyclopentanes [Eqs. (21) and (22)], *via* the reactions of metal carbonylates with 1,5-ditriflates (Scheme 32). The X-ray crystal structure of **100** has been determined (Fig. 30).[56]

Thermal decompositions of **99** or **100** yield 1-pentene, *cis* and *trans* 2-pentene as well as 1,4-pentadiene, however, no cyclopentane was observed. Decomposition in the presence of CO or PPh_3 provided cyclohexanone and either $M_3(CO)_{12}$ or $M(CO)_3(PPh_3)_2$, respectively.[56]

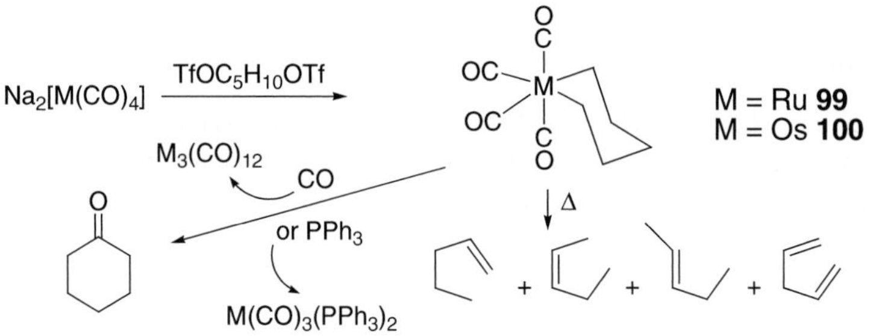

$Na_2[M(CO)_4]$ $\xrightarrow{TfOC_5H_{10}OTf}$ M = Ru **99**, M = Os **100**; CO or PPh_3 → cyclohexanone + $M_3(CO)_{12}$ or $M(CO)_3(PPh_3)_2$; Δ → pentenes + pentadiene

SCHEME 32.

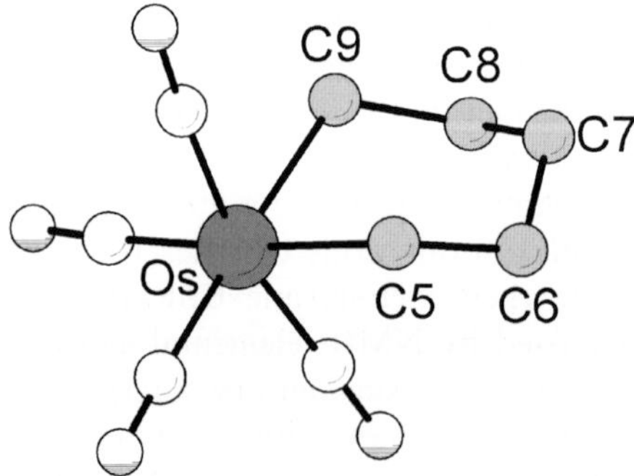

FIG. 30. Molecular structure of **100** (hydrogen atoms omitted).

Computational studies on the hypothetical 16-electron ruthenacyclohexane 'CpRuCl(C_5H_{10})' suggest that β-hydride elimination is feasible, as for the analogous ruthenacyclopentane.[65]

D. *Group 9 Metallacyclohexanes*

1. Rhodium

The rhodacyclohexane Cp*Rh(PPh_3)(C_5H_{10}) **101** was prepared by the reaction of Cp*Rh(PPh_3)I_2 in diethyl ether with a THF solution of MgBr(CH_2)$_5$BrMg at room temperature [Eq. (40)]. Compound **101** was isolated as orange crystals in 30% yield, with a melting point of 109 °C (decomp.).[67]

M—Cl, Ph$_3$P, Cl —BrMgC$_5$H$_{10}$MgBr→ M, Ph$_3$P; M = Rh **101**; M = Ir **102** (40)

2. Iridium

The iridacyclohexane Cp*Ir(PPh_3)(C_5H_{10}) **102** was prepared by the reaction of Cp*Ir(PPh_3)Cl_2 in THF with MgBr(CH_2)$_5$BrMg at room temperature [Eq. (40)]. Compound **102** was isolated as pale yellow crystals but only in 8% yield, and had a melting point of 188 °C (decomp.).[67]

E. *Group 10 Metallacyclohexanes*

The synthesis and characterisation of the platinahexacycle (PPh_3)$_2$Pt(C_5H_{10}) **103** was reported by Whitesides and co-workers in 1973.[80,81] The reaction of (COD)PtCl_2 with the appropriate di-Grignard produced the intermediate (COD)Pt (C_5H_{10}) **104**, which was then reacted with triphenylphosphine to give **103**. This complex had earlier been prepared in 1958 by Chatt and Shaw, *via* the reaction of *cis*-(PPh_3)$_2$PtCl_2 with 1,5 dilithiopentane (Scheme 33).[115] Compound **103** decomposes at 120 °C to give a mixture of 1-pentene (75%), 2-pentene (17%) and 1,4-pentadiene (8%).[81]

The closely related platinacyclohexanes (PMe_2Ph)$_2$Pt(CH_2CH_2)$_2$CR_2 (R = H **104a**, Me **104b**) were synthesised by the method reported by Whitesides *et al.* for compound **103**.[116] The structures of **105a** and **105b** (Fig. 31) were analysed by single crystal X-ray diffraction studies.[116] For **105a** the Pt–Cα bond lengths were found to be 2.195 and 2.097 Å while for **105b** the bond lengths were 2.089 and 2.102 Å. The Cα–Pt–Cα′ bond angle in **105a** was measured at 86.6° compared to 87.4° found in

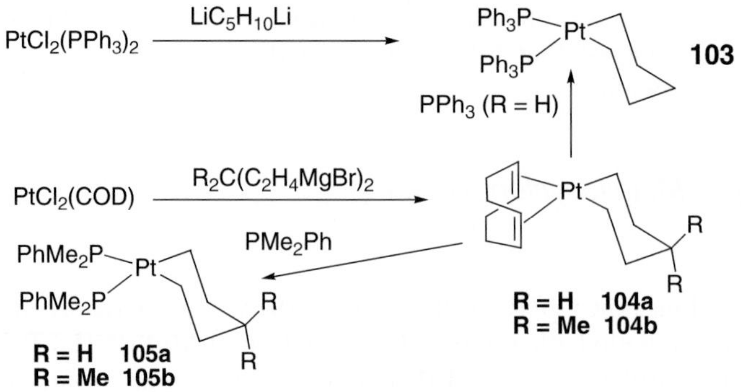

SCHEME 33.

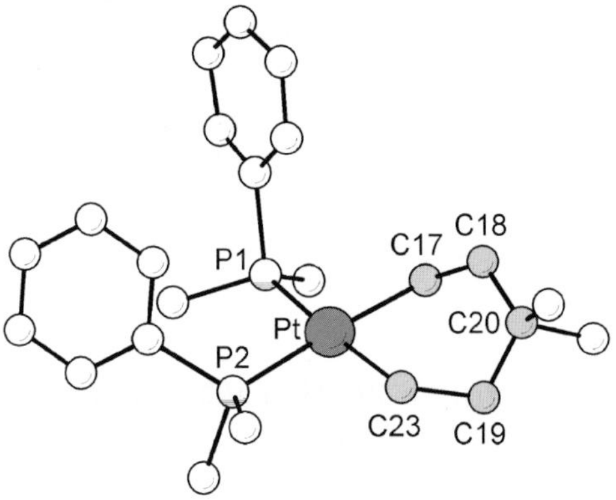

FIG. 31. Molecular structure of **105b** (hydrogen atoms omitted).

105b. Also detected was a difference in the conformation of the metallacyclic rings. The substituent methyl groups caused a change in the conformation of from half-chair in **105a** to skew in **105b** as well as a higher degree of puckering in the ring. This is similar behaviour to that found for metallacyclopentanes **63** and **64**.

Although the majority of metallacyclohexanes have been prepared either from 1,5-ditriflates, di-Grignards or dilithio derivatives, the report of a crystalline spiro-zincate $[Li(TMEDA)]_2[Zn(C_5H_{10})_2]$ **106** [Eq. (41)] (Fig. 32) may well serve as an alternative and valuable synthon for the development of metallacyclohexane chemistry.[117]

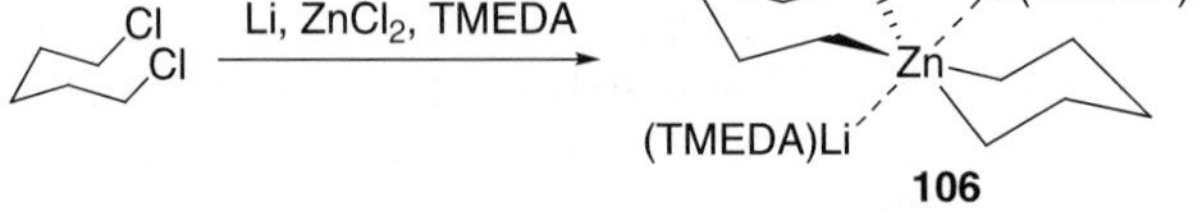

(41)

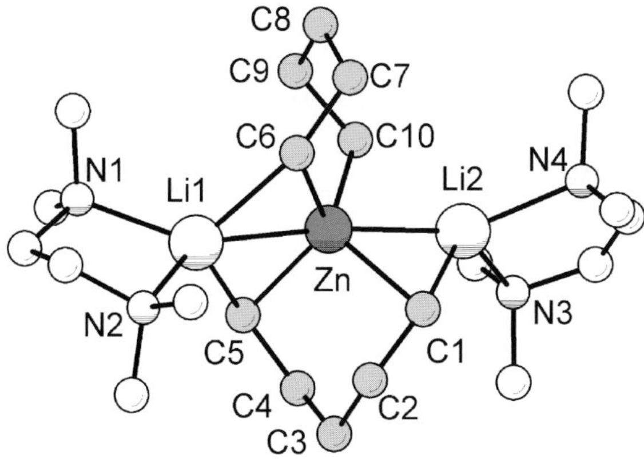

FIG. 32. Molecular structure of **106** (hydrogen atoms omitted).

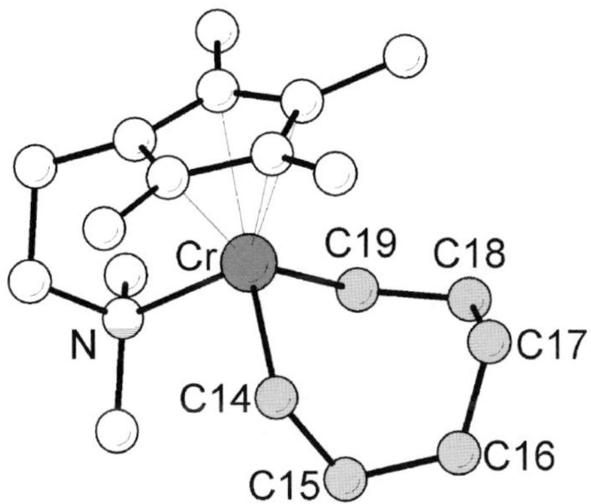

FIG. 33. Molecular structure of **107** (hydrogen atoms omitted).

IV
METALLACYCLOHEPTANES

A. *Groups 4, 5, 6, 7*

There are no reported metallacycloheptane compounds of group 4 or group 5 transition metals. There is, however, a report of a chromacycloheptane **107** by Emrich *et al.*[42] [Eq. (42)]. This metallacycle is prepared using the same methodology as was used to make the corresponding metallacyclopentane compound, but instead of using a dilithium reagent, a di-Grignard reagent was used. This chromacycloheptane was structurally characterised by X-ray crystallography (Fig. 33).

$$\xrightarrow{\text{ClMgC}_6\text{H}_{12}\text{MgCl}} \quad \mathbf{107} \qquad (42)$$

B. *Group 8 Metallacycloheptanes*

A rare example of a ferracycloheptane **108** was obtained as the product of the photochemical reaction of a Petitt's cyclobutadiene iron complex with dimethylmaleate [Eq. (43)].[118] The ferracycloheptane arises from the insertion of a maleate into each of two Fe–C bonds and might therefore be considered a special case of alkene trimerisation (*vide infra*). The cyclobutene fragment in the final metallacycle remains coordinated to iron, as established crystallographically (Fig. 34).

$$\xrightarrow[\text{MeO}_2\text{C–CH=CH–CO}_2\text{Me}]{h\nu} \quad \mathbf{108} \qquad (43)$$

Lindner *et al.*[119] have carried a detailed study of the reactions of $Na_2[Os(CO)_4]$ with a wide range of *bis*(triflates), $TfO(CH_2)_mOTf$ ($m = 5$–10, 12, 14, 16). They isolated the osmacycloheptane $(CO)_4Os(C_6H_{12})$ **109** as a colourless oil (m.p. + 18 °C), but only in 9% yield. The product was fully characterised by IR, ^1H and ^{13}C NMR, mass spectrometry and elemental analysis.[119]

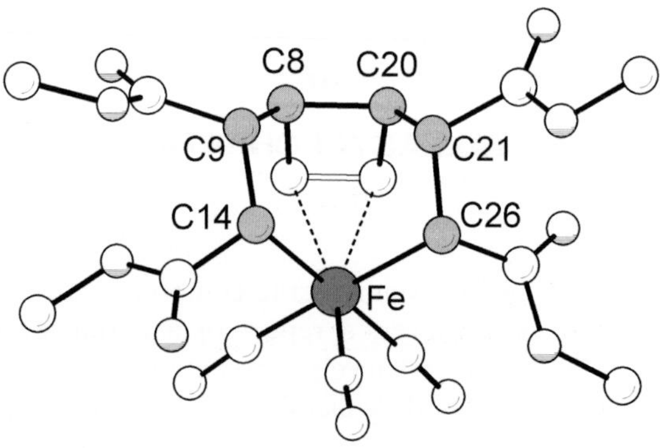

FIG. 34. Molecular structure of **108** (hydrogen atoms omitted).

C. *Group 9 Metallacycloheptanes*

1. Rhodium

The rhodaheptacycle $Cp^*Rh(PPh_3)(C_6H_{12})$ **110** was prepared by the reaction of $Cp^*Rh(PPh_3)I_2$ in diethyl ether with a THF solution of $BrMg(CH_2)_6MgBr$ at room temperature [Eq. (44)]. Compound **110** was isolated as orange crystals in 50% yield, with a melting point of 107 °C (decomp.).[67]

$$Cp^*Rh(PPh_3)I_2 \xrightarrow{BrMgC_5H_{10}MgBr} \mathbf{110} \quad (44)$$

Treatment of **110** with dry HCl gave a mixture of hexane (25%) and hexenes (75%) as well as the corresponding dihalide complex $Cp^*Rh(PPh_3)Cl_2$. Reaction of **110** with bromine yielded 1,6-dibromohexane in addition to $Cp^*Rh(PPh_3)Br_2$.[53]

The cyclotrimerisation of dimethylcyclopropene is effected by Wilkinson's catalyst $RhCl(PPh_3)_3$ to provide a rhodacycloheptane derivative **111** which has been structurally characterised (Fig. 35). The metallacycle is cleaved upon treatment with carbon monoxide to provide both cyclohexane (reductive eliminations) and cycloheptanone (CO insertion) products (Scheme 34).[120]

A new high yield route (>90%) to compound **112** has recently been reported by RCM on the bis-butenyl compound *cis*-$[Pt(PPh_3)_2(CH_2CH_2CH{=}CH_2)_2]$ using Grubbs 1st generation catalyst $RuCl_2({=}CHPh)(PCy_3)_2$. The 1st product is the novel platinacycloheptene which can then be hydrogenated (Pd/C) to give the platinacycloheptane.[123]

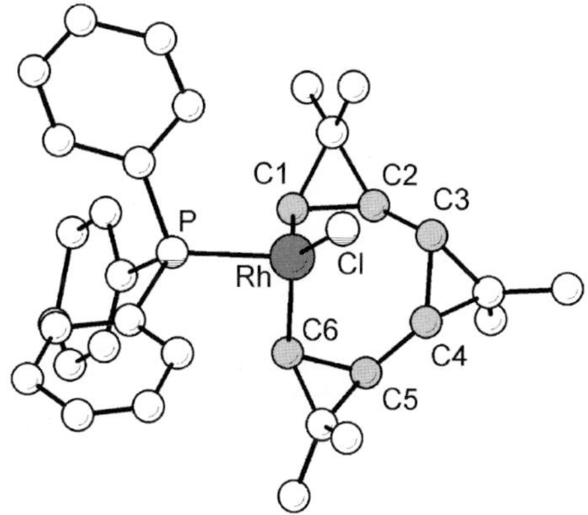

Fig. 35. Molecular structure of **111** (hydrogen atoms omitted).

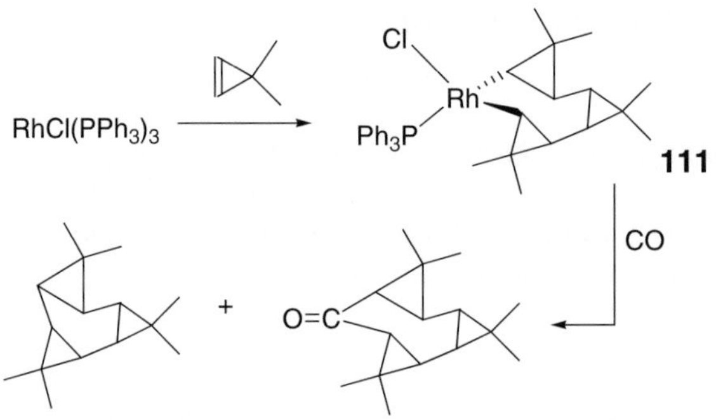

$RhCl(PPh_3)_3$ → 111 → CO

SCHEME 34.

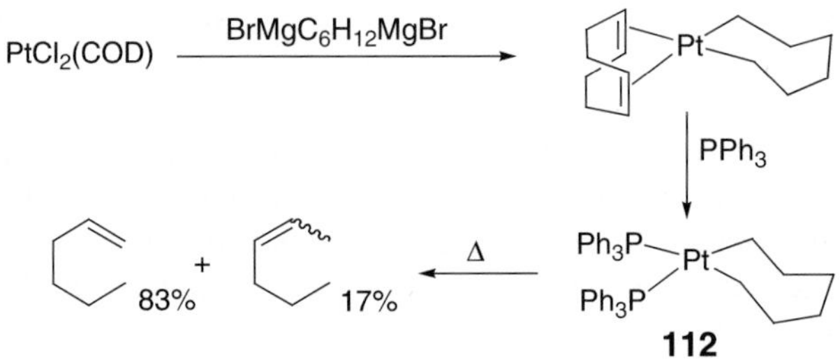

$PtCl_2(COD)$ → $BrMgC_6H_{12}MgBr$ → PPh_3 → 112 → Δ → 83% + 17%

SCHEME 35.

D. *Group 10 Metallacycloheptanes*

The platinacycloheptane **112** was synthesised in 3% yield by the reaction of (COD)$PtCl_2$ with a appropriate di-Grignard, and subsequent reaction with PPh_3 displaced the COD ligand (Scheme 35). The products of thermal decomposition of **112** at 60 °C were found to be 1-hexene (83%) and 2-hexene (17%). Interestingly, compound **112** is reported to be less thermally stable than the five- and six-membered ring analogues.[80]

A remarkably diastereoselective synthesis of a palladacycloheptane has been achieved *via* the oxidative cyclisation of the cyclopropene **113** by $Pd_2(dba)_3$ followed by treatment with 2,2-bipyridyl to provide the crystallographically characterised derivative **114** [Eq. (45)].

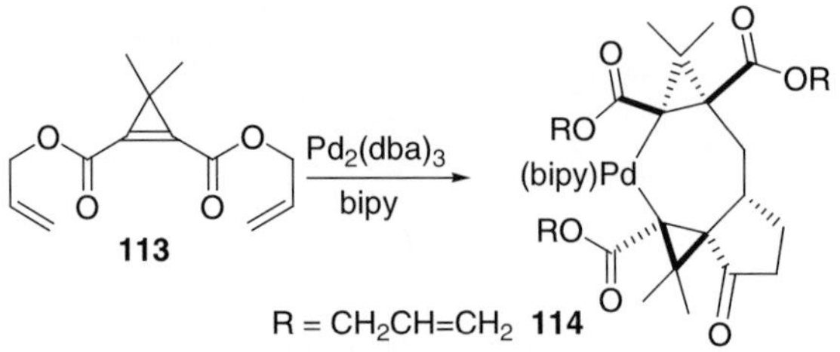

113 → $Pd_2(dba)_3$, bipy → 114, R = $CH_2CH{=}CH_2$ (45)

V

METALLACYCLOOCTANES AND LARGER RINGS

A. *Groups 4, 5, 6, 7 and 9*

There are no reported metallacycloalkanes higher than metallacycloheptanes reported for the groups 4–7 and 9 transition metals.

B. *Group 8 Metallacyclooctanes and Larger Rings*

Using the same methodology as they employed for $(CO)_4Os(C_6H_{12})$, **109** Lindner and co-workers carried out the reactions in attempts to make metallacyclooctanes and larger rings, according to Eq. (46). However, with $n = 7$, the yield of the expected osmacyclooctane dropped to 0%, similarly for $n = 8$ or 9. In fact, there are very few metallacycloalkanes with 8, 9 or 10 membered rings known. For Lindner and co-workers with $n = 10$, the yield increased to 1%, with $n = 12$, the yield increased to 13%, for $n = 14$, the yield was 21%, and for $n = 16$, the yield was 22%. Thus, using this synthetic route, the yield of product strongly depends on the metallacycloalkane ring size.

$$Na_2[Os(CO)_4] \xrightarrow{TfO(CH_2)_nOTf} (OC)_4Os\,(CH_2)_n \tag{46}$$

C. *Group 9 Metallacyclooctanes and Larger Rings*

The nine-membered palladium metallacycle **115** was formed by the oxidative coupling of 3,3-dimethylcyclopropene with a latent palladium(0) complex.[121,122] This is the first known example of an isolable metallanonacycle [Eq. (47)]. The structure of **115** (Fig. 36) was confirmed by X-ray crystallographic analysis. The geometry of the complex is distorted square-planar, with the conformation of the nonacyclic ring resembling the all-*cis* boat conformation of 1,3,5,7-cyclononatetraene, due to the steric rigidity of the four substituent cyclopropane rings.

Pd$_2$(dba)$_3$ → (3,3-dimethylcyclopropene; PMe$_2$Ph) → (PhMe$_2$P)$_2$Pd(metallacycle) **115** (47)

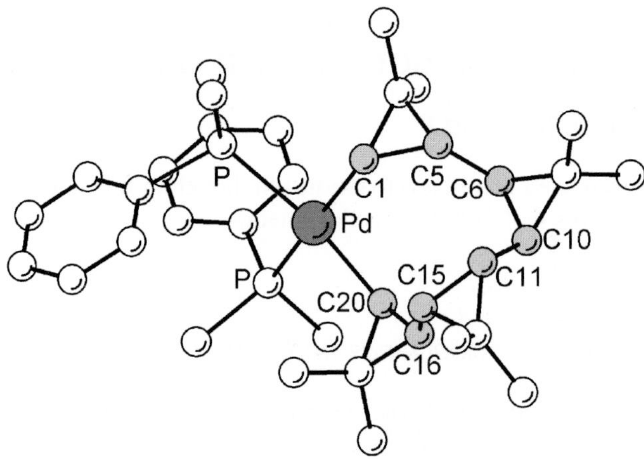

FIG. 36. Molecular structure of **116** (hydrogen atoms omitted).

VI
CONCLUSION

Owing to the realisation of the importance of medium to large metallacycles in selective catalytic oligomerisation processes, particularly ethylene trimerisation and tetramerisation, we envisage that the field of metallacycloalkanes will be a significant growth area. Clearly new routes will have to be devised for the synthesis of metallacycloheptane, -octane and -nonanes so that these novel ring compounds can be efficiently prepared and studied.

ACKNOWLEDGMENTS

We would like to thank the National Research Foundation and the University of Cape Town for support. Siyabonga Ngubane and Prinessa Chelan for help with some of the drawings and preparation of the manuscript and Tony Hill for help with the figures.

REFERENCES

(1) Adlhart, C.; Chen, P. *J. Am. Chem. Soc.* **2004**, *126*, 3496 and references therein.
(2) Grubbs, R. H. *Tetrahedron* **2004**, *60*, 7177.
(3) Schrock, R. R. *J. Mol. Cat. A Chem.* **2004**, *213*, 21.
(4) Grubbs, R. H. *Handbook of Olefin Metathesis*, Wiley–VCH, Weinheim, Germany, 2003.
(5) (a) McGuinness, D. S.; Wasserscheid, P.; Keim, W.; Dixon, J. T.; Grove, J. J. C.; Hu, C.; Englert, U. *Chem. Commun.* **2003**, 334. (b) McGuinness, D. S.; Wasserscheid, P.; Keim, W.; Morgan, D. H.; Dixon, J. T.; Bollmann, A.; Maumela, H.; Hess, F. M.; Englert, U. *J. Am. Chem. Soc.* **2003**, *125*, 5272.
(6) Agapie, T.; Schofer, S. L.; Labinger, J. A.; Bercaw, J. E. *J. Am. Chem. Soc.* **2004**, *126*, 1304.
(7) Dry, M. E. *Appl. Catal.* **1996**, *138*, 319.
(8) (a) Bollmann, A.; Blann, K.; Dixon, J. T.; Hess, F. M.; Killian, E.; Maumela, H.; McGuinness, D. S.; Morgan, D. H.; Neveling, A.; Otto, S.; Overett, M.; Slawin, A. M. Z.; Wasserscheid, P.; Kuhlmann, S. *J. Am. Chem. Soc.* **2004**, *126*, 14712. (b) Overett, M.; Blann, K.; Bollmann, A.; Dixon, J. T.; Hess, F.; Killian, E.; Maumela, H.; Morgan, D. H.; Neveling, A.; Otto, S. *Chem. Commun.* **2005**, 622.

(9) Puddephatt, R. J. *Comment. Inorg. Chem.* **1982**, *2*, 69.
(10) Chappel, S. D.; Cole-Hamilton, D. J. *Polyhedron* **1982**, *1*, 739.
(11) Collman, J. P.; Hegedus, L. S.; Norton, J. R.; Finke, R. G. *The Principles and Applications of Organotransition Metal Chemistry*, University Science Books, Mill Valley, CA, USA, 1987.
(12) Cámpora, J.; Palma, P.; Carmona, E. *Coord. Chem. Rev.* **1999**, *193*, 207.
(13) McDermott, J. X.; Wilson, M. E.; Whitesides, G. M. *J. Am. Chem. Soc.* **1976**, *98*, 6529.
(14) (a) Mashima, K.; Takaya, H. *Organometallics* **1985**, *4*, 1464. (b) Mashima, K.; Sakai, N.; Takaya, H. *Bull. Chem. Soc. Jpn.* **1991**, *64*, 2475.
(15) Grubbs, R. H.; Miyashita, A. *J. Am. Chem. Soc.* **1978**, *100*, 1300.
(16) (a) Lukesova, L.; Stepnick, P.; Fejfarova, K.; Gyepes, R.; Cisarova, I.; Horacek, M.; Kubista, J.; Mach, K. *Organometallics* **2002**, *21*, 2639. (b) Horacek, M.; Stepnicka, P.; Gyepes, R.; Cisarova, I.; Tislerova, I.; Zemanek, J.; Kubista, J.; Mach, K. *Chem. Eur. J.* **2000**, *6*, 2397.
(17) (a) Thorn, M. G.; Hill, J. E.; Waratuke, S. A.; Johnson, E. S.; Fanwick, P. E.; Rothwell, I. P. *J. Am. Chem. Soc.* **1997**, *119*, 8630. (b) Waratuke, S. A.; Johnson, E. S.; Thorn, M. G.; Fanwick, P. E.; Rothwell, I. P. *Chem. Commun.* **1996**, 2617.
(18) Balaich, G. J.; Hill, J. E.; Waratuke, S. A.; Fanwick, P. E.; Rothwell, I. P. *Organometallics* **1995**, *14*, 656.
(19) Hill, J. E.; Fanwick, P. E.; Rothwell, I. P. *Organometallics* **1991**, *10*, 15.
(20) (a) Graham, T. W.; Kickham, J.; Courtenay, S.; Wei, P.; Stephan, D. W. *Organometallics* **2004**, *23*, 3309. (b) Stephan, D. *Adv. Organomet. Chem.* **2005**, *54*, 275.
(21) Manriquez, J. M.; McAlister, D. R.; Sanner, R. D.; Bercaw, J. E. *J. Am. Chem. Soc.* **1978**, *100*, 2716.
(22) Takahasi, T.; Fischer, R.; Xi, Z.; Nakajima, K. *Chem. Lett.* **1996**, 357.
(23) Mansel, S.; Thomas, D.; Lefeber, C.; Heller, D.; Kempe, R.; Baumann, W.; Rosenthal, U. *Organometallics* **1997**, *16*, 2886.
(24) Sun, H.; Burlakov, V. V.; Spannenburg, A.; Baumann, W.; Arndt, P.; Rosenthal, U. *Organometallics* **2001**, *20*, 5472.
(25) Lee, L. W. M.; Piers, W. E.; Parvez, M.; Rettig, S. J.; Young, V. G. *Organometallics* **1999**, *18*, 3904.
(26) Hagadorn, J. R.; Arnold, J. *J. Chem. Soc. Dalton Trans.* **1997**, 3087.
(27) Warren, T. H.; Erker, G.; Fröhlic, R.; Wibbeling, B. *Organometallics* **2000**, *19*, 127.
(28) (a) Taber, D. F.; Louey, J. P.; Wang, Y.; Nugent, W. A.; Dixon, D. A.; Harlow, R. L. *J. Am. Chem. Soc.* **1994**, *116*, 9457. (b) Nugent, W. A.; Taber, D. F. *J. Am. Chem. Soc.* **1989**, *111*, 6435.
(29) (a) Knight KS Wang, D.; Waymouth, R. M.; Ziller, J. *J. Am. Chem. Soc.* **1994**, *116*, 1845. (b) Knight, K. S.; Waymouth, R. M. *J. Am. Chem. Soc.* **2006**, *54*, 275. (c) Wischmeyer, U.; Knight, K. S.; Waymouth, R. M. *Tetrahedron Lett.* **1992**, *33*, 7735.
(30) Fischer, R.; Walther, D.; Gebhardt, P.; Görls, H. *Organometallics* **2000**, *19*, 2532.
(31) Xi, Z.; Hara, R.; Takahashi, T. *J. Org. Chem.* **1995**, *60*, 4444.
(32) Kasatkin, A. N.; Whitby, R. J. *Tetrahedron Lett.* **2000**, *41*, 5275.
(33) Copéret, C.; Negishi, E.-I.; Xi, Z.; Takahashi, T. *Tetrahedron Lett.* **1994**, *35*, 695.
(34) Li, P.-X.; Xi, Z.-F.; Takahashi, T. *Chinese J. Chem.* **2001**, *19*, 45.
(35) Erker, G.; Dorf, U.; Rheingold, A. L. *Organometallics* **1988**, *7*, 138.
(36) Takahashi, T.; Tamura, M.; Saburi, M.; Uchida, Y.; Negishi, E.-I. *J. Chem. Soc. Chem. Commun.* **1989**, 852.
(37) Bercaw, J. E.; Moss, J. R. *Organometallics* **1992**, *11*, 639.
(38) Jonas, K.; Russeler, W.; Kruger, C.; Raabe, E. *Angew. Chem. Int. Ed. Engl.* **1986**, *25*, 925.
(39) McLain, S. J.; Wood, C. D.; Schrock, R. R. *J. Am. Chem. Soc.* **1977**, *99*, 3519.
(40) McLain, S. J.; Wood, C. D.; Schrock, R. R. *J. Am. Chem. Soc.* **1979**, *101*, 4558.
(41) (a) Churchill, M. R.; Young, W. J. *J. Am. Chem. Soc.* **1979**, *101*, 6462. (b) Churchill, M. R.; Youngs, W. J. *Inorg. Chem.* **1980**, *19*, 3106.
(42) Emrich, R.; Heinemann, O.; Jolly, P. W.; Krüger, C.; Verhovnik, G. P. J. *Organometallics* **1997**, *16*, 1511.
(43) Dohring, A.; Emrich, R.; Goddard, R.; Jolly, P. W.; Kruger, C. *Polyhedron* **1993**, *12*, 2671.
(44) Krausse, J.; Schödl, G. *J. Organomet. Chem.* **1971**, *27*, 59.
(45) Tsang, W. C. P.; Jamieson, J. Y.; Aeilts, S. L.; Hultzsch, K. C.; Schrock, R. R.; Hoveyda, A. H. *Organometallics* **2004**, *23*, 1997.

(46) (a) Cameron, T. M.; Ortiz, C. G.; Ghiviriga, I.; Abboud, K. A.; Boncella, J. M. *Organometallics* **2001**, *20*, 2032. (b) Wang, S.-Y. S.; VanderLende, S. D.; Abboud, K. A.; Boncella, J. M. *Organometallics* **1998**, *17*, 2628.
(47) Diversi, P.; Ingrosso, G.; Lucherini, A.; Porzio, W.; Zocchi, M. *J. Chem. Soc. Dalton Trans.* **1983**, 967.
(48) Chisholm, M. H.; Huffman, J. C.; Hampden-Smith, M. J. *J. Am. Chem. Soc.* **1989**, *111*, 5284.
(49) (a) Giannini, L.; Solari, E.; Floriani, C.; Chiesi-Villa, A.; Rizzoli, C. *J. Am. Chem. Soc.* **1998**, *120*, 823. (b) Giannini, L.; Guillemot, G.; Solari, E.; Floriani, C.; Re, N.; Chiesi-Villa, A.; Rizzoli, C. *J. Am. Chem. Soc* **1999**, *121*, 2797.
(50) For a review of organometallic derivatives of calix[4]arenes see Floriani, C.; Floriani-Moro, R. *Adv. Organomet. Chem.* **2001**, *47*, 167.
(51) (a) Jonas, K.; Häselhoff C-, C.; Goddard, R.; Krüger, C. *Inorg. Chim. Acta* **1992**, *198*, 533. (b) Jonas, K.; Burkart, G.; Haselhoff, C.; Betz, P.; Krüger, C. *Angew. Chem. Int. Ed. Engl.* **1990**, *29*, 322.
(52) Yang, G. K.; Bergman, R. G. *J. Am. Chem. Soc.* **1983**, *105*, 6500.
(53) Yang, G. K.; Bergamn, R. G. *Organometallics* **1985**, *4*, 129.
(54) Lindner, E.; Wassing, W. *Organometallics* **1991**, *10*, 1640.
(55) Lindner, E.; Jansen, R.-M.; Mayer, H. A. *Angew. Chem. Int. Ed.* **1986**, *25*, 1008.
(56) Lindner, E.; Jansen, R.-M.; Hiller, W.; Fawzi, R. *Chem. Ber.* **1989**, *122*, 1403.
(57) (a) Lindner, E.; Schauß, E.; Hiller, W.; Fawzi, R. *Chem. Ber.* **1985**, *118*, 3915. (b) Lindner, E.; Schauß, E.; Hiller, W.; Fawzi, R. *Angew. Chem. Int. Ed. Engl.* **1984**, *23*, 711.
(58) Manuel, T. A.; Stafford, S. L.; Stone, F. G. A. *J. Am. Chem. Soc.* **1961**, *83*, 249.
(59) Stockis, A.; Hoffmann, R. *J. Am. Chem. Soc.* **1980**, *102*, 2952.
(60) Fischer W Hembre, R. T.; Sidler, D. R.; Norton, J. R. *Inorg. Chim. Acta* **1992**, *198–200*, 57.
(61) Lindner, E.; Pabel, R.; Fawzi, R.; Steimann, M. *J. Organomet. Chem.* **1992**, *441*, 63.
(62) Lindner, E.; Jansen, R.-M.; Mayer, H. A.; Hiller, W.; Fawzi, R. *Organometallics* **1989**, *8*, 2355.
(63) Barabotti, P.; Diversi, P.; Ingrosso, G.; Lucherini, A.; Marchetti, F.; Sagramora, L.; Adovasio, V.; Nardelli, M. *J. Chem. Soc. Dalton Trans.* **1990**, 179.
(64) Nagashima, H.; Michino, Y.; Ara K-, I.; Fukahori, T.; Itoh, K. *J. Organomet. Chem.* **1991**, *406*, 189.
(65) Huang, Z.; Zhu, J.; Lin, Z. *Organometallics* **2004**, *23*, 4154.
(66) (a) Trepanier, S. J.; Sterenberg, B. T.; McDonald, R.; Cowie, M. *J. Am. Chem. Soc.* **1999**, *121*, 2613. (b) Trepanier, S. J.; Dennett, J. N. L.; Sterenberg, B. T.; McDonald, R.; Cowie, M. *J. Am. Chem. Soc.* **2004**, *126*, 8046.
(67) (a) Diversi, P.; Ingrosso, G.; Lucherini, A.; Porzio, W.; Zocchi, M. *J. Chem. Soc. Chem. Commun.* **1977**, 811. (b) Diversi, P.; Ingrosso, G.; Lucherini, A.; Porzio, W.; Zocchi, M. *Inorg. Chem.* **1980**, *19*, 3590.
(68) Wakatsuki, Y.; Nomura, O.; Tone, H.; Yamazaki, H. *J. Chem. Soc. Perkin. II* **1980**, 1344.
(69) Wakatsuki, Y.; Sakurai, T.; Yamazaki, H. *J. Chem. Soc. Dalton Trans.* **1982**, 1923.
(70) Bertani, R.; Diversi, P.; Ingrosso, G.; Lucherini, A.; Marchetti, F.; Adovasio, V.; Nardelli, M.; Pucci, S. *J. Chem. Soc. Dalton Trans.* **1988**, 2983.
(71) Barabotti, P.; Diversi, P.; Ingrosso, G.; Lucherini, A.; Nuti, F. *J. Chem. Soc. Dalton Trans.* **1984**, 2517.
(72) Cuccuru, A.; Diversi, P.; Ingrosso, G.; Lucherini, A. *J. Organomet. Chem.* **1981**, *204*, 123.
(73) Chen, J.; Angelici, R. J. *J. Organomet. Chem.* **2001**, *621*, 55.
(74) Grubbs, R. H.; Miyashita, A. *J. Am. Chem. Soc.* **1978**, *100*, 7418.
(75) Grubbs, R. H.; Miyashita, A. *Fund. Res. Homogeneous Catal.* **1979**, *3*, 151.
(76) Biefeld, C. G.; Eick, H. A.; Grubbs, R. H. *Inorg. Chem.* **1973**, *12*, 2166.
(77) Grubbs, R. H.; Miyashita, A.; Liu, M.; Burk, P. *J. Am. Chem. Soc.* **1978**, *100*, 2418.
(78) Grubbs, R. H.; Miyashita, A.; Liu, M.; Burk, P. *J. Am. Chem. Soc.* **1977**, *99*, 3863.
(79) Grubbs, R. H.; Miyashita, A. *J. Am. Chem. Soc.* **1978**, *100*, 7416.
(80) McDermott, J. X.; White, J. F.; Whitesides, G. M. *J. Am. Chem. Soc.* **1973**, *95*, 4451.
(81) McDermott, J. X.; White, J. F.; Whitesides, G. M. *J. Am. Chem. Soc.* **1976**, *98*, 6521.
(82) Young, G. B.; Whitesides, G. M. *J. Am. Chem. Soc.* **1978**, *100*, 5808.
(83) McKinney, R. J.; Thorn, D. L.; Hoffmann, R.; Stockis, A. *J. Am. Chem. Soc.* **1981**, *103*, 2595.

(84) Fröhlich, H.-O.; Wyrwa, R.; Görls, H. *J. Organomet. Chem.* **1992**, *441*, 169.
(85) Wyrwa, R.; Fröhlic, H.-O.; Görls, H. *J. Organomet. Chem.* **1995**, *491*, 41.
(86) Takahashi, S.; Suzuki, Y.; Sonogashira, K.; Hagihara, N. *J. Chem. Soc. Chem. Comm.* **1976**, 839.
(87) Binger, P.; Doyle, M. J.; Krüger, C.; Tsay, Y.-H. Z. *Naturforsch.* **1979**, *34b*, 1289.
(88) Matsunaga, P. T.; Mavropoulos, J. C.; Hillhouse, G. L. *Polyhedron* **1995**, *14*, 175.
(89) Diversi, P.; Ingrosso, G.; Lucherini, A.; Murtas, S. *J. Chem. Soc. Dalton Trans.* **1980**, 1633.
(90) Diversi, P.; Ingrosso, G.; Lucherini, A.; Lumini, T.; Marchetti, F.; Adovasio, V.; Nardelli, M. *J. Chem. Soc. Dalton Trans.* **1988**, 133.
(91) Canty, A. J.; Skelton, B. W.; Traill, P. R.; White, A. H. *Aust. J. Chem.* **1994**, *47*, 2119.
(92) Ozawa, F.; Yamamoto, A.; Ikariya, T.; Grubbs, R. H. *Organometallics* **1982**, *1*, 1481.
(93) Fröhlich, H.-O.; Wyrwa, R.; Görls, H. *J. Organomet. Chem.* **1993**, *456*, 7.
(94) Binger, P.; Doyle, M. J.; McMeeking, J.; Kruger, C.; Tsay, Y.-H. *J. Organomet. Chem.* **1977**, *135*, 405.
(95) Hashmi, A. S. K.; Grundl, M. A.; Bats, J. W.; Bolte, M. *Eur. J. Org. Chem.* **2002**, 1263.
(96) Hashmi, A. S. K.; Naumann, F.; Bolte, M. *Organometallics* **1998**, *17*, 2385.
(97) Hashmi, A. S. K.; Naumann, F.; Bats, J. W. *Chem. Ber.* **1997**, *130*, 1457.
(98) Hashmi, A. S. K.; Naumann, F.; Rivas Nass, A.; Bolte, M. *J. Prakt. Chem. – Chem. Zeitung* **1998**, *340*, 240.
(99) Hashmi, A. S. K.; Naumann, F.; Probst, R.; Bats, J. W. *Angew. Chem. Int. Ed. Engl.* **1997**, *36*, 104.
(100) Hashmi, A. S. K.; Bats, J. W.; Naumann, F.; Berger, B. *Eur. J. Inorg. Chem.* **1998**, 1987.
(101) Canty, A. J.; Jin, H.; Skelton, B. W.; White, A. H. *Inorg. Chem.* **1998**, *37*, 3975.
(102) Canty, A. J.; Traill, P. R. *J. Organomet. Chem.* **1992**, *435*, C8.
(103) Canty, A. J.; Patel, J.; Rodemann, T.; Ruan, J. H.; Skelton, B. W.; White, A. H. *Organometallics* **2004**, *23*, 3466.
(104) Canty, A. J.; Jin, H.; Roberts, A. S.; Skelton, B. W.; White, A. H. *Organometallics* **1995**, *14*, 199.
(105) Canty, A. J.; Jin, H.; Roberts, A. S.; Skelton, B. W.; White, A. H. *Organometallics* **1996**, *15*, 5713.
(106) Canty, A. J.; Jin, H.; Skelton, B. W.; White, A. H. *J. Organomet. Chem.* **1995**, *503*, C16.
(107) Barker, G. K.; Green, M.; Howard, J. A. K.; Spencer, J. L.; Stone, F. G. A. *J. Chem. Soc. Dalton Trans.* **1978**, 1839.
(108) Green, M.; Howard, J. A. K.; Mitprachachon, P.; Pfeffer, M.; Spencer, J. L.; Stone, F. G. A.; Woodward, P. *J. Chem. Soc. Dalton Trans.* **1978**, 306.
(109) Rashidi, M.; Esmaeilbeig, A. R.; Shahabadi, N.; Tangestaninejad, S.; Puddephatt, R. J. *J. Organomet. Chem.* **1998**, *568*, 53.
(110) Fröhlich, H.-O.; Wyrwa, R.; Görls, H.; Pieper, U. *J. Organomet. Chem.* **1994**, *471*, 23.
(111) Wyrwa, R.; Fröhlich, H.-O.; Görls, H. Z. *Anorg. Allg. Chem.* **2000**, *626*, 819.
(112) Fröhlic, H.-O.; Wyrwa, R.; Görls, H. *Angew. Chem. Int. Ed. Engl.* **1993**, *32*, 387.
(113) Cheetham, A. K.; Puddephatt, R. J.; Zalkin, A.; Templeton, D. H. *Inorg. Chem.* **1976**, *15*, 2977.
(114) Neilsen, W. D.; Larsen, R. D.; Jennings, P. W. *J. Am. Chem. Soc.* **1988**, *110*, 8657.
(115) Chatt, J.; Shaw, B. L. *J. Chem. Soc.* **1959**, 4020.
(116) Alcock, N. W.; Bryars, K. H.; Pringle, P. G. *J. Organomet. Chem.* **1990**, *386*, 399.
(117) Frölich, H-O.; Kosan, B.; Müller, B.; Hiller, W. *J. Organomet. Chem.* **1992**, *441*, 177.
(118) Riley, P. E.; Davis, R. E. *Inorg. Chem.* **1975**, *14*, 2507.
(119) Lindner, E.; Leibfritz, T.; Fawzi, R.; Steimann, M. *Chem. Ber.* **1997**, *130*, 347.
(120) Cetinkaya, B.; Binger, P.; Krüger, C. *Chem. Ber.* **1982**, *115*, 3414.
(121) Hashmi, S. K.; Nass, A. R.; Bats, J. W.; Molte, M. *Angew. Chem. Int. Ed. Engl.* **1999**, *38*, 3370.
(122) Büch, H. M.; Krüger, C. *Acta Cryst.* **1984**, C40, 28.
(123) Dralle, K.; Jaffa, N. L.; le Roux, T.; Moss, J. R.; Travis, S.; Watermeyer, N. D.; Sivaramakrishna, A. *Chem. Commun.* **2005**, 3865.

The Chemistry of Cyclometallated Gold(III) Complexes with C,N-Donor Ligands

WILLIAM HENDERSON*

Department of Chemistry, The University of Waikato, Private Bag 3105, Hamilton, New Zealand

I

INTRODUCTION

The direct activation of arene C–H bonds by gold(III) chloride to give aryl gold(III) compounds $(ArAuCl_2)_2$ has been known since 1931,[1] and has been reinvestigated on a number of occasions,[2] including a recent study.[3] The cyclometallation reaction is a fundamental and elegant route by which the C–H bond activation process can proceed by formation of a metallacyclic complex.[4] In contrast to the vast body of knowledge concerning cyclometallation reactions with palladium(II) and

*Corresponding author. Fax: +64 (7) 838 4219
E-mail: w.henderson@waikato.ac.nz (W. Henderson).

ADVANCES IN ORGANOMETALLIC CHEMISTRY
VOLUME 54 ISSN 0065-3055/DOI 10.1016/S0065-3055(05)54005-X

© 2006 Elsevier Inc.
All rights reserved.

platinum(II), relatively little is known concerning cyclometallation reactions of the isoelectronic gold(III). However, in the last 20 years or so there has been considerable and recently increased activity in this area. Cyclometallated C,N ligands stabilise the gold(III) centre towards reduction to gold(I), and complexes of the type (C,N)$AuCl_2$ are analogues of the widely studied isoelectronic d^8-square planar platinum(II) systems, L_2PtCl_2. The observation that many gold(III) complexes, like their platinum(II) counterparts, can show potent anti-tumour activity, has been a major impetus for the increased interest in this area.

Cycloauration reactions provide products containing five- or less commonly six-membered ring systems, often (but not always) containing a gold(III)–aryl bond; a limited number of examples of cycloauration reactions forming alkyl derivatives are now known. The ligand bears a strong influence on the conditions under which a cycloauration reaction proceeds. Reactions may occur directly, either at room temperature or with heating, in some cases they may require 'chemical assistance' such as the addition of a silver(I) salt. In other cases, the cycloaurated complex may only be obtained by a transmetallation reaction, typically involving an organomercury(II) precursor. Many examples of cycloauration reactions of these various types are now known.

A number of general reviews of organogold chemistry have been published,[5] including ones specifically covering aspects of the synthesis and structure,[6] reactions[7] and applications of organogold compounds.[8] In this review, the chemistry of gold(III) complexes containing cycloaurated ligands is described, with particular emphasis on the synthesis and reactivity of these compounds. Discussion is restricted to compounds where the anionic group is primarily a carbon donor, and the neutral group is a nitrogen donor ligand. A range of nitrogen donor ligand groups have been employed, and the coordinating ability of these can bear an influence on the chemistry of the resulting cycloaurated complex. For example, complexes containing the relatively strongly coordinating NMe_2 group are relatively 'robust' towards displacement reactions with phosphine ligands, but those containing the azo (–N=N–) group are more reactive. The nitrogen donor ability bears an influence on reactions such as biaryl formation, and C–H activation.

Owing to the substantial body of knowledge that has now been developed in this field (with seminal contributions from a number of research groups, including those of Vicente, Parish, Abram, Fuchita, Minghetti, Manassero and Cinellu, to name but a few) it has not been possible to be fully comprehensive. Instead, an attempt has been made to survey the main types of cycloaurated complexes that have been synthesised by various methods, and to report on the use of these complexes as substrates for further reactivity studies.

II
SYNTHESIS OF FIVE-MEMBERED RING CYCLOAURATED COMPLEXES

This section describes the various reaction pathways that lead to the formation of five-membered ring cycloaurated complexes; six-membered cycloaurated ring systems are described in Section III and complexes containing two fused rings are classified according to the size of the *cyclometallated* ring. Ligand substitution reactions – primarily of the ancillary (halide) ligands – together with reactions involving modification of the coordination mode, or displacement of, the cycloaurated ligand, are described in Section IV.

A. *Direct Cycloauration Reactions*

In the first work in this area, Constable and Leese[9] investigated the reactivity of 2-phenylpyridine towards $H[AuCl_4]$ or $Na[AuCl_4]$. The initial product isolated under a range of conditions was the yellow N-bonded complex **1**. When this compound was heated in aqueous acetonitrile, the white cyclometallated complex **2** (also termed $ppyAuCl_2$) was obtained.[9] Eisenberg and co-workers[10] have reported on the poor reproducibility of this solution thermolysis procedure, and investigated thermolysis in the solid state, by thermogravimetric analysis (TGA) under nitrogen. Decomposition of **1** to the cycloaurated species **2** began at 150 °C and continued up to 220 °C, suggesting that a higher reaction temperature is required for optimal conversion. The cycloaurated product was found to be stable up to 360 °C, at which point decomposition to metallic gold occurred. Preparative solid-state thermolysis has also been subsequently reported in the Russian literature.[11]

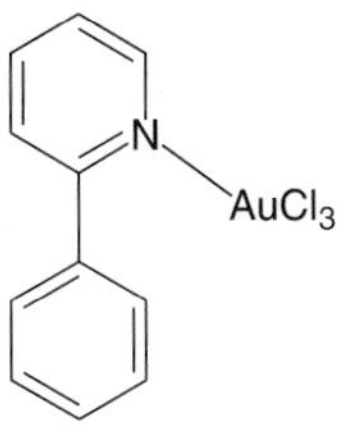

1 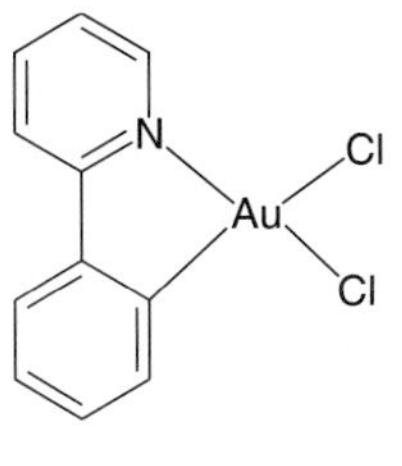 **2** 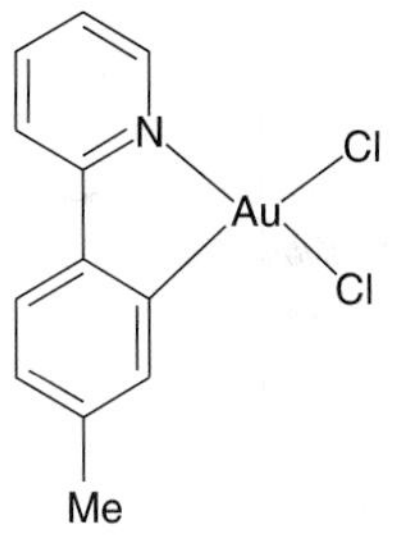 **3**

Because of the potential for phenyl and pyridyl rings of cycloaurated phenylpyridyl gold derivatives to be indistinguishable by X-ray diffraction methods,[12] an alternative complex has been developed using commercially available 2-*p*-tolyl)pyridine in place of 2-phenylpyridine.[13] The resulting cycloaurated complex **3** contains a methyl substituent on the aryl carbon, reducing any possibility of crystallographic disorder and eliminating indistinguishability.

Heterocyclic derivatives also undergo direct cycloauration reactions. Isomeric 2-(3-thienyl)pyridine **4** and 2-(2-thienyl)pyridine **5** react with $Na[AuCl_4]\cdot 2H_2O$ to form the $AuCl_3$ adducts **6** and **7**. The 3-thienyl complex undergoes direct cycloauration upon heating to 80 °C in $MeCN/H_2O$, giving **8** in 81% yield; this complex can also be conveniently obtained by the direct reaction of the ligand with $Na[AuCl_4]\cdot 2H_2O$ in $MeCN/H_2O$.[14] In contrast, the corresponding 2-thienyl derivative **7** does not give the analogous cycloaurated product **9** under the same conditions as for **8**; this complex was instead cycloaurated using Ag[BF_4] in refluxing dichloromethane (refer Section II.B).

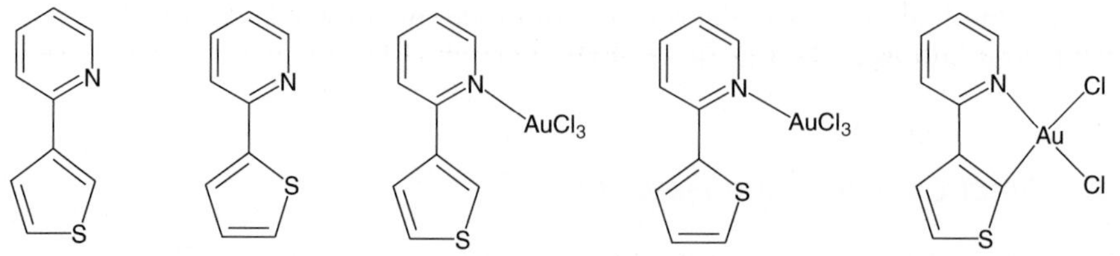

4 **5** **6** **7** **8**

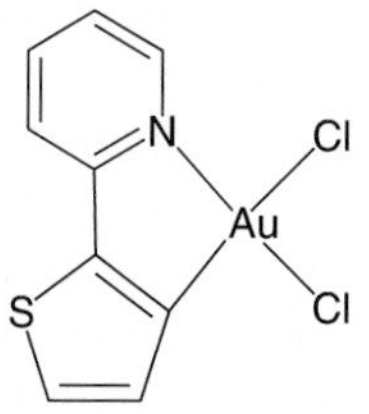

9

In a related system, reaction of 6-(2″-thienyl)-2,2′-bipyridine (HL) with [$AuCl_4$]$^-$ was initially reported by Constable *et al.*[15,16] to give a cycloaurated product, but subsequent work by the same group identified this assignment as incorrect. Reaction of the ligand with Na[$AuCl_4$] in aqueous MeCN gave the non-cycloaurated complex $AuCl_3$(HL).

Most examples of direct cycloauration reactions have concerned reaction of an sp^2 C–H bond, but there are a few examples of the activation of an sp^3 C–H bond. Reaction of Na[$AuCl_4$] with the 2-pyridyl derivatives **10** at room temperature gave the cycloaurated complexes **11** (R = Ph, Me or OMe).[17] The analogous bromo complexes were obtainable from **10** upon reaction with Na[$AuBr_4$], or by metathesis of the chloro complexes **11** with NaBr. 2-Acetylpyridine has also been found to undergo direct cycloauration with Na[$AuCl_4$] · $2H_2O$, giving **12** as a white solid. The reaction proceeds in water under mild conditions (60 °C, 3 days), making this a potentially useful precursor complex.[18]

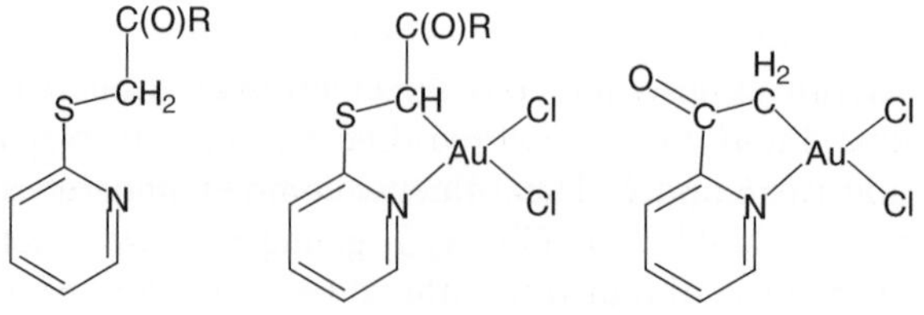

10 **11** **12**

B. *Silver Ion-Assisted Cycloauration Reactions*

In a number of cases, the use of silver(I) salts assists the cyclometallation process, by abstraction of chloride and formation of a reactive coordination site. As an illustrative example, reaction of 2-phenylthiazole (LH, **13**) with $H[AuCl_4]\cdot 4H_2O$ gave the salt $[H_2L][AuCl_4]$ **14**, while reaction with $Na[AuCl_4]\cdot 2H_2O$ or $AuCl_3\cdot 4H_2O$ gave the adduct $[AuCl_3(LH)]$ **15**. When the adduct was heated with $Ag[BF_4]$ in 1,2-dichloroethane, cycloauration occurred giving **16**.[19] Similar behaviour was observed with 1-ethyl-2-phenylimidazole, **17**;[20] reaction of this compound with $AuCl_3\cdot 4H_2O$ gave the protonated salt **18**, but reaction with $H[AuCl_4]\cdot 4H_2O$ gave the neutral $AuCl_3$ adduct **19**, which was cycloaurated to **20** on refluxing in dichloromethane with $Ag[BF_4]$. These were the first examples of cycloauration of aromatic rings other than pyridine derivatives.

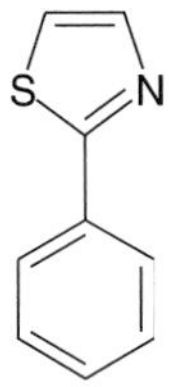

13 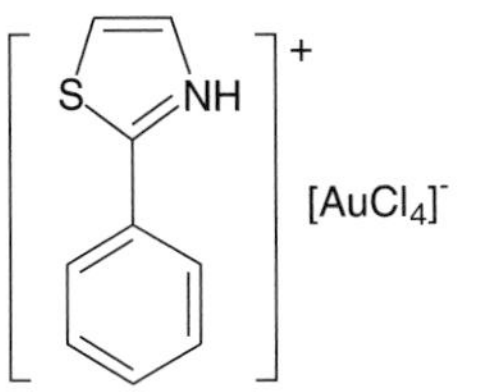 **14** 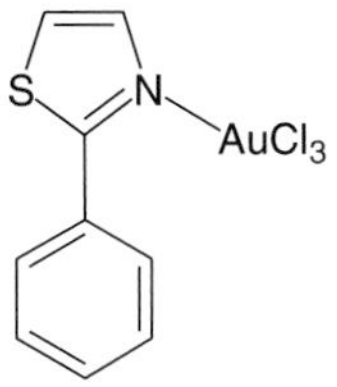 **15** 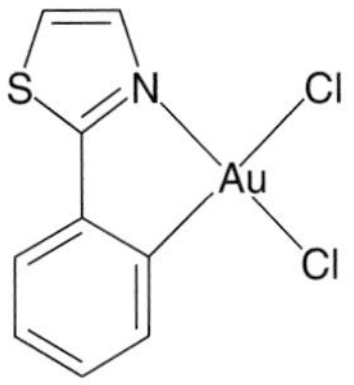 **16**

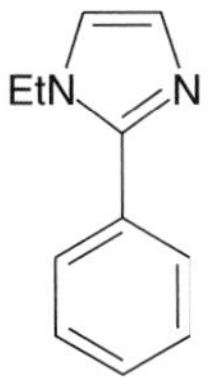

17 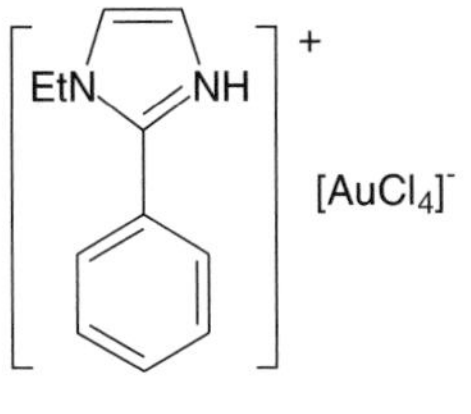 **18** 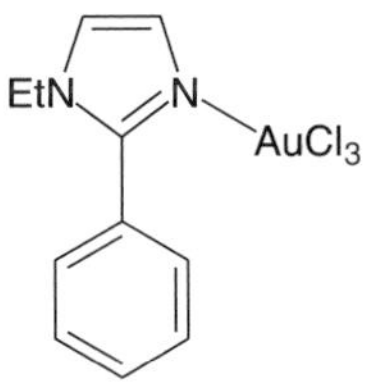 **19** 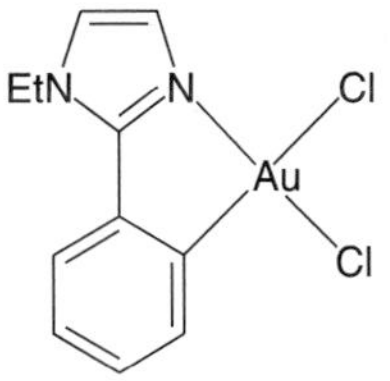 **20**

As described in Section II.A, the complex 2-(3-thienyl)pyridine$AuCl_3$ **6** undergoes direct cycloauration in refluxing $MeCN/H_2O$, but the 2-thienyl isomer **7** does not.[14] However, reaction of **7** with $Ag[BF_4]$ in dichloromethane did give the desired product **9**.

Ligands containing aryl-substituted 2,2′-bipyridine or 1,10-phenanthroline groups can also undergo silver ion-assisted direct cycloauration reactions, to give products containing two chelate rings. Reaction of the substituted bipyridine **21** with $K[AuCl_4]$ in the presence of $[CF_3SO_3]Ag$ in hot MeCN for 72 h gave the cycloaurated product **22**.[21] The corresponding reaction of $K[AuCl_4]$ with 2,9-diphenyl-1,10-phenanthroline and $[p\text{-}CH_3C_6H_4SO_3]Ag$ produced the complex **23**, which was obtained as the $[p\text{-}CH_3C_6H_4SO_3]^-$ or $[ClO_4]^-$ salts.[22]

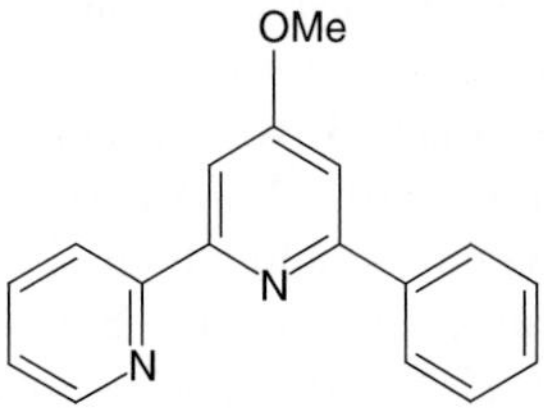

21 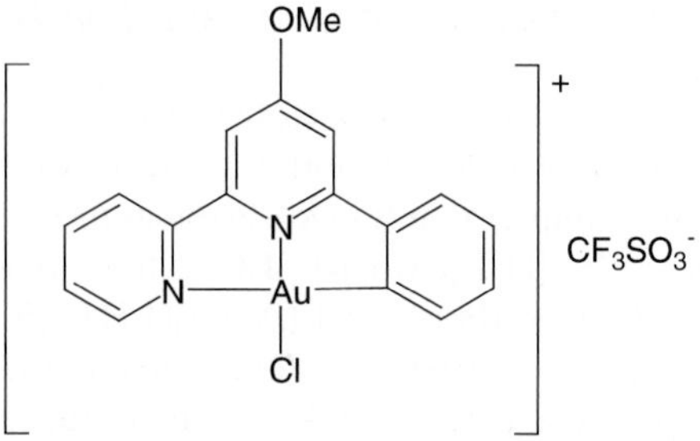 **22**

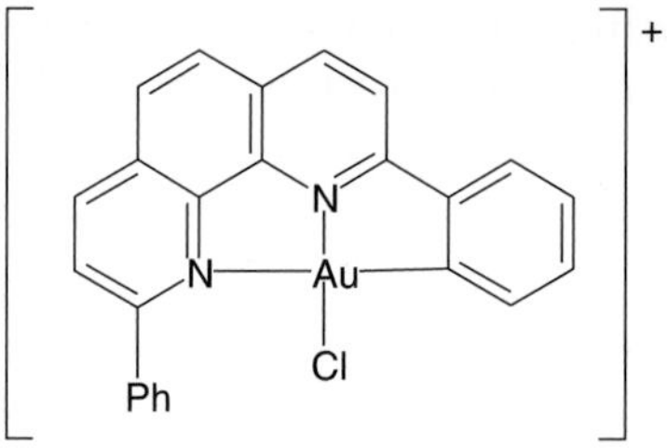

23

6-*t*-Butyl-2,2′-bipyridine, containing a *tert*-butyl substituent in the 6-position readily forms a neutral $AuCl_3$ adduct which undergoes $Ag[BF_4]$-assisted cycloauration in acetone solution to give complex **24**, which was structurally characterised.[23] This is another (see also **11**, **12**) relatively rare example of C–H activation of an sp^3 C–H bond. In the same paper, cycloauration of a range of related 6-benzylbipyridines was discussed, but C–H activation instead occurs at the phenyl substituent, giving six-membered rings; these complexes are discussed in Section III.

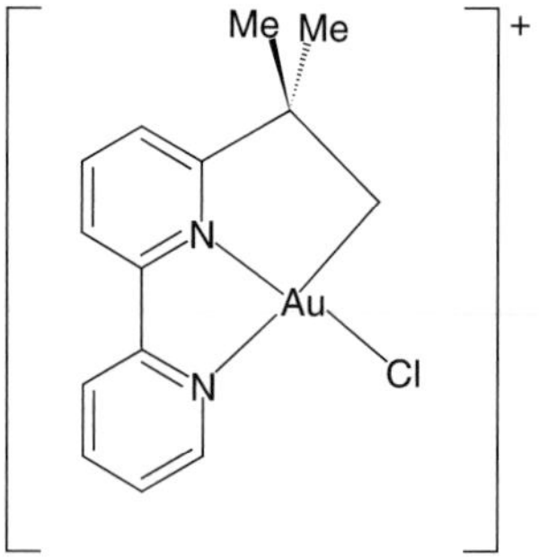

24

C. *Transmetallation Reactions*

The transmetallation reaction between a gold(III) precursor [invariably an $AuCl_4^-$ salt or an $LAuCl_3$ adduct] and an organomercury substrate is perhaps the most versatile methodology for the synthesis of cycloaurated complexes, by transfer of the organic group from Hg to Au. The method is particularly useful for substrates which do not undergo direct cycloauration reactions. The organomercury compounds, of the type RHgCl (or in some cases R_2Hg) are readily prepared by reaction of the starting organic compound with *n*-butyl lithium, followed by reaction of the resulting organolithium with $HgCl_2$. Transmetallation reactions are often carried out with added $[Me_4N]Cl$, which results in the formation of the relatively insoluble chloromercurate salt $[Me_4N]_2[Hg_2Cl_6]$, which is easily separated from the cycloaurated product.

The widely studied cycloaurated azobenzene complex **25** [papAuCl$_2$] has been known for many years, and is prepared by reaction of the organomercury complex papHgCl **26** with $[AuCl_3(tht)]$ (tht = tetrahydrothiophene, SC_4H_8), in the presence of added $[Me_4N]Cl$.[24,25] Previously, attempts to cycloaurate azobenzene (PhN =NPh) with gold(III) halides produced only adducts $[AuX_3(PhN{=}NPh)]$.[26,27] Analogous complexes such as **27** and **28** containing *para* butyl or butoxy substituents on the phenyl rings have been prepared from the corresponding di-aryl mercury compounds, for investigation of their potential liquid crystallinity properties.[28] Gold(III)–aryl complexes are also able to serve as precursors for reaction with cycloaurating reagents; reaction of the corresponding di-organomercury derivative $[Hg(C_6H_4NNPh)_2]$ **29** with the anionic gold(III) mono-aryl complex **30** gave the cycloaurated product **31**.[29]

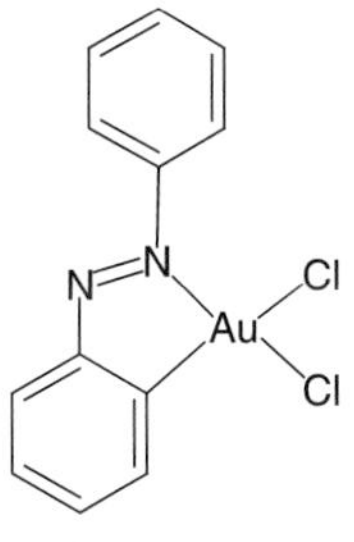

25 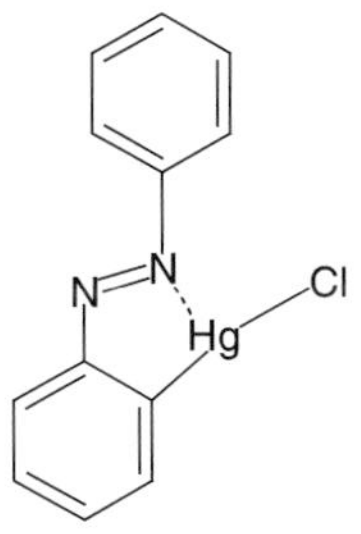 **26** 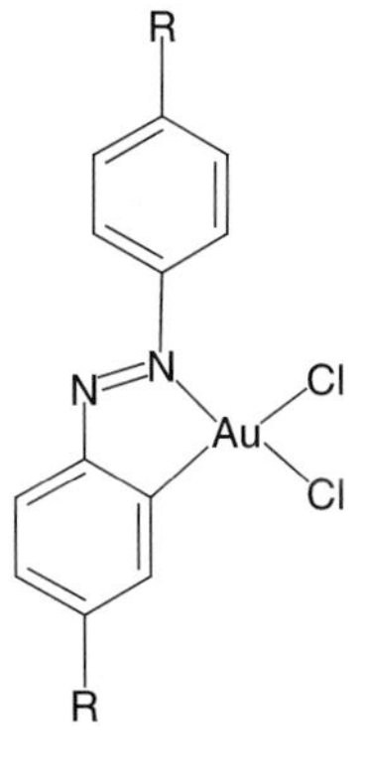 **27**, R = ^{n}Bu **28**, R = ^{n}BuO **29**

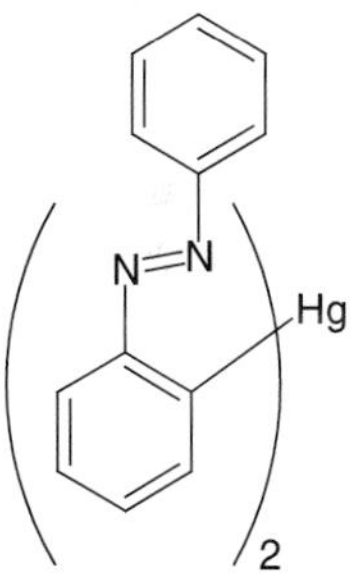

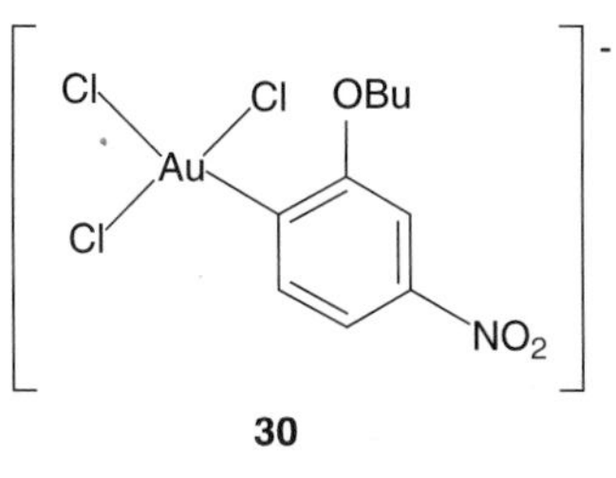

30 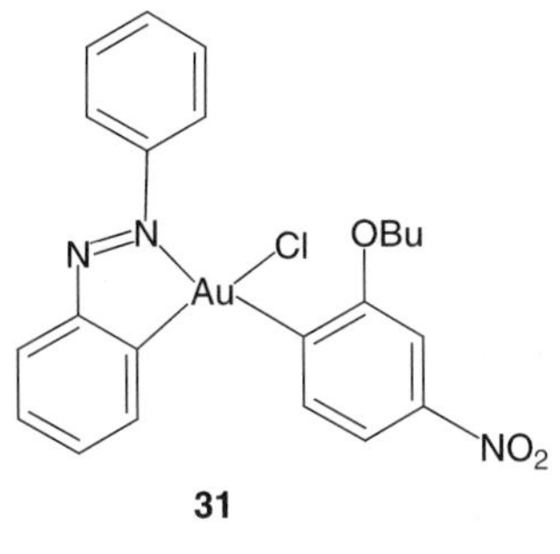 **31**

Related systems have also been cycloaurated *via* organomercury reagents; reaction of the imine-containing organomercury complex **32** with $[Me_4N][AuCl_4]$ gave **33**, the first example of a cycloaurated imine complex.[30] Similarly, reaction of $Ph_3P{=}NPh$ with $^{n}BuLi$ and $HgCl_2$ has recently been found to give the new organomercury complex **34**, which reacts with $[Me_4N][AuCl_4]$ to give the new derivative **35**.[31]

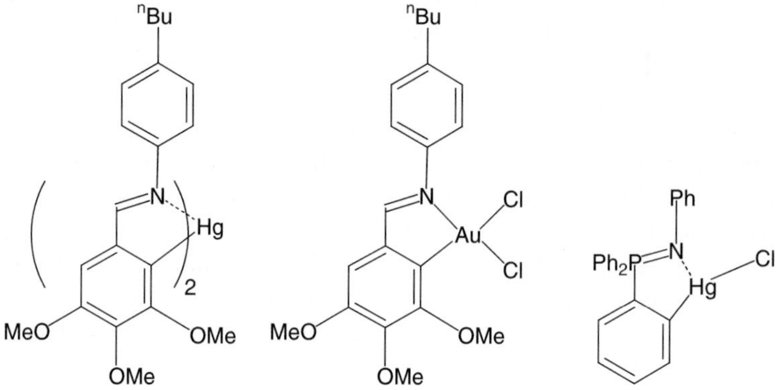

32 **33** **34**

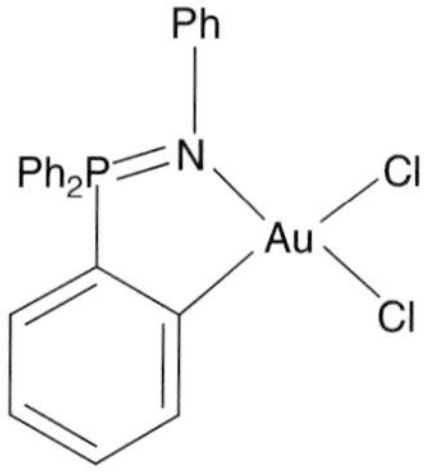

35

The 2-(*N*,*N*-dimethylaminomethyl)phenyl ($C_6H_4CH_2NMe_2$, damp) complex **36**, containing a coordinated tertiary amine donor, is perhaps the most widely studied of all cycloaurated complexes. It produces downstream products with excellent solubility and NMR spectroscopic characteristics, and in some cases excellent anti-tumour activity (Section V), which are probably the reasons for its widespread adoption. This complex has been synthesised by reaction of the organomercury precursor dampHgCl (**37**) with $Na[AuCl_4]$ alone,[32] with $Na[AuCl_4]/[Me_4N]Cl$[33] with $[Me_4N][AuCl_4]$[34] or with $AuCl_3(tht)/[Me_4N]Cl$.[34] The organomercury complex can be prepared by selective *ortho*-lithiation of $PhCH_2NMe_2$ with *n*-butyl lithium, followed by reaction with $HgCl_2$. Complex **36** can also be prepared by reaction of $[Me_4N][AuCl_4]$ with the bis(organo)mercury reagent $Hg(C_6H_4CH_2NMe_2)_2$.[34] However, attempted direct reaction of the organolithium reagent with $AuCl_3(tht)$ or $[Me_4N][AuCl_4]$ did not yield **36**. The molecular structure of **36** has been determined,[33] Fig. 1, and confirms the presence of the five-membered cycloaurated ring, and two *cis* chlorine atoms. As a result of the differing *trans*-influences of the aryl C- and N-donor groups, the two gold–chlorine bond distances are rather different; Au–Cl(1) *trans* to the aryl group is longer [2.387(2) Å] than Au–Cl(2) [2.285(2) Å], which is *trans* to the NMe_2 group.

36 **37**

The proposed mechanism of the transmetallation reaction is shown for the synthesis of dampAuCl_2 in Scheme 1, and involves initial coordination of the amine donor to gold, forming a bimetallic intermediate; the organomercury precursor generally has a weak (if any) interaction between the metal and amine donor. Transfer of the aryl group from mercury to gold, with concomitant elimination of $HgCl_2$ furnishes the cycloaurated product. The chloride ligands can readily be

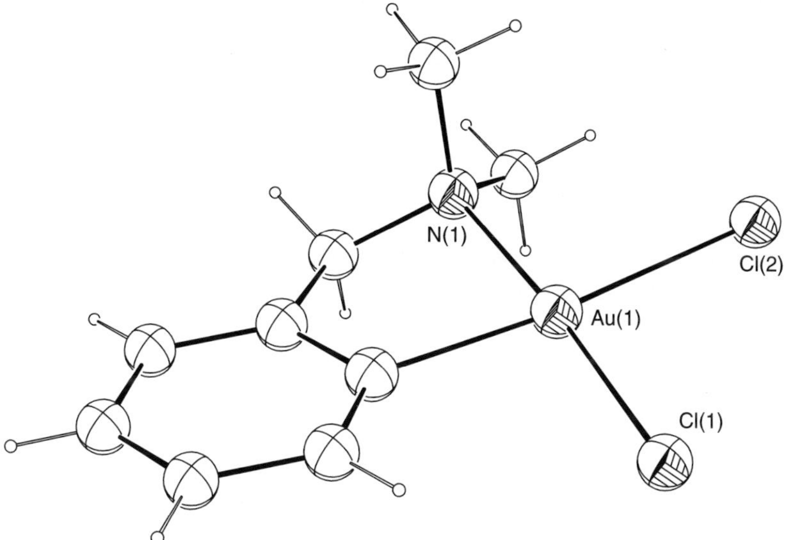

FIG. 1. Molecular structure of [AuCl_2($C_6H_4CH_2NMe_2$-N,C)], [dampAuCl_2, **36**]. Selected bond distances (Å); Au–C 2.046(8), Au–N 2.087(6), Au–Cl(1) 2.387(2), Au–Cl(2) 2.285(2).

SCHEME 1.

replaced by a wide range of other anionic ligands, including halides, pseudohalides, carboxylates etc (Section IV).

Bonnardel and Parish[35] have subsequently reported the syntheses and spectroscopic properties of a wide range of organomercury compounds, which can be used in the synthesis of new cycloaurated gold(III) dichloride complexes, by reaction with $[Me_4N][AuCl_4]$ in dry acetone or acetonitrile.[36] Compounds prepared included complexes with oxazoline nitrogen donors (**38**, **39** and **40**), and compounds with methoxy- and amino-functionalised damp ligands (**41a**, **42a**, **43a** and **44**). Analogous naphthyl derivatives **45a** and **45b** were also obtained, but without the addition of $[Me_4N]Cl$, which appeared to inhibit the synthesis. The 'pincer' complex **46** was also obtained as its $[Hg_2Cl_6]^{2-}$ salt **46a** in the absence of $[Me_4N]Cl$, but in aqueous solution the corresponding $[AuCl_4]^-$ salt **46b** was formed. The cycloaurated sulfonamide and amide compounds **47** and **48** were also formed (but not isolated) when dimethyl sulfoxide (DMSO) was used as the solvent, but not in acetone or acetonitrile, indicating that the solvent may play an important role in these transmetallation reactions.

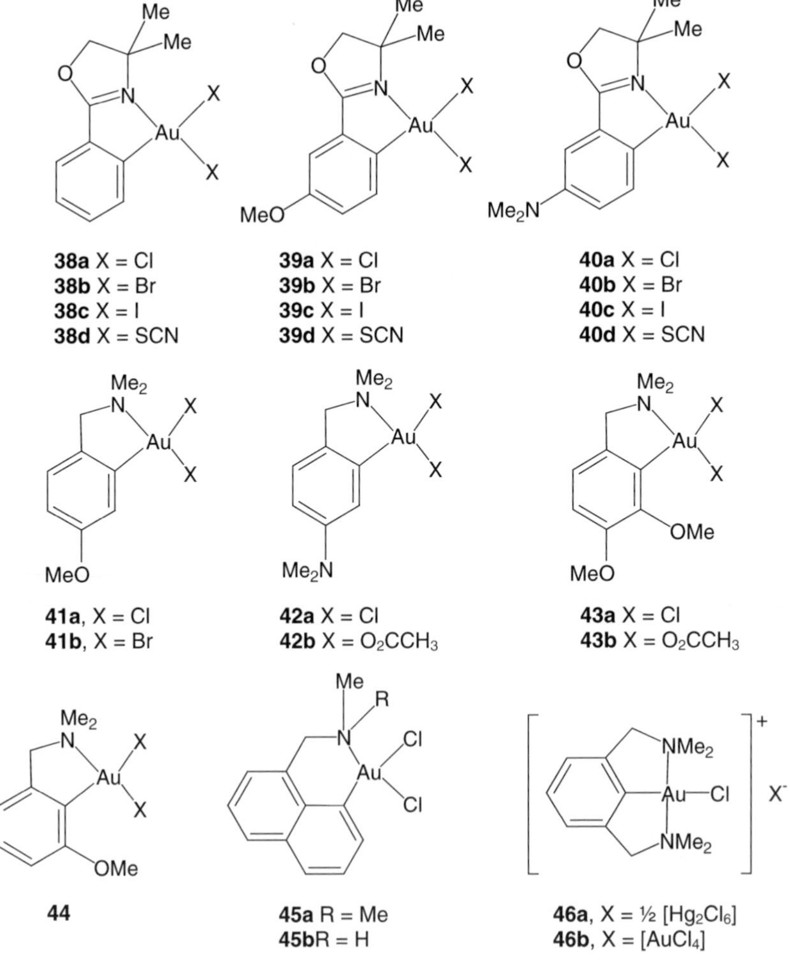

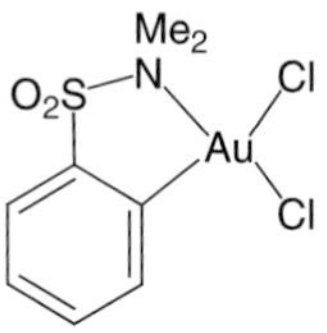

47

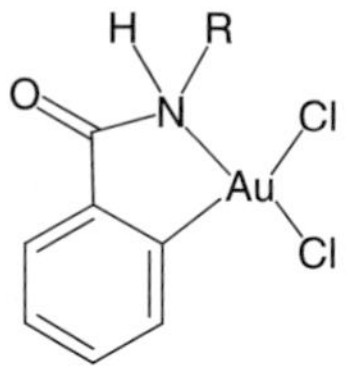

48a, R = H
48b, R = Et

The cycloaurated phenylpyridine complex **2**, which was synthesised in early work by Constable and Leese[9] by direct cycloauration can also be obtained by reaction of $H[AuCl_4]$ with LHgCl **49**. The organomercury chemistry has also been extended by Parish and co-workers to the synthesis of a number of cycloaurated 2-phenylpyridine compounds, including the parent complex **2**, but also **50–53** containing various substituents on the pyridine ring. The cycloaurated product derived from 2-phenyl-4-methylcarboxylato)quinoline **54** was also reported.[12] This route to phenylpyridyl derivatives is more efficient than the direct cycloauration route using $[AuCl_4]^-$, described in Section II.A.

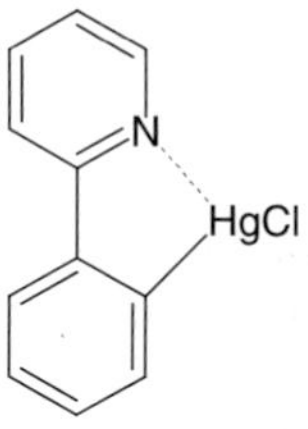

49

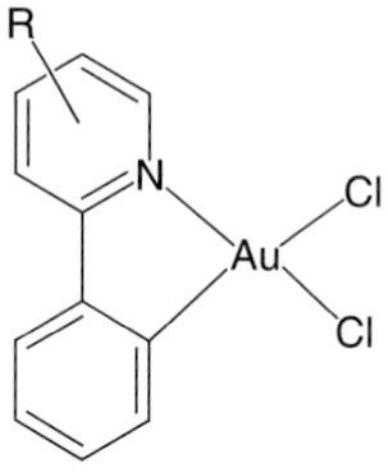

50, R = 3-Me
51, R = 3,5-Me_2
52, R = 4-^{n}Pr
53, R = 4-^{t}Bu

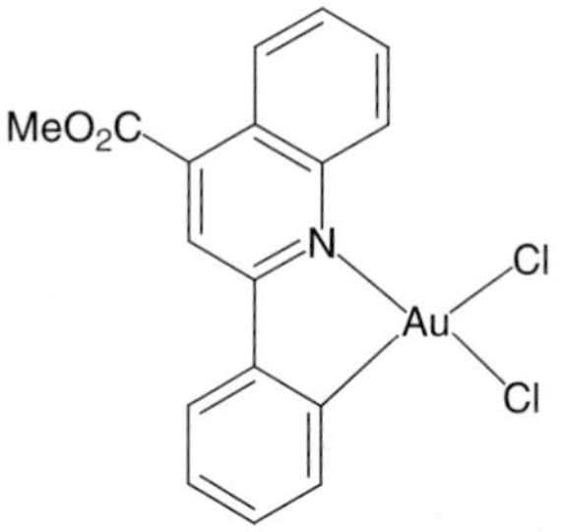

54

2,6-Diphenylpyridine has been developed as a tridentate C–N–C ligand through the formation of two five-membered cycloaurated rings in complex **55**. Again, this complex was developed using organomercury chemistry by the way of **56**.[37] The corresponding derivative with *t*-butyl substituents **57** has also been prepared.[38]

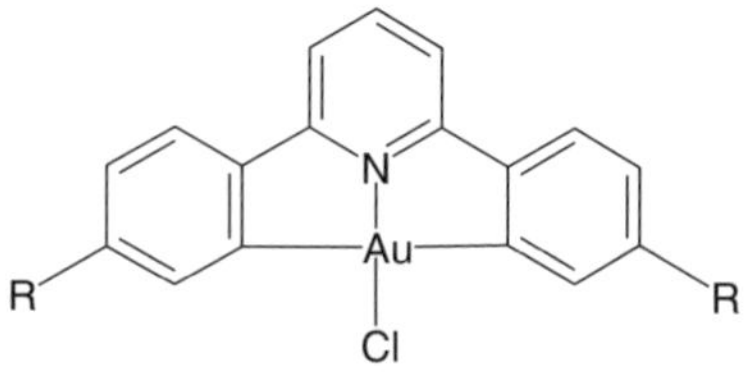

55; R = H
57; R = ^{t}Bu

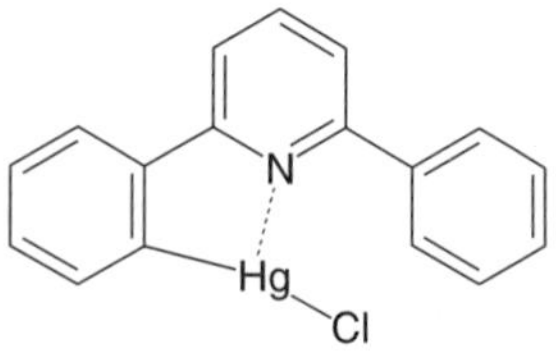

56

2,2′-Bipyridine-derived cycloaurated complexes have also been synthesised through the transmetallation route. Conversion of 6-phenyl-2,2′-bipyridine to its organomercury derivative **58**, followed by reaction with Na[$AuCl_4$] gave the cycloaurated complex **59** in 38% yield as its $[AuCl_4]^-$ salt.[39]

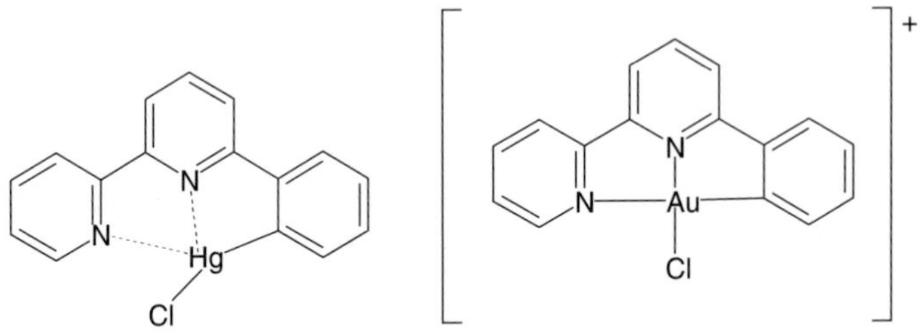

58 **59**

D. *Complexes Containing Two Potentially Cycloaurating Ligands*

As part of a search for penta-coordinated gold(III) complexes, the reactivity of cycloaurated gold(III) complexes with potentially cycloaurating organomercury reagents has been explored. Reactions of damp$AuCl_2$ **36** with either 0.5 equivalents of Hg(2-$C_6H_4CH_2NMe_2$)$_2$, or with 1 equivalent of dampHgCl **37**, both in the presence of excess [Me_4N]Cl, give the complex [Au(damp)$_2$Cl], **60**. This complex is a non-conductor, and IR and NMR spectroscopic data indicate the presence of both mono- and bidentate damp ligands, which undergo fluxional exchange *via* a four-coordinate bis(cyclometallated) intermediate **61a**, Scheme 2. Reaction of **60** with cyanide gave the corresponding cyano complex [Au(damp)$_2$(CN)], which also has one monodentate and one bidentate damp ligand, and is fluxional in solution. However, here the proposed interconversion proceeds through a five-coordinate intermediate because of the strength of the Au–CN bond compared to Au–Cl in **60**. Reaction of **60** with Ag[ClO_4] gave the ionic bis(cycloaurated) complex $[Au(damp)_2]^+ ClO_4^-$ **61b**.[40]

Other bis(cycloaurated) complexes related to **61** have been synthesised by closely related chemistry; reaction of equimolar amounts of **29** and [Me_4N][$AuCl_4$] gave [Au(azo)$_2$Cl] **62**, which with Ag[ClO_4] gave the corresponding cationic bis(cycloaurated)

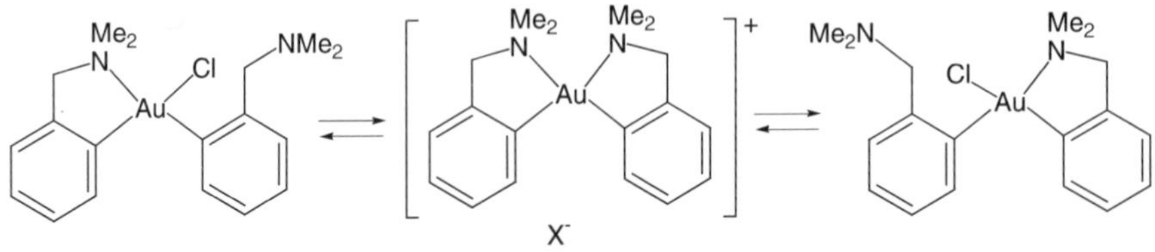

60

61a; X = Cl
61b; X = ClO_4

SCHEME 2.

complex **63**, with two cycloaurated azobenzene ligands.[41] Addition of a metal salt, M^+X^-, to **63** results in displacement of one of the azo donors and formation of a range of complexes **64** containing mono- and bidentate azobenzene ligands, plus a coordinated anion X. Similarly, reaction of the quinolinecarboxylate derivative **54** with one molar equivalent of the corresponding organomercury chloride gave complex **65**, which again contains one monodentate and one cycloaurated ligand. This product was characterised by an X-ray structure determination.[12]

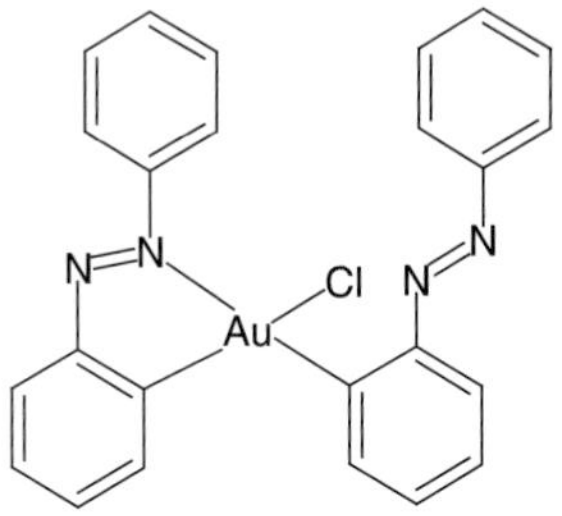

62

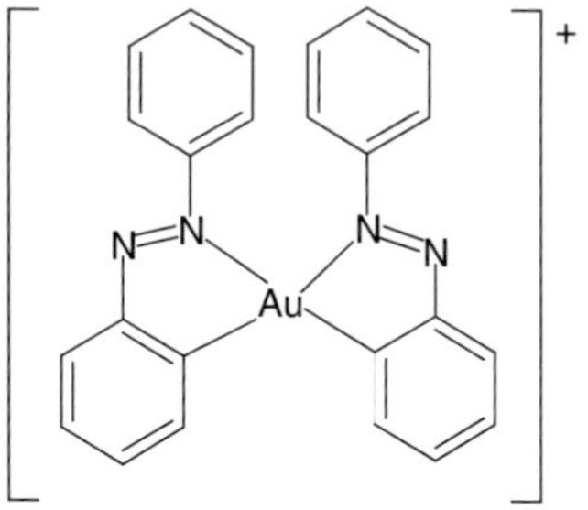

63

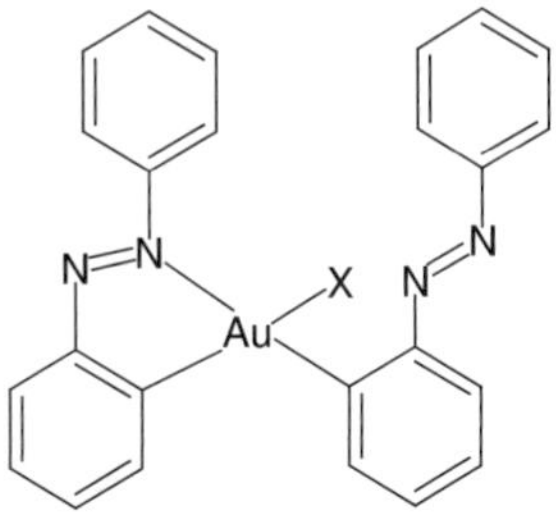

64; X = Br, I, CN, O_2CMe

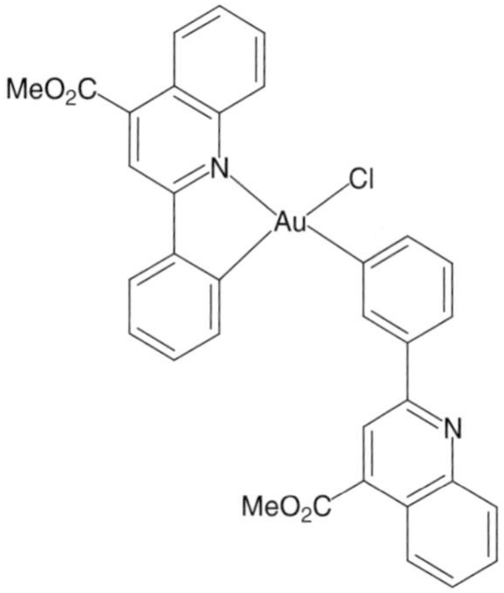

65

Compounds with two different cycloaurated ligands have also been reported. Reaction between dampAuCl_2 **36** and Hg(pap)$_2$ **29** gave complex **66a**, which has the more strongly coordinating NMe_2 group forming the cycloaurated ring.[42] This complex is also formed, but in lower yield, and with metallic gold as a by-product, from reaction of papAuCl_2 **25** with dampHgCl **37**. Removal of the chloride ligand of **66a** using Ag[ClO_4] gave the bis(cycloaurated) complex **67a** as its perchlorate salt. This could be converted into the tetrachloroaurate salt **67b** on reaction with [Me_4N][AuCl_4]. The molecular structure of the cation of **67** is shown in Fig. 2, which confirms the presence of two different cycloaurated ligands. Addition of

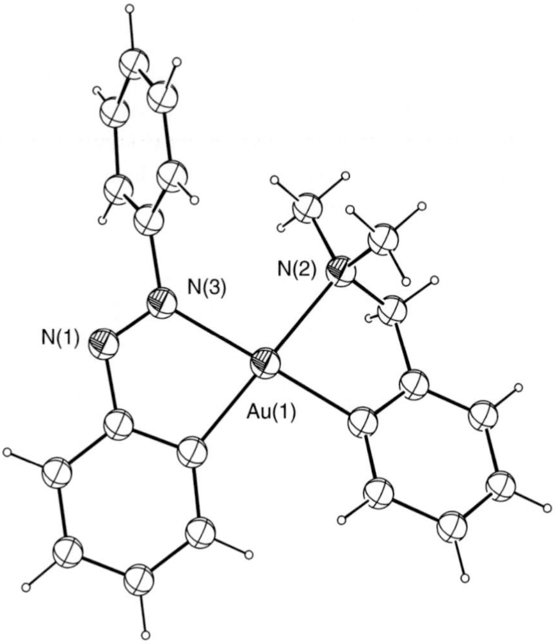

FIG. 2. Molecular structure of the cationic complex $[Au(pap)(damp)]^+$ **67**, showing the presence of two different cycloaurated ligands.

various ligands to the bis(cycloaurated) species **67a** resulted in displacement of the weaker azo donor ligand, resulting in the formation of four-coordinate complexes **66b** (by reaction with anions X^-) and **68** (by reaction with neutral ligands).[42]

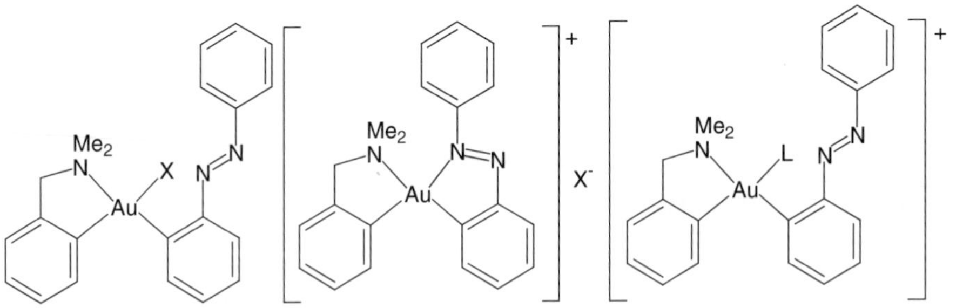

66a, X = Cl
66b, X = CN or O_2CCH_3

67a, X = ClO_4
67b, X = $AuCl_4$

68, L = PPh_3, pyridine

Electrochemical or chemical reduction has also been found to result in the formation of a bis(cycloaurated) complex, and is described in Section V.

E. *Miscellaneous Syntheses*

Reaction of [AuCl(tht)] with the organolithium reagent **69** gave the highly reactive dinuclear gold(I) complex **70**, which reacted with CH_2I_2 to give the unusual gold(III) complex **71**,[43] containing a five-membered cycloaurated ring analogous to the well-known damp systems.

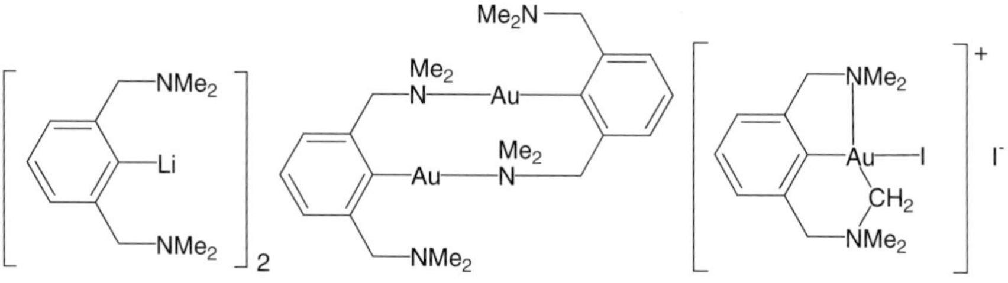

69 **70** **71**

F. *Unsuccessful Cycloauration Systems*

Relatively few 'unsuccessful' cycloauration reactions, potentially giving five-membered cycloaurated rings, have been reported in the literature; some illustrative examples are given in this section. Reaction of 6-(2″-thienyl)-2,2′-bipyridine with $Na[AuCl_4]$ in aqueous MeCN at 90 °C gave a red-brown dimeric product **72**, containing a thiophene group aurated at the (most electrophilic) C(5) position.[16] A similar C-auration of a thienyl ring occurred when the 2-thienylpyridine complex **7** was heated in dichloromethane, giving **73**.[14]

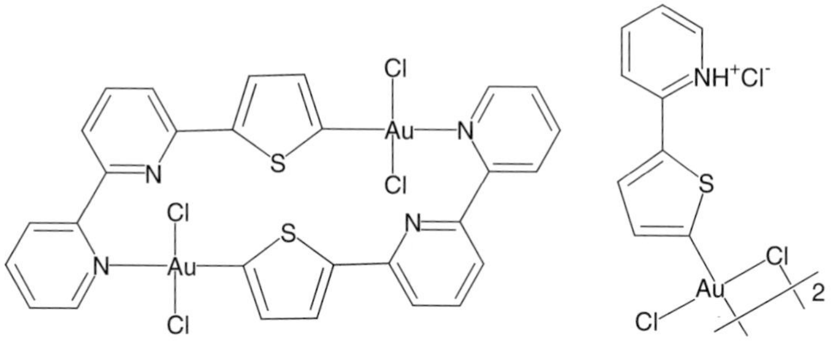

72 **73**

Another example comes from the reactions of 1-phenylpyrazole (HL, **74**) with gold(III) compounds, which do not give a C,N-cycloaurated product. The expected 1:1 adduct $[Au(HL)Cl_3]$ (where the deprotonated ligand is bonded through the nitrogen atom) is obtained from reactions of $Na[AuCl_4]\cdot 2H_2O$ or $AuCl_3\cdot xH_2O$ and **74** in water. However, reaction with Au_2Cl_6 in dichloromethane gave a complex $[Au(L)Cl_2\cdot HCl]_n$, where the gold is coordinated to the 4-C atom of the pyrazole ligand. Deprotonation using NaOH or KOH gave a material analysing as $[Au(L)Cl_2]_n$, which may be dimeric or oligomeric. Reaction instead with K_2CO_3 or proton sponge [1,8-bis(dimethylamino)naphthalene, B] gave the salts M[Au(L)Cl_3] **75** (M = BH or K), also containing the C-bonded phenylpyrazole ligand. Similarly, reaction of the $[Au(L)Cl_2]_n$ product with 3,5-dimethylpyridine or $P(C_6H_4Me\text{-}p)_3$ gave the neutral adducts **76** and **77**, respectively. The 3,5-dimethylpyridine adduct was characterised by an X-ray diffraction study, which confirmed the *trans* isomer and the C-bonded pyrazole ring.[44]

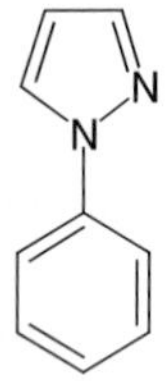

74 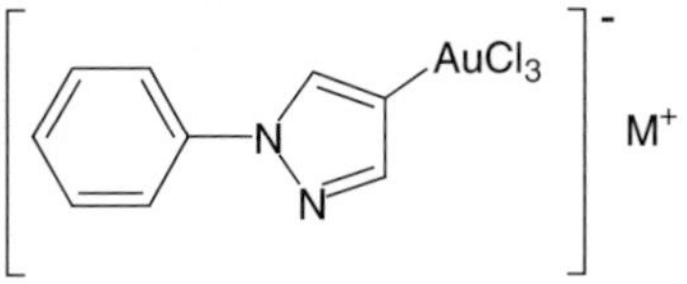 **75**

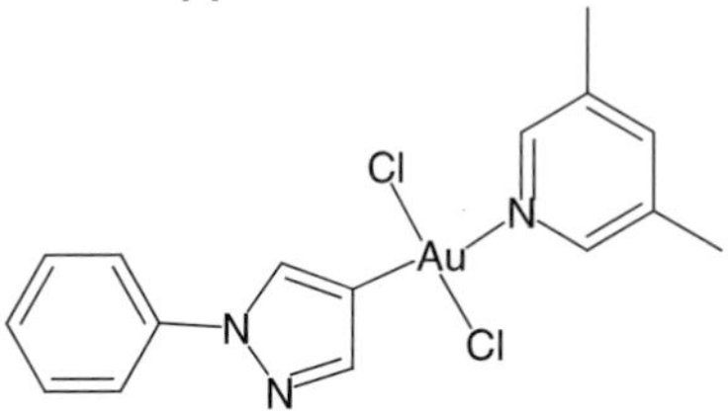

76 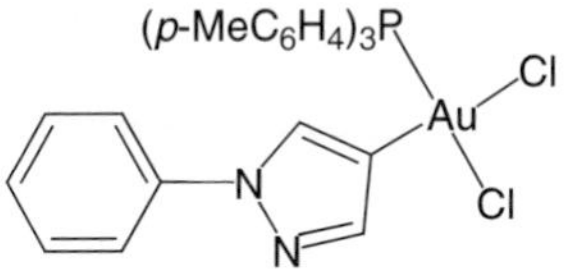 **77**

Given the widespread utility of complexes containing the cycloaurated damp and pap ligands, a direct route to such complexes would be highly desirable, though no such route has been found. Reaction of gold bromide with $PhCH_2NMe_2$ gave $[AuBr_3(Me_2NCH_2Ph)]$, $[PhCH_2NMe_2H]^+[AuBr_4]^-$ or $[PhCH_2NMe_2H]^+[AuBr_2]^-$ depending on the reaction conditions. Likewise, azobenzene gave [$AuBr_3$(PhNNPh)] and 2-vinylpyridine gave [$AuBr_3$(2-vinylpyridine)]. Attempts at cycloaurating these complexes were unsuccessful.[27]

Attempts at cycloaurating 1,4-benzodiazepin-2-ones (L) such as **78** (R = Me, CH_2 cyclopropyl) have also been unsuccessful, with only $LAuCl_3$ adducts being obtained through nitrogen coordination;[45] such ligands can, however, by cyclometallated with palladium(II) and platinum(II) substrates, demonstrating distinct differences in reactivity between the metals.[46]

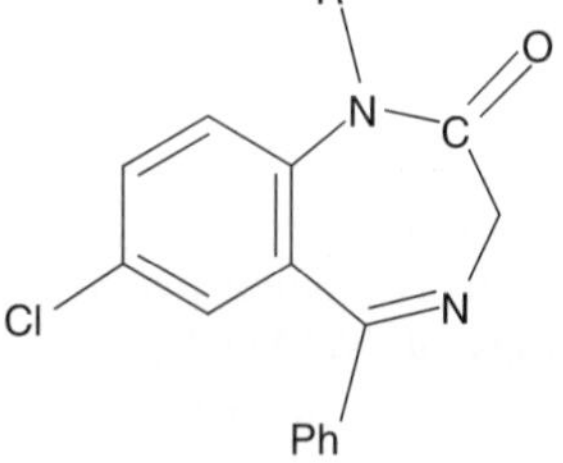

78

G. *Related Compounds with Anionic Nitrogen Donor Ligands*

Related complexes with anionic nitrogen donor ligands are also included briefly in this review, because of general similarities with their organometallic counterparts. For example, two reports appeared in 2003 concerning the reactivity of picolinamide towards $H[AuCl_4]$ or $Na[AuCl_4]$ to give the yellow cycloaurated complex **79** containing pyridine and deprotonated amide donor ligands.[47,48] The crystal structure of this complex has been reported.[47,48] The related pyridine–amide complex **80** $[AuCl_2(HL)]$ has been prepared by reaction of $K[AuCl_4]$ with the parent amide H_2L[49] and complexes containing chelated pyridine–amide complexes **81** and **82** have been prepared by reaction of the parent ligand with $H[AuCl_4]$ in methanol.[50]

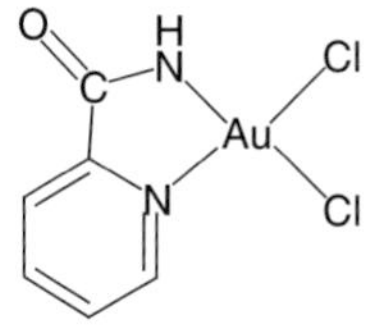

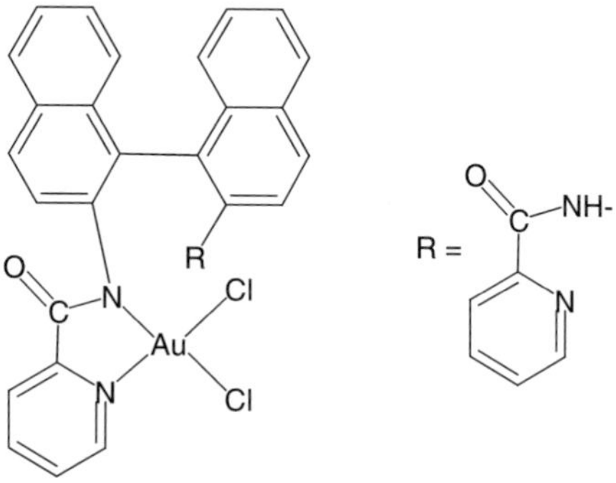

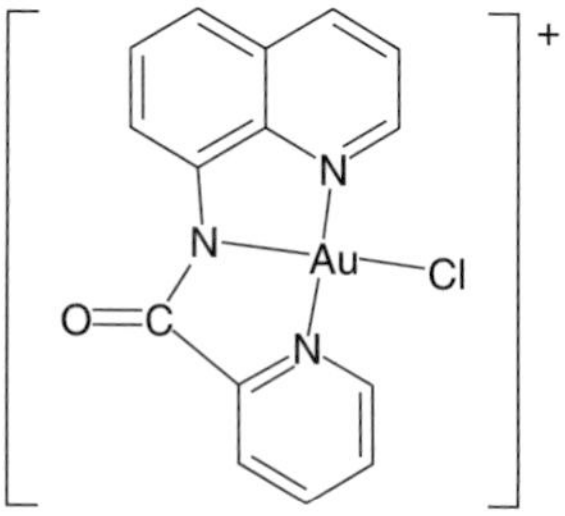

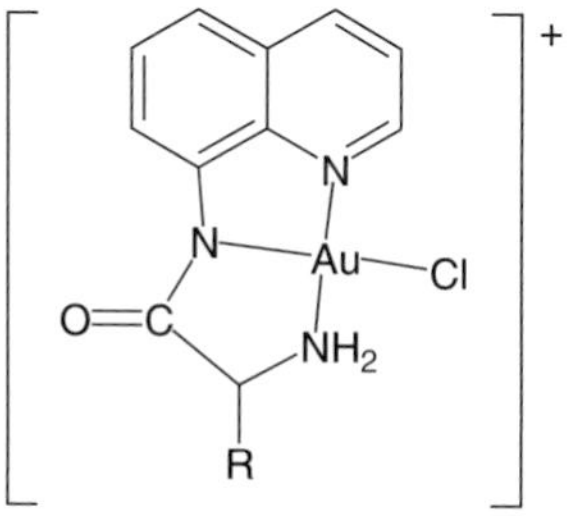

III
SYNTHESIS OF SIX-MEMBERED RING CYCLOAURATED COMPLEXES

A. *Direct Cycloauration Reactions*

One of the most facile, direct cycloauration reactions is that of an $[AuCl_4]^-$ source with (commercially available) 2-anilinopyridine. Reaction of $Na[AuCl_4]$ with

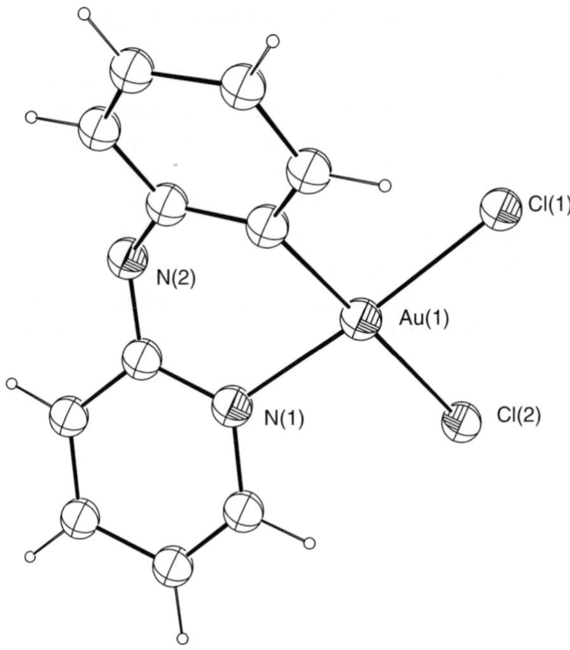

FIG. 3. Molecular structure of the anilinopyridine complex anp$AuCl_2$ **83**.

2-anilinopyridine in boiling water gave **83**, which was structurally characterized;[51] the molecular structure of this complex is shown in Fig. 3. Using similar conditions, the bromo complex was prepared from $AuBr_3 \cdot 2H_2O$, and related compounds [2-(*p*-toluidino)pyridine, 2-(*N*-methylanilino)pyridine and 2-(*p*-toluidino)pyrimidine] also produced cycloaurated products **84–86**, respectively. An improved procedure for the synthesis of **83** under very mild conditions was reported by Fuchita *et al.*,[52] and involved the reaction of either $H[AuCl_4] \cdot 4H_2O$ or $Na[AuCl_4] \cdot 2H_2O$ with excess 2-anilinopyridine in ethanol at room temperature. The excess ligand accelerates the cyclometallation reaction, probably by removing the byproduct HCl. Although this complex can be obtained simply, in excellent yields (88% and 93% from the two $[AuCl_4]^-$ compounds), the low solubility of **83** (and derivatives thereof) will probably preclude more widespread uptake as a precursor to other cycloaurated compounds.

83 **84** **85** **86**

Direct cycloauration is also observed for closely related naphthalene derivatives 1-(2-pyridylamino)naphthalene **87** and 1-(2-pyrimidinylamino)naphthalene **88**, as well as 2-(*p*-toluidino)quinoline **89** with Na[$AuCl_4$]·$2H_2O$ in refluxing aqueous ethanol, giving products **90–92**, respectively. Although these conditions are more forcing than required for 2-anilinopyridine, the direct cycloauration reaction still occurs quite readily.[53]

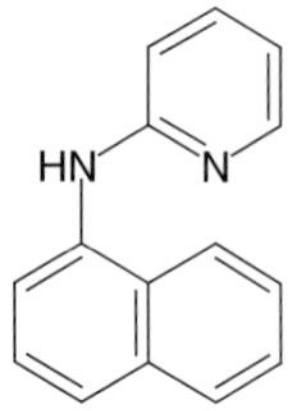

87 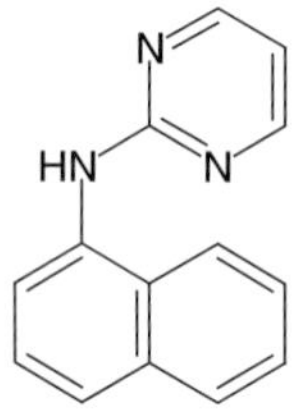 **88** 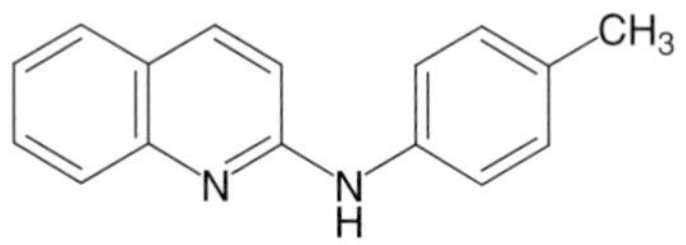 **89**

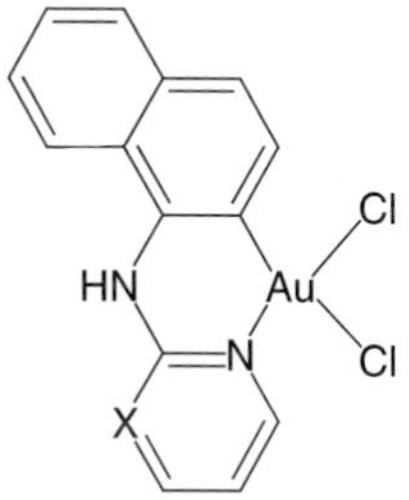

90, X = CH
91, X = N

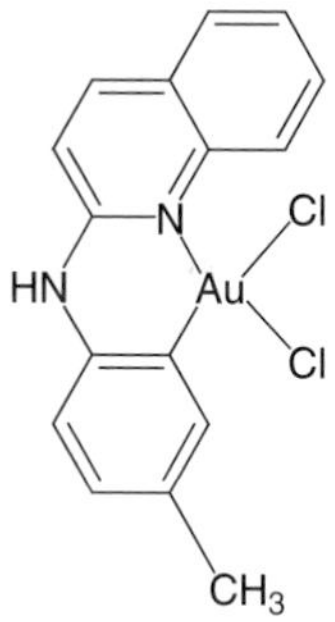

92

3-Phenyl-6-(*p*-toluidinyl)pyridazine (Hptp) **93** has been investigated as a potentially ambidentate cycloauration substrate. Direct cycloauration typically produces six-membered rings more readily, so this substrate was used to assess the preference for six- over five-membered rings. Direct reaction of **93** with Na[$AuCl_4$]·$2H_2O$ or $AuBr_3$·$2H_2O$ in refluxing aqueous ethanol overnight gave the cycloaurated products [Au(ptp)Cl_2] and [Au(ptp)Br_2], respectively. Double cycloauration of this substrate did not occur. Spectroscopic data indicated the formation of six-membered rings (**94**) in these complexes, which was confirmed by an X-ray structure determination on the dibromide complex. In contrast, cyclometallation with Li_2[$PdCl_4$] and $RhCl_3$·$3H_2O$ gave products containing five-membered rings formed by cyclometallation at the pyridazine phenyl substituents.[54]

93 **94**; X = Cl or Br

2-Benzylpyridines also directly form cycloaurated products, though the intermediates formed in the reactions are dependent on the substituents on the ligand. Reaction of either $AuCl_3 \cdot 2H_2O$ or $Na[AuCl_4]$ with **95** or **96** gave the $AuCl_3$ adducts **98** and **99**, but with the dimethyl derivative **97** under the same conditions, only the salt **100** was isolated. By warming suspensions of these adducts or salt in aqueous acetonitrile the cycloaurated products **101**–**103** were obtained as white solids. Alternatively, by refluxing $AuCl_3 \cdot 2H_2O$ with the pyridine directly in water, the cycloaurated species can be formed directly.[55] The cycloaurated benzylpyridyl derivative **101** is an attractive candidate for the further study of such gold(III) complexes; it is easily prepared in good yield (in a desirable solvent!) from commercially available 2-benzylpyridine, and derivatives with this ligand often have moderate solubilities. The cycloauration reaction is readily monitored by the change in colour of the bright yellow $AuCl_3$ adduct **98** to white **101**. An X-ray structure determination on the dimethyl derivative **103** (Fig. 4) reveals the presence of a boat conformation for the six-membered ring, with the two methyl substituents in different environments. The presence in the ^{1}H NMR spectrum of an AB multiplet for the CH_2 protons of **101**, and two CH_3 signals in **103** indicate that this conformation is retained in solution.[55] Papaverine **104** (an isoquinoline alkaloid found in opium) contains the 2-benzylpyridine structural unit, and can also be cycloaurated using $Na[AuCl_4]$ in refluxing ethanol–water, to give **105**.[56]

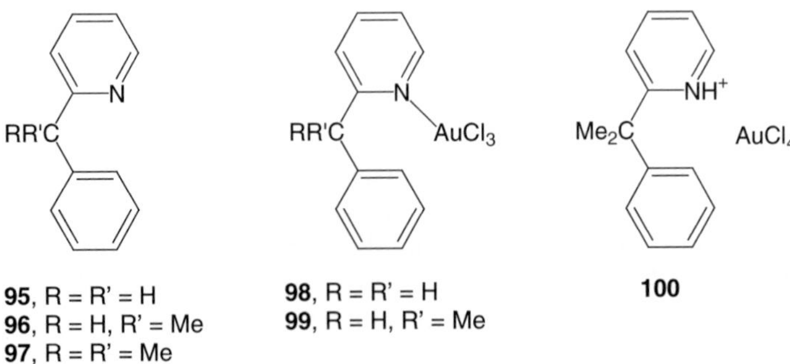

95, R = R' = H
96, R = H, R' = Me
97, R = R' = Me

98, R = R' = H
99, R = H, R' = Me

100

101, R = R' = H
102, R = H, R' = Me
103, R = R' = Me

104

105

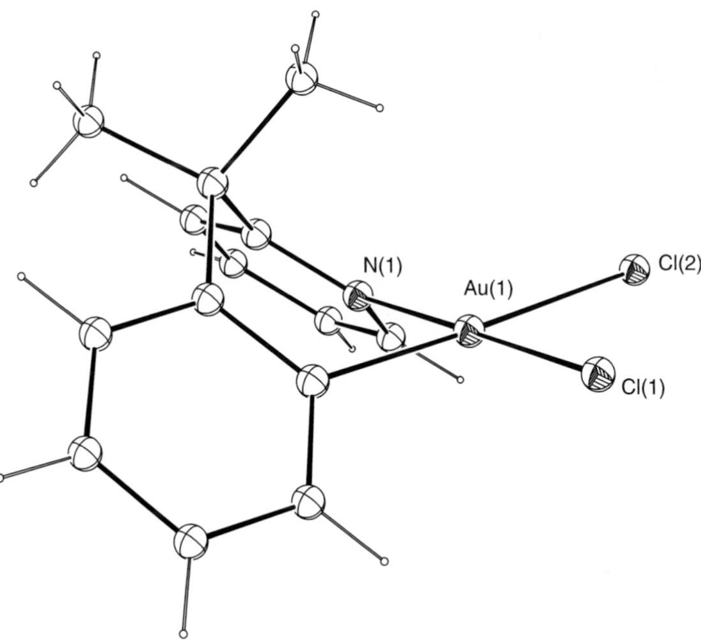

Fig. 4. Molecular structure of the dimethyl benzylpyridyl complex **103**, showing the presence of a puckered six-membered ring.

Closely related derivatives 2-phenoxypyridine (Hpop) and 2-(phenylsulfanyl)pyridine (Hptp) are not directly cycloaurated by $H[AuCl_4]\cdot 4H_2O$ at room temperature, instead producing the salts $[H_2pop][AuCl_4]$ and $[H_2ptp][AuCl_4]$.[52] In acetonitrile–water (1:5), the same reaction or the isolated salts produce the adducts $[AuCl_3(Hpop)]$ and $[AuCl_3(Hptp)]$. Upon heating to 105 °C in this solvent mixture, the salts or adducts produce cycloaurated products **106** and **107** in 70% and 20% yields, respectively. Direct cycloauration of Hpop and Hptp by $H[AuCl_4]\cdot 4H_2O$ in wet ethanol or 2-propanol was also found to occur at higher temperatures.[52] A wide range of 2-phenoxypyridine derivatives containing substituents in the 5-position of the pyridine ring have also been subjected to cycloauration reactions with $Na[AuCl_4]$.[57] The reactions proceed as expected *via* the formation of a yellow suspension of the phenoxypyridine–$AuCl_3$ adduct, which was refluxed in 1:5 MeCN–H_2O to directly produce the cycloaurated product **108**. The substituent groups were alkyl, substituted alkyl, phenyl, halo, ester and amido groups, but the presence of bulky, lipophilic groups decreased the yield of cycloaurated product, and favoured decomposition to gold metal. Ligands with electron-withdrawing nitrile or nitro groups did not coordinate, while the ligand with an electron-donating NMe_2 group was oxidised by the $[AuCl_4]^-$ anion.

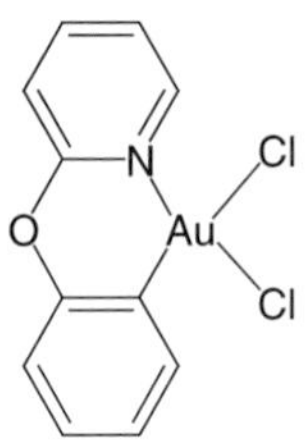

106 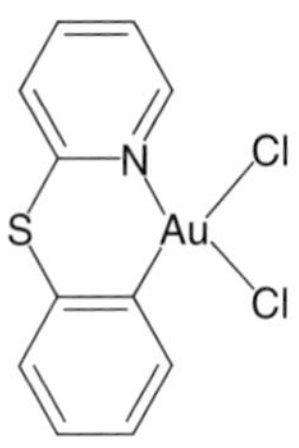 **107** 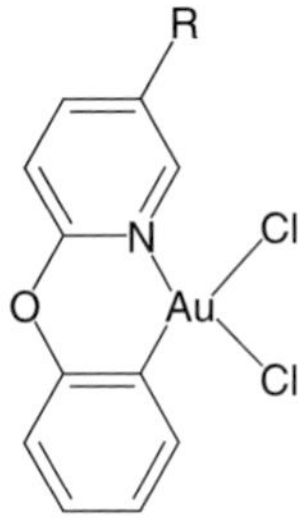 **108**

6-Benzyl-substituted 2,2′-bipyridines also readily form complexes containing six-membered cycloaurated rings. Reaction of $H[AuCl_4]$ with bipyridines HL **109**–**111** gave either neutral adducts $[AuCl_3(HL)]$ or salts $[H_2L][AuCl_4]$, depending on the experimental conditions.[23] The adducts can be readily converted into the cycloaurated products, either at room temperature in the case of **112** and **113**, or for **114**, by heating. The X-ray structure of the $[AuCl_4]^-$ salt of **114** (Fig. 5), like the parent benzylpyridyl complex **103**, shows puckering of the six-membered cycloaurated ring. Anion metathesis produced the $[BF_4]^-$ or $[PF_6]^-$ salts.[23] These complexes have been used as starting materials for the synthesis of other cycloaurated complexes, by ligand substitution of the halide, which is described in Section IV.

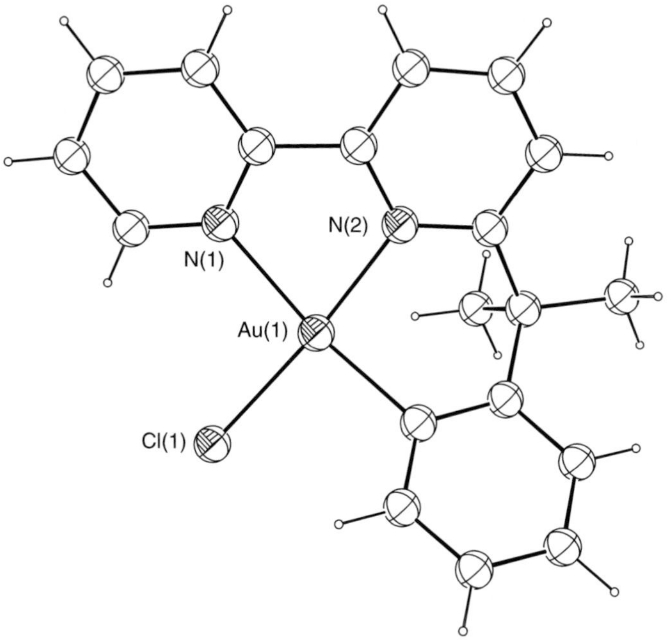

FIG. 5. Molecular structure of the cation **114**.

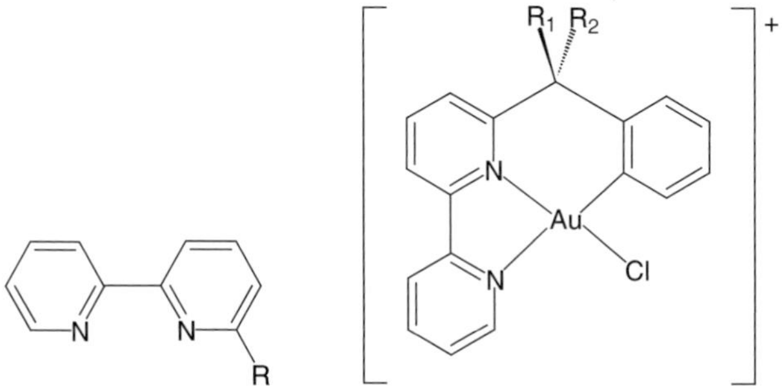

109, R = CH_2Ph
110, R = CHMePh
111, R = CMe_2Ph

112, $R_1 = R_2 = H$
113, $R_1 = H$, $R_2 = Me$
114, $R_1 = R_2 = Me$

B. *Chemically Assisted Cycloauration Reactions*

As discussed previously for the formation of five-membered ring cycloaurated complexes (Section II.B), a number of pyridine derivatives also form six-membered cycloaurated products when their $AuCl_3$ adducts are heated in the presence of added silver ions. For example, 2-benzoylpyridine (Hpcp) reacts with $H[AuCl_4]\cdot 4H_2O$ in ethanol to give $AuCl_3$(Hpcp), which gives the cycloaurated product **115** (albeit in 12% yield) on heating in propionitrile in the presence of silver trifluoroacetate.[58] No cycloauration occurred when acetonitrile or an acetonitrile–water mixed solvent was used, or when other silver salts, namely $Ag[BF_4]$ or $Ag[O_2CCH_3]$, were used.

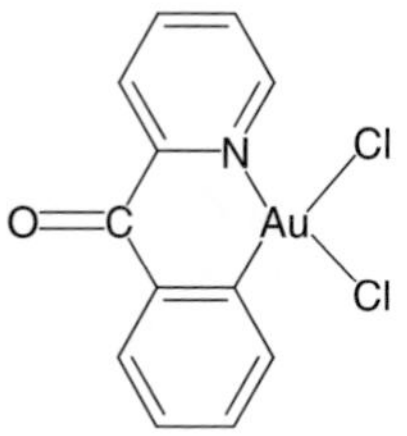

115

IV
REACTIVITY OF CYCLOAURATED COMPLEXES

A. *Phosphine Ligands*

Reactions of cycloaurated gold(III) dihalide complexes with tertiary phosphine ligands have been widely studied, because different types of behaviour can be observed. In some cases, where the Au–N bond is relatively strong, this is not cleaved by addition of 1 equivalent of phosphine, which instead replaces a halide ligand. In other cases where the Au–N bond is relatively weak, the nitrogen donor is displaced by the phosphine, giving a neutral phosphine adduct. Such behaviour can be a key indicator to other reactivity of these cycloaurated complexes. For example, complexes that contain a weak, phosphine-displaceable N donor are likely to undergo C–H activation reactions with weak carbon acids, specifically acetone, to give σ-acetonyl complexes.[30]

Early studies in this area were centred on the cycloaurated azobenzene derivatives, especially papAuCl_2 **25**. Addition of equimolar amounts of various ligands L, such as phosphines [PPh_3, PCy_3, $P(C_6H_4OMe\text{-}o)_3$, P^nBu_3] and also $AsPh_3$ to **25** in acetone gave the neutral complexes **116**.[25] Reaction with $Ph_2PCH_2CH_2PPh_2$ gave a corresponding phosphine-bridged dinuclear complex.[25] Reaction of the cycloaurated imine complex **33** with PPh_3 yielded an analogous neutral complex by displacement of the imine nitrogen donor.[30] When **25** is treated with 2 equivalents of phosphine together with $Na[ClO_4]$, cationic bis(phosphine) complexes **117** can be obtained, including derivatives with mixed phosphine ligands; these complexes have a *trans* disposition of phosphine ligands, as evidenced by IR and ^{31}P NMR spectroscopies.[25]

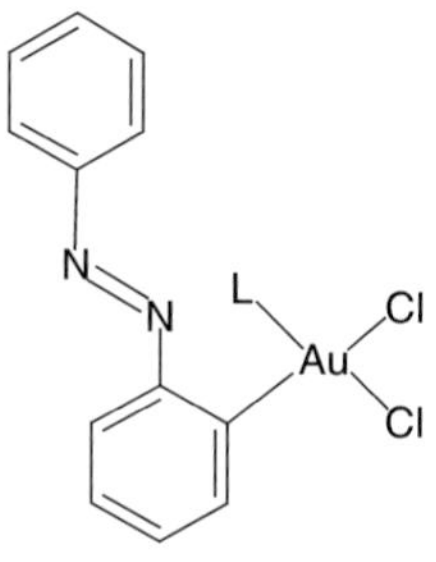

116 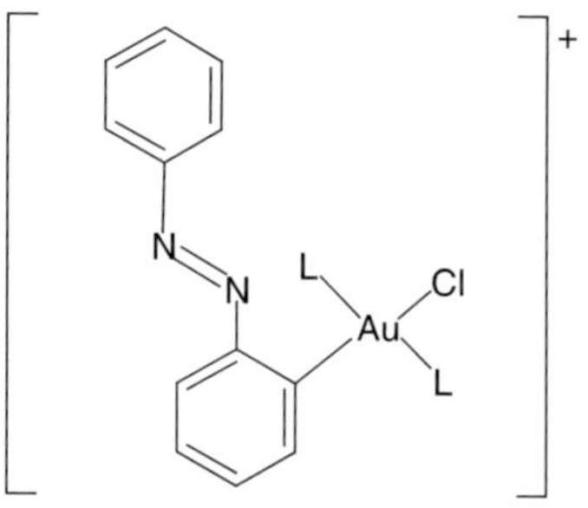 **117**

In contrast, however, as an example of the alternative type of behaviour, reaction of dampAuCl_2 **36** with 1 equivalent of PPh_3 in dichloromethane gave the cationic derivative **118** as its chloride or perchlorate salt.[34] The analogous complex with potentially bidentate 2-(diphenylphosphino)benzoic acid, $Ph_2PC_6H_4CO_2H$, has also been reported.[59,63] However, **36** does not react with poorer ligands such as $AsPh_3$, pyridine or tht alone, but in the presence of Na[ClO_4] the corresponding cationic derivatives **118** can be obtained. In these derivatives, there is no displacement of the NMe_2 donor.

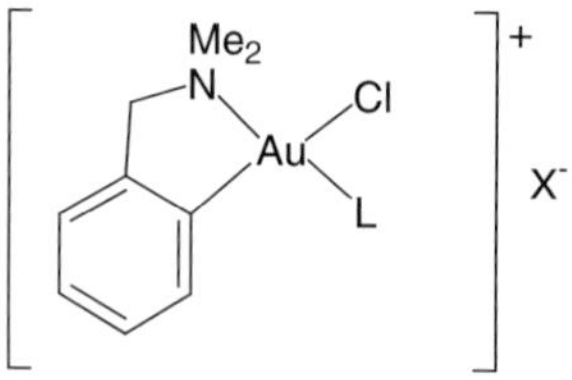

118, L = PPh_3, $Ph_2PC_6H_4CO_2H$, $AsPh_3$, pyridine or tht; X = Cl, ClO_4

A number of other cycloaurated complexes also show simple phosphine-for-halide substitution. Reaction of the cycloaurated 2- and 3-thienylpyridine complexes **8** and **9** with PPh_3 and Na[BF_4] yielded the cationic products **119** and **120**, respectively, with the PPh_3 ligands again *trans* to the nitrogen donor group.[14] Complexes **11** react with PPh_3 and Na[ClO_4] to give the mixture of isomers **121** and **122**, which are converted into the latter complex upon refluxing in chloroform.[17]

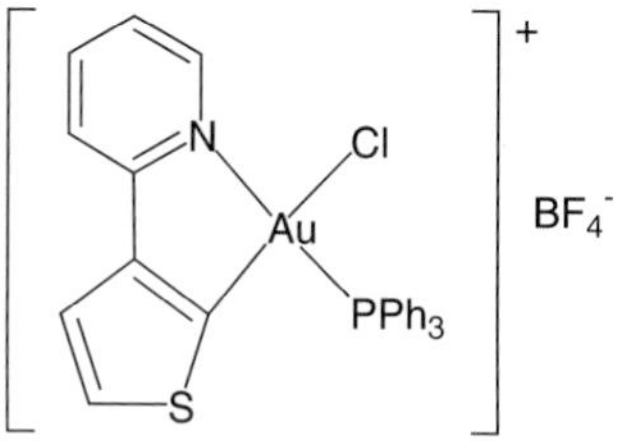

119 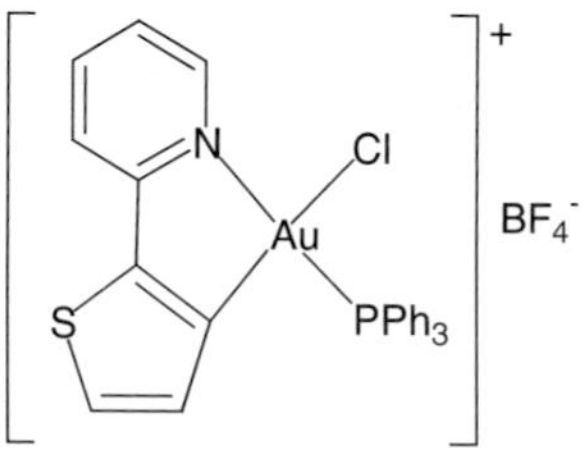 **120**

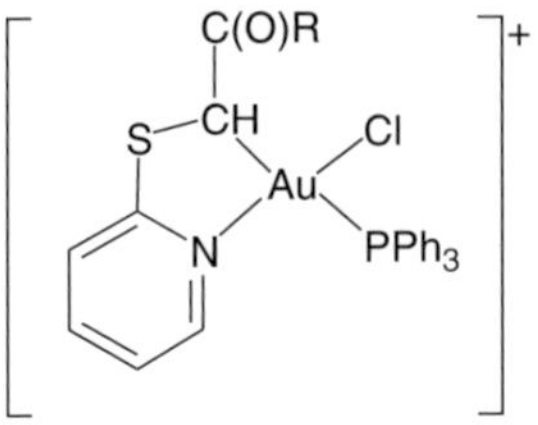

121 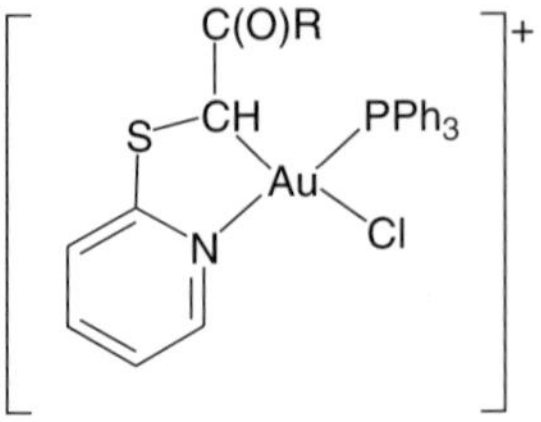 **122**

Cycloaurated complexes of benzylpyridines and closely related analogues provide some interesting comparisons of how the ring group (e.g., O, S, NH, CH_2) can influence reactivity. Reaction of the anilinopyridine complex **83** with $Ptol_3$ (tol = *p*-tolyl) also results in displacement of one chloride giving the cationic derivative **123**.[52] Similarly, reactions of **83** or the benzylpyridine derivatives **101**–**103** with equimolar PPh_3 in the presence of $Na[BF_4]$ gave the corresponding BF_4 salts **124** and **125**.[52,55] As expected, two diastereomers were observed for the monomethylbenzyl derivative.

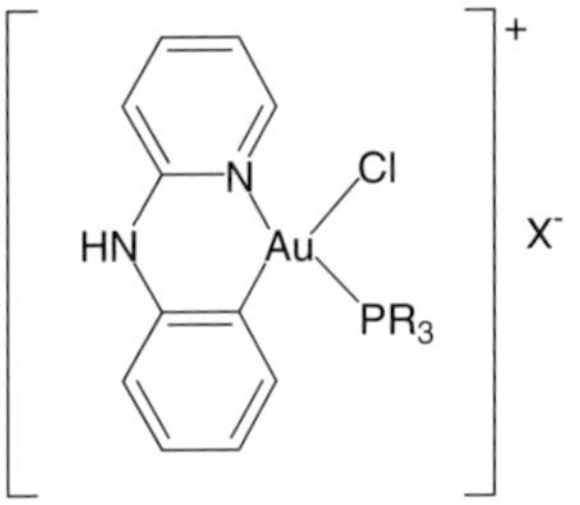

123, R = *p*-tol, X = Cl
124, R = Ph, X = BF_4

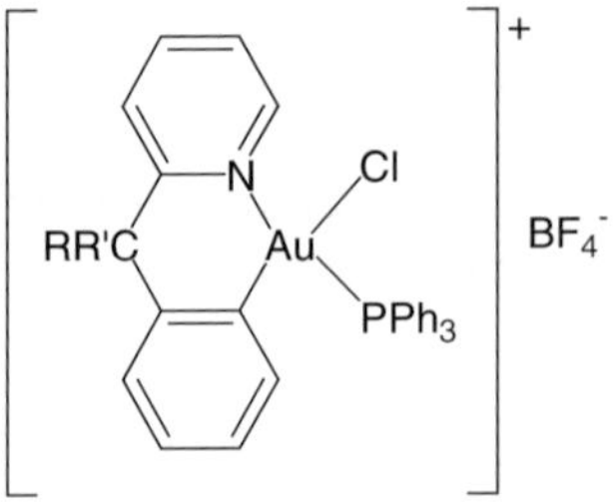

125a, R = R' = HH, Me
125b, R = H, R' = Me
125c, R = R' = Me

Addition of excess PPh_3 to **125a** does not result in further reaction. However, reaction of **101** or **103** with $Ph_2PCH_2CH_2PPh_2$ (dppe) in the presence of $Na[BF_4]$ leads to the cationic derivatives **126**, containing an uncoordinated pyridine ligand. Reaction of these complexes with $Ag[BF_4]$ removes the coordinated chloride, allowing re-coordination of the pyridine to give dicationic complexes **127**.[55] Reaction of **83** with 2 equivalents of PPh_3 also results in dissociation of the pyridine donor, giving the cationic complex **128**.[52] The related triethylphosphine complex **129** was also prepared and characterised by an X-ray diffraction study.

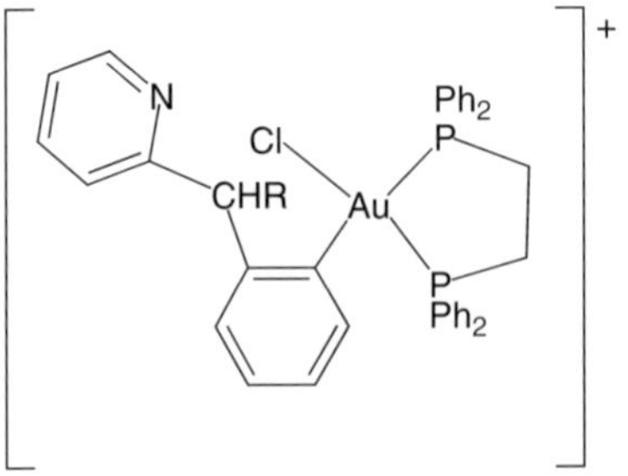

126; R = H, Me 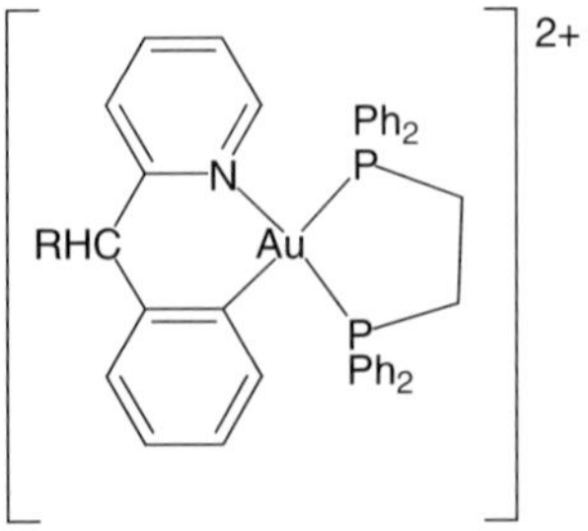 **127**

In contrast, for the oxy and thio analogues **106** and **107** only 1 equivalent of PPh_3 was necessary to produce the neutral, pyridine-displaced products **130** and **131**.[52] Reaction of the ketone derivative **115** with an equimolar amount of PPh_3 gave the neutral complex **132** formed by displacement of the pyridine ligand, while with 2 equivalents of PPh_3 in the presence of Na[BF_4] the cationic bis(phosphine) complex **133** is obtained.[58]

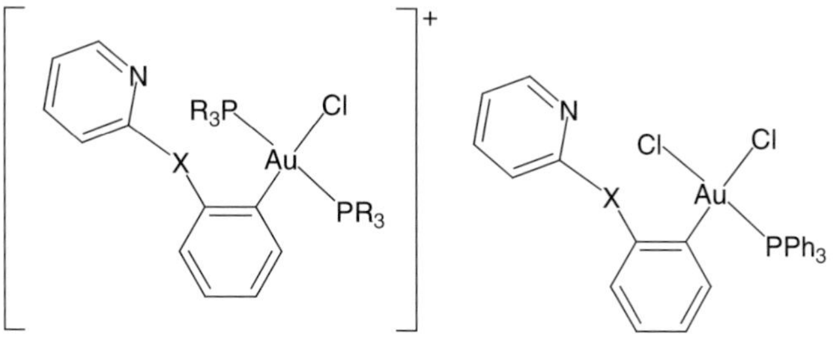

128; X = NH, R = Ph
129; X = NH, R = Et
133; X = CO, R = Ph

130; X = O
131; X = S
132; X = CO

Other behaviours can also be observed in reactions of the gold–halide complexes with phosphines. Reaction of **115** with PPh_3 and 3 equivalents of Ag[BF_4] gave the cationic bis(cycloaurated) product **134** by an unknown mechanism. In contrast, reaction of **101** with an equimolar amount of PPh_3 (in the absence of added counter-anion) gave a product which on the basis of spectroscopic data was assigned to the unusual five-coordinate complex **135** which gave complex **125a** with added Na[BF_4].[58]

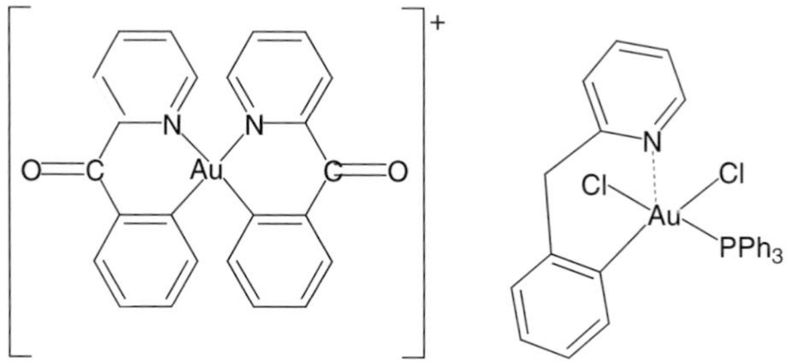

134

135

Reactions of the cycloaurated diphenylpyridine complex **55** with a number of phosphines have been described. The resulting cationic products with PPh_3 (**136**), dppm (**137**) and dppe (**138**) were all structurally characterised.[37]

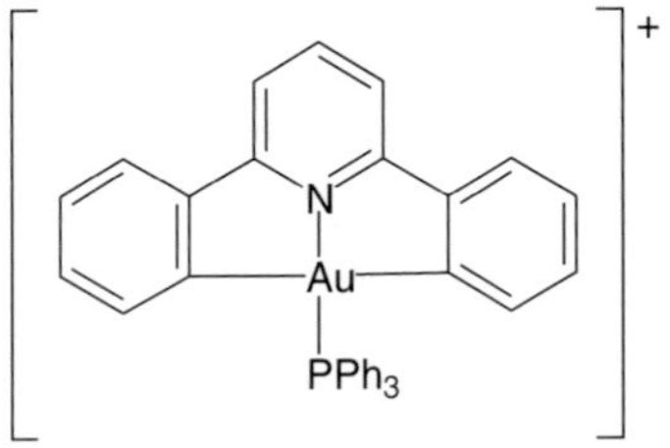

136

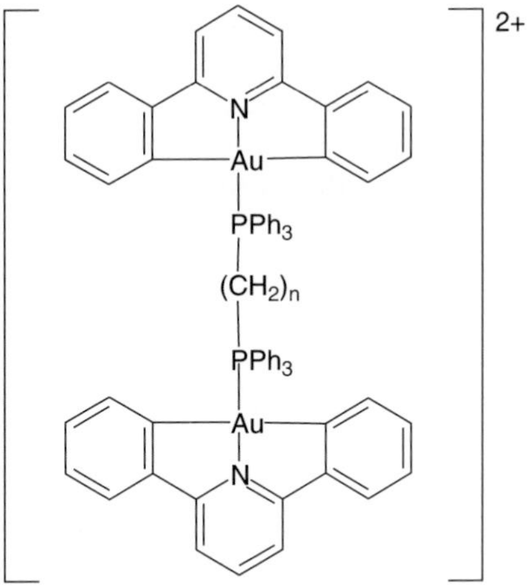

137, n = 1
138, n = 2

Reactions of cycloaurated complexes with phosphine ligands that lead to reductive coupling and the formation of biaryls are discussed separately in Section IV.K.

B. *Neutral Nitrogen Donor and Other Ligands*

Acetate derivatives (see Section IV.D) have provided a useful route into complexes containing pyridine and other nitrogen donor ligands. Reactions of the diacetate complex dampAu$(OAc)_2$ (**143**) with 2 mol equivalents of the pyridinium salts [pyH]ClO_4 or [pyH]BF_4 have been investigated, leading to the bis-pyridine complex **139**, through protonation of the acetate ligands. With 1 equivalent of [pyH]ClO_4 the monocationic mono-pyridine complex $[dampAu(OAc)py]^+$ was formed.[60] Similarly, papAuCl_2 **25** can be converted into the diacetate derivative with silver acetate, and this complex forms a parallel route into cationic mono- and bis-pyridine complexes by reaction with [pyH][ClO_4].[61] Ready displacement of the coordinated pyridine ligands occurs, giving a route to other dicationic complexes such as **140** with 1,10-phenanthroline, 2,2′-bipyridine, *N*,*N*,*N*′,*N*′-tetramethylethylenediamine or *o*-phenylenediamine ligands.[60,61] The dicationic phenanthroline complex **140** reacts with PPh_3 or $P(C_6H_4OMe\text{-}p)_3$ to give the rare five-coordinated complexes $[Au(damp)(phen)(PR_3)]^{2+}$ [62]

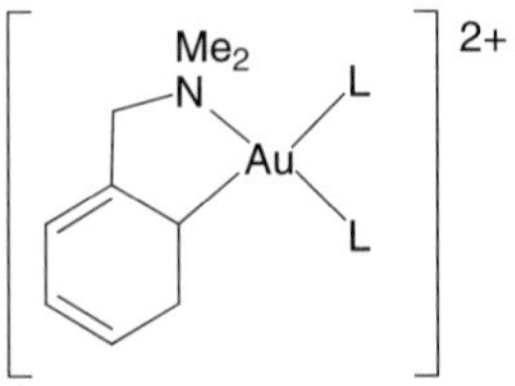

139 L = pyridine
140 L-L = bidentate N-donor ligand

Interactions of the biologically active damp complex dampAu$(O_2CCH_3)_2$ **143** towards nitrogen donor ligands of biological relevance have been investigated using NMR spectroscopy; caffeine and adenosine showed no reaction while guanosine and inosine showed partial reaction. Adenine however reacted quantitatively, and NMR data suggested the formation of a single monodentate adenine ligand, coordinated through the N(7) atom of the imidazole ring. Because of the absence of broadening of the acetate resonance upon addition of water or free acetate, it was assigned the position *trans* to the low *trans*-influence NMe_2 group, suggesting that the product has the structure **141**.[63]

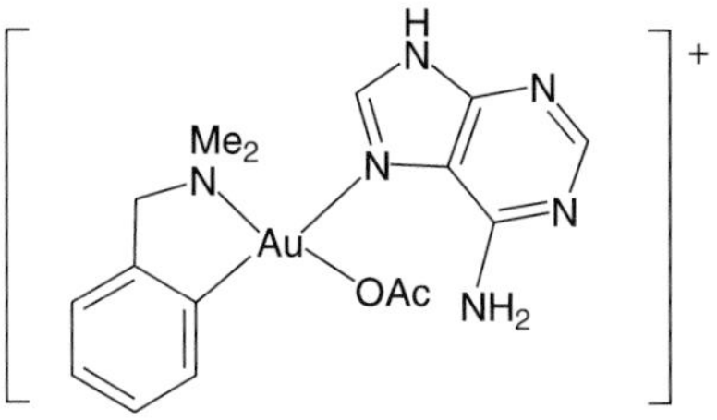

141

In related chemistry, the reaction of papAuCl_2 **25** with Ag[ClO_4] gave the putative perchlorato complex [papAuCl($OClO_3$)], which reacted with DMSO [S(O)Me_2] to give the first organogold complex containing a DMSO ligand, [papAu$(DMSO)_2$][ClO_4]$_2$.[64]

C. *Halide and Pseudohalide Ligands*

Generally, reactions of cycloaurated gold(III) dihalide complexes with simple anionic ligands proceeds readily, with formation of new complexes containing a wide range of anionic ligands with different donor atoms. For example, the reactions of dampAuCl_2 **36** (and closely related analogues) with KBr or KI result in displacement of the chloride ligands giving the corresponding complexes dampAuX_2; the iodo complex was unstable in solution.[34,36] The oxazoline-based complexes have also been converted into bromide and iodide analogues (**38b**, **38c**, **39b**, **39c**, **40b**, **40c**). Yields of the substituted oxazoline products were relatively low, and this was attributed to the susceptibility of the oxazoline group to nucleophilic attack, being often used as a protecting group in organic chemistry because of these characteristics.[36]

Reaction of the phenylpyridine complex **2** with NaSCN gave an interesting product, which was shown to be complex **142** by an X-ray diffraction study, and contains thiocyanate ligands displaying their ambidentate nature.[65] The molecular structure is illustrated in Fig. 6, which shows that the thiocyanate *cis* to the high *trans*-influence aryl carbon is S-bonded, while the other thiocyanate is N-bonded. Previously, thiocyanate derivatives of damp and oxazoline-based cycloaurated complexes have also been converted into the bis(thiocyanate) derivatives (**38d**, **39d** and **40d**), but in contrast to **142** it was suggested on the basis of IR spectroscopic data that the thiocyanate ligands were S-bonded.[32,36] Further structural studies into thiocyanate derivatives might shed more light on this matter.

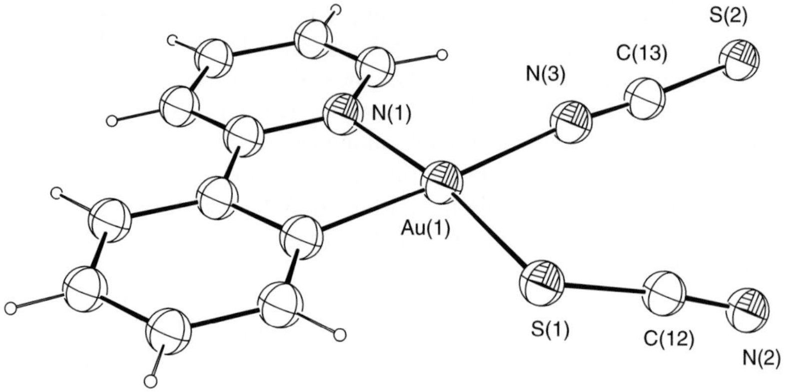

FIG. 6. Structure of the bis(thiocyanato) complex ppyAu(SCN)$_2$ **142**, showing the presence of N-bonded (*trans* to C) and S-bonded (*trans* to N) thiocyanate ligands.

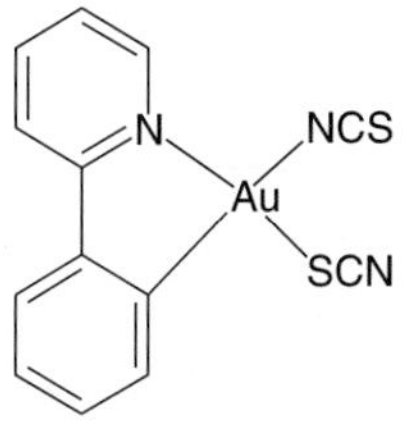

142

D. *Carboxylate Derivatives*

The most extensively studied carboxylate derivatives are the acetates. The damp complex **143** is prepared by reaction of dampAuCl$_2$ **36** with silver acetate.[32] This complex has been found to show promising anti-tumour activity, and the presence of the acetate groups improves the solubility characteristics of the complex. Crystals of the hemihydrate of **143** were obtained from moist CH_2Cl_2 solution and were characterised by an X-ray diffraction study which showed strong hydrogen-bonding between the water molecules and acetate groups.[33]

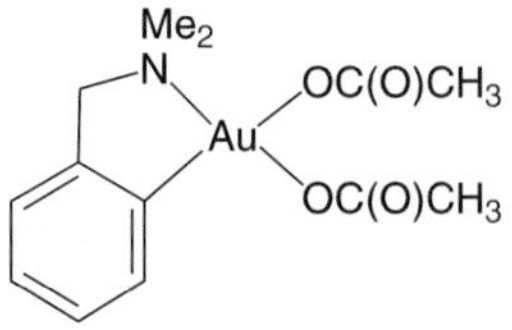

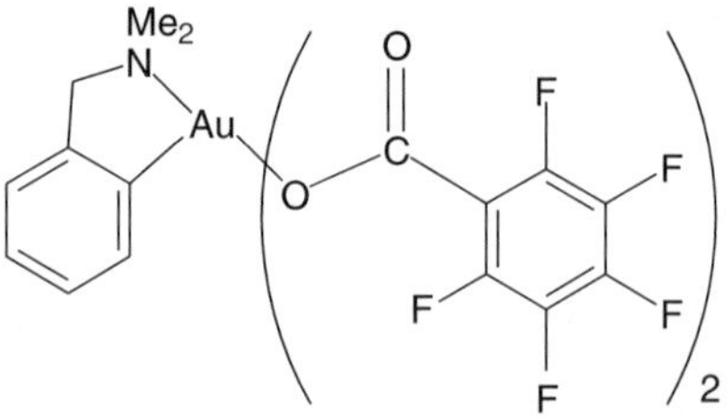

143 **144**

Complex **143** undergoes hydrolysis in aqueous solutions, or in organic solutions containing water.[63] One acetate ligand, presumably the most labile one *trans* to the

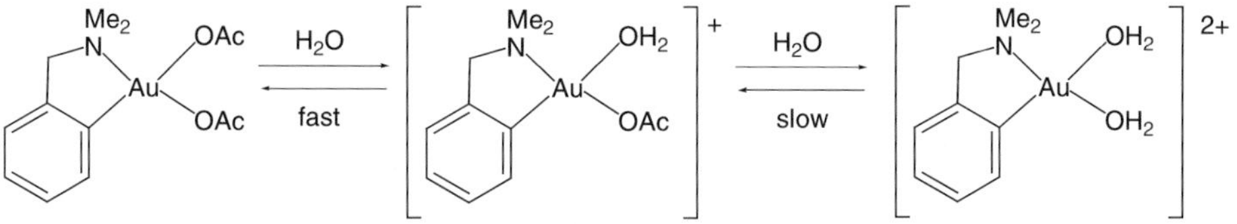

SCHEME 3. Hydrolysis of dampAu(OAc)$_2$ (OAc = O_2CCH_3). Modified from Parish *et al.*[63]

Au–C bond, exchanges rapidly on the NMR timescale. With higher concentrations of water, the second acetate undergoes hydrolysis, and in solutions that have been allowed to stand, a third hydrolysis species was tentatively assigned to the other isomer of $[dampAu(O_2CCH_3)(H_2O)]^+$; the hydrolysis process is summarised in Scheme 3.

Reaction of dampAuCl$_2$ with $[C_6F_5CO_2]Ag$ gave the pentafluorobenzoate complex **144**, which does not undergo decarboxylation up to 400 °C, with entire loss of the organic fragment occurring at 257 °C.

The azo complex **25** also reacts with silver(I) acetate to give the simple ligand-substituted product (azo)Au$(O_2CCH_3)_2$ **145**, which with HCl gave the monoacetate complex **146**.[61] Reaction of ppyAuCl$_2$ **2** with silver acetate, or with silver nitrate followed by sodium benzoate gave the corresponding bis(carboxylate) complexes **147**.[66] Other acetate complexes prepared using silver acetate include the bis(acetate) complexes derived from phenylpyridyl derivatives **50**, **51** and **53**[12] as well as **42b** and **43b**.[36]

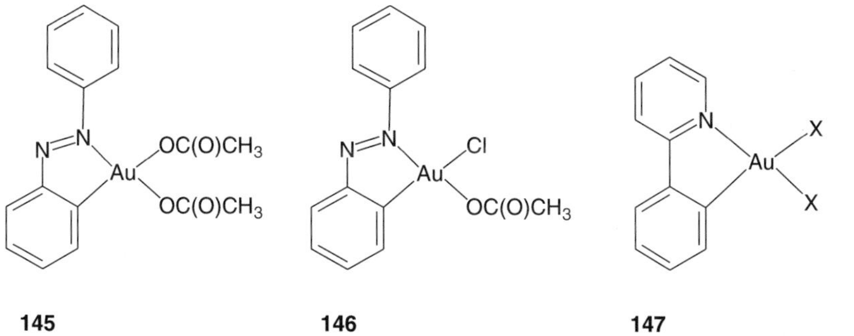

145 **146** **147**

X = OC(O)CH$_3$ or OC(O)Ph

E. *Complexes with Ancillary Aryl Ligands*

A number of aryl derivatives formed by replacement of coordinated halide ligands have been reported. Reaction between dampAuCl$_2$ **36** and HgPh$_2$ in chloroform solution gave the monophenyl complex dampAu(Ph)Cl **148** in 86% yield, which crystallises in two modifications, both with three independent molecules (differing in the orientation of the rings) in the asymmetric unit.[67] The structure of one of these molecules is shown in Fig. 7, and shows the two aryl groups in *cis* positions. The corresponding reaction of dampAuCl$_2$ **36** with Hg$(C_6H_4CF_3\text{-}o)_2$ gave the bis(aryl)

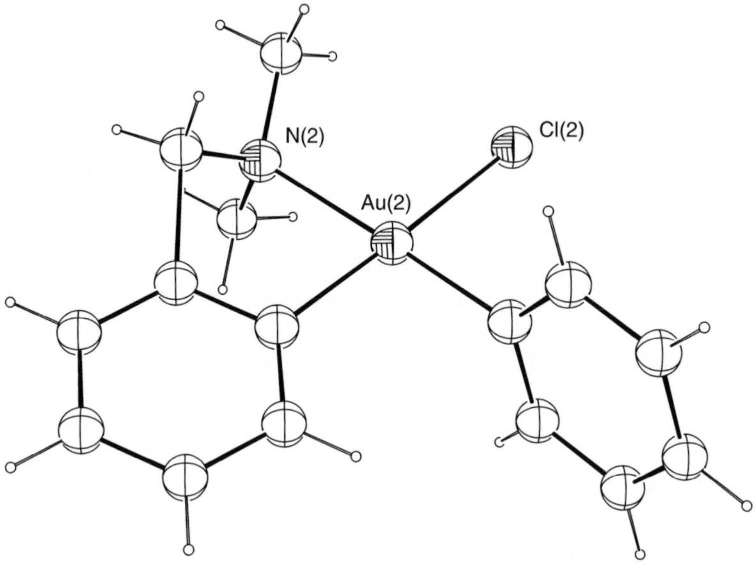

FIG. 7. Molecular structure of one of the six crystallographically independent cations of the complex dampAuCl(Ph), **148**.

complex **149** (which can also be obtained from dampHgCl **37** and *cis*-$[AuCl_2(C_6H_4CF_3\text{-}o)_2]^-$).[30] In contrast, reaction of **25** with $HgPh_2$ gave only PhHgCl and metallic gold.[67] By reaction with KBr, KI and KCN the chloride ligand of **148** can be replaced by other anionic ligands, giving **150–152**; the acetate complex **153** is best obtained by use of silver acetate.[67] Displacement of the chloride ligand by neutral ligands (L = pyridine, tetrahydrothiophene) in the presence of $[ClO_4]^-$ gave the cationic derivatives $[dampAu(Ph)L]^+[ClO_4]^-$.

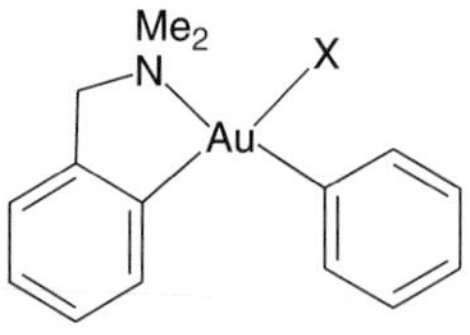

148, X = Cl
150, X = Br
151, X = I
152, X = CN
153, X = O_2CCH_3

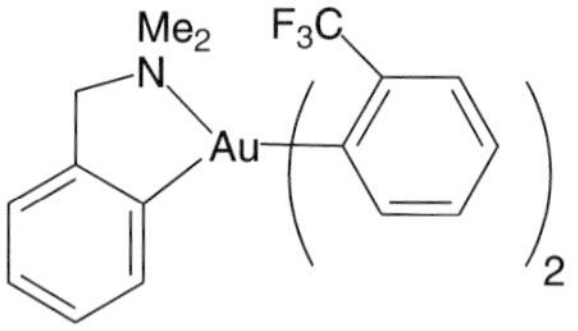

149

The mono(pentafluorophenyl) complex **154a** can likewise be obtained in moderate (41%) yield by reaction of $dampAuCl_2$ with $Hg(C_6F_5)_2$ and $[Me_4N]Cl$. Ligand replacement reactions of the chloride of **154a** with KBr or silver acetate gave the corresponding anionic complexes **154b** and **154c**, which were proposed to have the same geometry.[68]

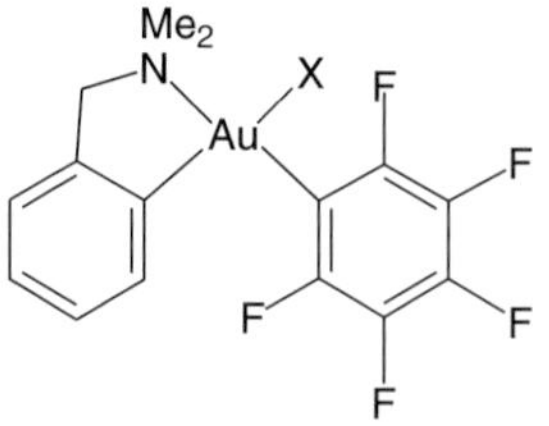

154a; X = Cl
154b ; X = Br
154c ; X = O_2CCH_3

F. *Acetonyl Complexes*

A number of cycloaurated complexes have been demonstrated to effect facile C–H activation reactions, particularly involving acetone. As commented in Section IV.A, lability of the nitrogen donor group (as evidenced by displacement with phosphine ligands) is a key indicator to likely reactivity in C–H activation.[30] The first work in this area was by Vicente *et al.*,[69,70] who reacted papAuCl$_2$ **25** with Tl[acac] [acac = CH_3-C(O)CHC(O)CH_3] in acetone at room temperature, which gave the σ-acetonyl complex **155**. When the reaction in acetone is carried out at 0 °C, or dichloromethane is used as the solvent at room temperature, the intermediate C-bonded acac complex **156** was isolated.[69,70] The acetonyl complex **155** is also formed in a number of other reactions involving papAuCl$_2$ **25** in acetone, including reactions with KCN, [Hg(C_6F_5)$_2$] plus Cl^-, Ag[ClO_4], 1,10-phenanthroline, or Hg(pap)$_2$ **29**.[70] Reactions of the butyl- and butoxy-substituted azobenzene complexes **27** and **28** with Ag[ClO_4] in acetone also gave the corresponding C-bonded acetonylactonato complexes **157** and **158**[28] and the cycloaurated imine complex **33** behaved similarly.[30]

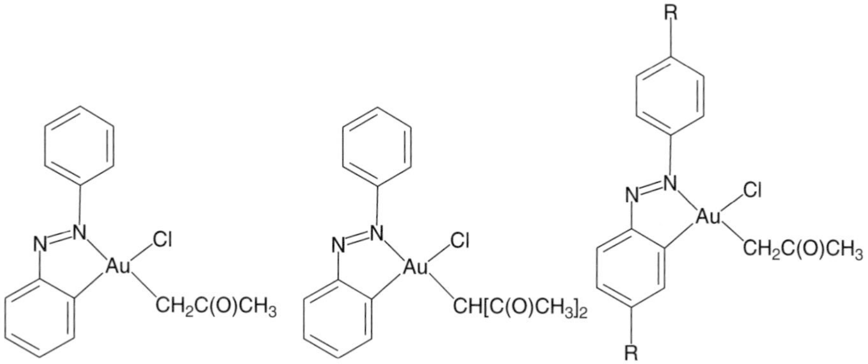

155 **156** **157**, R = n-Bu
158, R = n-BuO

C–H activation has been extended to a number of other methyl ketones MeC(O)R[64] and the reactivity of ketonyl gold complexes has been explored in some detail. One of the more interesting reactions was conversion of the C=O group of **155** to the C=NPh group by reaction with $PhNH_2$.[71] Other reactions involved replacement of the chloride ligand with other anions, and displacement of the azo nitrogen donor by reaction with phosphines, as is typical for this class of cycloaurated complex (refer Section IV.A).

In contrast, the complex damp$AuCl_2$ **36** shows rather different behaviour and (unlike pap$AuCl_2$ above) shows no reaction towards acetone in the presence of added KCN, $[Hg(C_6F_5)_2]$ plus Cl^-, or 1,10-phenanthroline. However, it does react with equimolar quantities of Tl[acac] to give the C-bonded acetylactonato complex **159**, which is the analogue of **156**.[70]

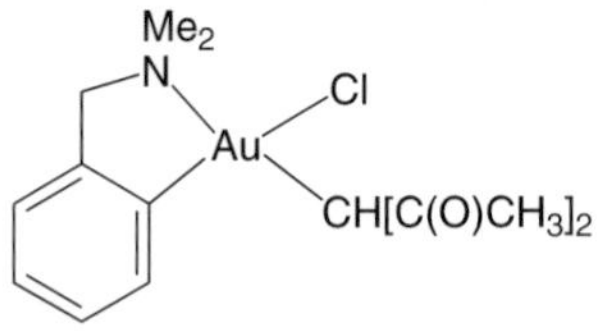

Me2
N
Au
Cl
CH[C(O)CH3]2

159

The reaction of ppy$AuCl_2$ **2** with silver lactate, in an attempt to synthesise a gold(III) lactate complex, resulted in crystallisation of the acetonyl complex **160** by C–H activation of the acetone solvent used for recrystallisation.[18] In a similar fashion, treatment of the nitrate complex ppy$Au(NO_3)_2$ (formed by reaction of ppy$AuCl_2$ with $AgNO_3$) with acetone for 72 h gave the analogous complex **161**, containing an acetonyl ligand and coordinated nitrate.[18]

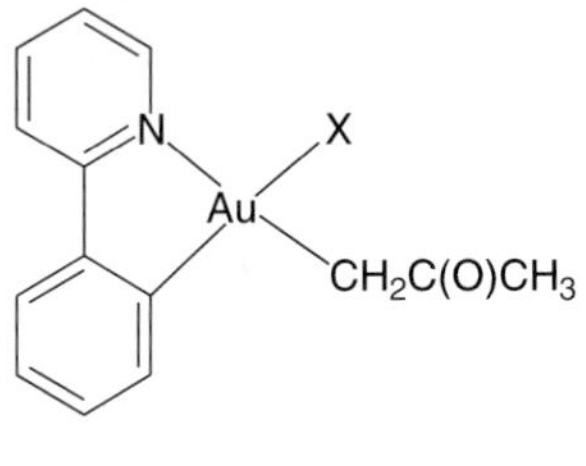

N
Au
X
CH2C(O)CH3

160, X = Cl
161, X = NO_3

Studies into the mechanism of C–H activation of acetone have been reported,[70] including isolation of some intermediates.[72] This area has been summarised in a review by Vicente and Chicote[73] on the 'acac method' for the synthesis of gold complexes.

G. *Ligand Displacement Reactions of Pincer Complexes*

The cycloaurated benzyl–bipyridine complexes **112–114** undergo ligand replacement reactions of the chloride ligand with a wide range of C-, N-, O- and S-donor ligands,[74] but probably the most versatile reactivity is displayed by the dimethyl derivative **114**. With NaOR (R = Me, Et), **114** gave the alkoxide complexes **162** and **163**; these were the first examples of isolated gold–methoxide and ethoxide complexes.[74] However, the monomethylbenzyl analogues cannot be obtained by simple alkoxide metathesis, because the benzylic C–H bonds in the starting chloro complex are reactive towards OR^-. Instead, alcoholysis of the acetate derivatives (such as **164**, prepared by reaction of **113** with $Ag[O_2CCH_3]$) also gives the alkoxide derivatives without reaction of the benzylic protons. The hydroxide complex **165** can be prepared as an air-stable and water-soluble white solid by reaction of an aqueous suspension of the chloro complex **114** with NaOH.[75] The methoxide and hydroxide complexes have been used as precursors to other derivatives.[74,75]

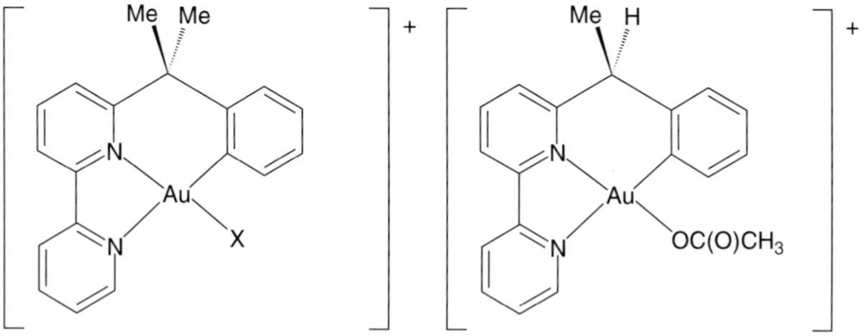

162; X = OMe
163; X = OEt
165; X = OH
166; X = $CH_2C(O)CH_3$
168; X = SPh
169; X = CCPh
170; X = acac

164

Reaction of the PF_6^- salt of **114** with $Ag[PF_6]$ in acetone resulted solely in the formation of the C-bonded acetonyl complex **166**.[74] However, when the BF_4^- salt of **114** is treated with $Ag[BF_4]$ in acetone, a mixture of **166** and the first example of its type of an oxo-bridged dinuclear gold(III) complex, **167**, was obtained.[76] The asymmetric stretch of the Au–O–Au moiety was observed in the IR spectrum at ca. 780 cm^{-1} An X-ray structure determination on **167** showed a C_2-symmetric dimer, with an Au–O bond distance of 1.96 Å and an Au–O–Au bond angle of 121.3(2)°. The oxo complex can also be obtained by condensation of the hydroxo complex **165** in refluxing tetrahydrofuran (THF), and the reaction can be reversed on refluxing in water. Surprisingly, the oxo dimer can be recovered unchanged from the reaction with aqueous $H[PF_6]$.[77,78] The chemistry of these and other gold(III) complexes with anionic oxygen donor ligands has recently been reviewed by Cinellu and Minghetti.[78]

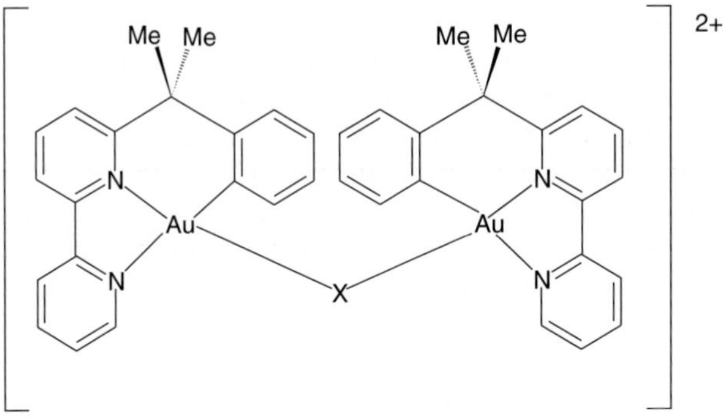

167

Reaction of the methoxo complex **162** with PhSH gave the thiolate complex **168**, while reaction with the weak carbon acid PhC≡CH, gave the corresponding alkynyl complex **169**.[74] Reaction of **114** with Tl(acac) gave the C-bonded acac complex **170**, which exists as a mixture of tautomers.

A number of amide derivatives of **114** have also been synthesised. The reactions of the methoxy and hydroxo complexes **162** and **165** with a range of primary and secondary aliphatic and aromatic amines resulted in formation of a number of air- and solution-stable amido derivatives **171**–**176**.[74,75] The related monoamido complex containing a cycloaurated benzylpyridine ligand, **177** was also prepared.[75] The stability of the toluidine **172** and xylidene **173** complexes in water or a physiological-like buffer solution has been investigated by UV–visible and NMR spectroscopies, and rapid hydrolysis occurred, probably giving the hydroxo complex **165**.[79]

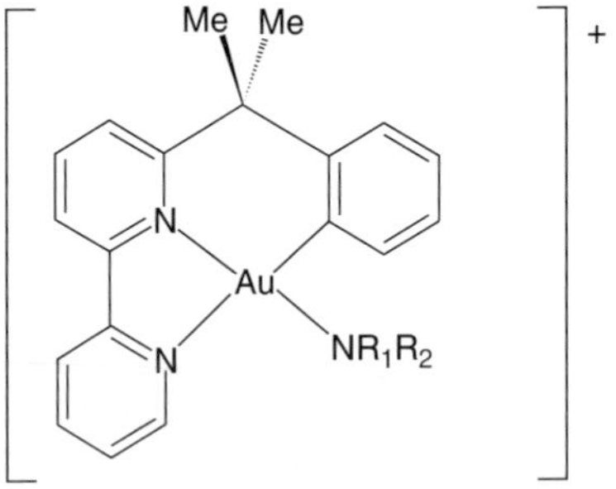

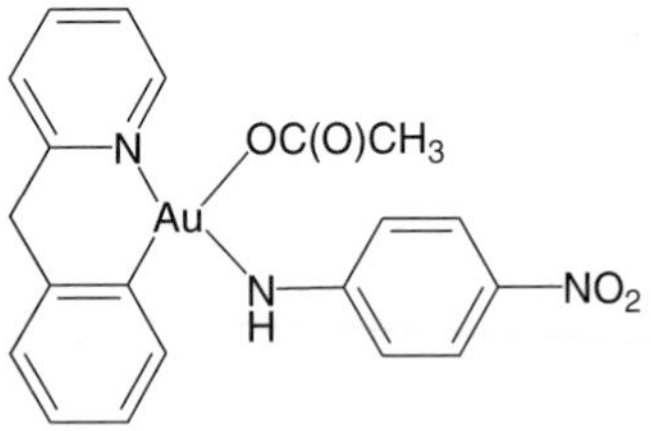

171; $R_1 = H$, $R_2 = C_6H_4NO_2$-4
172; $R_1 = H$, $R_2 = C_6H_4Me$-4
173, $R_1 = H$, $R_2 = C_6H_3Me_2$-2,6
174; $R_1 = H$, $R_2 = CH_2CHMe_2$
175; $R_1 = H$, $R_2 = CHMeEt$
176; $R_1 = R_2 = Et$

177

The diphenylpyridine complex **55** has been converted into a number of complexes that show interesting structural or photophysical properties; phosphine derivatives are covered in Section IV.A. Thus, the corresponding cyano complex, synthesised from **55**, forms a relatively weak supramolecular host–guest complex with a molecular receptor containing two parallel-disposed terpyridyl-Pt-Cl units. In contrast, the platinum complex, containing an isoelectronic Pt-CO group in place of Au-CN, formed a more stable supramolecular adduct, because of stronger Pt···Pt interactions.[80] Reaction of **55** or the *t*-butyl-substituted derivative **57** with terminal aryl acetylenes in CH_2Cl_2, with Et_3N and a catalytic amount of CuI gave a range of gold(III) alkynyl complexes **178**–**183**, which have interesting photophysical properties.[38,81]

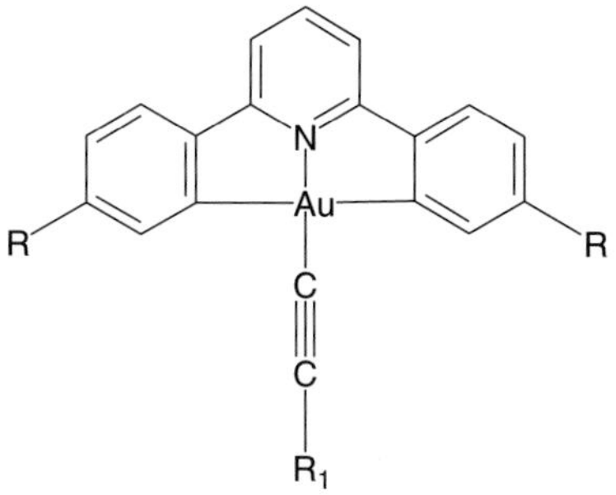

178; R = H, R_1 = Ph
179; R = H, R_1 = *p*-ClC_6H_4
180; R = H, R_1 = *p*-$MeOC_6H_4$
181; R = H, R_1 = *p*-$H_2NC_6H_4$
182 ; R = ^{t}Bu, R_1 = Ph
183; R = H, R_1 = *p*-$Me_2NC_6H_4$

H. *Bidentate Anionic Ligands with C-, N- and O-Donor Groups*

Cycloaurated complexes have been used as precursors for the synthesis of a wide range of auracyclic complexes, where the gold atom is also part of an additional metallacycle. Derivatives with C/C-, N/N-, O/O-, C/N- and N/O-donor atoms have all been reported.

A number of complexes containing bidentate carboxylate ligands have been prepared by metathesis reactions. For example, reaction of damp$AuCl_2$ **36** with silver malonate, or with silver nitrate and oxalic acid gave the malonate and oxalate derivatives **184** and **185**, respectively.[32] Related examples include the phenylpyridyl complexes **186** and **187**.[66] One example of a salicylate complex has been prepared using silver(I) oxide as a base; the initial product was a mixture of isomers **188a** and **188b**, in a 3:1 ratio, but on standing in solution, isomerisation occurred to give solely **188b**, which was structurally characterised.[82]

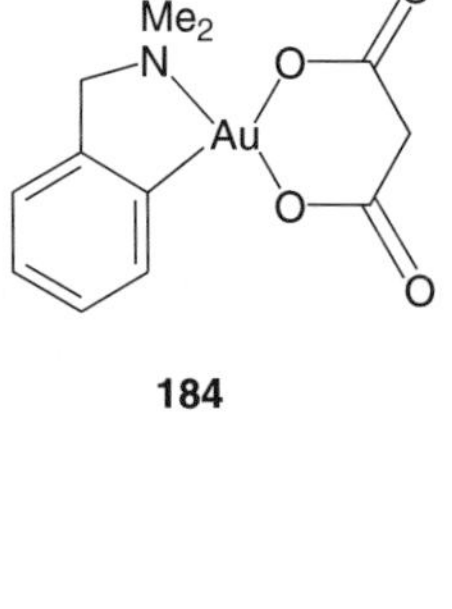

184 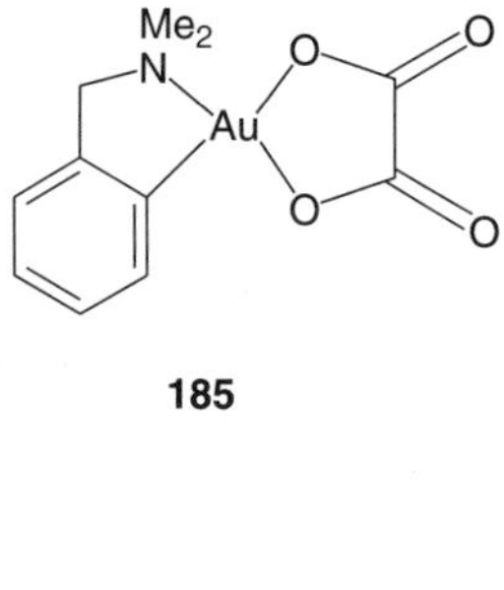 **185**

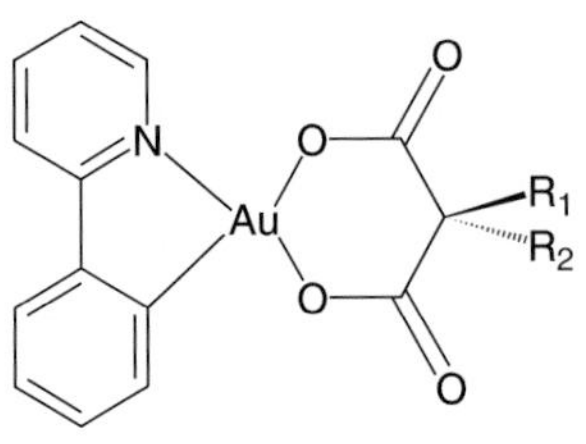

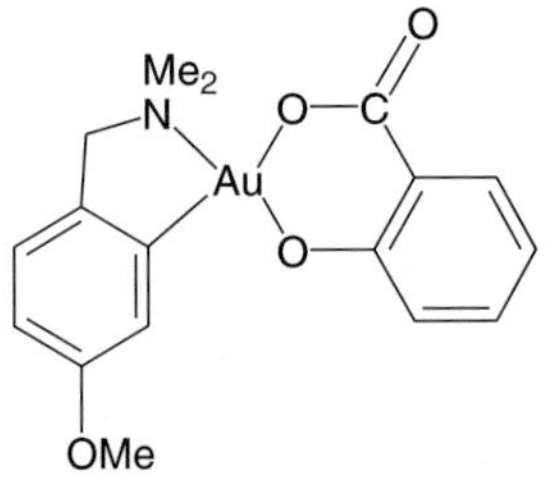

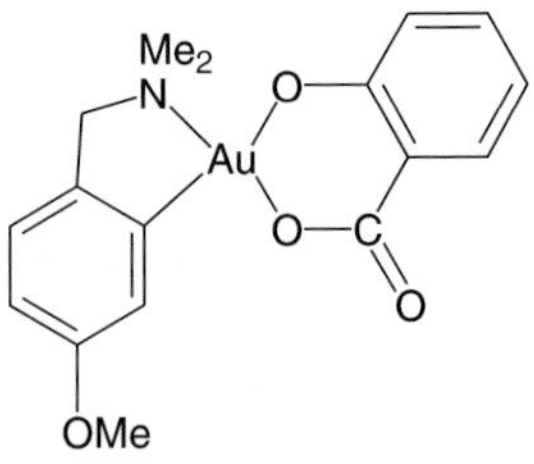

186; $R_1 = R_2 = H$
187; $R_1R_2 = (CH_2)_3$

188a

188b

A series of catecholate derivatives **189** and **190** were prepared from the parent dichloro complexes by reaction with *o*-catechol or tetrachloro-*o*-catechol and Me_3N in hot methanol.[83] The corresponding reaction with an unsymmetrical catechol, 3,5-di-*tert*-butylcatechol produced a mixture of isomers **191a** and **191b**. The related dioxolene complexes **192** and the amidophenolate complexes **193** were also prepared by the same method. An X-ray crystal structure on the latter complex shows that the phenolate oxygen is *trans* to the aryl carbon atom.

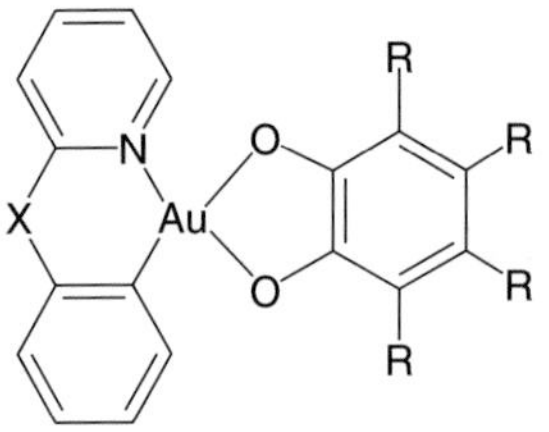

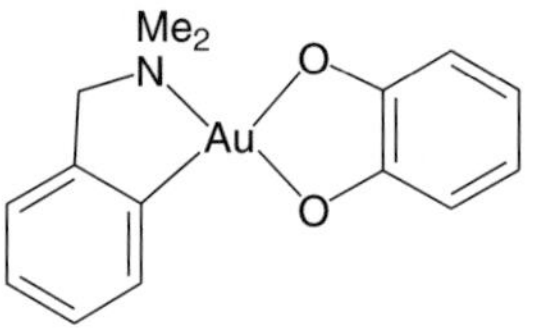

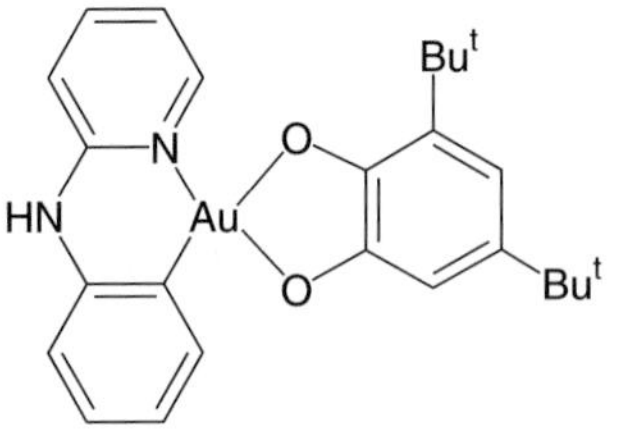

189
X = NH or CH_2; R = H, Cl

190

191a

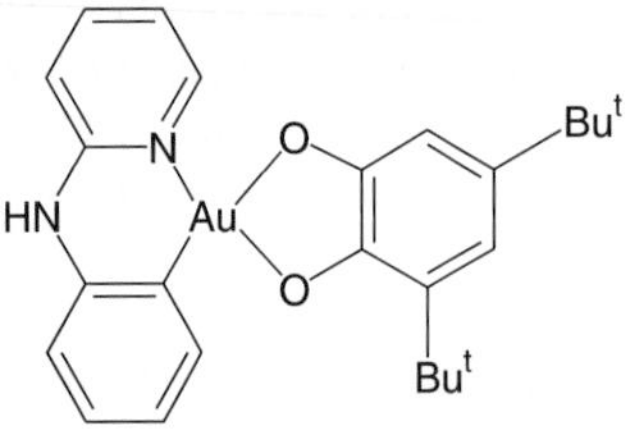

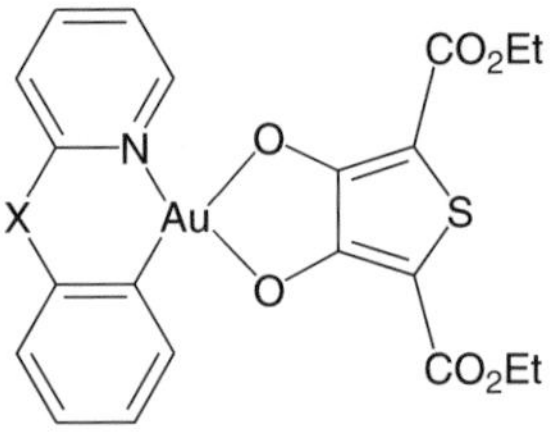

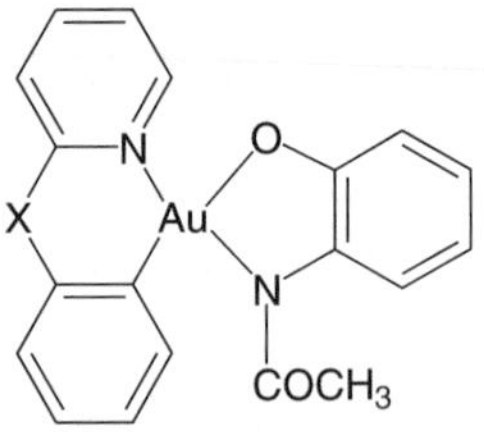

191b

192; X = NH or CH_2

193; X = NH or CH_2

The *O,O′*-bonded β-diketonate complex **194** has been prepared by reaction of the C-bonded acac complex **156** with $Ag[ClO_4]$; this was the first example of a cationic gold(III) acac complex.[64]

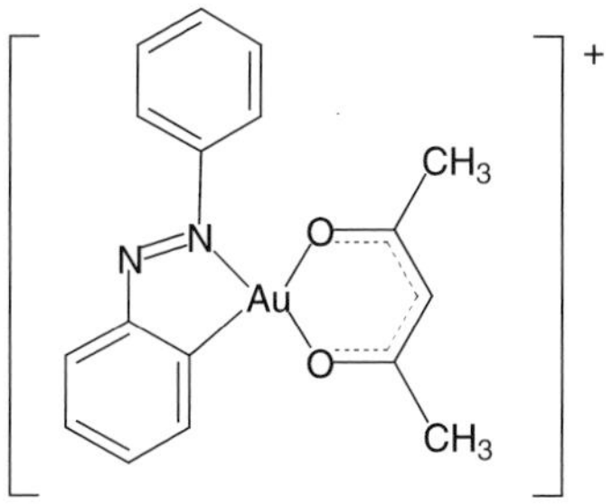

194

Four-membered ring ureylene complexes containing Au–NR–C(O)–NR rings **195** have been prepared by reaction of **41a** with PhNHC(O)NHPh or MeC(O)NH-C(O)NHC(O)Me and silver(I) oxide in refluxing dichloromethane; both complexes were structurally characterised.[84] The phenyl derivative underwent a ring expansion reaction with PhNCO, giving **196**. The isolated product was stable as a solid, but readily lost PhNCO in solution, regenerating the ureylene complex. This behaviour contrasts with analogous platinum(II) ureylene complexes, which formed stable insertion products with PhNCO.

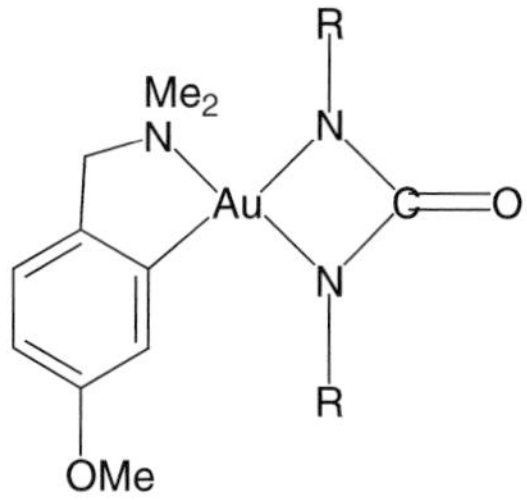

195
R = Ph or $C(O)CH_3$

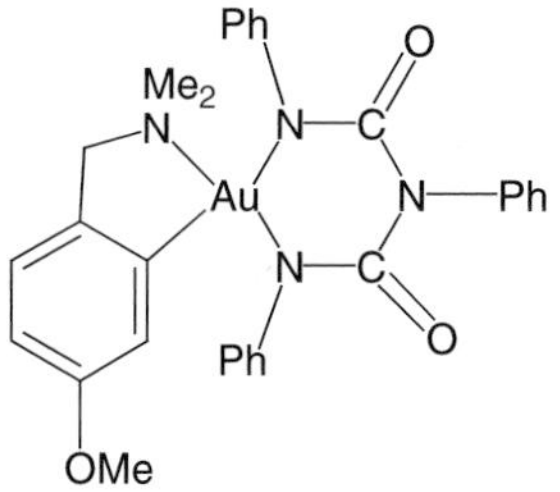

196

Tetrakis(pyrazol-1-yl)borate $[B(pz)_4]^-$ acts as a bidentate, monoanionic N-donor ligand with cycloaurated gold(III) fragments, but reactions of $Na[B(pz)_4]$ with dampAuCl$_2$ **36** and the analogue of papAuCl$_2$ **25** with methyl substituents in both *para* positions, both in the presence of $Na[ClO_4]$, produce different products.[85] With **36**, the cationic complex **197** is obtained, but with the azo complex, displacement of the coordinated azo ligand occurs, with retention of one chloride, giving **198**. A better route to this complex is obtained by omission of the $Na[ClO_4]$. Complex **197** was characterised by an X-ray structure determination; as expected, the Au–N bond *trans* to the aryl carbon was the longest of the three Au–N bonds in this complex. Despite the stabilisation of gold(III) provided by the cyclometallated ligand towards reducing agents the attempted preparation of a bis(pyrazolyl)borate complex from **36** was unsuccessful, producing only metallic gold, due to the greater reducing power of $[H_2B(pz)_2]^-$ compared with $[B(pz)_4]^-$.[85]

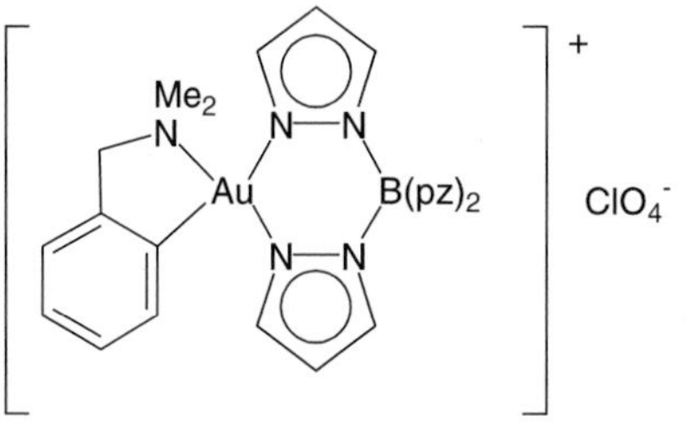

197

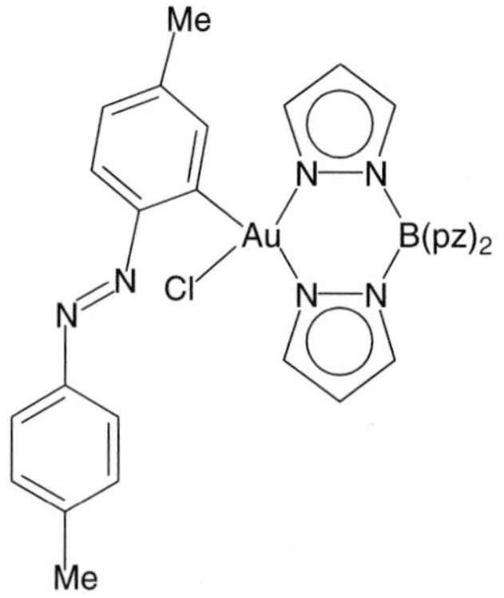

198

The chemistry of 'auralactam' complexes, containing four-membered Au–CHR–C(O)–NR′ ring systems, also shows similarities and differences to the corresponding platinum(II) chemistry. The reactions of **41a** with a range of amide precursors using the silver(I) oxide method gave the first examples of auralactam complexes **199**–**202**.[86] An X-ray crystal structure determination on **202** revealed the presence of a slightly puckered four-membered ring, with the two carbon donors in *cis* positions. When the cyanoacetylurethane product **199** was allowed to stand in $CDCl_3$ solution overnight, a novel dimerisation reaction occurred, to give the eight-membered ring product **203**, which has a much lower solubility, and crystallised from the solution. No such dimeric species have ever been observed in the analogous platinum(II) systems, and it is probable that the greater lability of gold(III) compared to platinum(II),[87] coupled with the insolubility of the gold dimer, accounts for the different behaviour.

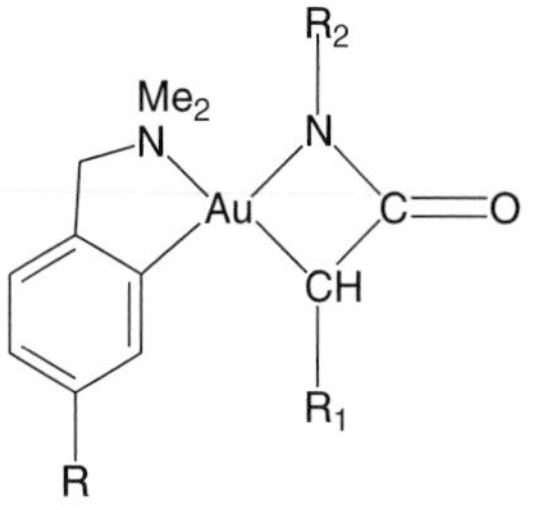

199; R = OMe, R_1 = CN, R_2 = CO_2Et
200; R = OMe, R_1 = C(O)Ph, R_2 = Ph
201; R = OMe, R_1 = C(O)Me, R_2 = Ph
202; R = H, R_1 = C(O)Me, R_2 = Ph

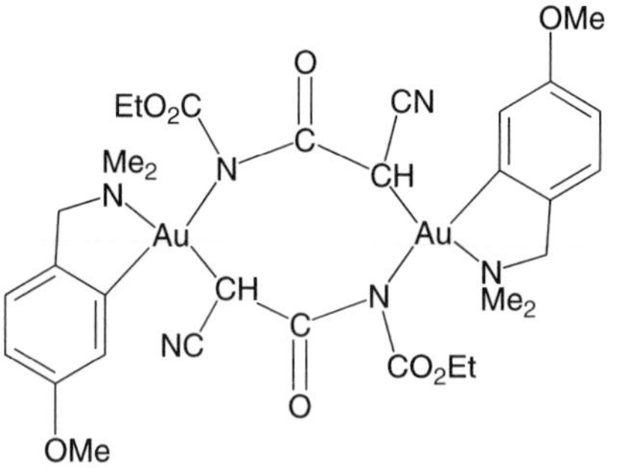

203

Cycloaurated gold halide complexes are also precursors to metallacyclic compounds with three gold–carbon bonds. The first examples of this type were the aurathietane-3,3-dioxide and auracyclobutan-3-one complexes **204** and **205**, prepared by reaction with $PhC(O)CH_2S(O)_2CH_2C(O)Ph$ or $MeO_2CCH_2C(O)CH_2CO_2Me$, respectively, again by reaction with Ag_2O in refluxing dichloromethane.[88] A range of related complexes derived from $PhC(O)CH_2S(O)_2CH_2C(O)Ph$ and $NCCH_2S(O)_2CH_2CN$ have been synthesised by reaction with cycloaurated gold dichloride complexes and Me_3N in refluxing methanol; the benzylpyridyl complex **206** undergoes a ring expansion reaction with ^tBuNC to give **207**.[89] The first example of an auracyclobutane ring system **208** was similarly prepared by reaction of **41a** with $(NC)_2CHCH_2CH(CN)_2$ and Ag_2O. The auracyclobutane ring was found to be slightly puckered (fold angle 20.1°).[90]

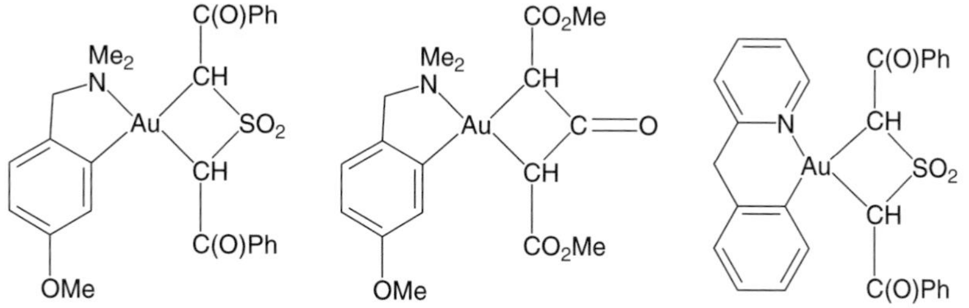

204 **205** **206**

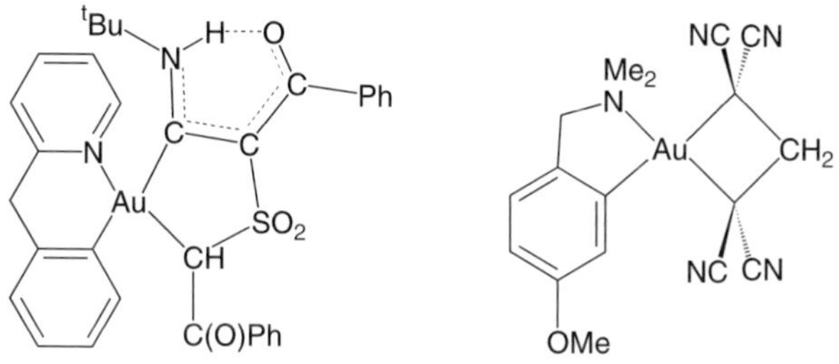

207 **208**

I. *Bidentate Ligands with S-Donor Groups*

Cyclometallated gold(III) dihalide complexes also act as useful precursors for the synthesis of gold(III) complexes with thiolate and other sulfur donor ligands. The cyclometallated ligand stabilises the gold(III) centre towards reduction, which is important due to the reducing potential of sulfur-based ligands.

A number of products formed by reaction with dithiol-based chelating ligands have been synthesised. Reaction of $ppyAuCl_2$ **2** with 3,4-toluenedithiolate gave the dithiolate complex as an orange solid, which was a mixture of isomers **209a** and **209b**[10] and reaction with $HSCH_2CH_2SH$ along with Et_3N in methanol gave the ethane-1,2-dithiolato complex **210**.[12] The damp Au maleonitriledithiolate complex **211** was readily prepared from **36** and $Na_2[S_2C_2(CN)_2]$ in aqueous acetone.[33] Related complexes **212** and **213** formed from $ppyAuCl_2$ **2** with sulfur-rich dithiolate ligands have also been described.[91] Both **212** and **213** undergo ready oxidation to the molecular cations with either I_2 (giving salts containing the I_3^- counterion) or with 7,7,8,8-tetracyano-*p*-quinodimethane (TCNQ) (to give salts containing the TCNQ radical anion). The oxidised complexes are electrical conductors in the solid state.[91] The 0.5-electron oxidised state of these materials has also been obtained for **212** by controlled current electrolysis of a benzonitrile solution of the complex with $[Et_4N][PF_6]$ as supporting electrolyte, giving **214**.[92] X-ray structure determinations on complexes **209a**,[10] **212**[91] and **214**[92] revealed that the Au–S bond *trans* to the Au–C bond is longer than that *trans* to the pyridine nitrogen, as a result of the high *trans*-influence of the aryl donor. The structure of **212** shows a zig-zag stacking arrangement of molecules, with alternating cycloaurated and dithiolate ligands (due to ligand polarisations). In contrast, the structure of **214** showed a columnar stacking of two independent cations, in the same molecular orientation.

Complex **2** also reacts directly with 2,3-dimercapto-1-propanol, or *meso*-2,3-dimercaptosuccinic acid to give **215** (as a mixture of *cis*/*trans* isomers) and **216**, respectively.[65]

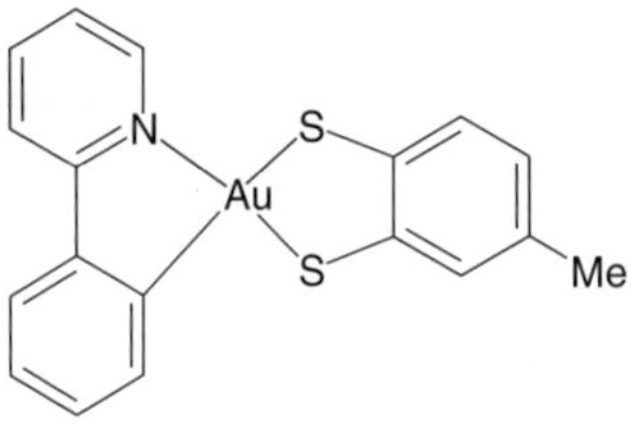

209a

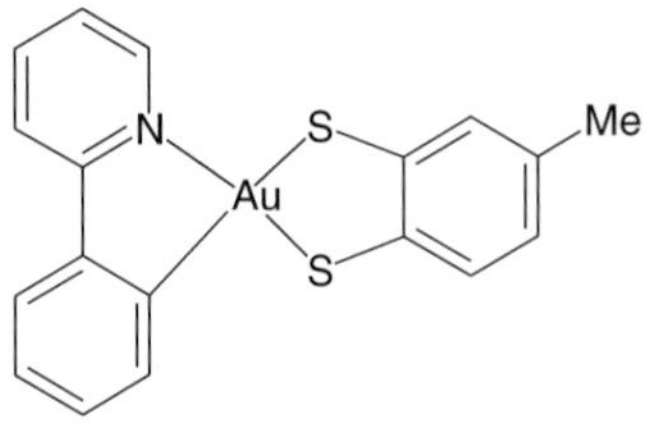

209b

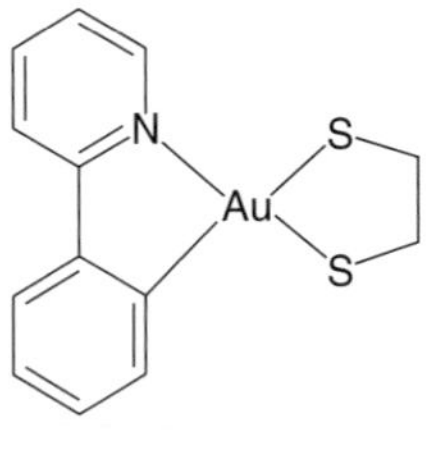

210

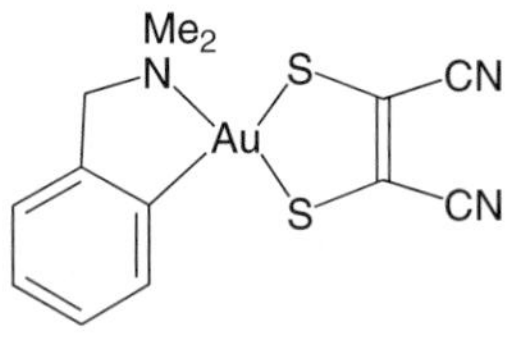

211

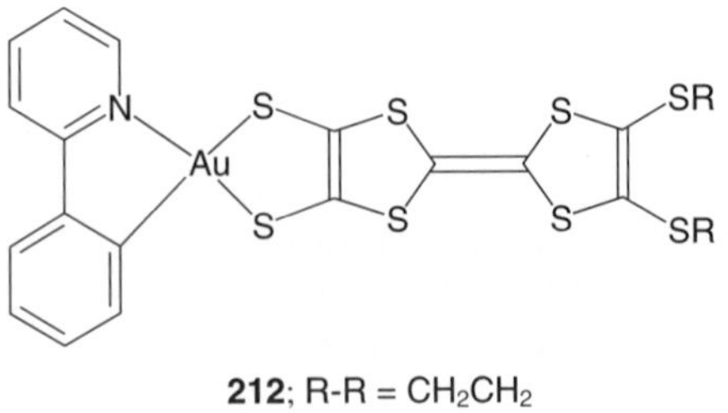

212; R-R = CH_2CH_2
213; R = n-$C_{10}H_{21}$

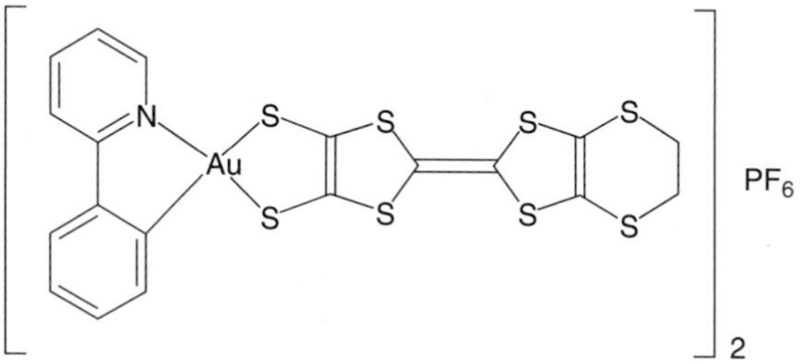

214

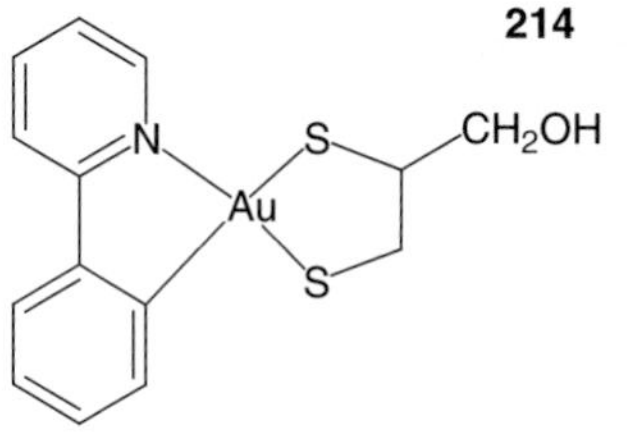

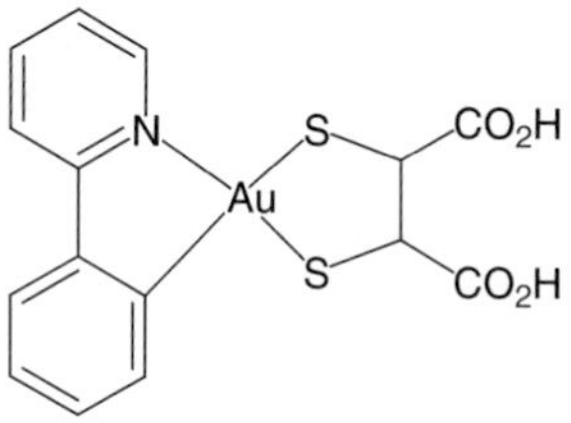

215

216

Mixed thiolate–carboxylate ligands also led to the formation of cyclic products. Reaction of **36** and **41a** with thiosalicylic acid (2-mercaptobenzoic acid) in the presence of silver(I) oxide gave the thiosalicylate complexes **217**.[82] A selection of other thiosalicylate complexes have been synthesised by reaction of the cyclometallated gold(III) dichloride complexes **2**, **3**, **83** and **101** with thiosalicylic acid and base.[13,65] Other sulfur-based chelating ligands behave similarly; reaction of ppy$AuCl_2$ **2** with $AgNO_3$ produced the intermediate nitrato complex ppy$Ag(NO_3)_2$, which reacts with thiolactic acid to give **218**. Reaction of damp$AuCl_2$ **36** with silver mercaptosuccinate gave complex **219**, characterised crystallographically[93] and reaction of **2** with mercaptosuccinic acid in acetone produced the ppy analogue **220**.[12]

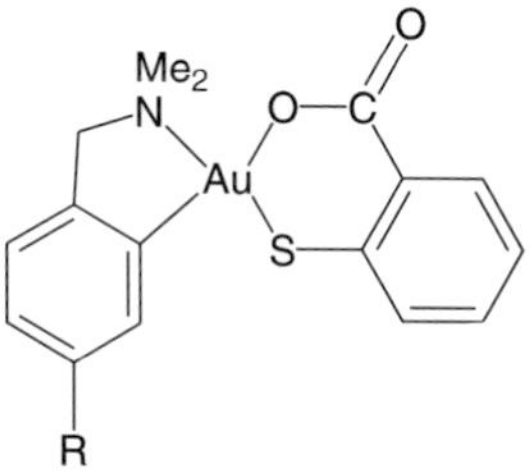

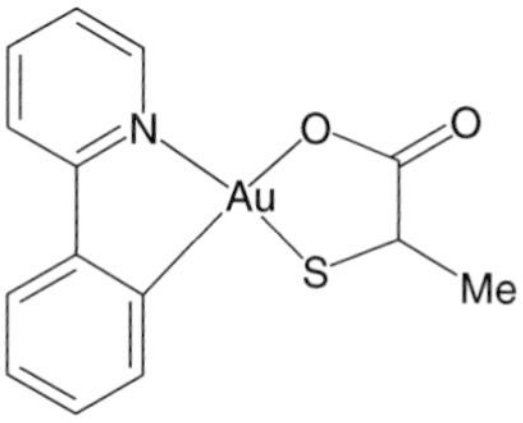

217a; R = H
217b; R = OMe

218

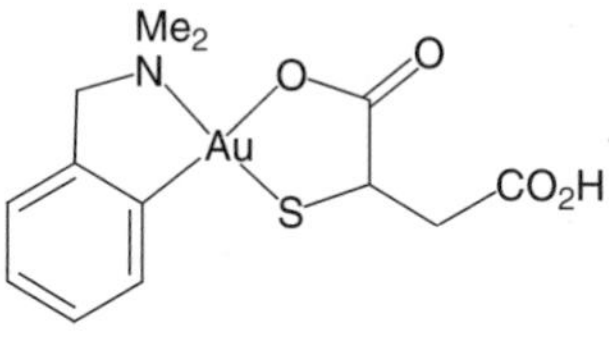

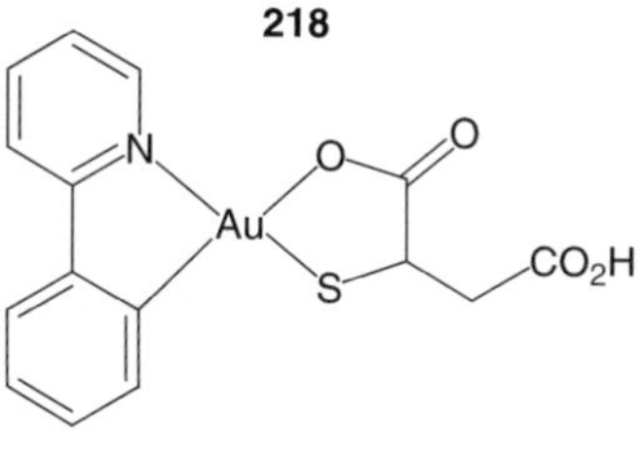

219

220

Complexes with S/P and S/N chelating ligands are also known. When the dicationic pyridine complex $[dampAu(py)_2]^{2+}$ **139** is reacted with 2-aminothiophenol, the chelate complex **221** is obtained; the X-ray structure determination of this complex located the two highest *trans*-influence ligands in the usual *cis* arrangement.[62] Similarly, reaction of **2** with $HSCH_2CH_2NH_2$ and Et_3N in methanol gave the neutral intermediate monosubstitution product **222**, as well as the cationic chelate **223** on addition of $Na[BPh_4]$.[12] The phosphino-thiol $Ph_2PC_6H_4SH$ with **2** in THF likewise gave **224**.[12] In this complex, it was less easy to predict which isomer had been formed, due to similar ligand characteristics of the PPh_2 and thiolate groups. The structure was determined by X-ray crystallography, though there was still some ambiguity due to the difficulty in distinguishing phenyl and pyridyl groups but the isomer with *cis* PPh_2 and aryl groups gave the best fit.

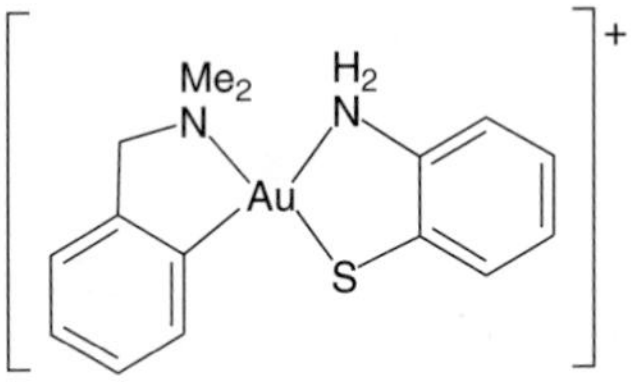

221 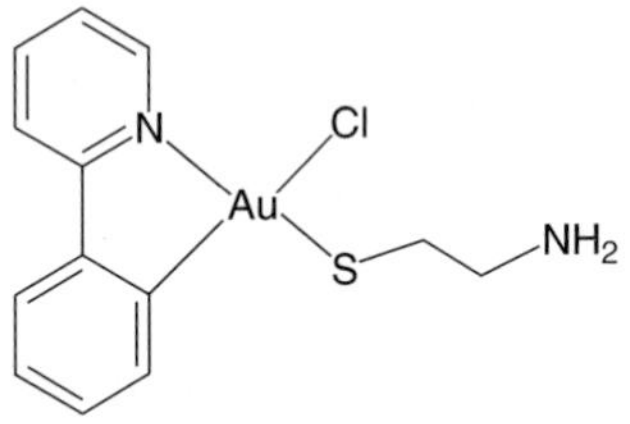 **222**

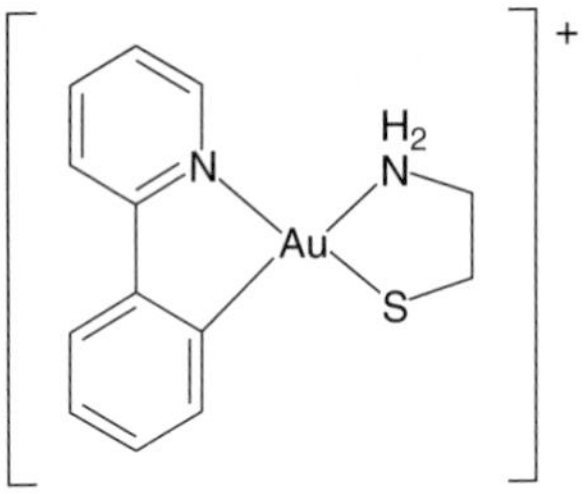

223 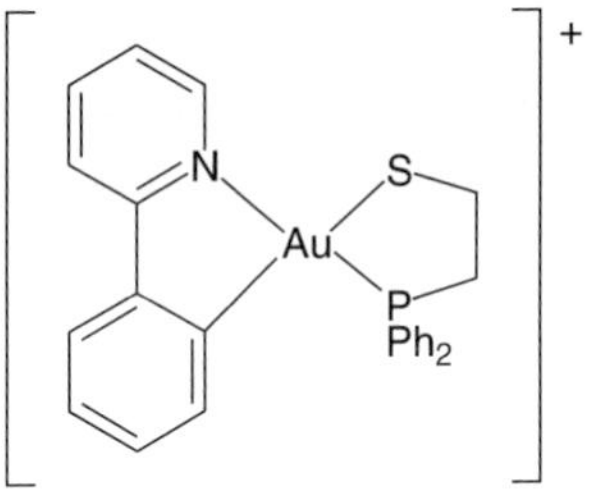 **224**

Reaction of the acetate complex **143** with the biologically important thiol-containing peptide glutathione has been probed using NMR spectroscopy.[63] Both acetate ligands are displaced giving a single product containing a chelated glutathione ligand; NMR shifts suggested coordination through thiolate sulfur and the amide nitrogen, giving the species **225**. The amide nitrogen may bind in a deprotonated form, though this was not suggested in the initial report. Reaction with L-cysteine was also investigated, and likewise gave a single product, which on the basis of NMR data was considered to be bound through sulfur and nitrogen. The affinity of the dampAu moiety for biologically relevant thiolate ligands, and its lack of affinity for DNA bases (Section IV.B), has implications for the mode of action of cycloaurated complexes displaying anti-tumour activity (Section V).

225

Reactions of the complexes **36** and **41b** with MeNHC(S)NHMe and silver(I) oxide in refluxing dichloromethane resulted in desulfurisation of the thiourea and the formation of novel gold(III)–silver(I) –sulfide–halide aggregate cations of the type **226**.[94] Electrospray ionisation mass spectrometry was a useful technique for the characterisation of these species. X-ray structures on two derivatives were reported, and the structure of the core of the cation with R = H and X = Br is shown in Fig. 8. The structure contains three $\{Au_2S_2\}$ groups (with each gold bearing a cycloaurated damp ligand – only N- and C-donor atoms are illustrated) coordinated to three silver(I)

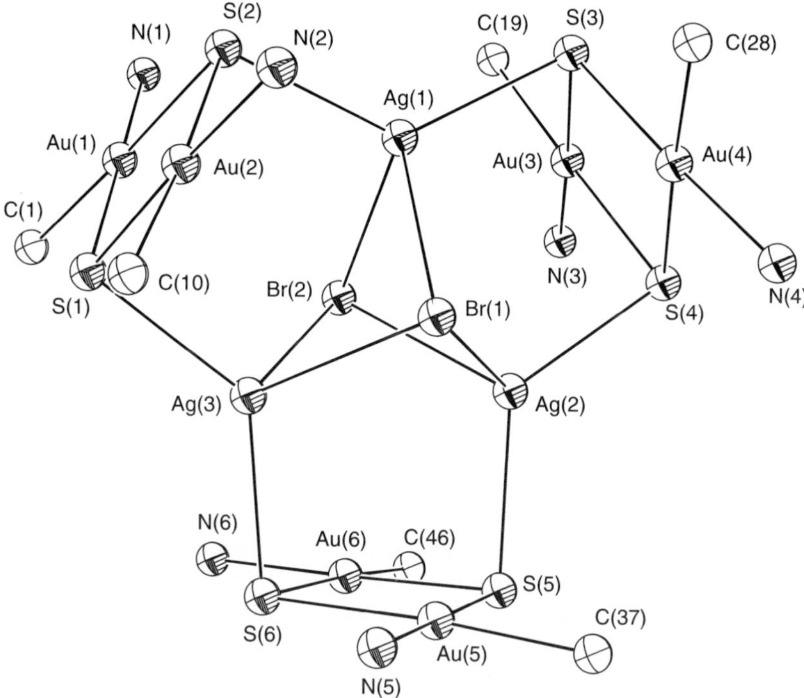

Fig. 8. Molecular structure of the gold–sulfur–silver–bromine core of the cationic complex **226**. Only the N- and C-donor atoms of the cycloaurated damp ligands are shown.

ions, which in turn are triply bridged by two bromide ions. In contrast, the analogous reaction of MeNHC(S)NHMe with *cis*-[$PtCl_2(PPh_3)_2$] and Ag_2O led to isolation of the mononuclear thiourea dianion complex [Pt{SC(NMe)NMe}($(PPh_3)_2$)], indicating significant differences in the analogous chemistry of gold(III) and platinum(II) systems on this occasion.[94] Reaction of anilinopyridine complex **83** with Ph_2NNHC(S)NHPh and Et_3N base in hot methanol gave a sparingly soluble product which was tentatively assigned the structure **227** based on the analogous product obtained and fully characterised with *cis*-[$PtCl_2(PPh_3)_2$].[95]

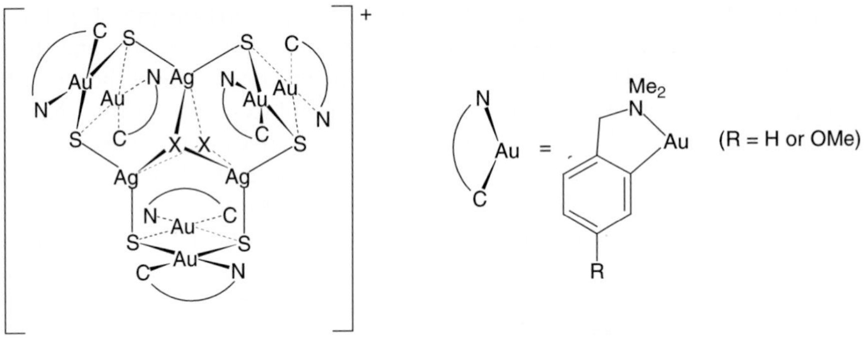

226

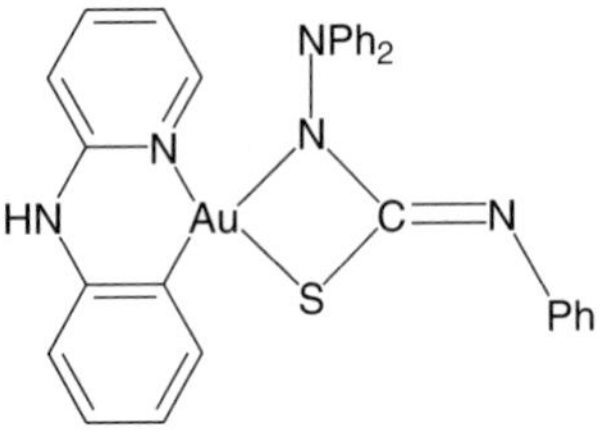

(R = H or OMe)

227

A selection of cycloaurated derivatives have been appended onto the metalloligand [$Pt_2(\mu\text{-}S)_2(PPh_3)_4$]. Thus, reaction of [$Pt_2(\mu\text{-}S)_2(PPh_3)_4$] with the tolylpyridine complex **3**, the anilinopyridine complex **83** and the isomeric thienylpyridine compounds **8** and **9** resulted in formation of the dicationic species **228**–**231**, characterised by electrospray ionisation mass spectrometry, and in the case of the tolylpyridine and anilinopyridine analogues, by X-ray structure determinations.[96] Reaction of [$Pt_2(\mu\text{-}Se)_2(PPh_3)_4$] with **83** has also been described, giving the selenide analogue of **229**.[97] Interestingly, the well-established {Pt_2S_2} core is isoelectronic with the {Au_2S_2} core isolated as a silver halide aggregate in complexes **226**, however attempts at the synthesis of the free {Au_2S_2} system, with each gold complexed by a cyclometallated ligand, have to date been unsuccessful.

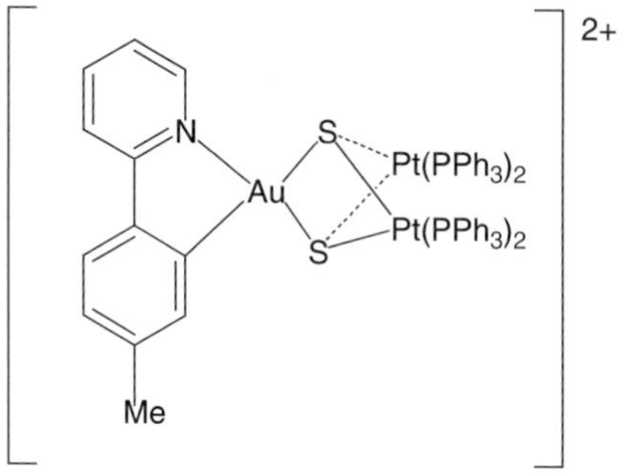

228

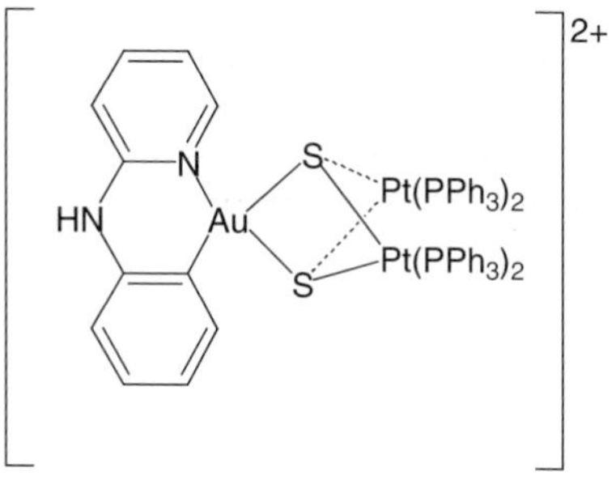

229

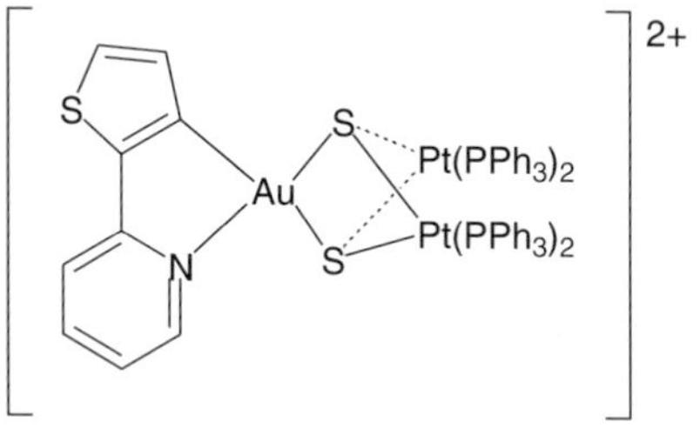

230

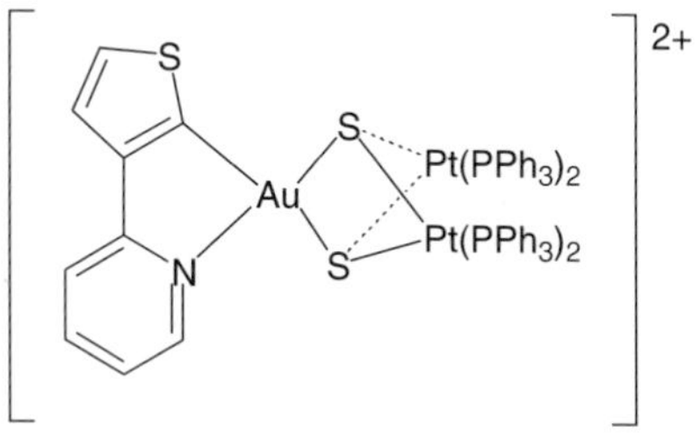

231

J. *Reactions Involving Displacement of the Coordinated N-Donor Ligand*

In much of the chemistry of cycloaurated complexes, the cycloaurated ligand retains its bidentate nature. However, in a number of reactions the nitrogen donor ligand can be displaced from the metal coordination sphere by another, generally more strongly binding ligand. Reactions with phosphines, described in Section IV.A, have already been discussed in this context. For complexes containing the relatively weakly binding azo donor, such as papAuCl_2 **25**, the N-donor ligand is often displaced from the gold centre, even by a relatively poorly donating chloride ligand.

1. With Halide and Cyanide Ligands

With simple halides and pseudohalides, although ligand substitution is generally the preferred process, in one instance involving the azo complex **25** the anionic complex $[papAuCl_3]^-$ was formed by *addition* of chloride (from $[PhCH_2PPh_3]^+Cl^-$). The product, which presumably contains a C-bonded monodentate aryl ligand, i.e. **232**, was stabilised by the presence of the bulky cation; use of $[Me_4N]Cl$ led only to recovery of unreacted **25**.[25] The corresponding complex with n-$C_{10}H_{21}$ substituents in *para* positions on the azobenzene can be prepared directly from reaction of the organomercury chloride with $AuCl_3$(tht) and $[Me_4N]Cl$.[28] In this system, even a relatively weak donor ligand (chloride) is able to displace the azo donor group under certain conditions.

Reactions of dampAuCl_2 **36** with cyanide ions have also been reported. The simple ligand-replacement complex dampAu$(CN)_2$ can be prepared by metathesis and has been characterised spectroscopically.[93] However, reaction of **36** with excess KCN in acetone gave the water-soluble anionic complex $[Au(damp)(CN)_3]^-$ **233**, containing a C-bonded damp ligand, as its potassium salt.

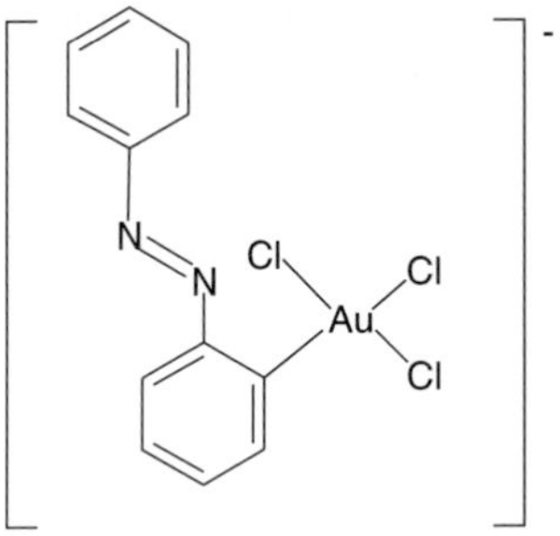

232 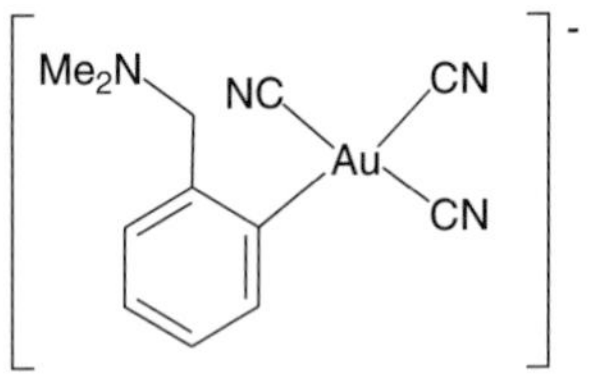 **233**

2. With Dithiocarbamate Ligands

Displacement of the coordinated nitrogen donor ligand can often occur in reactions of cycloaurated gold(III) dihalide complexes towards dithiocarbamate ($R_2NCS_2^-$) ligands. The nature of the products formed is dependent on the mole ratio used, with higher ratios of dithiocarbamate resulting in displacement of the nitrogen group. For example, with 1 equivalent of dithiocarbamate the cationic complex $[dampAu(S_2CNMe_2)]^+$ is formed, which contains a bidentate damp ligand together with a bidentate, but (because of the differing *trans*-influences of the NMe_2 and C ligands) somewhat asymmetrically bonded [Au–S 2.340(9) and 2.179(9) Å] dithiocarbamate ligand. The structure is therefore approximated by **234**, which places the two highest *trans*-influence ligands mutually *cis*.[93] Reactions of the oxazoline derivatives **38a**, **39a** and **40a** with $Na[S_2CNEt_2]$ with added $Na[PF_6]$ or $Na[BPh_4]$ also gave salts containing the cationic complexes **235**–**237**. Corresponding complexes formed from the naphthyl derivative **45a**[36]and **2**[12] are also known.

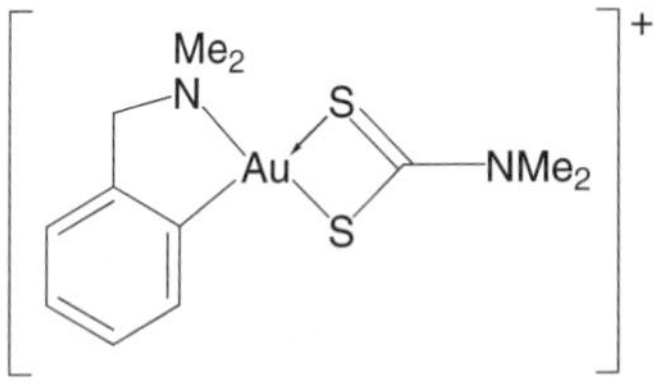

234

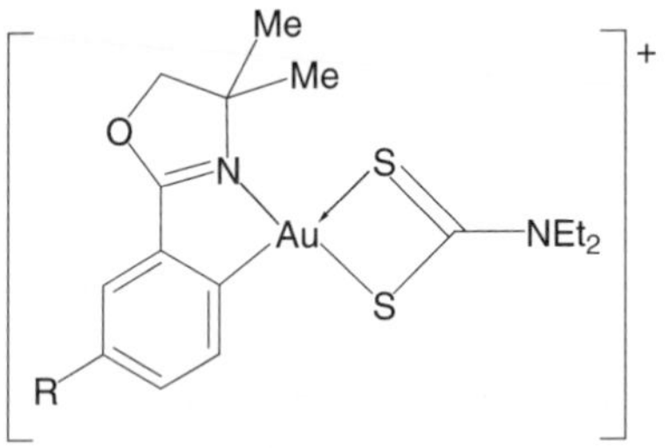

235, R = H
236, R = OMe
237, R = NMe_2

When 2 equivalents of dithiocarbamate are used, complexes of composition [Au(N–C)(S_2CNR_2)$_2$] are obtained.[12,93] The X-ray structure determination on the bis(dithiocarbamate) complex [Au(S_2CNEt_2)$_2$(damp)] reveals the presence of structure **238**.[93] The complex contains a C-bonded monodentate damp ligand, one monodentate dithiocarbamate and a bidentate dithiocarbamate that is more symmetrically bound than in [Au(S_2CNEt_2)(damp)]$^+$ **234**. Complex **238** shows a single set of dithiocarbamate resonances in solution, and fluxional behaviour shown in Scheme 4 was proposed. Analogous products were obtained by reaction of the oxazoline complex **39a**[36] and **2**[12] with 2 mol equivalents of Na[S_2CNEt_2].

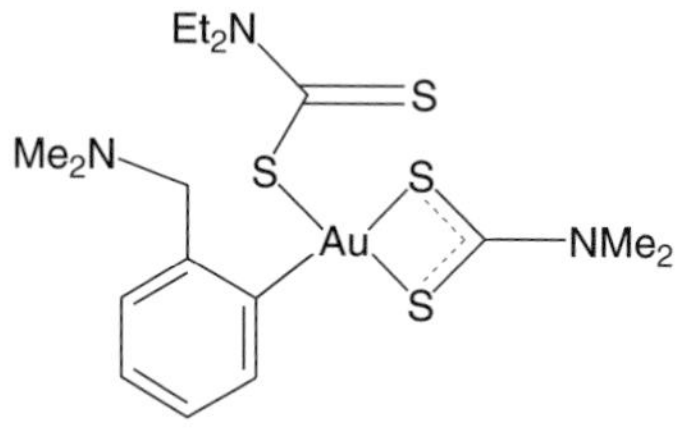

238

3. With Thiosemicarbazone Ligands

The reactions of dampAuCl_2 **36** with a range of thiosemicarbazones also leads to N-donor decomplexation and additionally, protonation.[98,99] In some other related cases, complete loss of the cycloaurated ligand occurs, and these systems are described in the next section. All products were obtained by displacement of chloride ligand(s), cleavage of the Au–N bond and protonation of the NMe_2 group, resulting in cationic products containing five-membered Au–S–C–N–N rings. As examples of the complexes, reaction of **36** with salicylaldehyde thiosemicarbazone gave complex

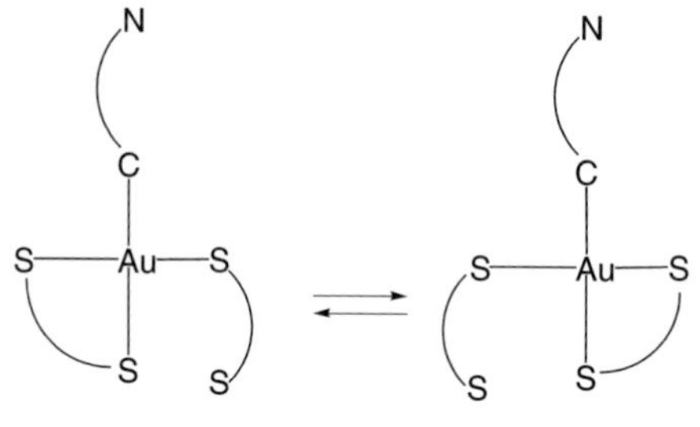

SCHEME 4.

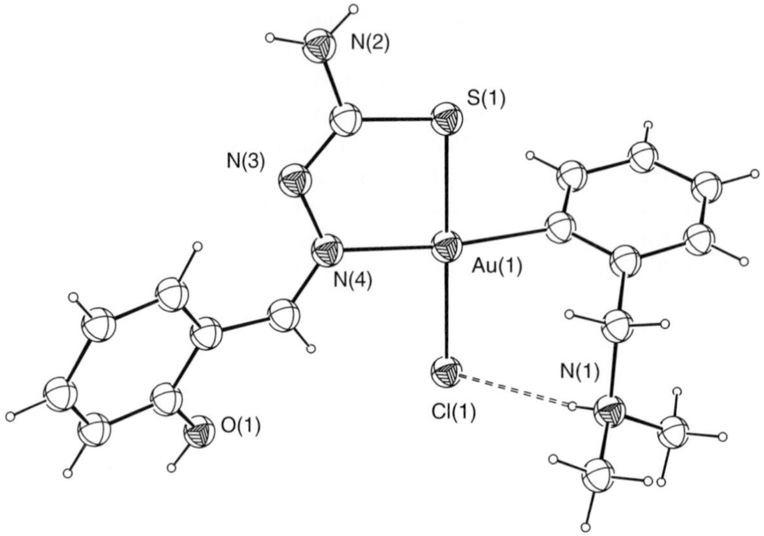

FIG. 9. The molecular structure of the salicylaldehyde thiosemicarbazone complex **239** clearly shows displacement of the NMe_2 group of the damp ligand, which is monodentate and C-bonded. The N–H proton shows an intramolecular hydrogen bond to the chlorine atom.

239 with one five-membered ring while reaction with 2-acetylpyridine thiosemicarbazide gave **240** containing two five-membered rings. The displacement of the NMe_2 group is illustrated in Fig. 9, which shows the molecular structure of the cation of **239**. Following on from these studies, a series of related derivatives were prepared from formylferrocene thiosemicarbazone ligands (**241**)[100] and from 2-pyridineformamide thiosemicarbazones (**242**).[101] In all complexes, the two highest *trans*-influence ligands (aryl carbon and sulfur) are mutually *cis*. The reaction of **36** with diphenylthiocarbazone [dithizone, PhNHNHC(S)N=NPh] produces an analogous complex **243**; the chloride ligand can be replaced by the thiolate anion, 1-methyl mercaptotetrazolate, which occurs with formation of the coordinated $[PhNNC(S)NNPh]^{2-}$ dianion.[102]

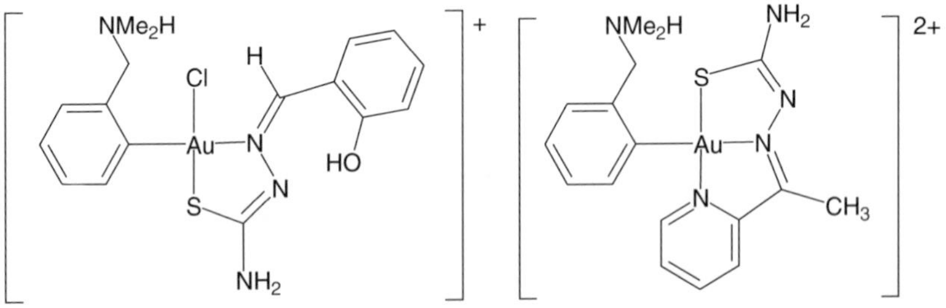

239 **240**

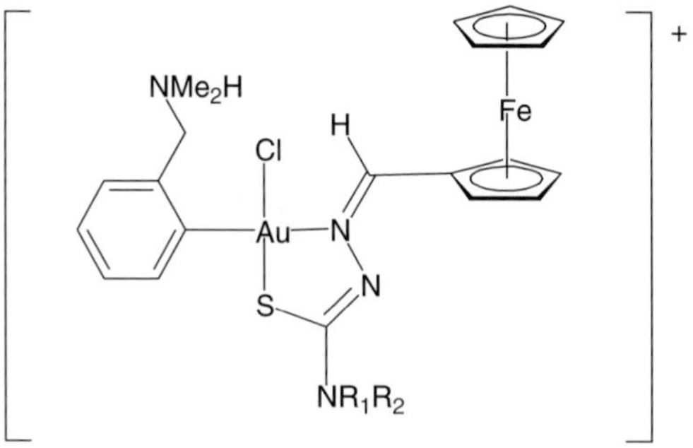

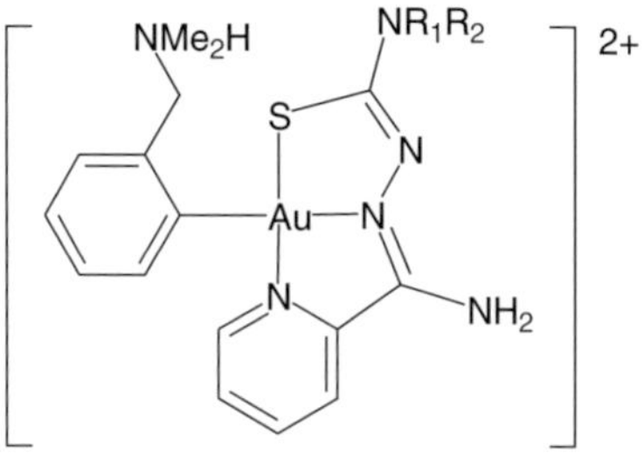

241

242: R_1R_2 = HH, MeMe, HMe or $(CH_2)_6$

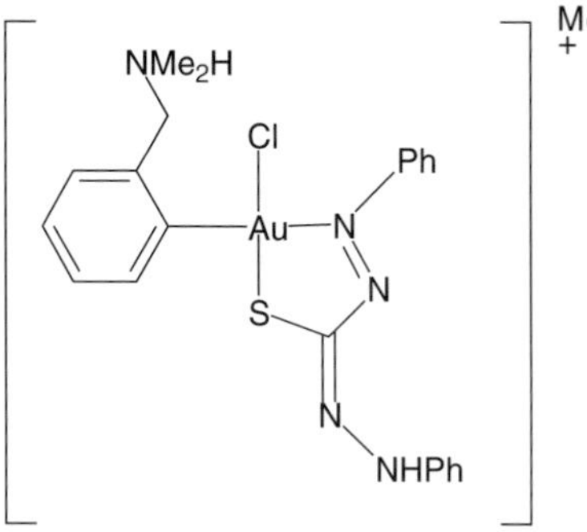

243

Decoordination and protonation of the NMe_2 group also occurs in reactions of damp$AuCl_2$ with heterocyclic thiols.[103] For example, reaction with 2-mercapto-1,3-thiazoline gave dicationic complex **244** containing a coordinated chloride ligand; 2-mercaptopyridine behaved in the same way. With sodium 1-methyl mercaptotetrazolate, displacement of both chlorides occurred, and the heterocycle binds as an anionic ligand, giving neutral **245**. More complex behaviour was observed with 4-methyl-3-mercapto-1,2,4-triazole, which gave different products, including a gold(I) complex formed by reduction.[103]

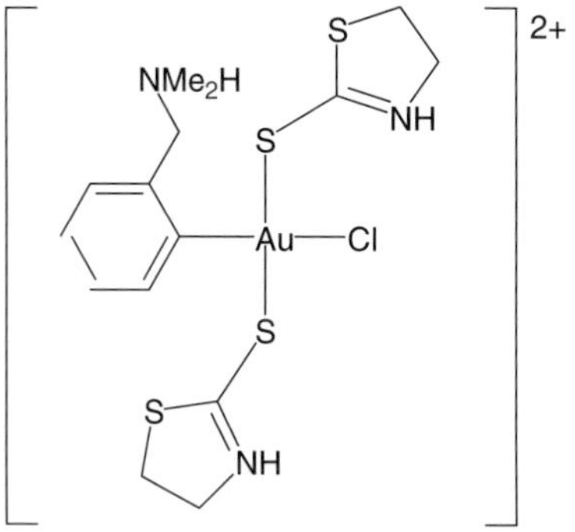

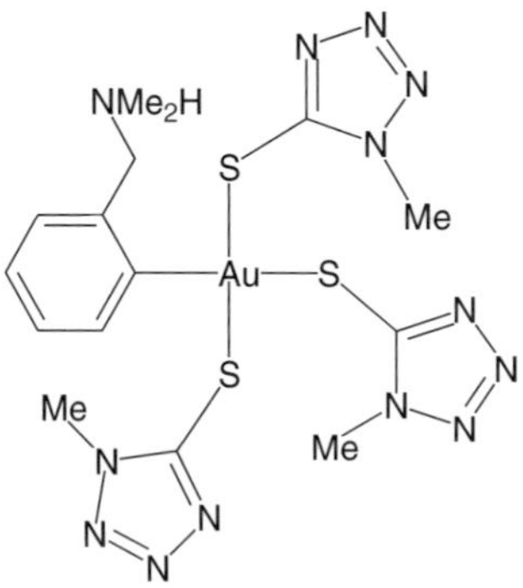

244

245

K. *Reactions Involving Complete Displacement of the Cycloaurated Ligand*

A few studies have been reported where the organic cycloaurated ligand is completely displaced from the metal centre; generally these are rare, because the Au–C bond is relatively strong. One example is given by the reaction of dampAuCl_2 **36** with the multidentate phosphinothiolate ligands **246** (R = H, $SiMe_3$),[104,105] which gave mononuclear products **247**, together with a mixed valence (non-organometallic) Au(I)–Au(III) product. Mass spectrometry provided evidence for an intermediate species such as **248**, containing a C-bonded damp ligand together with a phosphinothiolate ligand. Complex **247** is stabilised towards reduction to linear gold(I) species through the formation of fused chelate rings. Another example where complete substitution of the damp ligand occurs under mild conditions is reaction of **36** with 4-methyl-3-mercapto-1,2,4-triazole (HSmetriaz), which gave the gold(I) complex [Au(HSmetriaz)$_2$]Cl as one of the isolated products.[103]

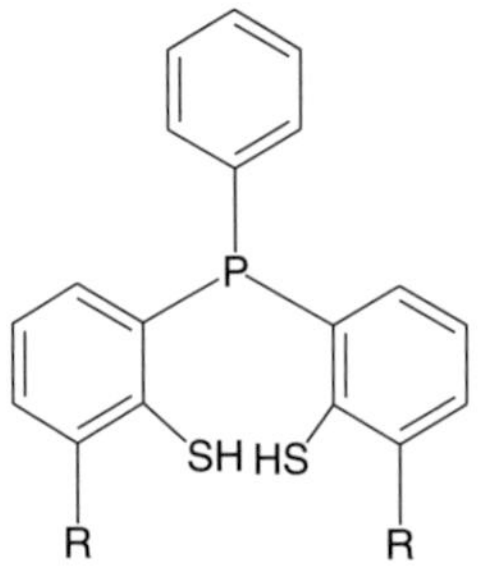

246 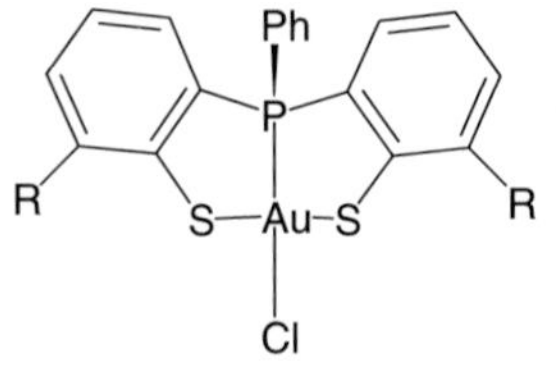 **247** 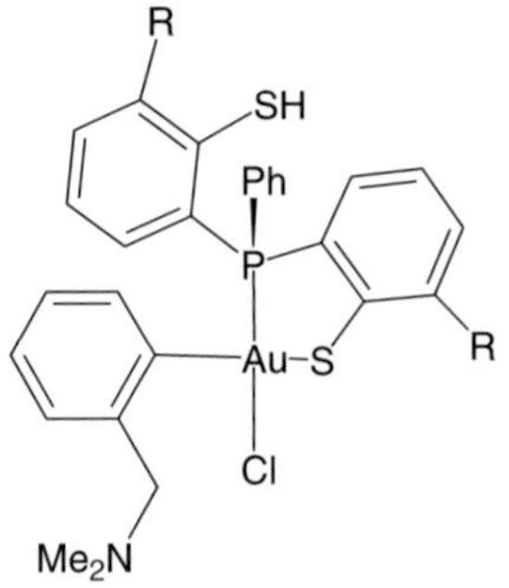 **248**

Relatively little work has concerned the use of cycloaurated complexes as reagents or intermediates for the synthesis of organic products, but it might be expected that such gold complexes could possess some unique reactivity. Diarylgold(III) compounds can undergo reductive elimination reactions when promoted by the addition of PPh_3, producing either symmetric or asymmetric biaryl compounds, containing a range of substituents, in good yields.[29,42] A prerequisite for reaction is that one of the aryl ligands must be cycloaurated, and preferably of the azo type (containing a relatively weak Au–N bond), which stabilises the gold–aryl.[106] As an example, reaction of the nitrophenyl complex **249** with PPh_3 gave the biaryl **250** together with Ph_3PAuCl. Use of an additional equivalent of PPh_3 together with added Na[ClO_4] produces [Au(PPh_3)$_2$]ClO_4 by-product, which is more easily removed from the biaryl.

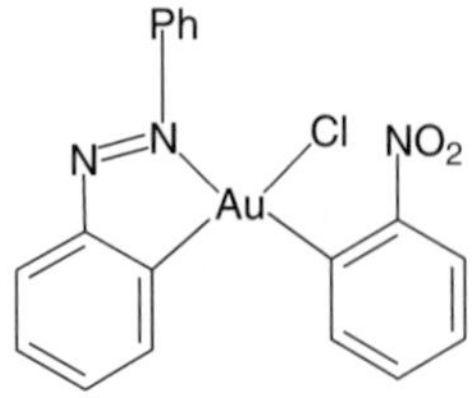

249 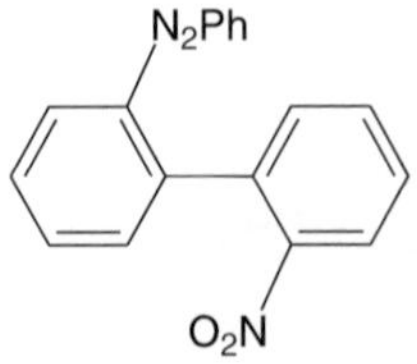 **250**

The benzylpyridine-derived complex **101** is relatively stable towards reaction with CO; at room temperature, no reaction occurs when CO is bubbled through a solution of the complex. However, under forcing conditions (50 atmosphere pressure of CO at 60 °C), bright orange-yellow organic products are formed, and the gold is reduced to give a gold mirror.[55] Organic products were isolated by chromatography and analysed by NMR and mass spectrometry. Compound **251** was considered to be the first product of reaction, formed by insertion of CO into the Au–C bond, followed by intramolecular nucleophilic attack of the pyridine nitrogen on the C=O group. It was believed that this was the first example of CO insertion into a gold(III)–carbon bond. Compound **252** was believed to be formed by oxidative chlorination of **251** by $AuCl_3$, and the dimeric species **253** formed by activation of a C–Cl bond of this species.

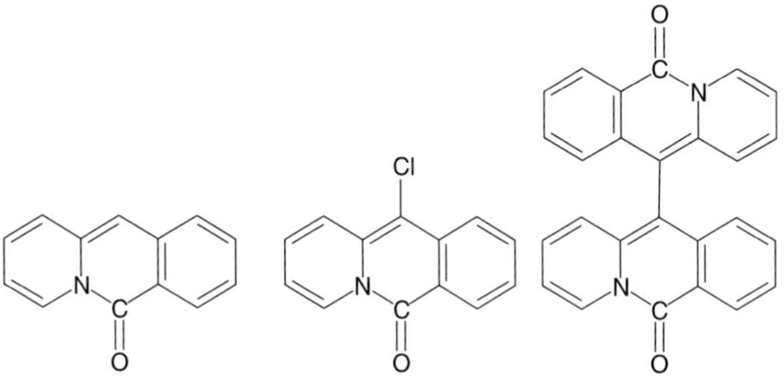

251 **252** **253**

V
PROPERTIES AND APPLICATIONS

A. *Biological Activity*

The 'medicinal' use of gold dates back to antiquity, and gold(I) compounds have been used clinically for the treatment of rheumatoid arthritis for some time. In recent years, considerable effort has been extended to the investigation of the anti-tumour properties of gold(III) compounds. One of the main driving forces for this is the extensive studies and clinical use of a range of platinum-based drugs; *cis*-$[PtCl_2(NH_3)_2]$ (*cisplatin*) is the most widely used anti-cancer drug in the world. Gold(III) is isoelectronic with platinum(II) and because both readily form square–planar complexes, they might be expected to show similar characteristics. However, a key factor that needs to be overcome is the tendency for many gold(III) complexes to be readily reduced in biological media, making them unsuitable. The stabilisation of the gold(III) centre by chelating ligands, especially cyclometallated ligands, has resulted in increased activity in recent years. A detailed discussion of the wide range of gold complexes that have been screened is beyond the scope of this review, and the reader is directed to excellent and comprehensive reviews on this topic.[107,108] Instead, we here briefly summarise some of the key discoveries, and provide an update on some of the more recent studies. A detailed review of the biochemistry of gold (in I and III oxidation states) has also appeared.[109]

Some of the most substantial discoveries in this area concern the anti-tumour properties of damp complexes dampAuX_2 (e.g., **36**), which have been studied in some detail and the developments have been reviewed.[110,111] A range of complexes having X = Cl, SCN, O_2CCH_3, and X_2 = oxalato, malonato have been screened, with the acetato and malonato complexes generally showing the best activity. These complexes showed selective *in vitro* cytotoxicity, as well as *in vivo* activity against human carcinoma xenografts.[32] In these complexes, the cycloaurated ligand stabilises the gold(III) state. For the acetate complex dampAu$(OAc)_2$ **143** hydrolysis occurs in solution (Section IV.D), which provides a comparison with *cisplatin*, which binds with its cellular target DNA more strongly after hydrolysis. However, the gold(III) cyclometallated complexes almost certainly have a different mode of action to *cisplatin*. For the dampAu complexes, studies using the technique of alkaline elution (where the formation of DNA crosslinks hinders the flow of DNA strands through a filter) have indicated that the gold complexes do not induce DNA crosslinks, as does *cisplatin*.[111] From chemical studies, the cycloaurated fragments have been found to have a strong affinity for thiolate ligands (Section IV.I), which (as discussed below) suggests that proteins may be an important target. Additionally, for *cisplatin,* hydrogen-bonding occurs between N–H groups and phosphate residues on the DNA backbone, but this hydrogen-bonding is not possible for complexes containing the dampAu moiety, unless displacement and protonation of the NMe_2 group occurs.

A number of recent studies have investigated the anti-tumour activity of benzyl–pyridine and benzyl–bipyridine-derived complexes. The complex bpAu$(OAc)_2$, and the xylidene amido complex **173** have been found to show significant activity against two ovarian carcinoma cell lines (A2780/S and A2780/R), with the toluidine analogue **172** and the gold–hydroxy complex **165** (the likely hydrolysis product of the amido complexes in aqueous media) also showing good activity.[79] Another study found that bpAu$(OAc)_2$, **165** and **173** are extremely efficient inhibitors of mitochondrial thioredoxin reductase, but do not interact strongly with DNA.[112] These results suggest that mitochondria could be a potential target for gold-based anti-tumour compounds.

Little information is known concerning the interactions of gold(III) complexes with proteins, though such adducts may be important in the distribution and activity of the gold species. The reactivity towards bovine serum albumin (BSA) of the cycloaurated complex [Au(6-(1,1-dimethylbenzyl)-2,2′-bipyridine-H)(OH)]PF_6 **165** (together with a number of non-cycloaurated gold(III) complexes with N-donor ligands) has been investigated by various spectroscopic methods (including circular dichroism and visible absorption) and separation techniques.[113] A strong metal–protein adduct was obtained, with considerable stability under physiological conditions, and it was suggested that adduct formation occurs through surface histidines. Complete removal of the gold from the BSA occurred only with excess KCN.

The promising results obtained to date indicate that there is considerable scope for the development of novel gold(III) complexes with potent anti-tumour activity;

the stabilisation of the gold(III) centre by cyclometallated ligands is certain to play a major role in this field.

B. *Electrochemical Properties*

Only one paper has reported the electrochemical behaviour of gold(III) complexes containing cyclometallated N,C ligands.[114] The complexes studied, in a range of solvent systems and using different techniques (cyclic voltammetry, controlled-potential Coulometry) were **59**, **101** and the three benzyl–bipyridine complexes **112–114**. All show an irreversible 1-electron reduction in the range −0.9 to −1.2 V Versus ferrocene/ferrocenium, using a platinum electrode in cyclic voltammetry. Complex decomposition of the electrogenerated species afforded metallic gold and various gold by-products. In the case of **59** it was possible to identify and characterise the bis(cyclometallated) product **254**, formed by a transcyclometallation process, which could also be obtained in low yield by chemical reduction of **59** with methanolic sodium borohydride.

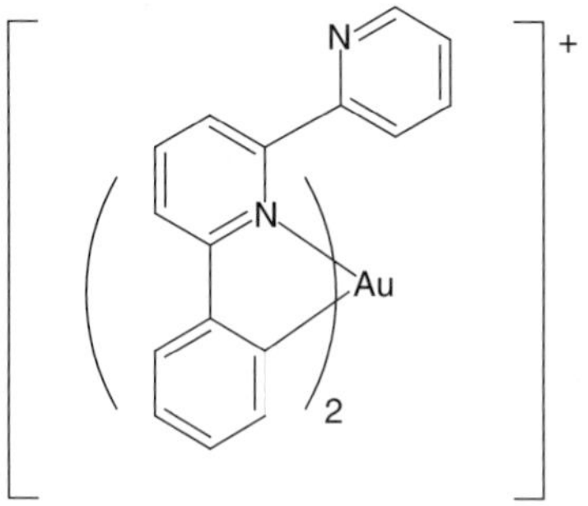

254

C. *Photophysical Properties*

Relatively few investigations have been carried out concerning the photophysical properties of gold(III) complexes, in contrast with the large number of studies on gold(I) complexes, but in recent years there has been developing interest into luminescence properties of cycloaurated gold(III) complexes. Complexes containing tridentate cycloaurated ligands have shown the most promise in this area. Che and co-workers[37] have investigated the photophysical properties of a number of 2,6-diphenylpyridine-derived complexes, which were found to be emissive at 77 K. However, some of the best results have been obtained by Yam *et al.*[38] who found that the acetylide complexes **178–182** show luminescence in various media at both low and ambient temperature; most other gold(III) complexes only exhibit luminescence at low temperature or are non-emissive. The lack of luminescence in many gold(III) systems has been attributed to the presence of low-energy d–d ligand field states, and the electrophilicity of gold(III).[38] The presence of the strongly σ-bonding alkynyl (acetylide) ligand makes the gold(III) centre more electron rich, increases the d–d splitting, resulting in enhanced luminescence by increasing the probability of populating the emissive state. Complexes **178** and **183** have been incorporated as electrophosphorescent emitters into multilayer organic light-emitting diodes (OLEDs), which show intense electroluminescence (EL) on application of a DC voltage.[81] The colour of the EL can be tuned by varying the DC voltage and the dopant concentration.

In contrast, the gold(III) dichloride complex ppyAuCl_2 **2** has been found to produce an intense emission only in a low temperature glassy matrix.[10] The toluenedithiolate complex **109** possessed mild solvatochromic absorptions in the visible region, but was not emissive, unlike the platinum(II) counterparts with diimine ligands such as 2,2-bipyridines.[10] It was suggested that the gold(III) orbitals are stabilised compared to those in platinum(II), leading to reduced intensity in the solvatochromic absorption, and no emission.[115]

The diphenylphenanthroline complex **23** is emissive in fluid solutions, with quantum yields of ca. 10^{-4} and lifetimes of 0.4–0.7 μs. The estimated excited state reduction potential of 2.2 V (vs. normal hydrogen electrode (NHE)) suggested that the complex is a strong photooxidant, which was demonstrated with the formation of the 1,4-dimethoxybenzene radical cation (DMB^+) upon UV–visible irradiation of an MeCN solution of the complex with DMB.[22]

Finally, several patents have covered the photophysical properties (electroluminescence, phosphorescence) of metal complexes containing C,N cyclometallated ligands.[116]

Acknowledgments

Support from the University of Waikato is gratefully acknowledged. The contribution of various coworkers is greatly appreciated, in particular my colleague Professor Brian Nicholson, who also prepared the X-ray structure diagrams for this article.

References

(1) Kharasch, M. S.; Isbell, H. S. *J. Am. Chem. Soc.* **1931**, *53*, 3053.
(2) Liddle, K. S.; Parkin, C. *J. Chem. Soc., Chem. Commun.* **1972**, 26; de Graaf, P. W. J.; Boersma, J.; van der Kerk, G. J. M. *J. Organomet. Chem.* **1976**, *105*, 399.
(3) Fuchita, Y.; Utsunomiya Y.; Yasutake, M. *J. Chem. Soc., Dalton Trans.* **2001**, 2330.
(4) Dehand, J.; Pfeffer, M. *Coord. Chem. Rev* **1976**, *18*, 327. Bruce, M. I. *Angew. Chem. Int. Ed. Engl* **1977**, *16*, 73. Omae, I. *Chem. Rev* **1979**, *79*, 287. *Coord. Chem. Rev.* **1980**, *32*, 235.
(5) Grohmann, A.; Schmidbaur, H. *Comprehensive Organometallic Chemistry II*, Elsevier, UK, 1995; Schmidbaur, H.; Grohmann, A.; Olmos, M. E. (H. Schmidbaur, Ed.), *Gold: Progress in Chemistry, Biochemistry and Technology*, Wiley, New York, 1999.
(6) Parish, R. V. *Gold. Bull.* **1997**, *30*, 3.
(7) Parish, R. V. *Gold. Bull.* **1997**, *30*, 55.
(8) Parish, R. V. *Gold. Bull.* **1998**, *31*, 14.
(9) Constable, E. C.; Leese, T. A. *J. Organomet. Chem.* **1989**, *363*, 419.
(10) Mansour, M. A.; Lachicotte, R. J.; Gysling, H. J.; Eisenberg, R. *Inorg. Chem.* **1998**, *37*, 4625.
(11) Ivanov, M. A.; Puzyk, M. V. *Russian J. Gen. Chem.* **2001**, *71*, 1660 (Transl. from *Zh. Obsch. Khim.* **2001**, *71*, 1751).
(12) Parish, R. V.; Wright, J. P.; Pritchard, R. G. *J. Organomet. Chem.* **2000**, *596*, 165.

(13) Henderson, W.; Nicholson, B. K.; Faville, S. J.; Fan, D.; Ranford, J. D. *J. Organomet. Chem* **2001**, *631*, 41.
(14) Fuchita, Y.; Ieda, H.; Wada, S.; Kameda, S.; Mikuriya, M. *J. Chem. Soc., Dalton Trans.* **1999**, 4431.
(15) Constable, E. C.; Henney, R. P. G.; Leese, T. A. *J. Organomet. Chem.* **1989**, *361*, 277.
(16) Constable, E. C.; Henney, R. P. G.; Raithby, P. R.; Sousa, L. R. *Angew. Chem. Int. Ed. Engl.* **1991**, *30*, 1363. Constable, E. C.; Henney, R. P. G.; Raithby, P. R.; Souza, L. R. *J. Chem. Soc., Dalton Trans.* **1992**, 2251.
(17) Vicente, J.; Chicote, M. T.; Lozano, M. I.; Huertas, S. *Organometallics* **1999**, *18*, 753.
(18) Fan, D.; Meléndez, E.; Ranford, J. D.; Lee, P. F.; Vittal, J. J. *J. Organomet. Chem.* **2004**, *689*, 2969.
(19) Ieda, H.; Fujiwara, H.; Fuchita, Y. *Inorg. Chim. Acta* **2001**, *319*, 203.
(20) Fuchita, Y.; Ieda, H.; Yasutake, M. *J. Chem. Soc., Dalton Trans.* **2000**, 271.
(21) Liu, H.-Q.; Cheung, T.-C.; Peng, S.-M. Che, C.-M. *J. Chem. Soc., Chem. Commun.* **1995**, 1787.
(22) Chan, C.-W.; Wong, W.-T.; Che, C.-M. *Inorg. Chem.* **1994**, *33*, 1266.
(23) Cinellu, M. A.; Zucca, A.; Stocoro, S.; Minghetti, G.; Manassero, M.; Sansoni, M. *J. Chem. Soc., Dalton Trans.* **1996**, 4217.
(24) Vicente, J.; Chicote, M. T. *Inorg. Chim. Acta* **1981**, *54*, L259.
(25) Vicente, J.; Chicote, M. T.; Bermúdez, M. D. *Inorg. Chim. Acta* **1982**, *63*, 35.
(26) Hüttel, R.; Konietzny, A. *Chem. Ber.* **1973**, *106*, 2098. Calderazzo, F.; Dell'Amico, D. B. *J. Organomet. Chem.* **1974**, *76*, C59.
(27) Monaghan, P. K.; Puddephatt, R. J. *Inorg. Chim. Acta* **1975**, *15*, 231.
(28) Vicente, J.; Bermúdez, M. D.; Carrión, F. J.; Martínez-Nicolás, G. *J. Organomet. Chem.* **1994**, *480*, 103.
(29) Vicente, J.; Bermúdez, M.-D.; Carrión, F.-J.; Jones, P. G. *Chem. Ber.* **1996**, *129*, 1395.
(30) Vicente, J.; Bermudez, M. D.; Carrion, F.-J.; Jones, P. G. *Chem. Ber.* **1996**, *129*, 1301.
(31) Nicholson, B. K.; Brown, S., personal communication.
(32) Buckley, R. G.; Elsome, A. M.; Fricker, S. P.; Henderson, G. R.; Theobald, B. R. C.; Parish, R. V.; Howe, B. P.; Kelland, L. R. *J. Med. Chem.* **1996**, *39*, 5208.
(33) Mack, J.; Ortner, K.; Abram, U.; Parish, R. V. *Z. Anorg. Allg. Chem.* **1997**, *623*, 873.
(34) Vicente, J.; Chicote, M. T.; Bermúdez, M. D. *J. Organomet. Chem.* **1984**, *268*, 191.
(35) Bonnardel, P. A.; Parish, R. V. *J. Organomet. Chem.* **1996**, *515*, 221.
(36) Bonnardel, P. A.; Parish, R. V.; Pritchard, R. G. *J. Chem. Soc., Dalton Trans.* **1996**, 3185.
(37) Wong K-, H.; Cheung, K.-K.; Chan, M. C.-W.; Che, C.-M. *Organometallics* **1998**, *17*, 3505.
(38) Yam, V. W.-W.; Wong, K. M.-C.; Hung, L.-L.; Zhu, N. *Angew. Chem. Int. Ed. Engl.* **2005**, *44*, 3107.
(39) Constable, E. C.; Henney, R. P. G.; Leese, T. A.; Tocher, D. A. *J. Chem. Soc., Dalton Trans.* **1990**, 443.
(40) Vicente, J.; Bermúdez, M. D.; Sánchez-Santana, M. J.; Payá, J. *Inorg. Chim. Acta* **1990**, *174*, 53.
(41) Vicente, J.; Chicote, M.-T.; Bermúdez, M. D.; Soláns, X.; Font-Altaba, M. *J. Chem. Soc., Dalton Trans.* **1984**, 557.
(42) Vicente, J.; Chicote M-, T.; Bermudez, M. D.; Sanchez-Santano, M. J. *J. Organomet. Chem.* **1986**, *310*, 401.
(43) Contel, M.; Nobel, D.; Spek, A. L.; van Koten, G. *Organometallics* **2000**, *19*, 3288.
(44) Minghetti, G.; Cinellu, M. A.; Pinna, M. V.; Stoccoro, S.; Zucca, A.; Manassero, M. *J. Organomet. Chem.* **1998**, *568*, 225.
(45) Minghetti, G.; Ganadu, M. L.; Foddai, C.; Cinellu, M. A.; Cariati, F.; Demartin, F.; Manassero, M. *Inorg. Chim. Acta* **1984**, *86*, 93.
(46) Cinellu, M. A.; Ganadu, M. L.; Minghetti, G.; Cariati, F.; Demartin, F.; Manassero, M. *Inorg. Chim. Acta* **1988**, *143*, 197. Stoccoro, S.; Cinellu, M. A.; Zucca, A.; Minghetti, G.; Demartin, F. *Inorg. Chim. Acta* **1994**, *215*, 17. Cinellu, M. A.; Gladiali, S.; Minghetti, G.; Stoccoro, S.; Demartin, F. *J. Organomet. Chem.* **1991**, *401*, 371.
(47) Hill, D. T.; Burns, K.; Titus, D. D.; Girard, G. R.; Reiff, W. M.; Mascavage, L. M. *Inorg. Chim. Acta* **2003**, *346*, 1.
(48) Fan, D.; Yang, C.-T.; Ranford, J. D.; Vittal, J. J. *J. Chem. Soc., Dalton Trans.* **2003**, 4749.
(49) Cheung, T.-C.; Lai, T.-F.; Che, C.-M. *Polyhedron* **1994**, *13*, 2073.

(50) Yang, T.; Tu, C.; Zhang, J.; Lin, L.; Zhang, X.; Liu, Q.; Ding, J.; Xu, Q.; Guo, Z. *Dalton Trans.* **2003**, 3419.
(51) Nonoyama, M.; Nakajima, K.; Nonoyama, K. *Polyhedron* **1997**, *16*, 4039.
(52) Fuchita, Y.; Ieda, H.; Kayama, A.; Kinoshita-Nagaoka, J.; Kawano, H.; Kameda, S.; Mikuriya, M. *J. Chem. Soc., Dalton Trans.* **1998**, 4095.
(53) Nonoyama, M.; Nakajima, K. *Transition Met. Chem.* **1999**, *24*, 449.
(54) Nonoyama, M.; Nakajima, K.; Nonoyama, K. *Polyhedron* **2001**, *20*, 3019.
(55) Cinellu, M. A.; Zucca, A.; Stoccoro, S.; Minghetti, G.; Manassero, M.; Sansoni, M. *J. Chem. Soc., Dalton Trans.* **1995**, 2865.
(56) Nonoyama, M. *Synth. React. Inorg. Met-Org. Chem* **1999**, *29*, 119.
(57) Zhu, Y.; Cameron, B. R.; Skerlj, R. T. *J. Organomet. Chem* **2003**, *677*, 57.
(58) Fuchita, Y.; Ieda, H.; Tsunemune, Y.; Kinoshita-Nagaoka, J.; Kawano, H. *J. Chem. Soc., Dalton Trans.* **1998**, 791.
(59) Howe, B. P.; Parish, R. V.; Pritchard, R. G. *Quimica Nova* **1998**, *21*, 564.
(60) Vicente, J.; Chicote, M.-T.; Bermudez, M.-D.; Jones, P. G.; Sheldrick, G. M. *J. Chem. Res. (S)* **1985**, 72.
(61) Vicente, J.; Chicote, M. T.; Bermudez, M. D.; Garcia-Garcia, M. *J. Organomet. Chem.* **1985**, *295*, 125.
(62) Vicente, J.; Chicote, M. T.; Bermúdez, M. D.; Jones, P. G.; Fittschen, C.; Sheldrick, G. M. *J. Chem. Soc., Dalton Trans.* **1986**, 2361.
(63) Parish, R. V.; Mack, J.; Hargreaves, L.; Wright, J. P.; Buckley, R. G.; Elsome, A. M.; Fricker, S. P.; Theobald, B. R. C. *J. Chem. Soc., Dalton Trans.* **1996**, 69.
(64) Vicente, J.; Bermúdez, M.-D.; Carrillo, M.-P.; Jones, P. G. *J. Chem. Soc., Dalton Trans.* **1992**, 1975.
(65) Fan, D.; Yang, C.-T.; Ranford, J. D.; Vittal, J. J.; Lee, P. F. *Dalton Trans.* **2003**, 3376.
(66) Fan, D.; Yang, C.-T.; Ranford, J. D.; Lee, P. F.; Vittal, J. J. *Dalton Trans.* **2003**, 2680.
(67) Vicente, J.; Chicote, M. T.; Bermudez, M. D.; Sanchez-Santano, M. J.; Jones, P. G. *J. Organomet. Chem* **1988**, *354*, 381.
(68) Vicente, J.; Bermúdez, M. D.; Chicote, M. T.; Sanchez-Santano, M. J. *J. Organomet. Chem.* **1989**, *371*, 129.
(69) Vicente, J.; Bermúdez, M.-D.; Chicote, M.-T.; Sanchez-Santano, M.-J.; *J. Chem. Soc., Chem. Commun.* **1989**, 141.
(70) Vicente, J.; Bermúdez, M.-D.; Chicote, M.-T.; Sanchez-Santano, M.-J. *J. Chem. Soc., Dalton Trans.* **1990**, 1945.
(71) Vicente, J.; Bermúdez, M-D.; Carrillo, M-P.; Jones, P. G. *J. Organomet. Chem.* **1993**, *456*, 305.
(72) Vicente, J.; Bermudez, M. D.; Escribano, J.; Carrillo, M. P.; Jones, P. G. *J. Chem. Soc., Dalton Trans.* **1990**, 3083.
(73) Vicente, J.; Chicote, M. T. *Coord. Chem. Rev.* **1999**, *193–195*, 1143.
(74) Cinellu, M. A.; Minghetti, G.; Pinna, M. V.; Stoccoro, S.; Zucca, A.; Manassero, M. *J. Chem. Soc., Dalton Trans.* **1999**, 2823.
(75) Cinellu, M. A.; Minghetti, G.; Pinna, M. V.; Stoccoro, S.; Zucca, A.; Manassero, M. *Eur. J. Inorg. Chem.* **2003**, 2304.
(76) Cinellu, M. A.; Minghetti, G.; Pinna, M. V.; Stoccoro, S.; Zucca, A.; Manassero, M. *Chem. Commun.* **1998**, 2397.
(77) Pinna, M. V. Ph.D. Thesis, University of Sassari, **2000**.
(78) Cinellu, M. A.; Minghetti, G. *Gold Bull.* **2002**, *35*, 11.
(79) Messori, L.; Marcon, G.; Cinellu, M. A.; Coronnello, M.; Mini, E.; Gabbiani, C.; Orioli, P. *Bioorg. Med. Chem.* **2004**, *12*, 6039.
(80) Crowley, J. D.; Steele, I. M.; Bosnich, B. *Inorg. Chem.* **2005**, *44*, 2989.
(81) Wong, K. M.-C.; Zhu, X.; Hung, L.-L.; Zhu, N.; Yam, V. W.-W.; Kwok, H.-S. *Chem. Commun.* **2005**, 2906.
(82) Dinger, M. B.; Henderson, W. *J. Organomet. Chem.* **1998**, *560*, 233.
(83) Goss, C. H. A.; Henderson, W.; Wilkins, A. L.; Evans, C. *J. Organomet. Chem.* **2003**, *679*, 194.
(84) Dinger, M. B.; Henderson, W. *J. Orgaomet. Chem.* **1998**, *557*, 231.
(85) Vicente, J.; Chicote, M. T.; Guerrero, R.; Herber, U.; Bautista, D. *Inorg. Chem.* **2002**, *41*, 1870.

(86) Henderson, W.; Nicholson, B. K.; Oliver, A. G. *J. Organomet. Chem.* **2001**, *620*, 182.
(87) Peseck, J. J.; Mason, W. R. *Inorg. Chem.* **1983**, *22*, 2958.
(88) Dinger, M. B.; Henderson, W. *J. Organomet. Chem.* **1997**, *547*, 243.
(89) Henderson, W.; Nicholson, B. K.; J Wilkins, A. L. *Organomet. Chem.* **2005**, *690*, 4971.
(90) Dinger, M. B.; Henderson, W. *J. Organomet. Chem.* **1999**, *577*, 219.
(91) Kubo, K.; Nakano, M.; Tamura, H.; Matsubayashi, G.; Nakamoto, M. *J. Organomet. Chem.* **2003**, *669*, 141.
(92) Kubo, K.; Nakano, M.; Tamura, H.; Matsubayashi, G. *Eur. J. Inorg. Chem.* **2003**, 4093.
(93) Parish, R. V.; Howe, B. P.; Wright, J. P.; Mack, J.; Pritchard, R. G.; Buckley, R. G.; Elsome, A. M.; Fricker, S. P. *Inorg. Chem.* **1996**, *35*, 1659.
(94) Dinger, M. B.; Henderson, W.; Nicholson, B. K.; Robinson, W. T. *J. Organomet. Chem.* **1996**, *560*, 169.
(95) Henderson, W.; Rickard, C. E. F. *Inorg. Chim. Acta* **2003**, *343*, 74.
(96) Fong, S.-W. A.; Vittal, J. J.; Henderson, W.; Hor, T. S. A.; Oliver, A. G.; Rickard, C. E. F. *Chem. Commun.* **2001**, 421; Fong, S.-W. A.; Yap, W. T.; Vittal, J. J.; Hor, T. S. A.; Henderson, W.; Oliver, A. G.; Rickard, C. E. F. *J. Chem. Soc., Dalton Trans.* **2001**, 1986.
(97) Yeo, J. S. L.; Vittal, J. J.; Henderson, W.; Hor, T. S. A. *J. Chem. Soc., Dalton Trans.* **2002**, 328.
(98) Ortner, K.; Abram, U. *Inorg. Chem. Commun.* **1998**, *1*, 251.
(99) Abram, U.; Ortner, K.; Gust, R.; Sommer, K. *J. Chem. Soc. Dalton Trans.* **2000**, 735.
(100) Casas, J. S.; Castaño, M. V.; Cifuentes, M. C.; García-Monteagudo, J. C.; Sánchez, A.; Sordo, J.; Abram, U. *J. Inorg. Biochem.* **2004**, *98*, 1009.
(101) Santos, I. G.; Hagenbach, A.; Abram, U. *Dalton Trans.* **2004**, 677.
(102) Ortner, K.; Abram, U. *Polyhedron* **1999**, *18*, 749.
(103) Abram, U.; Mack, J.; Ortner, K.; Müller, M. *J. Chem. Soc., Dalton Trans.* **1998**, 1011.
(104) Ortner, K.; Hilditch, L.; Dilworth, J. R.; Abram, U. *Inorg. Chem. Commun.* **1998**, *1*, 469.
(105) Ortner, K.; Hilditch, L.; Zheng, Y.; Dilworth, J. R.; Abram, U. *Inorg. Chem.* **2000**, *39*, 2801.
(106) Vicente, J.; Bermúdez, M. D.; Escribano, J. *Organometallics* **1991**, *10*, 3380.
(107) Tiekink, E. R. T. *Gold Bull.* **2003**, *36*, 117.
(108) Tiekink, E. R. T. *Crit. Rev. Oncol./Hematol.* **2002**, *42*, 225.
(109) Shaw III, C. F. (H. Schmidbaur, Ed.), Gold: Progress in Chemistry, Biochemistry and Technology, Wiley, New York, 1999.
(110) Parish, R. V. *Metal-Based Drugs* **1999**, *6*, 271.
(111) Fricker, S. P. *Metal-Based Drugs* **1999**, *6*, 291.
(112) Rigobello, M. P.; Messori, L.; Marcon, G.; Cinellu, M. A.; Bragadin, M.; Folda, A.; Scutari, G.; Bindoli, A. *J. Inorg. Biochem.* **2004**, *98*, 1634.
(113) Marcon, G.; Messori, L.; Orioli, P.; Cinellu, M. A.; Minghetti, G. *Eur. J. Biochem.* **2003**, *270*, 4655.
(114) Sanna, G.; Pilo, M. I.; Spano, N.; Minghetti, G.; Cinellu, M. A.; Zucca, A.; Seeber, R. *J. Organomet. Chem.* **2001**, *622*, 47.
(115) Paw, W.; Cummings, S. D.; Mansour, M. A.; Connick, W. B.; Geiger, D. K.; Eisenberg, R. *Coord. Chem. Rev.* **1998**, *171*, 125.
(116) Thompson, M. E.; Djurovich, P. I.; Li, J. PCT Int. Appl. WO 2003-US25936 (*Chem. Abstr.* **2004**, 162866; Burn, P. L.; Samuel, I. D. W.; Lo, S.-C. PCT Int. Appl. WO 2004-GB2127 (*Chem. Abstr.* **2004**, 1019964).

Sterically Demanding Phosphinimides: Ligands for Unique Main Group and Transition Metal Chemistry

DOUGLAS W. STEPHAN*

Chemistry and Biochemistry, University of Windsor, Windsor, ON, Canada N9B 3P4

I
INTRODUCTION

Much of the advancement of inorganic and organometallic chemistry has been based on the placement of a reactive center in a judiciously chosen ligand field. As a consequence, a good deal of effort has been devoted to the design and synthesis of ancillary ligands and the subsequent study of the geometries, characteristics and reactivities of the derived complexes. Such efforts in some cases also have led to the unearthing of new applications in stoichiometric chemistry, catalysis and material science. Throughout these studies, it has been demonstrated that perturbations to the steric and electronic characters of ancillary ligands permit tweaking of the structural features and as a consequence, alter the resulting chemistry. In seeking to exploit this tenet of organometallic chemistry to develop new reactivity, we sought anionic ancillary ligands where the synthetic route was readily amenable to such "tweaking." In this regard, we noted the considerable volume of structural studies reported for systems incorporating phosphinimide ligands.[1,2] These anionic ligands (R_3PN^-) are readily derived from neutral phosphinimine precursors which are prepared using the facile and long-known Staudinger reaction [Eq. (1)].[3] Thus, a

*Corresponding author. Fax: 519-973-7098.
E-mail: Stephan@uwindsor.ca (D.W. Stephan).

ADVANCES IN ORGANOMETALLIC CHEMISTRY
VOLUME 54 ISSN 0065-3055/DOI 10.1016/S0065-3055(05)54006-1

© 2006 Elsevier Inc.
All rights reserved.

variety of ligands with varied steric and electronic properties are readily accessible. Moreover, the high oxidation state of phosphorus in these ligands affords good thermal stability, while the presence of the phosphorus nucleus offers the convenience of monitoring reactions by ^{31}P NMR spectroscopy.

$$R_3P \xrightarrow[-N_2]{R'-N_3} R_3P{=}N{-}R' \quad (1)$$

Early studies of phosphinimide ligand chemistry employed simple alkyl and aryl derivatives such as R_3PN^- (R = Me, Et, Ph etc.) to prepare a variety of main group and transition metal systems. Although such studies have been reviewed in 1997[2] and 1999,[1] respectively, similar work has continued.[4–21] More recently, the application of phosphinimide ligands in lanthanoid chemistry has also been reviewed.[22] While much of this previous work has focused on the structural diversity of these derivatives, our approach to the chemistry of such compounds has been to focus on systems incorporating sterically demanding phosphinimides. Since this often involves the use of phosphinimides derived from bulky trialkyl phosphine, there is often an attendant practical advantage in the solubility these can impart. In this review, we limit the discussion to recent studies of complexes of bulky anionic phosphinimide ligands. Attention is drawn to the impact that results from perturbations to the steric and electronic properties of these ancillary ligands.

II

GROUP 1: LITHIUM

The only other known homoleptic Li-phosphinimide derivative was $[Li(\mu\text{-}NPPh_3)]_6$ previously reported by Dehnicke and coworkers.[23] Recently, the Li-salt of the conjugate base of $^tBu_3PNH$ was prepared by reaction with an alkyl lithium reagent. The X-ray structural data revealed that the Li salt is a distorted tetramer cuboid $[Li(\mu\text{-}NP^tBu_3)]_4$ (Fig. 1).[24] The steric bulk of this phosphinimide ligand presumably precludes an expanded M–N core. This species has proved to be an invaluable synthon for other phosphinimide complexes (vide infra).

III

GROUP 2: MAGNESIUM

The first homoleptic magnesium phosphinimide complexes $[Mg_2(\mu\text{-}NP^tBu_3)_2(NP^tBu_3)_2]$ and $[Mg_3(\mu\text{-}NP^iPr_3)_4(NP^iPr_3)_2]$ were obtained from the reaction of 2 equivalents of the phosphinimines $HNPR_3$ with Mg^nBu_2.[25] While both of these compounds contain three coordinate trigonal magnesium centers, the latter also includes a central tetrahedral four coordinate magnesium center (Fig. 2). The structural difference is thought to result from the greater steric demands in the ^tBu derivatives.

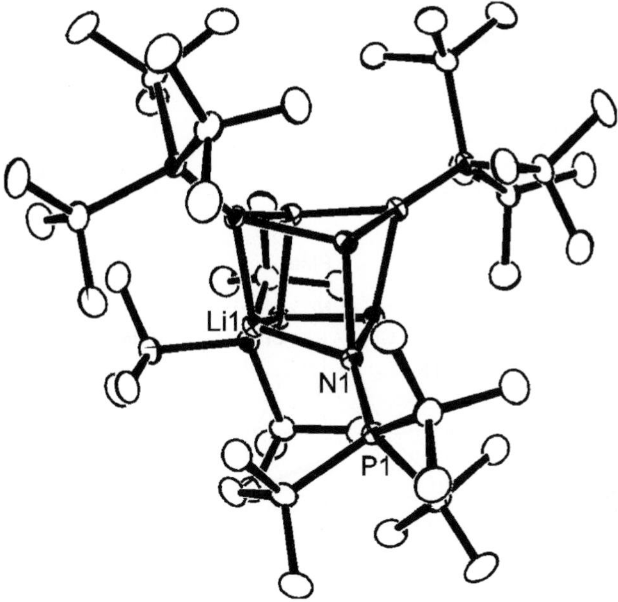

FIG. 1. Molecular structure of $[LiN{=}P^tBu_3]_4$ (hydrogen atoms omitted).

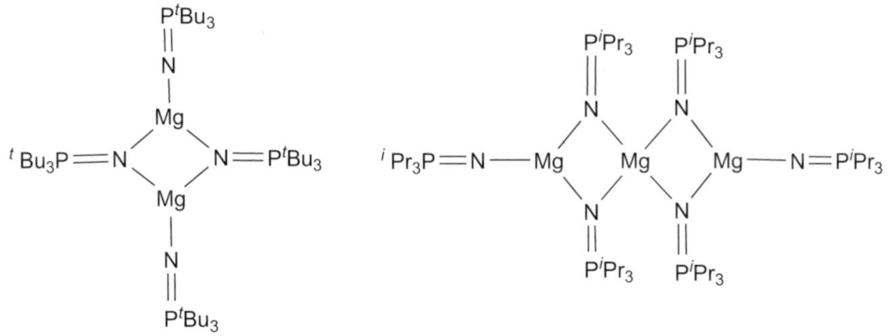

FIG. 2. Bi- and tri-nuclear magnesium phosphinimides.

IV
GROUP 4

A. *Titanium*

Several titanium complexes of relatively small phosphinimide ligands such as $[Ti(NPMe_3)_4]$, $[TiCl_3(\mu\text{-}NPR_3)]_2$ and $[Ti_3Cl_6(\mu\text{-}NPR_3)_5]^+$ (Fig. 3) were prepared and structurally characterized some years ago.[1] The structures adopted by these complexes suggest that the degree of aggregation may be determined by the steric demands of the substituents.

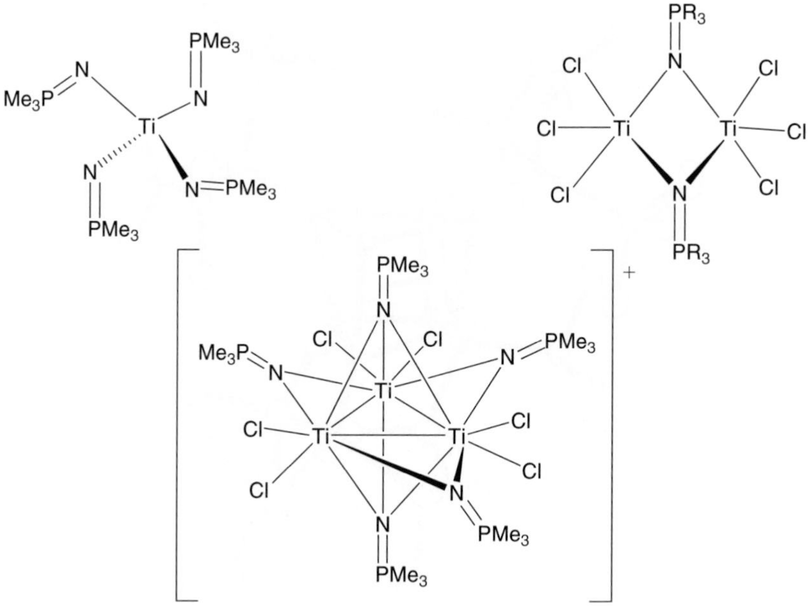

FIG. 3. Examples of titanium phosphinimide complexes.

More recently, the reactions of trimethylsilylphosphinimines with [$CpTiCl_3$] were used to prepare a series of complexes of the form [$CpTiCl_2(NPR_3)$] in high yields [Eq. (2)].[26,27] This synthetic route is versatile and works well for analogs with a variety of substituents on phosphorus or the cyclopentadienyl rings.[26,28] Similarly, the tropidinyl ($C_5H_5(C_2H_4)NMe$) complex, [(trop)$TiCl_2(NPR_3)$] can be prepared [Eq. (3)].[29] Chloride-phosphinimide metathesis is achieved using the lithium phosphinimide [$Li(\mu\text{-}NP^tBu_3)]_4$[24] in reactions with titanium halides. In an analogous manner, the species [($^tBu_3PN)TiCl_3$], [($^tBu_3PN)_2TiCl_2$] and [($^tBu_3PN)_3TiCl$] [Eq. (4)] were prepared.[30,31]

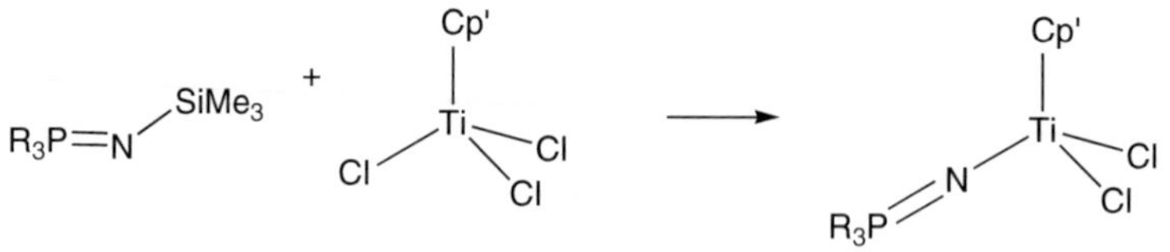

(2)

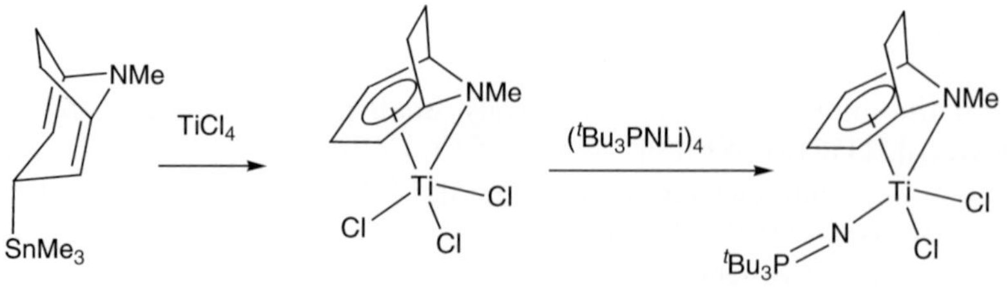

(3)

$TiCl_4$ —($^tBu_3PNLi)_4$→ $[^tBu_3P{=}N{-}TiCl_3]$ —($^tBu_3PNLi)_4$→ $[(^tBu_3P{=}N)_2TiCl_2]$ —($^tBu_3PNLi)_4$→ $[(^tBu_3P{=}N)_3TiCl]$ (4)

Complexes of the type [$CpTiCl_2(NPR_3)$] were screened for utility as ethene polymerization catalysts. In the presence of 500 equivalents of methylalumoxane (MAO) and 1 atm of ethene at 25 °C (Table I),[26,27] the data suggested a strong correlation of the activity with the steric bulk of the phosphinimide ligand substitutents. For example, the catalysts derived from [$CpTiCl_2(NPCy_3)$] and [$CpTiCl_2(NP^iPr_3)$] exhibit relatively low activity, whereas the catalyst from [$CpTiCl_2(NP^tBu_3)$] exhibited a polymerization activity of 652 g PE $mmol^{-1}\,h^{-1}\,atm^{-1}$. This compares with an activity of 895 g PE $mmol^{-1}\,h^{-1}\,atm^{-1}$ obtained for the catalyst derived from [Cp_2ZrCl_2] under the same conditions. This latter phosphinimide catalyst gave polyethene of molecular weight 89,000 with a polydispersity of 1.6, consistent with single-site behavior.[26,27] Interestingly, in contrast, the catalyst derived from [$(^tBu_3PN)_2TiX_2$] and MAO activation showed rather poor activity.

Generally, alkylation of CpTi- and cyclopentadienyl-free titanium phosphinimides proceeded readily with the use of alkyl lithium or Grignard reagents although steric crowding appeared to limit the alkylation of [$(^tBu_3PN)_3TiCl$] to methylation.[30] In a similar fashion, for the species [$LL'Ti(NP^tBu_3)Cl$] and [$L_2L'Ti(NP^tBu_3)$] (L, L′ = Cp or indenyl), the cyclopentadienyl and indenyl rings adopted either η^5 or η^1 bonding modes depending on temperature and degree of steric crowding. In the case where it was determined, NMR data were consistent with an η^5–η^1 exchange process with a barrier of 8–9 kcal mol^{-1} (Fig. 4).[28]

Table I

Selected Activities of Ethene Polymerization Catalysts (MAO Activation)[a]

Cat.	Act.	M_w	PDI	Pre-cat.	Act.	M_w	PDI
$CpTi(NPCy_3)Cl_2$	42	3590	1.8	$^tBuCpTi(NPCy_3)Cl_2$	46	7410	2.1
		336,000	2.2			893,500	3.4
$CpTi(NP^iPr_3)Cl_2$	49	18,700	2.8	$^tBuCpTi(NP^iPr_3)Cl_2$	16	7580	1.9
		578,500	2.4			910,200	2.5
$CpTi(NP^tBu_3)Cl_2$	652	89,900	1.6	$^tBuCpTi(NP^tBu_3)Cl_2$	881	65,400	2.4
$CpTi(NPPh_3)Cl_2$	34	109,200	2.6	$^nBuCpTi(NP^tBu_3)Cl_2$	2000		

[a]MAO 500 equiv.; 1 atm ethene and 25 °C, 5 min. Activity units: g PE $mmol^{-1}\,h^{-1}\,atm^{-1}$.

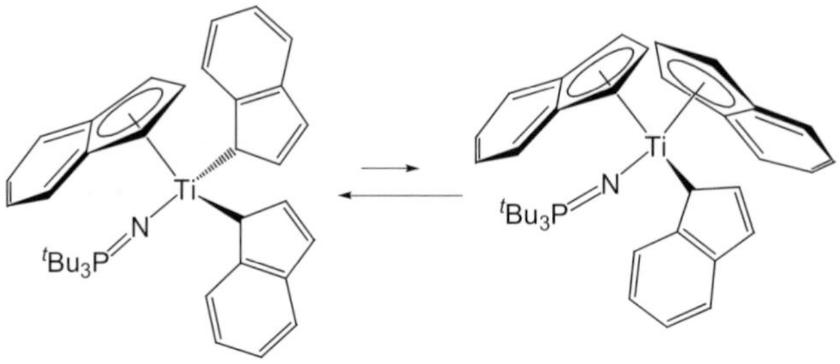

FIG. 4. η^5–η^1 Exchange in (indenyl)$_3$Ti(NP^{t}Bu$_3$).

TABLE II

SELECTED ACTIVITY FOR ETHENE POLYMERIZATION CATALYSTS ([Ph_3C][$B(C_6F_5)_4$] ACTIVATION)[a]

Cat.	Act.	M_w	PDI	Pre-cat.	Act.[1]	M_w	PDI
CpTi(NPCy$_3$)Me$_2$	231	134,600	2.8	^{t}BuCpTi(NPCy$_3$)Me$_2$	1807	310,200	7.5
CpTi(NP*i*Pr$_3$)Me$_2$	225	163,800	3.9	^{t}BuCpTi(NP^{i}Pr$_3$)Me$_2$	1193	259,200	9.9
CpTi(NP^{t}Bu$_3$)Me$_2$	401	165,800	3.4	^{t}BuCpTi(NP^{t}Bu$_3$)Me$_2$	1296	321,300	12.3

[a]1 atm ethylene, 25 °C, co-catalyst: trityl tetrakis(perfluorophenyl)borate [Ph_3C][$B(C_6F_5)_4$]. Activity units: g PE $mmol^{-1}\,h^{-1}\,atm^{-1}$.

Alkylated complexes also served as convenient precursors to ethene polymerization catalysts. For example, the zwitterionic complex [CpTiMe(NP^{t}Bu$_3$)(μ-MeB(C_6F_5)$_3$)], which was readily generated from the reaction of [CpTiMe$_2$(NP^{t}Bu$_3$)] and B(C_6F_5)$_3$, was shown to be a single-component catalyst.[27] Thus, simple dissolution of this species in hexane under an ethene atmosphere led to the production of polyethylene with an activity of 459 g PE $mmol^{-1}\,h^{-1}\,atm^{-1}$. Similarly, treatment of dialkyl precursors with [Ph_3C][$B(C_6F_5)_4$] generated more active polymerization catalysts (Table II). The role of steric and electronic factors in determining the level of activity was evident from the data. Larger substituents on phosphorus and the cyclopentadienyl ligand such as in ^{t}BuC$_5$H$_4$TiMe$_2$(NPR$_3$) (R = Cy, ^{i}Pr, ^{t}Bu) gave rise to dramatically higher activities of 1807, 1193 and 1296 g PE $mmol^{-1}\,h^{-1}\,atm^{-1}$, respectively (Table II). In the absence of ethene, the active Ti-cationic intermediates could be intercepted by donor ligands. For example, the salts [CpTiMe(NPR$_3$)(L)][MeB(C_6F_5)$_3$] and [CpTiMe(NPR$_3$)(L)][B(C_6F_5)$_4$] were prepared in reactions of [CpTiMe$_2$(NPR$_3$)] with B(C_6F_5)$_3$ or [Ph_3C][$B(C_6F_5)_4$] in the presence of donor ligands.[32]

An initial attempt to improve catalyst activity involved the use of alternative sterically demanding phosphinimide ligands. To this end, the sterically demanding adamantane-like cage phosphines PR($C_6H_4O_3Me_4$) developed by Pringle *et al.*[33] were oxidized with Me_3SiN_3 to the corresponding phosphinimines and the latter product used to prepare titanium dihalide and dimethyl complexes by established methods.[34] Screening of these complexes for catalytic activity in ethene polymerization revealed only very low activity (< 10 g PE $mmol^{-1}\,h^{-1}\,atm^{-1}$). Reactions of the "cage-phosphinimide" derivatives with $AlMe_3$ gave a complex mixture of products. One of the minor products was isolated and characterized as

[$CpTiMe_2(NPCy(C_6H_4O_3Me_5)(\mu\text{-}AlMe_2)(AlMe_3))$] [Eq. (5)]. This observation suggested that the Lewis acid induced ligand rupture resulting in a product that was inactive in catalysis.

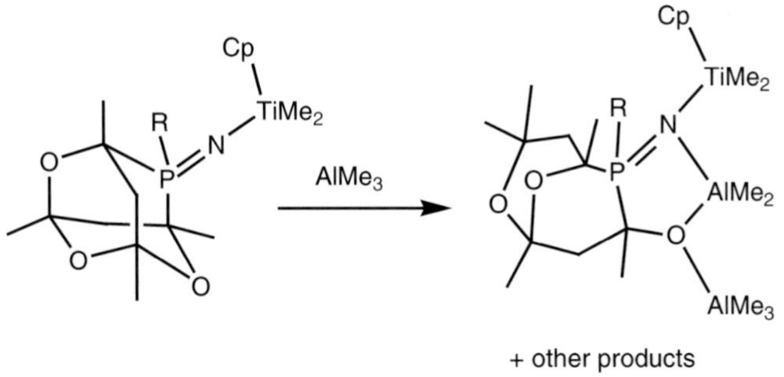

(5)

Another approach to a bulky ancillary phosphinimide ligand involved compounds of the type $Me_3SiNPPh_2(NPR_3)$ in which one of the substituents on a phosphinimide ligand is itself a phosphinimide fragment. These ligands and the resulting complexes of the form [$CpTiCl_2\{NPPh_2(NPR_3)\}$] were readily prepared by established procedures in high yield.[35] Subsequent alkylation was also straightforward (Scheme 1).

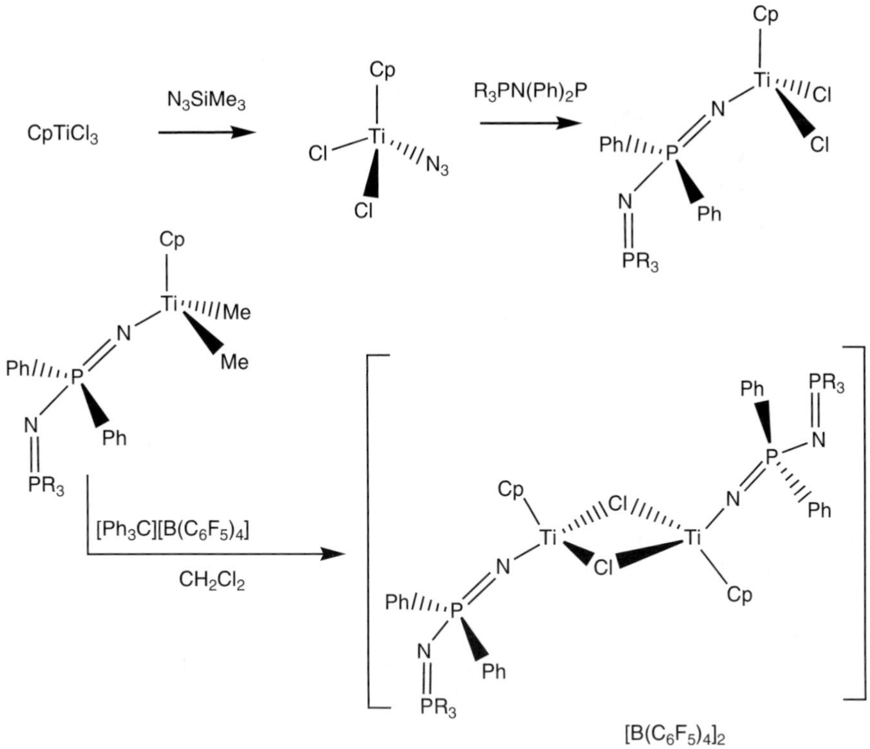

Scheme 1.

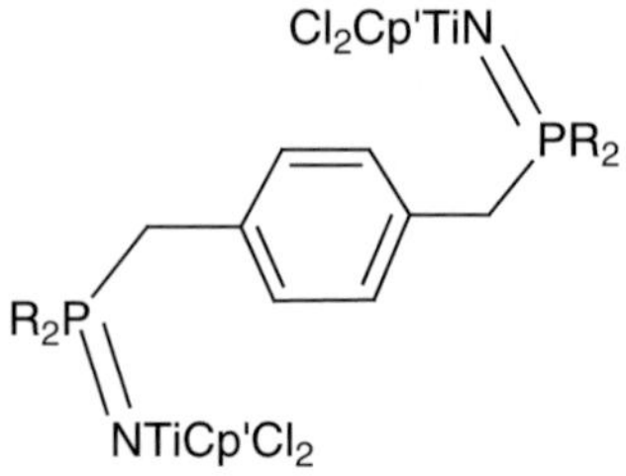

FIG. 5. Bimetallic phosphinimide derivatives.

Evaluation of these compounds in olefin polymerization catalysis revealed only intermediate levels of catalyst activity.[35] For example, [CpTiCl$_2${NPPh$_2$(NP^{i}Pr$_3$)}], upon activation with MAO produced polyethene with an activity of 299 g PE mmol^{-1} h^{-1} atm^{-1} at 25 °C. Attempts to isolate zwitterionic or cationic derivatives in the reactions of complexes CpTiMe$_2${NPPh$_2$(NPR$_3$)} with B(C$_6$F$_5$)$_3$ or [Ph$_3$C][B(C$_6$F$_5$)$_4$] yielded highly reactive products. In one such case, while a complex mixture of products was obtained, the isolation and characterization of [CpTi(μ-Cl){NPPh$_2$(NPR$_3$)}]$_2$[B(C$_6$F$_5$)$_4$]$_2$ was evidence of solvent activation.[35]

Bimetallic phosphinimide-based titanium catalysts have also been prepared and evaluated. Derived from diphosphines, the complexes *p*-C$_6$H$_4${CH$_2$PR$_2$NTiCp′Cl$_2$}$_2$ (R = ^{t}Bu, Cy; Cp′ = Cp, Cp*) (Fig. 5) were prepared and the methylated analogs obtained employing conventional methodologies.[36] Activation of the dihalide or dialkyl derivatives with MAO or [Ph$_3$C][B(C$_6$F$_5$)$_4$], respectively, yielded catalysts for the polymerization of ethene. However, the activities were lower than the corresponding monometallic analogs, presumably a result of differences in solubility. Interestingly, the bimetallic species gave rise to polyethylenes that showed broad bimodal molecular weight distributions.[36] Interestingly, related monometallic catalysts also showed broader molecular weight distributions, prompting the suggestion that the acidic benzylic protons reacted to give modified catalyst sites. This view is supported by the metallation of the methylene protons of {CH$_2$PPh$_2$NTiCp*Cl$_2$}$_2$ reported by Bochmann and coworkers.[37]

A contrived approach for ligand design was based on a computational examination of the mechanism.[38] Ziegler and coworkers[46] described the mechanism of insertions of ethene using the model precursor, CpTiMe$_2$NPH$_3$ and the model activator, BCl$_3$. Subsequent computations for the series CpTiMe$_2$(NPR$_3$), (R = Me, NH$_2$, Cl, F) were consistent with the domination of the energy profiles by the ion-pair separation energies.[38] These data also showed that electron-donating substituents gave significantly smaller ion-pair separation energy thus enhancing catalyst activity. These findings prompted the synthesis of the series of titanium complexes of the form [CpTiCl$_2${NP(NR$_2$)$_3$}] and the corresponding methylated species (Scheme 2).[38] Activation of the analogous dimethyl derivatives with B(C$_6$F$_5$)$_3$ gave consistently high ethene polymerization activity for the more sterically demanding systems (1200–2600 g PE mmol^{-1} h^{-1} atm^{-1}). The facile synthesis of these systems together with high activity demonstrates the ability to judiciously tune steric and electronic properties of the phosphinimide ligands to achieve a desired outcome.

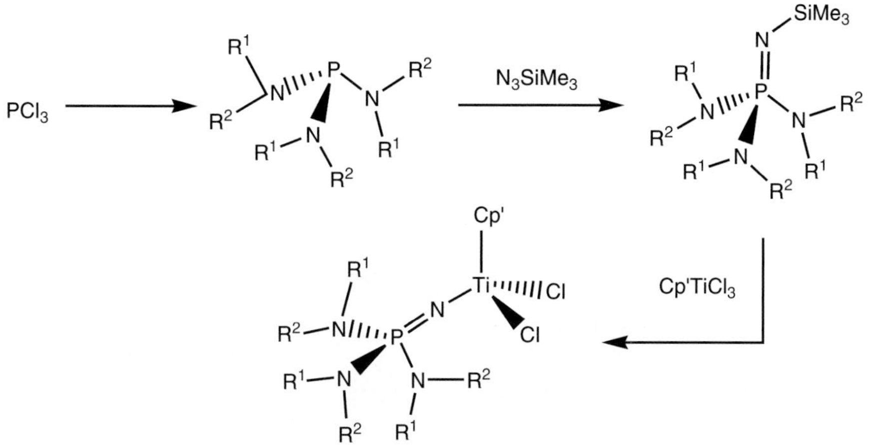

SCHEME 2.

TABLE III

ETHENE POLYMERIZATION CATALYSTS UNDER INDUSTRIALLY RELEVANT CONDITIONS ($[Ph_3C][B(C_6F_5)_4]$ ACTIVATION)[a]

Cat.	Act.	M_W	PDI
$(^tBu_3PN)_2TiMe_2$	62,310	77,500	1.9
$(^tBuNSiMe_2C_5Me_4)TiMe_2$	16,130	134,500	2.5
Cp_2ZrMe_2	8850	12,300	3.8

[a] 160 °C, flow conditions. Activity units: g PE $mmol^{-1} h^{-1} atm^{-1}$.

Activation of $[(^tBu_3PN)_2TiMe_2]$ with $B(C_6F_5)_3$ or $[Ph_3C][B(C_6F_5)_4]$ provided very active single-site catalysts.[31] In fact, under industrial conditions of higher temperatures (160 °C) and pressures of ethene, the catalysts derived from Cp_2ZrMe_2, $(C_5Me_4SiMe_2N^tBu)TiMe_2$ and $(^tBu_3PN)_2TiMe_2$ activated by $[Ph_3C][B(C_6F_5)_4]$ were tested and compared (Table III). The zirconocene catalyst that served as our standard gave rise to polyethylene formed at a rate of 8800 g PE $mmol^{-1} h^{-1} atm^{-1}$. The activity was approximately doubled by use of catalysts based on the Dow/Exxon constrained geometry catalyst $[(C_5Me_4SiMe_2N^tBu)TiMe_2]$, whereas the system based on $[(^tBu_3PN)_2 TiMe_2]$ gave an activity approximately eight times that of the zirconocene.[31]
Stoichiometric reactions of $[(^tBu_3PN)_2TiMe_2]$ with $B(C_6F_5)_3$ were shown to give the expected zwitterionic complex $[(^tBu_3PN)_2TiMe(\mu\text{-}Me)B(C_6F_5)_3]$, however, in the presence of an excess of borane, the dizwitterion $[(^tBu_3PN)_2Ti(\mu\text{-}MeB(C_6F_5)_3)_2]$ was formed (Fig. 6).[39] Remarkably, this compound exhibited negligible polymerization activity and was stable with respect to comproportionation with $[(^tBu_3PN)_2TiMe_2]$. These observations revealed that while $B(C_6F_5)_3$ is an activator, it also can act as catalyst poison.

These "single-site phosphinimide catalysts" showed generally better activity with B-based activators and poor activity with MAO activation.[31,40,41] To probe this

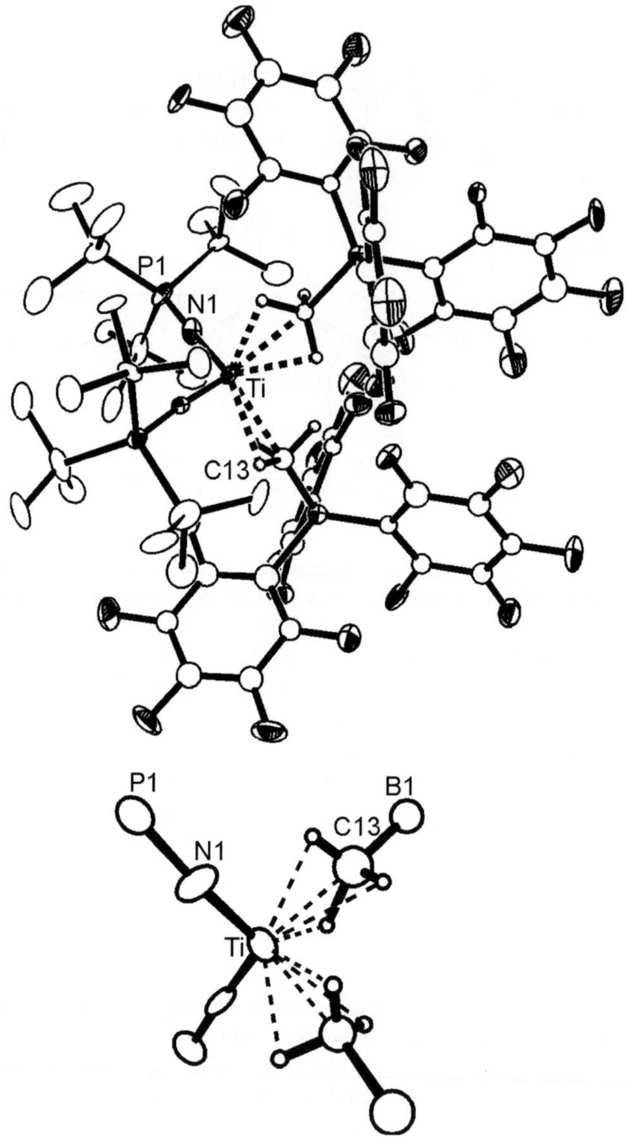

FIG. 6. ORTEP representation of [(^{t}Bu$_3$PN)$_2$Ti(μ-MeB(C$_6$F$_5$)$_3$)$_2$].

curious observation, reactions of phosphinimide complexes with MAO were modeled by reactions with AlMe$_3$. For example, [CpTi(NPR$_3$)Me$_2$] or [(indenyl)(^{i}Pr$_3$PN)TiMe$_2$] reacted with AlMe$_3$ smoothly at room temperature to give [Cp′Ti(μ_2-Me)(μ_2-NPR$_3$)(μ_4-C)(AlMe$_2$)$_3$] (Cp′ = Cp, indenyl). The formation of these Ti–Al carbide complexes involved triple C–H bond activation.[42] In addition, an excess of AlMe$_3$ established an equilibrium with the five coordinate, pseudo-trigonal bipyramidal carbide species [Cp′Ti(μ_2-Me)(μ_2-NPR$_3$)(μ_5-C)(AlMe$_2$)$_3$(AlMe$_3$)]

[Eq. (6)] with an equilibrium constant of 129 M^{-1} at 260 K [41] and ΔG (260 K) = $-10.5\,kJ\,mol^{-1}$, $\Delta H = -57.9\,kJ\,mol^{-1}$ and $\Delta S = -0.183\,kJ\,mol^{-1}\,K^{-1}$.[42,43] In similar reactions, steric crowding was shown to limit the degree of C–H activation as the reactions of [Cp(^{i}Pr$_3$PN)Ti(CH$_2$SiMe$_3$)$_2$] and [Cp*(^{i}Pr$_3$PN)TiMe$_2$] with AlMe$_3$ gave [CpTi(μ_2-Me) (μ_2-NP^{i}Pr$_3$)(μ_3-CSiMe$_3$)(AlMe$_2$)$_2$] and [Cp*Ti(μ_2-Me) (μ_2-NP^{i}Pr$_3$)(μ_3-CH)(AlMe$_2$)$_2$], respectively.[42]

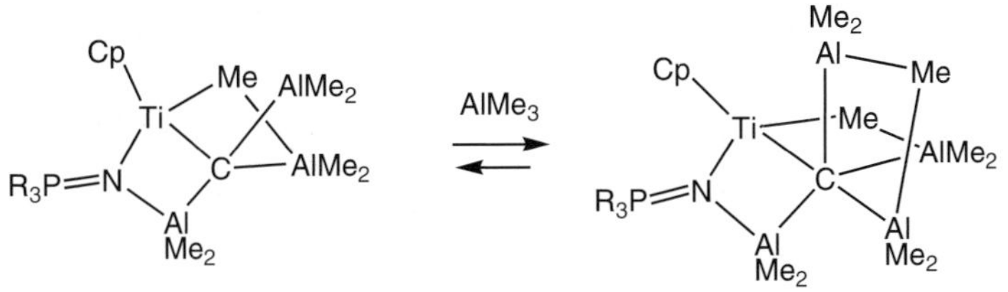

(6)

In an analogous manner, reactions of [(^{t}Bu$_3$PN)$_2$TiMe$_2$] with AlMe$_3$ gave the two titanium-based products [(μ_2-^{t}Bu$_3$PN)Ti(μ-Me)(μ_4-C)(AlMe$_2$)$_2$]$_2$ and [(^{t}Bu$_3$PN)Ti (μ_2-^{t}Bu$_3$PN)(μ_3-CH$_2$)$_2$(AlMe$_2$)$_2$(AlMe$_3$)] (Fig. 7).[44] A detailed kinetic study revealed that the mechanistic pathway to these products involved competitive methyl-phosphinimide metathesis and C–H activation.[44] Metathesis generated [(^{t}Bu$_3$PN)-TiMe$_3$] that underwent three sequential C–H activations with rate constants $k_1 = 3.9(5) \times 10^{-4}\,M^{-1}\,s^{-1}$, $k_2 = 1.4(2) \times 10^{-3}\,s^{-1}$ and $k_3 = 7(1) \times 10^{-3}\,s^{-1}$ giving the carbide product [(μ_2–^{t}Bu$_3$PN)Ti(μ-Me)(μ_4-C)(AlMe$_2$)$_2$]$_2$.[44] The competitive pathway proceeded by sequential C–H activations to give [(^{t}Bu$_3$PN)Ti(μ_2-^{t}Bu$_3$PN)(μ_3-CH$_2$)$_2$(AlMe$_2$)$_2$(AlMe$_3$)]. The rate constants of the initial C–H activation and ligand metathesis k_{met}, were found to be similar, $k_{obs} = 6(1) \times 10^{-5}\,s^{-1}$ and $6.1(5) \times 10^{-5}\,s^{-1}$, thus accounting for the divergent pathways.[44] In addition, transient concentrations of the intermediates were consistent with non-productive equilibria involving AlMe$_3$ and the intermediates in each pathway. This view was supported by stochastic-kinetic simulations.[44]

Efforts to enhance the stability of *bis*-phosphinimide complexes by the incorporation of chelating ligands prompted the synthesis of titanium phosphinimide complexes derived from the diphosphines *m*-C$_6$H$_4$(CH$_2$PR$_2$)$_2$ (R = ^{t}Bu, Cy). The amido, halo and alkyl derivatives [*m*-C$_6$H$_4$(CH$_2$R$_2$PN)$_2$Ti(NMe$_2$)$_2$], [*m*-C$_6$H$_4$

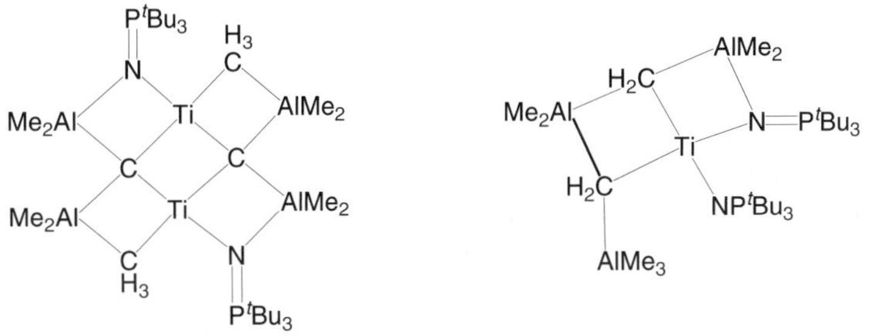

FIG. 7. Structures of [(μ_2-^{t}Bu$_3$PN)Ti(μ-Me)(μ_4-C)(AlMe$_2$)$_2$]$_2$ and (^{t}Bu$_3$PN)Ti(μ_2-^{t}Bu$_3$PN)(μ_3-CH$_2$)$_2$(AlMe$_2$)$_2$(AlMe$_3$).

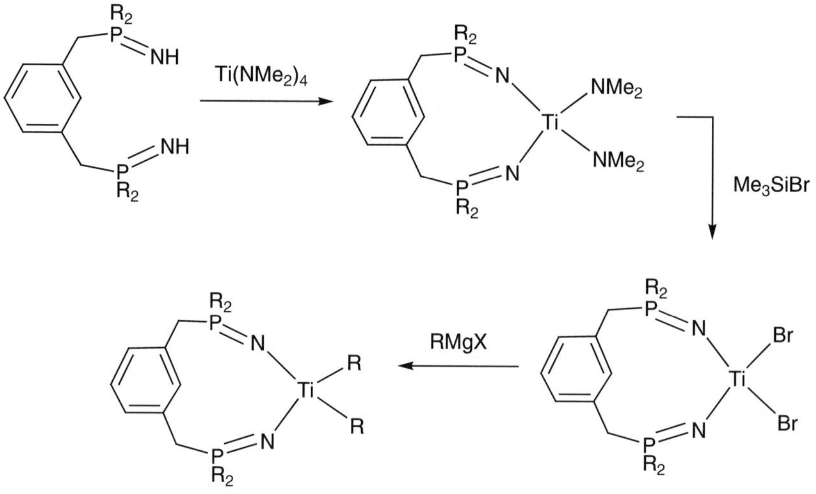

SCHEME 3.

$(CH_2R_2PN)_2TiX_2]$ and $[m\text{-}C_6H_4\{CH_2(^tBu)_2PN\}_2TiR_2]$ (R = Me, CH_2Ph) (Scheme 3) were prepared and characterized.[45] These compounds showed poor activity as ethene polymerization catalysts irrespective of the activation methodology used. It was inferred that the acidity of the benzylic protons was problematic.[37]

Ziegler and coworkers[46] have studied these phosphinimide catalysts by computational studies. DFT calculations predicted that the cation $[(R_3PN)_2TiMe]^+$ has a lower ion separation energy than $[Cp(R_2CN)TiMe]^+$, $[(CpSiR_2NR')TiMe]^+$, $[CpOSiR_3TiMe]^+$ and $[Cp(R_3PN)TiMe]^+$ and that this was further decreased with increasing size of the R group. The implication was that bulky phosphinimide ligands should provide active polymerization catalysts. Further computational studies have revealed that a β-hydrogen atom is transferred to an incoming olefin to terminate polymerization yielding a metal-hydride complex. Computations revealed that steric bulk and electron-donating phosphinimide ligands in the cations $[(R_3PN)_2TiR]^+$ or $[Cp(R_3PN)TiR]^+$ suppressed such hydrogen transfer.[47] In related synthetic chemistry, Piers and coworkers[48] recently have reported that the treatment of $[Cp(R_3PN)TiMe]^+$ with H_2 in ClC_6H_5 rompted reduction of Ti(IV) to Ti(III) and yielded the dimeric, dication $[Cp(R_3PN)Ti]_2^{2+}$, whereas the same reaction in toluene followed by addition of THF gave the cationic hydride species $[Cp(R_3PN)TiH(THF)]^+$ (Scheme 4).

Synthetic strategies involving thiolate-chloride metathesis or protonolysis of metal–carbon bonds by thiols provided facile routes to titanium phosphinimide thiolate complexes. Monometallic and dimeric complexes such as $[CpTi(NPR_3)(SR')_2]$ (R = ^{i}Pr, ^{t}Bu; R′ = Bn, Ph, ^{t}Bu), $[CpTi(NP^iPr_3)(S_2R)]$ (R = $(CH_2)_3$, $1,2\text{-}(CH_2)_2C_6H_4$ and $[CpTi(NP^iPr_3)\{S_2(CH_2)_2\}]_2$ were prepared using these methods (Scheme 5).[49,50] The latter dimeric species contrasted with the monomeric propanedithiolate derivative, presumably a result of subtle steric effects. The mixed akyl thiolate species $[CpTi(NP^tBu_3)Me(SPh)]$ and $[(^tBu_3PN)_2Ti(Me)(S^tBu)]$ were isolated although the

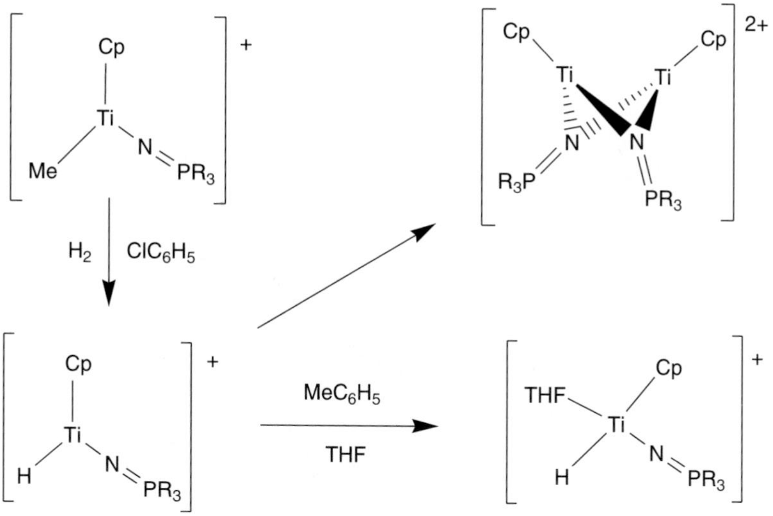

SCHEME 4.

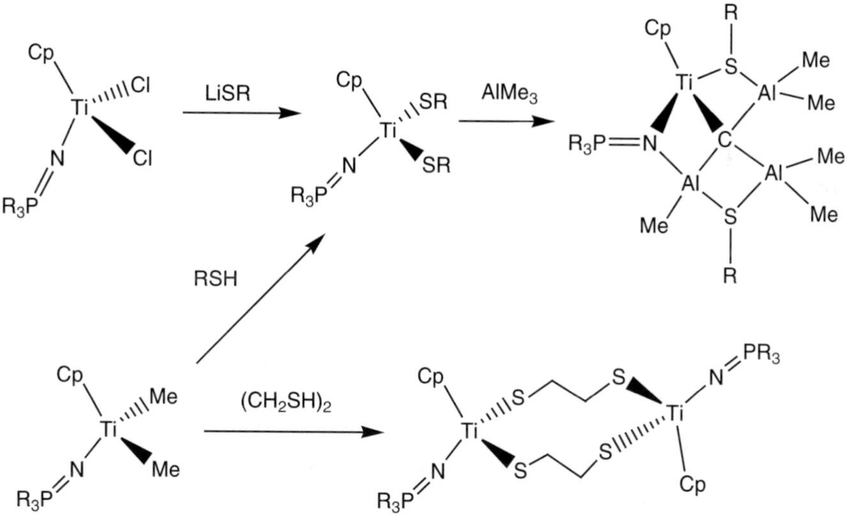

SCHEME 5.

related species $[({}^t\mathrm{Bu}_3\mathrm{PN})_2\mathrm{Ti(Me)(SCH_2Ph)}]$ was unstable, evolving CH_4 to give the thiobenzaldehyde complex $[({}^t\mathrm{Bu}_3\mathrm{PN})_2\mathrm{Ti}(\eta^2\text{-S}{=}\mathrm{CHPh})]$. The dithiolate complex $[\mathrm{CpTi(NPR_3)(SR)_2}]$ reacted with $AlMe_3$ by methyl thiolate exchange and gave the carbide product $[\mathrm{CpTi}(\mu\text{-SR})(\mu\text{-NP}^i\mathrm{Pr}_3)(\mu_4\text{-C})(\mathrm{AlMe}_2)_2(\mu\text{-SR})\mathrm{AlMe}]$ (Scheme 5).[50] The corresponding reaction of chelating dithiolate complexes with Lewis acids gave adducts such as $[\mathrm{Cp}({}^i\mathrm{Pr}_3\mathrm{PN})\mathrm{Ti(SRS)}]\cdot(\mathrm{AlMe}_3)_3$ as methyl thiolate exchange was inhibited by the chelate effect.[49]

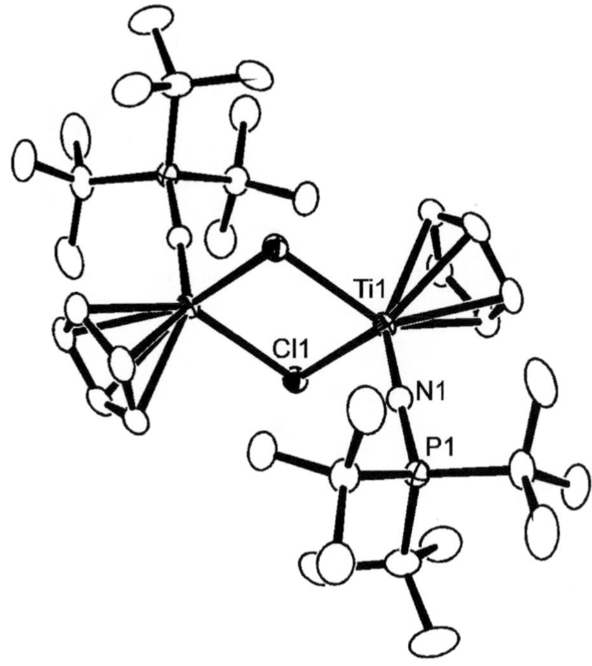

FIG. 8. ORTEP representation of $[CpTi(NP^{t}\text{-}Bu_3)(\mu\text{-}Cl)]_2$ (hydrogen atoms omitted).

Amide-substituted phosphinimide complexes such as $[Cp(R_3PN)Ti(NMe_2)_2]$, $[Cp(^tBu_3PN)Ti(NHC_6H_3^iPr_2\text{-}2,6)Cl]$, $[(^tBu_3PN)Ti(NMe_2)_3]$ and $[(^tBu_3PN)Ti(NHC_6H_3^iPr_2\text{-}2,6)_3]$ were prepared *via* salt metathesis of halide by amide. The complexes incorporating primary amido groups did not eliminate amine to form imido derivatives upon thermolysis, however, they reacted with $AlMe_3$ to effect methyl for amide exchange to $[Cp(^tBu_3PN)TiMeCl]$ or $[Cp(^tBu_3PN)TiMe_2]$ and the byproduct $[Al_2(\mu\text{-}NHC_6H_3^iPr_2\text{-}2,6)_2Me_4]$.[51]

The species $[Cp(^tBu_3PN)TiCl_2]$ is reduced by magnesium powder in benzene to give the reddish-brown product, $[Cp(^tBu_3PN)Ti(\mu\text{-}Cl)]_2$ (Fig. 8).[52] A solid single-crystal X-band EPR study confirmed the structure and revealed that the *g* matrix and zero-field parameters were consistent with EHMO calculations, which implied a HOMO comprised principally of metal d orbitals orthogonal to the Ti–Ti vector. This study showed that while phosphinimide and cyclopentadienyl ligands are sterically similar, they are electronically distinct.

In marked contrast to $[Cp(^tBu_3PN)Ti(\mu\text{-}Cl)]_2$, reduction of $[Cp(R_3PN)TiCl_2]$ ($R = Me$, ^iPr) with magnesium in THF gave the Ti(III)-phosphinimide bridged dimers $[CpClTi(\mu\text{-}NPR_3)]_2$ ($R = Me$, ^iPr) (Fig. 9).[53] This structure was similar to $[TiBr_2(\mu\text{-}NPPh_3)]_2$.[54] In further contrast, a transient Ti(II) was generated by reduction of $[Cp(^tBu_3PN)TiCl_2]$ by magnesium in THF. This species was intercepted with dienes, acetylenes or olefins.[53] Similarly, the *titanacyclopentane* derivatives $[Cp^*(^tBu_3PN)Ti(C_4H_8)]$ and $[(\eta^5\text{-}C_9H_7)(^tBu_3PN)Ti(C_4H_8)]$ were obtained by reduction in the presence of ethene. The formation of metallacyclopentanes by alkene

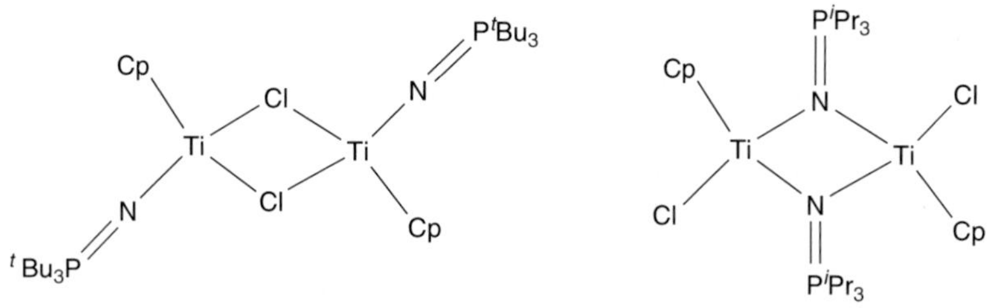

FIG. 9. Structures of $[CpTiCl(\mu\text{-}NP^{i}Pr_3)]_2$ and $[CpTi(NP^{t}\text{-}Bu_3)(\mu\text{-}Cl)]_2$.

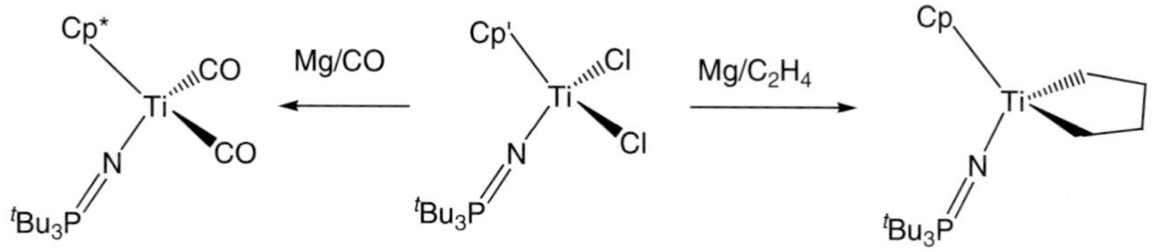

SCHEME 6.

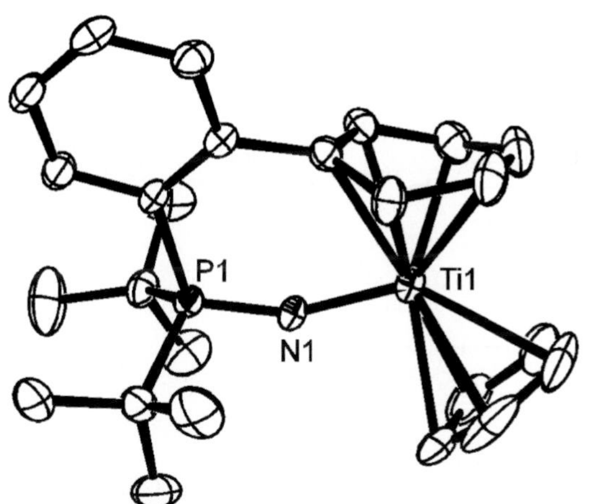

FIG. 10. ORTEP representation of $[CpTi(NP^{t}Bu_2(C_6H_4Ph\text{-}2))]$ (hydrogen atoms omitted).

coupling on low-valent group 4 metal centers is detailed elsewhere in this volume. Reduction of $[Cp^*(^{t}Bu_3PN)TiCl_2]$ in the presence of CO afforded $[Cp^*(^{t}Bu_3PN)Ti(CO)_2]$ which confirmed the role of Ti(II) in this chemistry (Scheme 6).[53]

In related chemistry, the complexes $[Cp'Ti(NP^{t}Bu_2C_6H_4Ph\text{-}2)Cl_2]$ ($Cp' = Cp$, Cp^*) were prepared in conventional manner and reduced to give $[Cp'Ti(NP^{t}\text{-}Bu_2C_6H_4Ph\text{-}2)]$ ($Cp' = Cp$, Cp^*). In these products, the phenyl substituents interact with Ti in a manner that is best formulated as a Ti(IV) cyclohexa-1,4-dien-3,6-di-yl complex (Fig. 10).[53]

B. *Zirconium*

Complexes of the form [$CpZr(NPR_3)Cl_2$] and [$Cp^*Zr(NPR_3)Cl_2$] were prepared *via* protonolysis in a manner similar to those described for titanium. It is noteworthy that attempts to employ metathesis using phosphinimide lithium salts were unsuccessful.[40,55] However, mixed-ring derivatives such as [$Cp^*Zr(NPR_3)(Cp)Cl$] were cleanly obtained from the reaction between NaCp and the precursor. Alternatively, alkylation or arylation proceeded in a straightforward manner to provide [$CpZr(NP^tBu_3)R_2$], [$Cp^*Zr(NP^tBu_3)R_2$] and [$Cp_2Zr(NP^tBu_3)R$] (R = Me, Bn) in high yield. The supine isomer of the butadiene complex $Cp^*Zr(NPR_3)(CH_2CMeC MeCH_2)$ was obtained by reduction of the zirconium(IV) dihalide with magnesium powder in the presence of 2,3-dimethylbutadiene.[55]

The zirconium-*bis*-phosphinimide species [$Zr(NP^tBu_3)_2Cl_2$] was obtained in a two-step process involving protonolysis of [$Zr(NEt_2)_4$] with 2 equivalents of HNP^tBu_3 to give [$Zr(NP^tBu_3)_2(NEt_2)_2$], followed by treatment with Me_3SiCl (Scheme 7).[55] Attempts to alkylate this product employing a number of methods were unsuccessful.

Screening of these zirconium species for ethene polymerizations using MAO activation gave only moderate activities (Table IV).[40] It was suggested that interactions of aluminum with the phosphinimide nitrogen atom led to ligand abstraction

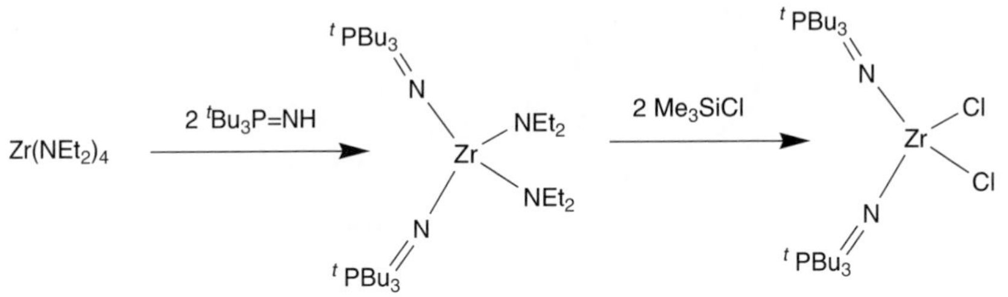

$$Zr(NEt_2)_4 \xrightarrow{2\ {}^tBu_3P{=}NH} Zr(N{=}P^tBu_3)_2(NEt_2)_2 \xrightarrow{2\ Me_3SiCl} Zr(N{=}P^tBu_3)_2Cl_2$$

SCHEME 7.

TABLE IV

SELECTED ACTIVITIES OF ETHENE POLYMERIZATION ZIRCONIUM CATALYSTS[a]

Cat.	Co-cat	Act.	M_n	M_w	PDI
$Cp^*Zr(NP^t\text{-}Bu_3)Cl_2$	MAO	170	3070	5930	1.9
			783,900	1,261,000	1.6
$CpZr(NP^t\text{-}Bu_3)Cl_2$	MAO	100	3450	234,300	67.9
$Cp^*Zr(NP^t\text{-}Bu_3)Me_2$	MAO	70			
$Cp^*Zr(NP^t\text{-}Bu_3)Me_2$	$B(C_6F_5)_3$	270	59,900	161,300	2.7
$CpZr(NP^t\text{-}Bu_3)Me_2$	$B(C_6F_5)_3$	400	6640	94,900	14.3
$Cp^*Zr(NP^t\text{-}Bu_3)Me_2$	$[Ph_3C][B(C_6F_5)_4]$	1010	54,200	177,000	3.3
$CpZr(NP^t\text{-}Bu_3)Me_2$	$[Ph_3C][B(C_6F_5)_4]$	1270	7450	24,400	3.3

[a]Toluene, 30 °C, 1.82 atm C_2H_4, 50 μM catalyst, 30 min; MAO activation Zr:Al ratio 1:500; $B(C_6F_5)_3$ or $Ph_3CB(C_6F_5)_4$ activation: Zr:Al ratio 1:20 (i-Bu_3Al), 10 min. Activity units: g PE $mmol^{-1}\,h^{-1}\,atm^{-1}$.

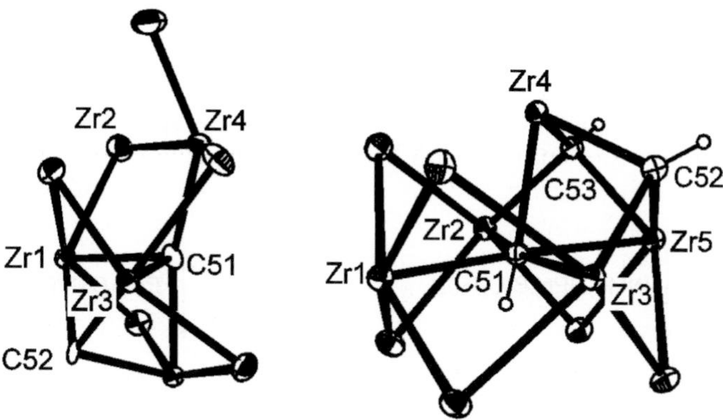

FIG. 11. ORTEP representations of core structures of the clusters, $[(Cp^*Zr)_4(\mu\text{-}Cl)_5(Cl)(\mu\text{-}CH)_2]$ and $[(Cp^*Zr)_5(\mu\text{-}Cl)_6(\mu\text{-}CH)_3]$.

or C–H activation processes similar to those seen with titanium. This view was supported by Gaussian calculations on $[CpZr(NP^tBu_3)Cl_2]$ and $[Zr(NP^tBu_3)_2Cl_2]$ that showed the HOMO, HOMO-1 and HOMO-2 orbitals were N–Zr in character and sterically accessible to Lewis acids.[40] Model reactions of zirconium phosphinimide complexes with $AlMe_3$ gave complex mixtures of products. Two clusters $[(Cp^*Zr)_4(\mu\text{-}Cl)_5(Cl)(\mu\text{-}CH)_2]$ and $[(Cp^*Zr)_5(\mu\text{-}Cl)_6(\mu\text{-}CH)_3]$ (Fig. 11) were identified.[39] These species confirmed that ligand metathesis and C–H bond activation account for the generally poor activity of zirconium phosphinimide complexes with aluminum-based activators.

Reactions of zirconium phosphinimides with $B(C_6F_5)_3$ led to the rapid formation of $[CpZr(NP^tBu_3)(C_6F_5)_2]$. Similar methyl-for-aryl exchange reactions have been reported for titanium amido species.[56] Computational studies showed that such aryl group transfers were more facile for zirconium than titanium, and that electron-rich and sterically bulky ligands precluded this process.[57] Interestingly, activation by $B(C_6F_5)_3$ or $[Ph_3C][B(C_6F_5)_4]$ of zirconium phosphinimide precursors afforded high polymerization activities. Under these conditions, $[CpZr(NP^tBu_3)(C_6F_5)_2]$ was not formed, clearly showing that olefin insertion was rapid with respect to methyl-for-C_6F_5 exchange.[40]

Donor ligands were used to intercept the zirconium cations. For example, the reaction of $[CpZr(NP^tBu_3)Me_2]$ with $B(C_6F_5)_3$ in THF gave $[CpZr(NP^tBu_3)Me(THF)][MeB(C_6F_5)_3]$, while the analogous reaction with $[HNMe_2Ph][B(C_6F_5)_4]$ afforded $[CpZr(NP^tBu_3)Me(NMe_2Ph)][B(C_6F_5)_4]$. Similarly, the salt $[CpZr(NP^tBu_3)_2][PhCH_2B(C_6F_5)_3]$ was formed from reaction of $[CpZr(NP^tBu_3)_2CH_2Ph]$ with $B(C_6F_5)_3$.[55] In the analogous reaction of the zwitterionic complex $[Cp'Zr(NP^tBu_3)(Me)\{\mu\text{-}MeB(C_6F_5)_3\}]$ with diisopropylcarbodiimide, the insertion product $[Cp'Zr(NP^tBu_3)((^iPrN)_2CMe)][MeB(C_6F_5)_3]$ was obtained (Scheme 8).[53] Finally, the reaction of a zirconium butadiene complex with $B(C_6F_5)_3$ gave the product $[Cp^*Zr(NP^tBu_3)\{CH_2CMeCMeCH_2B(C_6F_5)_3\}]$.[55] In this compound, the interaction of the cationic zirconium center with an ortho-fluorine of one of the arene rings was evidenced by NMR data.

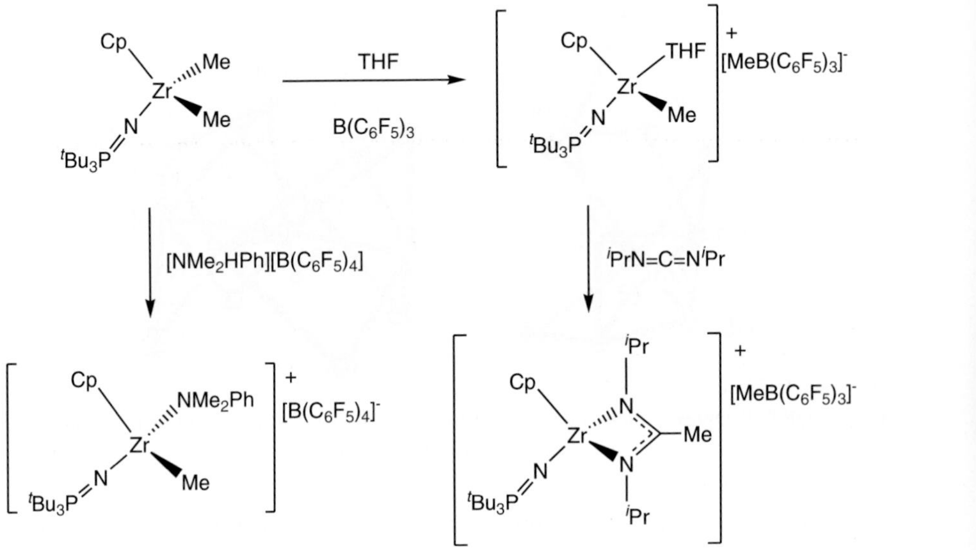

SCHEME 8.

V
GROUP 5

A. *Vanadium*

The synthesis of vanadium phosphinimide complexes was challenging. Reactions of $VOCl_3$ with Me_3SiNPR_3 or Li[NPR_3] led to unresolved mixtures of products. A few crystals of "$VCl_2(NPPh_3)_3$" correctly formulated as the salt [$VCl(NPPh_3)_3$]Cl were isolated in one such effort.[58] In contrast, reactions of [$VCl_3(NAr)$] with $R_3PNSiMe_3$ gave the complexes [$VCl_2(NAr)(NPR_3)$] in a clean fashion (Scheme 9). Subsequent alkylation or arylation gave [$VR_2(NAr)(NP^tBu_3)$], (R = Me, Ph),[58] while metathesis with amide or thiolate gave [$VCl\{N(SiMe_3)_2\}(NPh)(NP^tBu_3)$] or [$V(SBn)_2(NC_6H_3^iPr_2\text{-}2,6)(NP^tBu_3)$]. Preliminary tests of these vanadium phosphinimide compounds for polymerization activity showed only very low activities.[58]

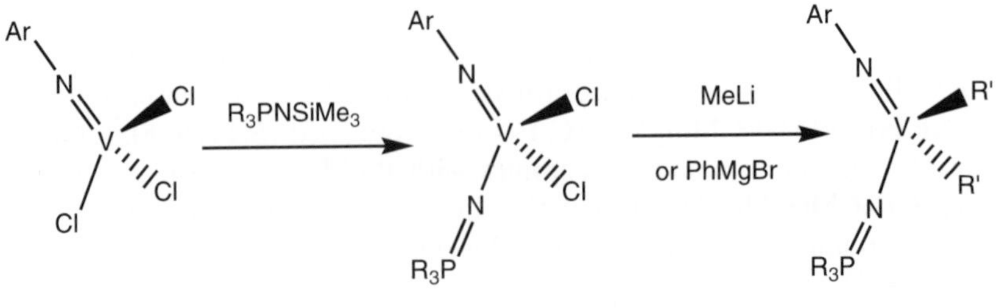

SCHEME 9.

B. *Tantalum*

Tantalum complexes of the form [Cp*Ta(NPR$_3$)Cl$_3$] were readily prepared by conventional means.[57] In the case of [Cp*Ta(NP^{t}Bu$_3$)Cl$_3$], variable temperature ^{1}H NMR studies showed that rotation about the Ta–N bond was restricted at low temperature due to a "cogwheel" effect with the cyclopentadienyl ligand. Alkylation of these species was straightforward giving such compounds as [Cp*Ta(NPR$_3$)RCl$_2$] (R = Me, Bn) and [Cp*Ta(NP^{t}Bu$_3$)(CH$_2$Ph)Me$_2$].[59] The reaction with an excess of benzyl-Grignard reagents resulted in the formation of the alkylidene complex [Cp*Ta(NPR$_3$)(=CHPh)(CH$_2$Ph)] (Scheme 10). Subsequent reaction with MeI gave the tantalacycle [Cp*Ta(NP^{t}Bu$_3$)(η^2-H$_2$CCHPh)(CH$_2$Ph)]. Alkylation with EtMgCl led to β-hydrogen elimination yielding [Cp*Ta(NP^{t}Bu$_3$)(η^2-C$_2$H$_4$)Cl], while treatment with an excess of EtMgCl afforded [Cp*Ta(NP^{t}Bu$_3$)(η^2-C$_2$H$_4$)(CH$_2$CH$_3$)].[59]

[Cp*Ta(NP^{t}Bu$_3$)Cl$_3$] reacted with B(C$_6$F$_5$)$_3$ to give the salt [Cp*Ta(NP^{t}Bu$_3$)Cl$_2$][ClB(C$_6$F$_5$)$_3$], while [Cp*Ta(NP^{t}Bu$_3$)Cl$_2$][B(C$_6$F$_5$)$_4$] was formed in the corresponding reaction with Ph$_3$C[B(C$_6$F$_5$)$_4$].[59] The complexes [Cp*Ta(NPR$_3$)Me$_2${MeB(C$_6$F$_5$)$_3$}], [Cp*Ta(NPR$_3$)Me$_2$][B(C$_6$F$_5$)$_4$] and [Cp*Ta(NP^{t}Bu$_3$)(CH$_2$Ph)Cl][ClB(C$_6$F$_5$)$_3$] were

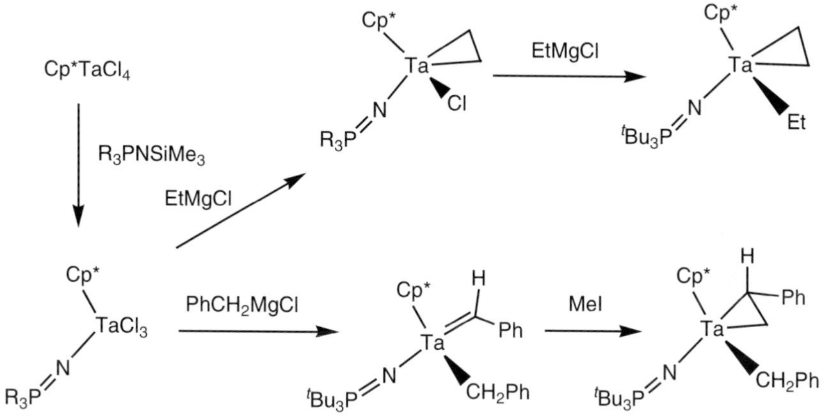

SCHEME 10.

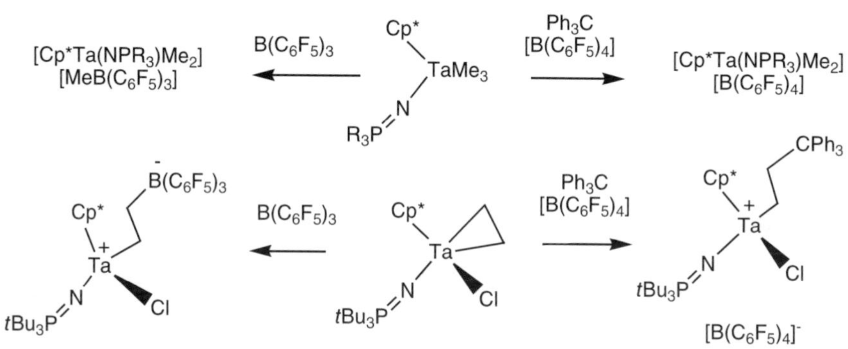

SCHEME 11.

prepared under similar reactions. Similarly, treatment of [Cp*Ta(NPR$_3$)(η^2-C$_2$H$_4$)Cl] with [Ph$_3$C][B(C$_6$F$_5$)$_4$] gave [Cp*Ta(NP^{t}Bu$_3$)Cl(CH$_2$CH$_2$CPh$_3$)] [B(C$_6$F$_5$)$_4$], while reaction of [Cp*Ta(NPR$_3$)(η^2-C$_2$H$_4$)Cl] with B(C$_6$F$_5$)$_3$ formed the zwitterionic complex [Cp*Ta(NP^{t}Bu$_3$)Cl(CH$_2$CH$_2$B(C$_6$F$_5$)$_3$)] (Scheme 11). These tantalum species were shown not to be active in ethene polymerization.[59]

VI

GROUP 8: IRON

The mixed-valent Fe(III)–Fe(II) complex [Cl$_3$Fe$_2$(μ-NP^{t}Bu$_3$)$_2$] was prepared by the reaction of [LiNP^{t}Bu$_3$]$_4$ with FeCl$_2$.[60] This species was reversibly oxidized by one-electron generating [Cl$_3$Fe$_2$(μ-NP^{t}Bu$_3$)$_2$]$^+$, while chemical oxidation with [Cp$_2$Fe]PF$_6$ afforded the Fe(III)–Fe(III) dimer [Cl$_2$Fe(μ-NP^{t}Bu$_3$)]$_2$ (Scheme 12). The species [Br$_4$Fe$_2$(μ-NP^{t}Bu$_3$)$_2$], formed from the reaction of [LiNP^{t}Bu$_3$]$_4$ with FeBr$_3$, underwent hydrolysis to give [(^{t}Bu$_3$PNH)Br$_2$Fe]$_2$(μ-O). The species [Br$_4$Fe$_2$ (μ-NP^{t}Bu$_3$)$_2$] was found to be a low-activity, single-site catalyst for the polymerization of ethene. Interestingly, [(^{t}Bu$_3$PNH)Br$_2$Fe]$_2$(μ-O) afforded polymer similar to that produced by [Br$_4$Fe$_2$(μ-NP^t-Bu$_3$)$_2$] suggesting that MAO dehydrates [(^{t}Bu$_3$PNH)Br$_2$Fe]$_2$(μ-O) generating a catalyst similar to that derived from [Br$_4$Fe$_2$ (μ-NP^t-Bu$_3$)$_2$] (Scheme 12).[60]

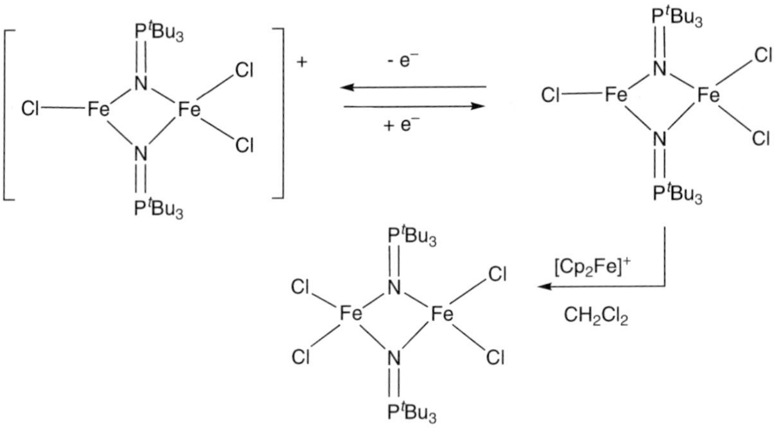

SCHEME 12.

VII

GROUP 13

A. *Boron*

The dimeric dication [(Et$_3$PN)$_2$B$_2$]$^{2+}$,[61] the cyclic trications [(Et$_3$PNBH)$_3$]$^{3+}$ [62] and [(Et$_3$PNBH)$_4$NPEt$_3$]$^{3+}$ [63] and the neutral borane [(Ph$_3$PN)$_3$B] have been previously reported.[64] More recently, the reaction of [Li(μ-NP^{t}Bu$_3$)]$_4$ with [BCl$_3$] was shown to give the borinium salt [(^{t}Bu$_3$PN)$_2$B]Cl (Fig. 12).[65] This salt exhibited a unique

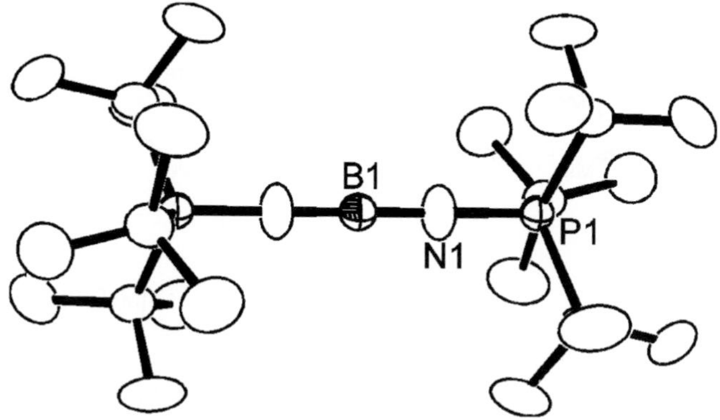

FIG. 12. ORTEP representation of the cation: $[(^t Bu_3PN)_2B]^+$.

extended linear PNBNP linkage, presumed to be a direct result of the steric demands of the phosphinimide ligands. The related species $[(^t Bu_3PN)_2BH]$ containing three coordinate boron was prepared in the reaction of BH_3SMe_2 with HNP^tBu_3. Subsequent reaction with $[Ph_3C][B(C_6F_5)_4]$ gave the salt $[(^t Bu_3PN)_2B]$ $[B(C_6F_5)_4]$.[65]

B. *Aluminum*

A variety of silylphosphinimine adducts was formed *via* the reactions of $R_3PNSiMe_3$ with the aluminum alkyl/halides $AlCl_3$, $AlMeCl_2$, $AlMe_2Cl$ and $AlMe_3$. In contrast, reactions of $(^t Bu_3PNH)$ and $AlMe_3$ gave the phosphinimide dimers $[X_2Al(\mu\text{-}NP^tBu_3)]_2$ (X = Me, Cl).[66] Alternatively, these compounds could be prepared by nucleophilic substitution reactions involving $[Li(\mu\text{-}NP^tBu_3)]_4$ and aluminum halides. Stoichiometric reaction of the alkylated dimer $[Me_2Al(\mu\text{-}NP^tBu_3)]_2$ with the borane $B(C_6F_5)_3$ gave the zwitterionic species $[Me_2Al(\mu\text{-}NP^tBu_3)_2AlMe][MeB(C_6F_5)_3]$ (Scheme 13). This latter product reacted with trimethylphosphine to give the salt $[Me_2Al(\mu\text{-}NP^tBu_3)_2AlMe(PMe_3)][\text{-}MeB(C_6F_5)_3]$.[66]

VIII
GROUP 14

While silyl phosphinimides of the form $(R_3PN)SiMe_3$ were prepared by conventional Staudinger oxidation of the phosphine by N_3SiMe_3, the germanium and tin analogs $(^t Bu_3PN)MMe_3$ (M = Ge, Sn) were prepared by stoichiometric reaction of $[LiNP^tBu_3]_4$ and Me_3MCl.[67] Similarly, the di- and tri-substituted products $(^t Bu_3PN)_2MMe_2$ (M = Si, Ge, Sn) and $(^t Bu_3PN)_3SnMe$ were prepared. Analogous metathesis reactions afforded $R_3PNSnPh_3$ (R = ^{t}Bu, ^{i}Pr), $(^t Bu_3PN)_2SnPh_2$ and $(^t Bu_3PN)_2Sn(^t Bu)_2$. The reaction of $(Me_3PN)SiMe_3$ with $B(C_6F_5)_3$ gave the adduct $Me_3PN(SiMe_3)B(C_6F_5)_3$, whereas bulkier phosphinimines reacted with $B(C_6F_5)_3$ to

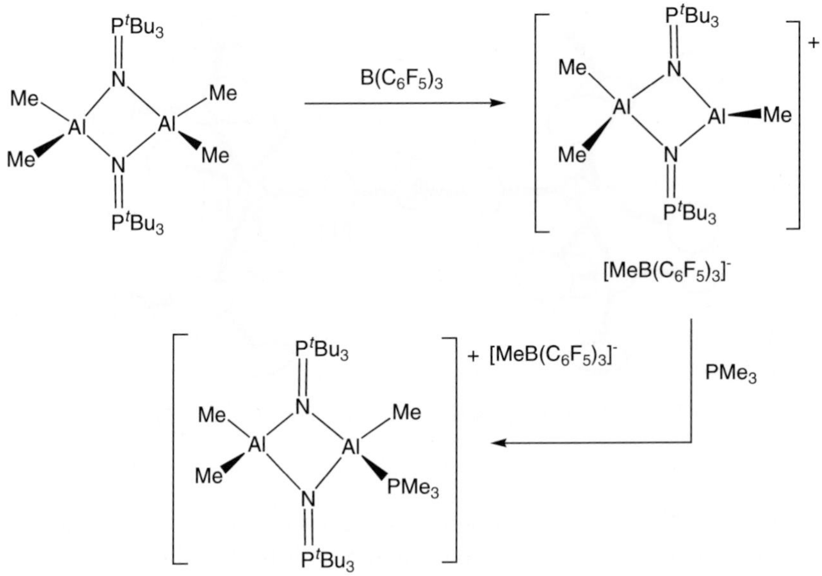

SCHEME 13.

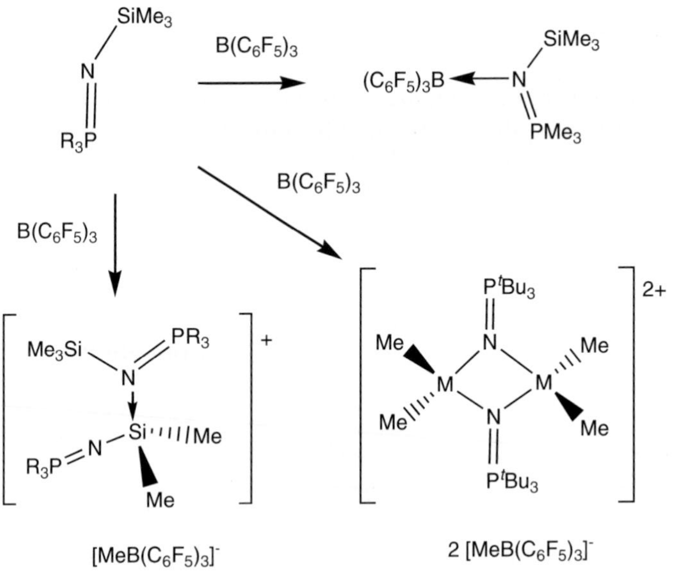

SCHEME 14.

give $[R_3PNSiMe_2(N(PR_3)SiMe_3)][MeB(C_6F_5)_3]$ ($R = {}^iPr$, Ph).[67] In further contrast, reaction of still bulkier phosphinimide derivatives $({}^tBu_3PN)MMe_3$ with $B(C_6F_5)_3$ gave the dimeric dication salt $[((\mu\text{-}{}^tBu_3PN)MMe_2)_2][MeB(C_6F_5)_3]_2$ (M = Si, Sn) (Scheme 14).[67]

IX
SUMMARY AND FUTURE DIRECTIONS

The work described herein demonstrates a variety of systems that exhibit unique structural and reactivity characteristics as a result of the incorporation of sterically demanding phosphinimide ligands. These ligands act as strong electron donors and yet provide, in the presence of sufficient steric bulk, an envelope around the M–N bond, thus protecting it from reaction. In such systems, the central main group or transition metal center has second-sphere protection provided by the steric bulk of the ligand substituents and yet the displacement of this steric protection from the central atom, may generate a highly reactive environment. While the main group chemistry described herein has focused on unique structural features, the work on early transition metal polymerization catalysts has illustrated that the features provided by sterically demanding phosphinimide ligands can be exploited for a "real-world" application. We view this as the first of many applications of complexes derived from sterically demanding phosphinimides as organometallic chemists begin to incorporate these ligands into their arsenal for the design of new systems for applications in unique stoichiometric or catalytic chemistry.

Acknowledgments

I wish to acknowledge the enthusiastic and very bright scientists I have had the pleasure of working with during the course of the studies described herein. They include Jeff Stewart, Fred Guerin, Jim Kickham, Charles Carraz, Chris Ong, Chad Beddie, Emily Hollink, Steve Clemens, Lourisa Cabrera, Silke Courtenay, Todd Graham, Chris Frazer and Pingrong Wei. NOVA Chemicals Corporation, NSERC of Canada and the University of Windsor are thanked for financial support.

References

(1) Dehnicke, K.; Krieger, M.; Massa, W. *Coord. Chem. Rev.* **1999**, *182*, 19.
(2) Dehnicke, K.; Weller, F. *Coord. Chem. Rev.* **1997**, *158*, 103.
(3) Staudinger, H.; Meyer, J. *Helv. Chim. Acta.* **1919**, *2*, 635.
(4) Ackermann, H.; Leo, R.; Massa, W.; Neumuller, B.; Dehnicke, K. Z. *Anorg. Allg. Chem.* **2000**, *626*, 284.
(5) Ackermann, H.; Seybert, G.; Massa, W.; Bock, O.; Muller, U.; Dehnicke, K. Z *Anorg. Allg. Chem.* **2000**, *626*, 2463.
(6) Ackermann, H.; Weller, F.; Dehnicke, K. Z. *Naturforsch* **2000**, *55b*, 448.
(7) Chitsaz, S.; Harms, K.; Neumuller, B.; Dehnicke, K. Z. *Anorg. Allg. Chem.* **1999**, *625*, 939.
(8) Chitsaz, S.; Neumuller, B.; Dehnicke, K. Z. *Anorg. Allg. Chem.* **2000**, *626*, 813.
(9) Chitsaz, S.; Neumuller, B.; Dehnicke, K. Z. *Anorg. Allg. Chem.* **2000**, *626*, 1535.
(10) Dietrich, A.; Neumuller, B.; Dehnicke, K. Z. *Anorg. Allg. Chem.* **2000**, *626*, 1837.
(11) Grob, T.; Geiseler, G.; Harms, K.; Greiner, A.; Dehnicke, K. Z. *Anorg. Allg. Chem.* **2002**, *628*, 217.
(12) Jaeschke, B.; Jansen, M. Z. *Anorg. Allg. Chem.* **2002**, *628*, 2000.
(13) Krieger, M.; Gould, R. O.; Dehnicke, K. Z. *Anorg. Allg. Chem.* **2002**, *628*, 1289.
(14) Krieger, M.; Gould, R. O.; Harms, K.; Greiner, A.; Dehnicke, K. Z. *Anorg. Allg. Chem.* **2001**, *627*, 747.
(15) Mueller, U.; Bock, O.; Sippel, H.; Groeb, T.; Dehnicke, K.; Greiner, A. Z. *Anorg. Allg. Chem.* **2002**, *628*, 1703.
(16) Neumueller, B.; Dehnicke, K. Z. *Anorg. Allg. Chem.* **2004**, *630*, 369.
(17) Neumueller, B.; Dehnicke, K. Z. *Anorg. Allg. Chem.* **2004**, *630*, 799.
(18) Neumueller, B.; Dehnicke, K. Z. *Anorg. Allg. Chem.* **2004**, *630*, 1360.

(19) Raab, M.; Sundermann, A.; Schick, G.; Nieger, M.; Schoeller, W. W.; Niecke, E. *Phosphorus, Sulfur Silicon Rel. Elements* **2002**, *177*, 2153.
(20) Raab, M.; Tirree, J.; Nieger, M.; Niecke, E. Z. *Anorg. Allg. Chem.* **2003**, *629*, 769.
(21) Sundermann, A.; Schoeller, W. W. *J. Am. Chem. Soc.* **2000**, *122*, 4729.
(22) Dehnicke, K.; Greiner, A. *Angew. Chem., Int. Ed. Engl.* **2003**, *42*, 1340.
(23) Anfang, S.; Seybert, G.; Harms, K.; Geiseler, G.; Massa, W.; Dehnicke, K. Z. *Anorg. Allg. Chem.* **1998**, *624*, 1187.
(24) Courtenay, S.; Wei, P. *Stephan D W Can. J. Chem.* **2003**, *81*, 1471.
(25) Hollink, E.; Wei, P.; Stephan, D. W. *Can. J. Chem.* **2005**, *83*, 430.
(26) Stephan, D. W.; Stewart, J. C.; Guerin, F.; Courtenay, S.; Kickham, J.; Hollink, E.; Beddie, C.; Hoskin, A.; Graham, T.; Wei, P.; Spence, R. E. v. H.; Xu, W.; Koch, L.; Gao, X.; Harrison, D. G. *Organometallics* **2003**, *22*, 1937.
(27) Stephan, D. W.; Stewart, J. C.; Guerin, F.; Spence, R. E. v. H.; Xu, W.; Harrison, D. G. *Organometallics* **1999**, *18*, 1116.
(28) Guerin, F.; Beddie, C. L.; Stephan, D. W.; Spence, R. E. v. H.; Wurz, R. *Organometallics* **2001**, *20*, 3466.
(29) Brown, S. J.; Gao, X.; Kowalchuk, M. G.; Spence, R. E. v. H.; Stephan, D. W.; Swabey, J. *Can. J. Chem.* **2002**, *80*, 1618.
(30) Guerin, F.; Stewart, J. C.; Beddie, C.; Stephan, D. W. *Organometallics* **2000**, *19*, 2994.
(31) Stephan, D. W.; Guerin, F.; Spence, R. E. v. H.; Koch, L.; Gao, X.; Brown, S. J.; Swabey, J. W.; Wang, Q.; Xu, W.; Zoricak, P.; Harrison, D. G. *Organometallics* **1999**, *18*, 2046.
(32) Cabrera, L.; Hollink, E.; Stewart, J. C.; Wie, P.; Stephan, D. W. *Organometallics* **2005**, *24*, 1091.
(33) Gee, V.; Orpen, A. G.; Phetmung, H.; Pringle, P. G.; Pugh, R. I. *Chem. Commun.* **1999**, 901–902.
(34) Carraz, C.-A.; Stephan, D. W. *Organometallics* **2000**, *19*, 3791.
(35) Yue, N. L. S.; Stephan, D. W. *Organometallics* **2001**, *20*, 2303.
(36) Hollink, E.; Wei, P.; Stephan, D. W. *Can. J. Chem.* **2004**, *82*, 1304.
(37) Said, M.; Hughes, D. L.; Bochmann, M. *Dalton Trans.,* **2004**, 359–360.
(38) Beddie, C.; Hollink, E.; Wie, P.; Gauld, J.; Stephan, D. W. *Organometallics* **2004**, *23*, 5240.
(39) Guerin, F.; Stephan, D. W. *Angew. Chem., Int. Ed. Engl.* **2000**, *39*, 1298.
(40) Yue, N.; Hollink, E.; Guerin, F.; Stephan, D. W. *Organometallics* **2001**, *20*, 4424.
(41) Stephan, D. W.; Stewart, J. C.; Harrison, D. G. Supported phosphinimine-cyclopentadienyl catalysts for polymerization of olefins. **1999**, 890581, USA.
(42) Kickham, J. E.; Guerin, F.; Stewart, J. C.; Urbanska, E.; Ong, C. M.; Stephan, D. W. *Organometallics* **2001**, *20*, 1175.
(43) Kickham, J. E.; Guerin, F.; Stewart, J. C.; Stephan, D. W. *Angew. Chem., Int. Ed. Engl.* **2000**, *39*, 3263.
(44) Kickham, J. E.; Guerin, F.; Stephan, D. W. *J. Am. Chem. Soc.* **2002**, *124*, 11486.
(45) Hollink, E.; Stewart, J. C.; Wie, P.; Stephan, D. W. *Dalton Trans.*, **2003**, 3968–3974.
(46) Xu, Z.; Vanka, K.; Firman, T.; Michalak, A.; Zurek, E.; Zhu, C.; Ziegler, T. *Organometallics* **2002**, *21*, 2444.
(47) Wondimagegn, T.; Vanka, K.; Xu, Z.; Ziegler, T. *Organometallics* **2004**, *23*, 2651.
(48) Ma, K.; Piers, W. E.; Gao, Y.; Parvez, M. *J. Am. Chem. Soc.* **2004**, *126*, 5668.
(49) Ong, C.; Kickham, J.; Clemens, S.; Guerin, F.; Stephan, D. W. *Organometallics* **2002**, *21*, 1646.
(50) Guerin, F.; Stephan, D. W. *Angew. Chem. Int. Ed. Engl.* **1999**, *38*, 3698.
(51) Hollink, E.; Wei, P.; Stephan, D. W. *Can. J. Chem.* **2004**, *82*, 1634.
(52) Sung, R. C. W.; Courtenay, S.; McGarvey, B. R.; Stephan, D. W. *Inorg. Chem.* **2000**, *39*, 2542.
(53) Graham, T. W.; Kickham, J.; Courtenay, S.; Wie, P.; Stephan, D. W. *Organometallics* **2004**, *23*, 3309.
(54) Li, J.-S.; Weller, F.; Schmock, F.; Dehnicke, K. Z. *Anorg. Allg. Chem.* **1995**, *621*, 2097.
(55) Hollink, E.; Wie, P.; Stephan, D. W. *Organometallics* **2004**, *23*, 1562.
(56) Scollard, J. D.; McConville, D. H.; Rettig, S. J. *Organometallics* **1997**, *16*, 1810.
(57) Wondimagegn, T.; Xu, Z.; Vanka, K.; Ziegler, T. *Organometallics* **2004**, *23*, 3847.
(58) Hawkeswood, S. B.; Stephan, D. W. *Inorg. Chem.* **2003**, *42*, 5429.
(59) Courtenay, S.; Stephan, D. W. *Organometallics* **2001**, *20*, 1442.
(60) Le Pichon, L.; Stephan, D. W.; Gao, X.; Wang, Q. *Organometallics* **2002**, *21*, 1362.

(61) Moehlen, M.; Neumueller, B.; Faza, N.; Mueller, C.; Massa, W.; Dehnicke, Z. *Anorg. Allg. Chem.* **1997**, *623*, 1567.
(62) Moehlen, M.; Neumueller, B.; Dehnicke, K. Z. *Anorg. Allg. Chem.* **1999**, *625*, 197.
(63) Moehlen, M.; Neumueller, B.; Harms, K.; Krautscheid, H.; Fenske, D.; Diedenhofen, M.; Frenking, G.; Dehnicke, K. Z. *Anorg. Allg. Chem.* **1998**, *624*, 1105.
(64) Moehlen, M.; Neumueller, B.; Dehnicke, K. Z. *Anorg. Allg. Chem.* **1998**, *624*, 177.
(65) Courtenay, S.; Mutus, J. Y.; Schurko, R. W.; Stephan, D. W. *Angew. Chem., Int. Ed. Engl.* **2002**, *41*, 498.
(66) Ong, C. M.; McKarns, P.; Stephan, D. W. *Organometallics* **1999**, *18*, 4197.
(67) Courtenay, S.; Ong, C. M.; Stephan, D. W. *Organometallics* **2003**, *22*, 818.

Transition Metal Complexes of Tethered Arenes

JOANNE R. ADAMS[†] and MARTIN A. BENNETT*

Research School of Chemistry, Institute of Advanced Studies, Australian National University, Canberra, ACT 0200, Australia

I

INTRODUCTION

Tethered cyclopentadienyl (Cp) and arene ligands can be regarded as a special class of heterobidentate or hemi-labile ligand, i.e., ligands in which the connected donors differ in binding ability and lability.[1–3] Tethered and funtionalized Cp complexes have been investigated extensively over the last 15 years as part of an effort to improve the range of applications of Cp complexes as catalysts and as reagents for stereoselective synthesis. A strongly binding donor atom helps to anchor the Cp group to the metal centre and thus inhibits its dissociation, a feature that may be especially important for main group elements; on the other hand, a relatively weak donor may protect a vacant site at a transition metal centre until it is required by an incoming ligand [Eq. (1)]:

L' / − L' (1)

The chelate effect of tethered Cp ligands may be used to modulate the coordination sphere in order to stabilize reactive metal centres and may also serve to introduce chirality at a metal centre. A number of reviews are available.[4–11]

*Corresponding author. Tel.: +61-2-61253639; fax: +61-2-61253216.
E-mail: bennett@rsc.anu.edu.au (M.A. Bennett).
[†]Previously Joanne R. Harper (Refs. 59, 60).

ADVANCES IN ORGANOMETALLIC CHEMISTRY
VOLUME 54 ISSN 0065-3055/DOI 10.1016/S0065-3055(05)54007-3

© 2006 Elsevier Inc.
All rights reserved.

It is well known that, although the Cp anion and benzene usually both behave as 6π-electron donors to d-block elements, benzene complexes are generally less stable and less numerous than their Cp counterparts, in part because benzene, as a neutral molecule, is lost or displaced by other ligands more readily than the Cp anion. We cite just two examples: the ruthenium(II) complex $[RuMe_2(\eta^6\text{-}C_6H_6)(PMe_2Ph)]$ decomposes, presumably with loss of benzene, in the solid state above –40 °C,[12] whereas the isoelectronic rhodium(III) complex $[RhMe_2(\eta^5\text{-}C_5H_5)(PMe_2Ph)]$ is stable at room temperature.[13] Likewise, the coordinated benzene in $[Fe(\eta^6\text{-}C_6H_6)_2]^{2+}$ is easily displaced by acetonitrile and halide ions,[14] in contrast to the stability of ferrocene. This trend holds even when stabilizing methyl groups are present on the carbocyclic ring.

As a consequence of the chelate effect, the arene in a tethered arene might be more effectively anchored to a metal centre, perhaps resulting in a stabilization of η^6-arene coordination for a greater variety of metals and oxidation states than is available for non-tethered arenes. Conversely, the lability of a coordinated, tethered arene might be used to protect potentially catalytic sites at a metal centre. Both effects could be modulated by variation in the length and nature of the tether, the nature of the other donor atom, and by the introduction of substituents on the aromatic ring.

In this review, we attempt to summarize progress in this emerging field. It will be noted that in most of the reported tethered arene complexes the metal atoms have the stable electronic configurations nd^6 [such as Cr(0) and Ru(II)], or nd^8 [mainly Rh(I) and Ir(I)], in part because, as weak σ-donors, arenes require π-back donation from the metal, which is less favourable for metals in high-oxidation states. The tripodal arrangement of auxiliary ligands in the half-sandwich geometry is also highly favoured for nd^6 organometallic compounds.

II
PREPARATIVE METHODS

A. *General Considerations*

In principle, the preparative procedures for tethered arene complexes fall into four distinct classes, which are represented in over-simplified form in Scheme 1. The

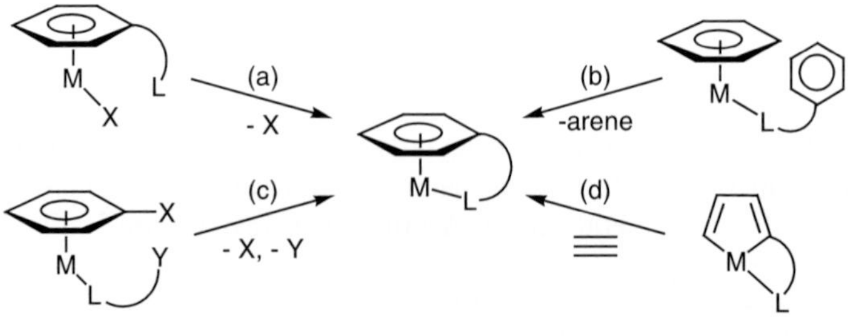

SCHEME 1.

first involves η^6-coordination to the metal centre of a functionalized arene connected by a group of atoms to a potential donor, and subsequent formation of the tether by displacement of a ligand (X) and coordination of the donor atom to the metal centre (Scheme 1a). In the second, a labile ligand, which may itself be an arene, is displaced by a second arene, which is attached by a tether to the already coordinated donor atom (Scheme 1b). Routes (a) and (b) depicted in Scheme 1 thus differ according to whether the donor atom L is coordinated after or before the arene. The third procedure involves an intramolecular condensation reaction between a substituent on the η^6-arene and a functional group on the donor atom (L) (Scheme 1c). The final method involves construction of the arene by an alkyne condensation reaction in the coordination sphere (Scheme 1d). The metal atom is already coordinated to the donor atom (L) that is connected to a metalla-1,3-diene. The alkyne reacts with the diene to construct the six-membered ring, thus giving rise to the tethered species.

We now illustrate these procedures according to the routes outlined in Scheme 1 but note that in some cases the reaction sequence for the formation of the complexes is unclear.

B. *Syntheses via Initial Arene Coordination (Scheme 1a)*

This general approach has been used to synthesize tethered arene complexes of chromium(0) and ruthenium(II). The complexes $[Cr(CO)_2(\eta^2{:}\eta^6\text{-}CH_2=CHXC_6H_5)]$ $[X = CH_2, OCH_2, (CH_2)_2, (CH_2)_3, (CH_2)_4]$ have been made by photolysis of the appropriate $[Cr(CO)_3(\eta^6\text{-}C_6H_5XCH=CH_2)]$ precursor.[15] A related, more extensive series of compounds has been obtained by tethering the arene to one or more phosphite groups, the metal–ligand bonds of which are expected to be more stable than those of an alkene. Photolysis of $[Cr(CO)_3\{\eta^6\text{-}3,5\text{-}Me_2C_6H_3(CH_2)_nOP(OPh)_2\}]$ ($n = 1$–3) forms the tethered complex **1** [Eq. (2)].[16] A second strap can be formed by initially coordinating arenes of the type 1,2-, 1,3-, or $1,4\text{-}C_6H_4\{(CH_2)_nOP(OPh)_2\}_2$ ($n = 2,3$) to the $Cr(CO)_3$ unit; photolysis of the resulting complexes causes consecutive elimination of two CO groups and intramolecular cyclization to give complexes of type **2** [Eq. (3)]. In the case of $n = 1$, only the mono-tethered complex is formed.[17] Photolysis of the $Cr(CO)_3$ complex of the difluorophosphite triester $1,3,5\text{-}C_6H_3\{(CH_2)_2OPF_2\}_3$ removes all three CO groups giving the triply tethered complex **3** [Eq. (4)].[18]

(2)

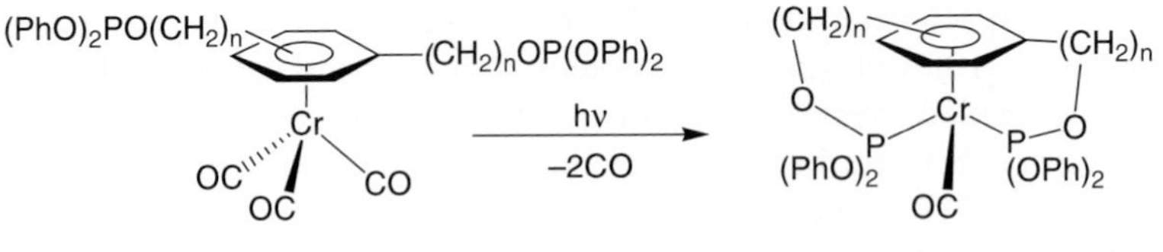

(3)

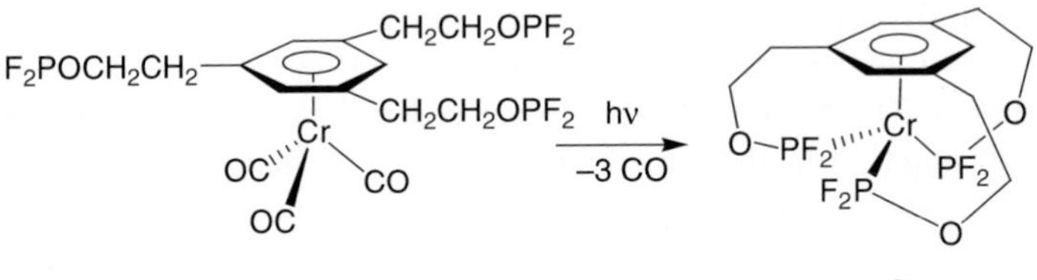

(4)

A tethered arene complex may also be an intermediate in the reaction of the $Cr(CO)_3$ derivative of $C_6H_5CH_2CD_2I$ with $AgBF_4$ (Scheme 2).[19] The resulting carbocation $[Cr(CO)_3(\eta^6\text{-}C_6H_5CH_2CD_2)]^+BF_4^-$ (**4**) can be stabilized by donation of an electron pair from chromium to the coordinatively and electronically unsaturated carbon atom, to form the tethered intermediate (**5**). This immediately undergoes nucleophilic attack by methanol to give a single methoxy-substituted product $[Cr(CO)_3(\eta^6\text{-}C_6H_5CH_2CD_2OMe)]$ (**6**). Additional experiments with substrates that contain two chiral centres in the side-chain indicate that the nucleophile enters exclusively from the remote side of the metal centre. The reaction takes a significantly different course in the absence of the $Cr(CO)_3$ group. The reaction of the isotopically labelled halide $C_6H_5CH_2CD_2I$ (**7**) with $AgBF_4$ generates a phenonium ion intermediate that yields a 1:1 ratio of $C_6H_5CH_2CD_2OMe$ (**8**) and $C_6H_5CD_2CH_2OMe$ (**9**) on treatment with methanol [Eq. (5)].[19] The two products arise by rearrangement of the intermediate phenonium ion, producing two different carbocations that are attacked by methanol.

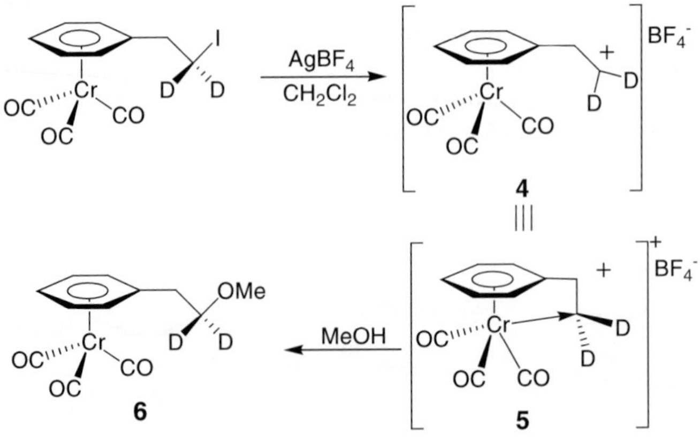

SCHEME 2.

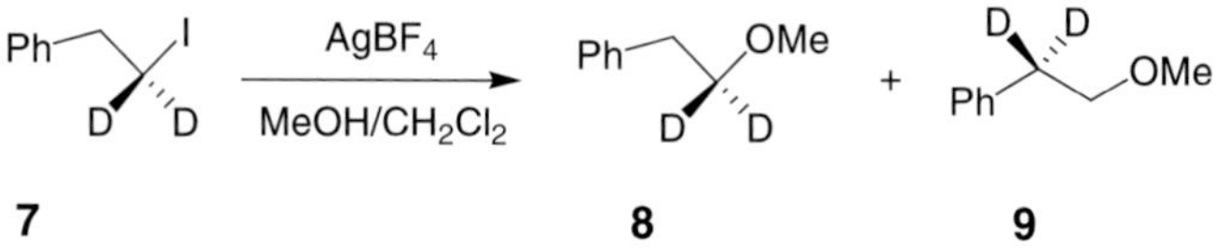

(5)

7 **8** **9**

Tethered arene–ruthenium(II) complexes with oxygen as the second donor atom have been derived from the 3-phenylpropanol complex $[RuCl_2\{\eta^6\text{-}C_6H_5(CH_2)_3OH\}]_2$ (**10**).[20–24] The tertiary phosphine complexes $[RuCl_2\{\eta^6\text{-}C_6H_5(CH_2)_3OH\}L]$ [L = PEt_3 (**11a**), PPh_3 (**11b**)], on treatment with $AgBF_4$, lose one of the chloride ions, generating the tethered alcohol complexes **12a** and **12b** [Eq. (6)]. Similarly, the mono-cationic, bidentate N-donor complexes $[RuCl(N\cap N)\{\eta^6\text{-}C_6H_5(CH_2)_3OH\}]BF_4$ (**13**) give the dicationic salts **14a**–**14c**, which can be deprotonated to the corresponding alkoxo complexes **15a**–**15c** (Scheme 3). When the tether is shortened by one carbon atom, similar derivatives are not obtained. However, in the case of the corresponding amines $C_6H_5(CH_2)_nNH_2$, tethered complexes are obtained for both $n = 2$ and $n = 3$. Reaction of **12** with R_2PCl (R = Ph, ^{i}Pr) in the presence of $^iPr_2NEt$ gives tethered phosphinous ester complexes $[RuCl_2\{\eta^6\text{-}C_6H_5(CH_2)_3OPR_2\text{-}\kappa P\}]$ (**16a**, **16b**). The tethered alcohol complex $[RuCl\{\eta^6\text{-}O(H)(CH_2)_3C_6H_5\text{-}\kappa O\}(PCy_3)]BF_4$ has been shown to catalyse ring-closing metathesis (RCM) reactions.[22]

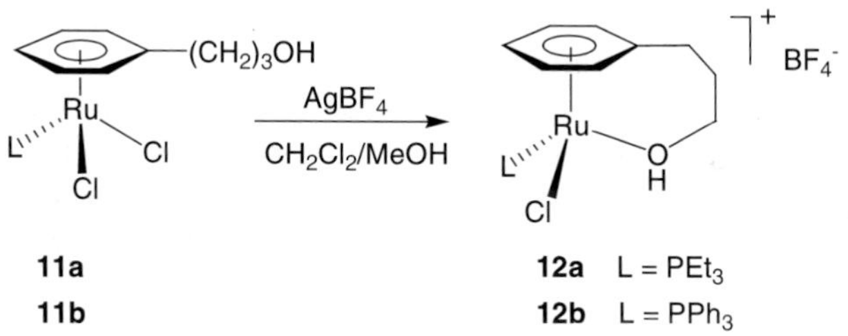

(6)

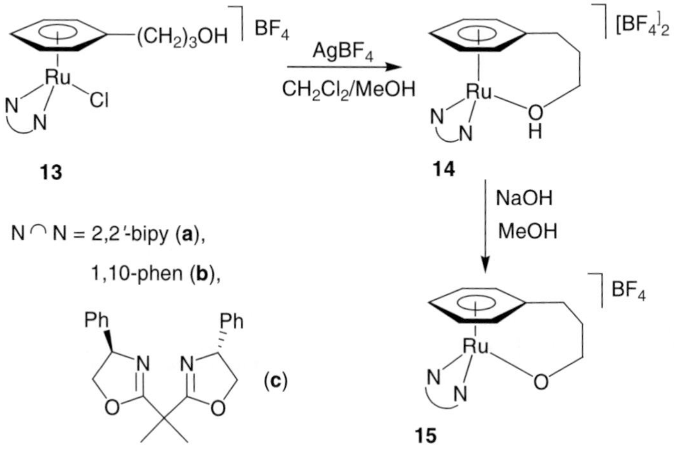

11a **11b** **12a** L = PEt_3 **12b** L = PPh_3

13 **14** **15**

$N\cap N$ = 2,2′-bipy (**a**), 1,10-phen (**b**), (**c**)

Scheme 3.

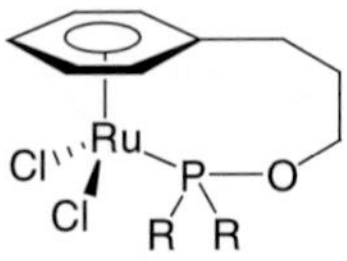

16a R = Ph
16b R = ^{i}Pr

Dinuclear tethered arene–ruthenium complexes can also be prepared from **10** (Scheme 4).[23,24] Treatment with the base 2-aminoethanol in the presence of $NaBF_4$ affords the dinuclear, tri-bridged species **17** in which the two alkoxide oxygen atoms are on the same side of the Ru–Ru axis. The bridging chloride is abstracted by $AgBF_4$ in acetonitrile to give complex **18** in which the oxygen atoms are on opposite sides of the Ru–Ru axis. The acetonitrile ligands can be replaced by bridging bidentate ligands. For example, on irradiation ($300 < \lambda < 400$ nm) in the presence of *trans*-azobenzene, the purple μ-*cis*-azobenzene complex **19** is obtained in which the alkoxide oxygen atoms have reverted to the *syn*-arrangement with respect to the Ru–Ru axis. Irradiation of **19** in CD_3CN ($\lambda > 510$ nm) forms a photostationary mixture of **19**, **18**, and *trans*-azobenzene in a ratio of 8:92:92, whereas under the same conditions free azobenzene (*cis*:*trans* ratio 50:50) is converted into a photostationary mixture in which the *cis*:*trans* ratio is 22:78. Complexes similar to **19** are formed by diphenyl disulfide and pyridazine.

A tridentate (η^6:η^1:η^1) strapped arene complex of ruthenium(II), **20**, containing a pair of auxiliary nitrogen donor atoms derived from (*R*,*R*)-1,2-diphenylethylenediamine, has been prepared by a sequence in which the functionalized arene is first coordinated to ruthenium(II), as shown in Scheme 5.[25] The structure of **20** has been

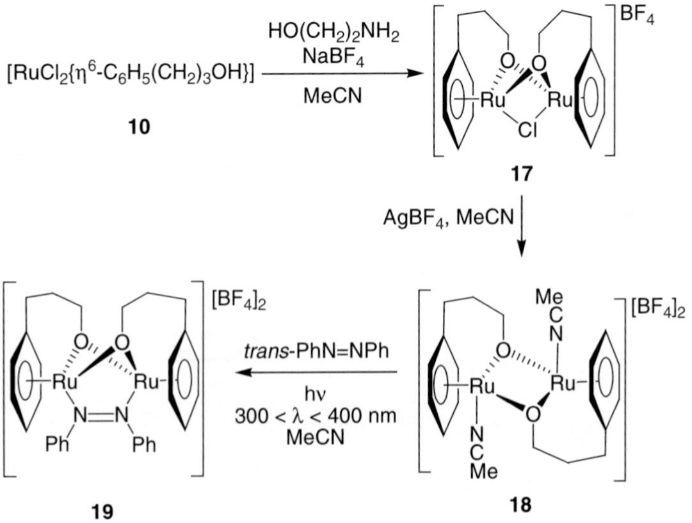

Scheme 4.

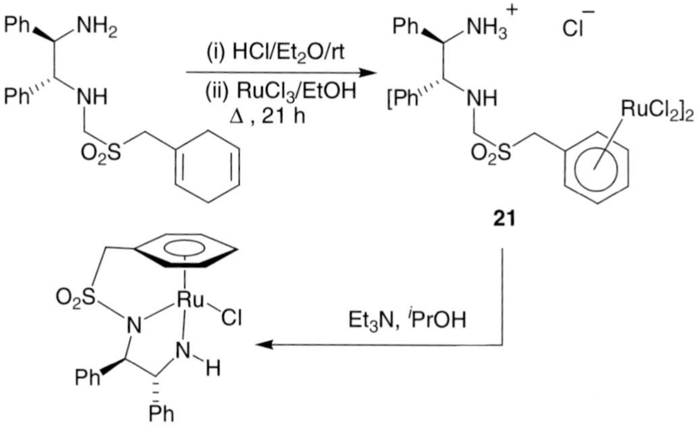

SCHEME 5.

confirmed by X-ray crystallography. Complex **20**, either isolated or generated *in situ* from its precursor **21**, is a long-lived Noyori-type catalyst for the reduction of acetophenone in HCO_2H/Et_3N to give PhCH(OH)Me in >99% yield and 96% ee (*R*) after ca. 21 h at 28 °C. Complex **22** derived from (1*R*,2*S*)-norephedrine can also be generated *in situ* and used as a catalyst for transfer hydrogenation of aromatic ketones in ^{i}PrOH/KOH, though the enantioselectivities are generally lower than those obtained by use of **20** in HCO_2H/Et_3N.

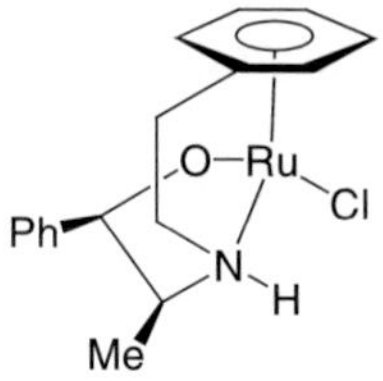

22

C. *Syntheses via Initial Donor Atom Coordination (Scheme 1b)*

This approach, in which an arene appended to a coordinated donor atom replaces a labile ligand, is by far the most generally useful procedure and has been applied to the preparation of tethered arene complexes of Ti(IV), Cr(0), Mo(0), Mo(II), W(II), Ru(II), Rh(I), and Ir(I). The labile ligand itself is often an arene, as shown in Scheme 1b, in which case the replacement is an intramolecular version of the intermolecular exchange between free and coordinated arenes, which has been studied for complexes of the type [M(CO)$_3$(η^6-arene)] (M = Cr, Mo),[26–31] [Ir(η^6-arene)(η^4-1,5-COD)]BF$_4$,[28,32] and [Ru(η^6-arene)(η^4-1,5-COD)][28] (COD = cyclooctadiene).

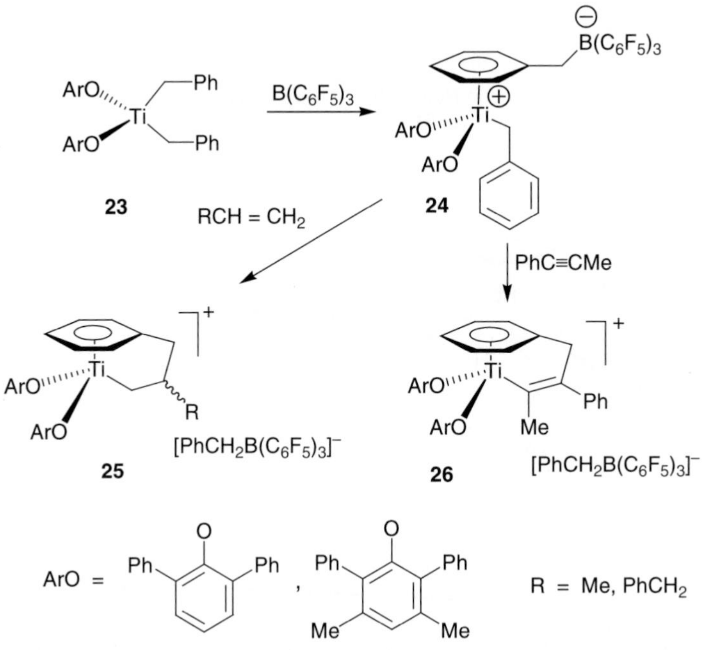

SCHEME 6.

The only examples of tethered arene complexes of titanium have been made by an approach in which a donor atom (carbon, in this case) is first attached to the metal, although it was then necessary to construct the tether by a procedure like that in Scheme 1c. As shown in Scheme 6, addition of $B(C_6F_5)_3$ to the titanium bis(benzyl) complexes **23** generates zwitterionic η^1-benzyl complexes **24** containing non-tethered $[\eta^6\text{-}C_6H_5CH_2B(C_6F_5)_3]^-$. These react with primary alkenes, such as propene or allylbenzene, to give the tethered complexes **25**. Spectroscopic studies show that these are formed by insertion of the alkene into the Ti-η^1-CH_2Ph bond and displacement of $[\eta^6\text{-}C_6H_5CH_2B(C_6F_5)_3]^-$ by the dangling arene of the resulting carbon chain.[32,33] A similar tethered complex **26** is obtained by use of phenylmethylacetylene in place of the 1-alkene.

The complexes of 1,1-bis(diphenylarsino)methane (dpam, $Ph_2AsCH_2AsPh_2$) with Cr(0) and Mo(0) were the first tethered arene complexes to be described (Scheme 7). The first products of heating $Cr(CO)_6$ or $Mo(CO)_6$ with dpam in decane are the arsenic-coordinated complexes $[M(CO)_5(Ph_2AsCH_2AsPh_2\text{-}\kappa As)]$ (**27a**, **27b**), which then lose CO to form the tethered species $[M(CO)_2\{\eta^6\text{-}C_6H_5As(Ph)CH_2AsPh_2\text{-}\kappa As\}]$ [M = Cr (**28a**), Mo (**28b**)].[34,35] Complex **28b** is also formed on heating $[Mo(CO)_4(Ph_2AsCH_2AsPh_2)_2\text{-}\chi^2 \cdot As]$ (**29**) in decane or *o*-xylene. After the first arsenic atom has coordinated, the phenyl group attached to the second arsenic atom evidently competes successfully with that arsenic atom for access to the metal atom. In contrast, under similar conditions, the stronger donor 1,1-bis(diphenylphosphino)methane (dppm, $Ph_2PCH_2PPh_2$) gives the P-coordinated

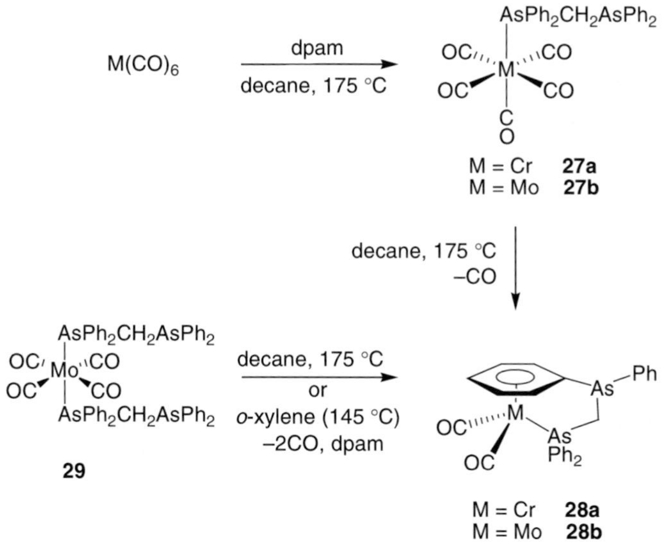

SCHEME 7.

chelate complex [M(CO)$_4$(Ph$_2$PCH$_2$PPh$_2$-$\kappa^2 P$)], and no tethered arene complexes are formed.[36]

Some examples have been reported of tethered η^6-arene complexes of molybdenum(II) and tungsten(II) that incorporate oxygen as the auxiliary donor atom.[37–40] Reaction of neat 2,6-diphenylphenol with [MH$_4$(PMePh$_2$)$_4$] (M = Mo, W) at 150 °C gives the tethered hydrido(aryloxo)metal(II) complexes **30** and **31**.[36] A metal–hydride bond is probably first cleaved by the phenolic oxygen with loss of H$_2$; coordination of the pendant arene then induces loss of another equivalent of H$_2$.

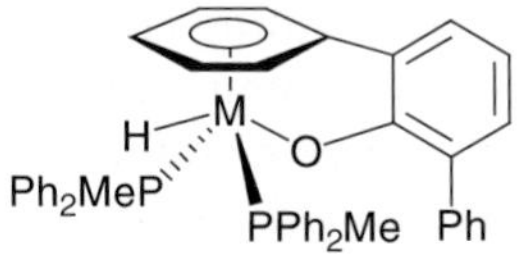

An alternative approach has been adopted with more highly substituted phenols. Reaction of WCl$_6$ with 2,3,5,6-tetraphenylphenol gives the tungsten(VI) aryloxide derivative **32**. As shown in Scheme 8, on reduction with sodium amalgam in the presence of PMe$_2$Ph or PMePh$_2$, one of the phenyl substituents of the aryloxide coordinates to the metal, thus forming the tethered arene–metal(II) complexes **33a** (L = PMe$_2$Ph) and **33b** (L = PMePh$_2$).[38] Under an atmosphere of dihydrogen, two of the formal double bonds of the *ortho*-phenyl group of the non-tethered aryloxide

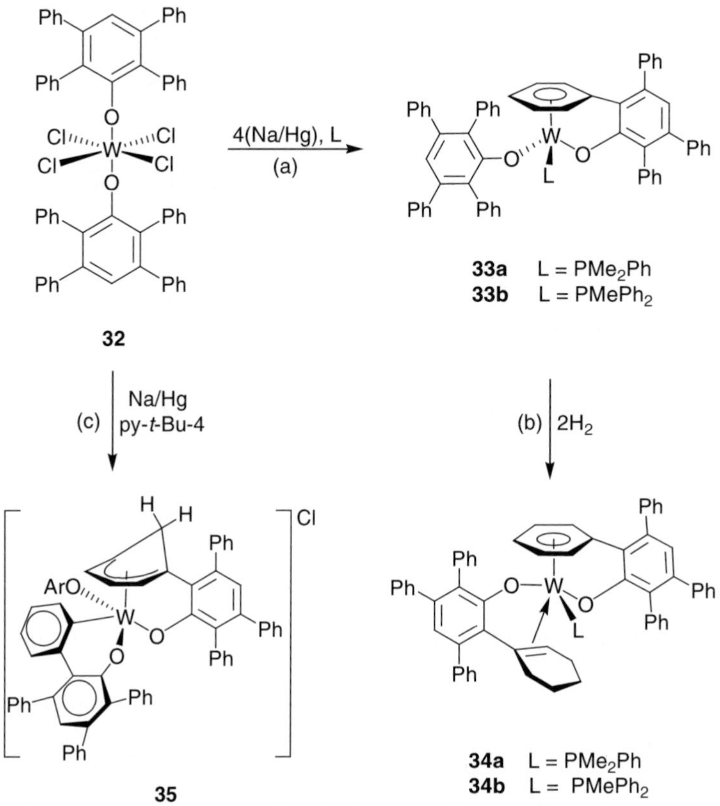

SCHEME 8.

are hydrogenated to give the corresponding chelate η^2-cyclohexene complexes **34a** and **34b**. When **32** is reduced with sodium amalgam in the presence of 4-*t*-butylpyridine, a second complex is obtained in addition to the expected analogue of **33a** and **33b** having L = 4-^{t}BuC_5H_4N. The additional product is the tethered η^5-cyclohexadienyl tungsten(VI) complex **35**, which contains three differently bound aryloxides: one terminally bound, one cyclometalated, and one chelated *via* the η^5-cyclohexadienyl interaction.[40] This is thought to be formed *via* a cyclometalated intermediate [WHCl(η^1-C_6H_4-2,3,5-C_6HPh_3O-κO)(OC_6HPh_4-2,3,5,6)], which undergoes hydride transfer to one of the *ortho*-phenyl rings.

The ability of an appended arene to compete with another potential donor in the same molecule is also illustrated by the coordination behaviour of the 2,6-diphenylbenzenethiolate anion. This reacts with [$MoBr_2(CO)_4$]$_2$ to give the tethered arene–molybdenum(II) complex **36**[41] and with [$RuCl_2(PPh_3)_3$] to give the similar ruthenium(II) complex **37**.[42] Coordination of thiolate anion probably precedes that of the arene in both cases. The triply strapped arene–ruthenium(II) complexes **38** and **39** incorporating tripodal, heterocyclic N-donor ligands have been prepared by reaction of the ligands with [$RuCl_2(DMSO)_4$] in the presence of NH_4PF_6.[43] In the first step, it seems likely that one or more of the nitrogen atoms enters the coordination sphere before the arene.

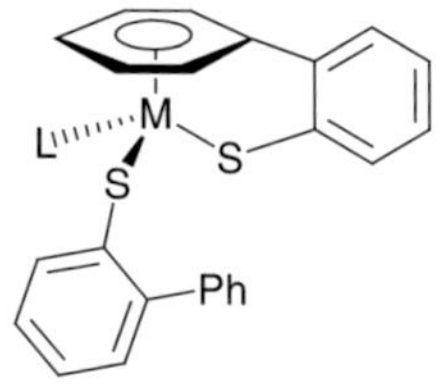

M = Mo, L = CO **36**
M = Ru, L = PPh_3 **37**

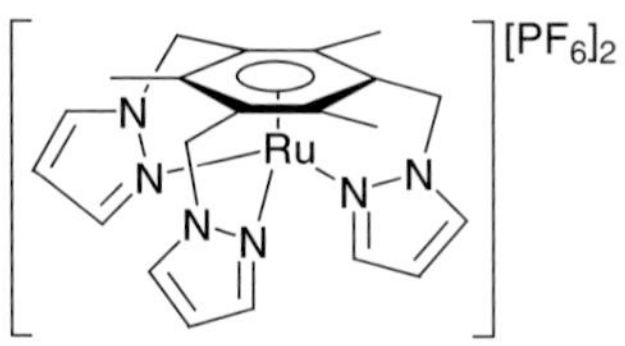

38

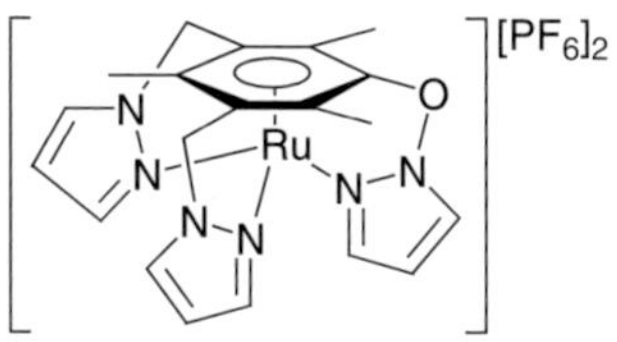

39

The many tethered arene complexes in the literature that contain phosphorus as the auxiliary donor atom have usually been made by the synthetic route of Scheme 1b. The ligands $Ph_2P(CH_2)_2X$-*p*-C_6H_4Y (X = O, Y = H; X = CH_2, Y = H, F) react with the labile rhodium(I) salt $[Rh(THF)_2(\eta^2\text{-}C_8H_{14})_2]BF_4$ to form the tethered arene–rhodium(I) salts **40–43**.[44,45] For the ether complex **40**, 2D ^1H NMR EXSY studies in the temperature range 10–30 °C in CD_2Cl_2 have shown that the free and bound arenes exchange, the activation parameters at 293 K being $\Delta G^{\neq} = 74.1\,kJ\,mol^{-1}$, $\Delta H^{\neq} = 83.3\,kJ\,mol^{-1}$ and $\Delta S^{\neq} = 25.5\,J\,mol^{-1}\,K^{-1}$. Since complex **41** having no ether function does not behave similarly, it is likely that the exchange proceeds *via* an ether-coordinated intermediate, [*cis*-$Rh(Ph_2PCH_2CH_2OPh)_2$-$\kappa O,\kappa P$]$^+$, in which both arenes are uncoordinated. Complex **42** does show arene-for-arene exchange at 60 °C in CD_2ClCD_2Cl, the $\Delta G^{\neq}$ value being ca. $23\,kJ\,mol^{-1}$ greater than for **40**, presumably as a result of the weakening of arene–Rh^+ binding by the electrophilic substituent. Complex **43** containing the redox-active ferrocenyl (Fc, $C_5H_5FeC_5H_4$) substituent also undergoes arene-for-arene exchange.[46,47] It can be oxidized at the iron centres in two successive one-electron steps, the ferrocenyl group attached to the dangling arene being oxidized before that on the complexed arene. The $\Delta G^{\neq}$ values for exchange of the arene groups are not greatly different for **43**, [**43**]$^+$, and [**43**]$^{2+}$, though this feature masks significant differences in the $\Delta H^{\neq}$ and $\Delta S^{\neq}$ values, especially for [**43**]$^{2+}$. The coordinated arenes of **40**–**43** are readily displaced by acetonitrile to give $[Rh(NCMe)_2(Ph_2PCH_2CH_2XPh)_2\text{-}\kappa P]^+$ or $[Rh(NCMe)_2(Ph_2PCH_2CH_2OFc)_2]^+$ and by CO to give similar P-coordinated tricarbonylrhodium(I) cations, which are in equilibrium with the corresponding dicarbonyls in solution. Labilization of the weakly bound arene in [**43**]$^{2+}$ increases the affinity of the rhodium centre for acetonitrile and also increases the catalytic activity for the isomerization of allyl ethyl ether. All the tethered arene–rhodium(I) cations undergo reversible one-electron oxidation by cyclic voltammetry with $E_{1/2}$-values in the range 515–630 mV vs. FcH/FcH^+, which suggests that the corresponding strapped arene–rhodium(II) cations may be chemically accessible.

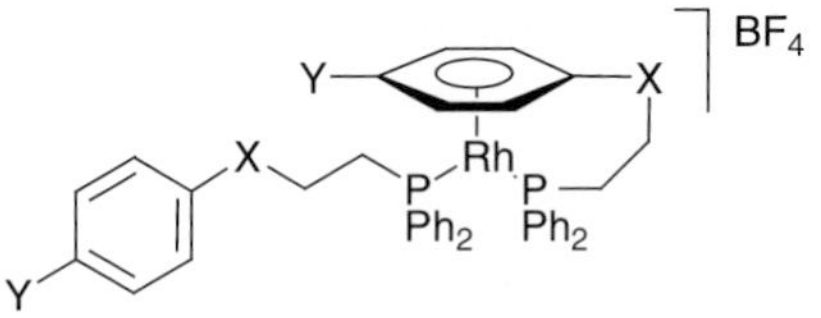

X = O, Y = H **40**
X = CH_2, Y = H **41**
X = CH_2, Y = F **42**
X = O, Y = Fc **43**

A similar synthetic approach has been adopted to synthesize rhodium(I) complexes of tethered arenes having bulky substituents on phosphorus. The labile rhodium(I) cation $[Rh(acetone)_2(\eta^2\text{-}C_8H_{14})_2]^+$ reacts with $C_6H_5CH_2CH_2P^iPr_2$ to give successively a mono-complex **44** containing cyclooctene and a bis-complex, the latter being analogous to **41**. The cyclooctene in **44** is displaced by ethene, maleic anhydride, or ethyl propiolate to give complexes **45a**–**45c** without displacement of the η^6-arene [Eq. (7)].[48] Ethene also displaces cyclooctene from the complex $[Rh(\eta^6\text{-}C_6H_5CH_2CH_2CH_2P^iPr_2\text{-}\kappa P)(\eta^2\text{-}C_8H_{14})]PF_6$. The corresponding complexes formed by the *t*-butyl substituted ligands $C_6H_5CH_2CH_2P^tBu_2$ and $C_6H_5OCH_2CH_2P^tBu_2$ are more stable in solution and as solids than their isopropyl analogues, probably owing to greater steric shielding of the rhodium centre.

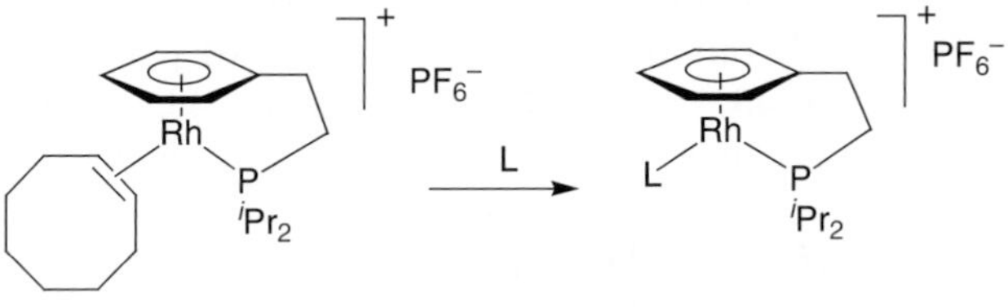

(7)

44

45a L = C_2H_4
45b L = maleic anhydride
45c L = $HC{\equiv}CCO_2Et$

The presence of bulky substituents on phosphorus can also induce metalation of the aromatic ring in the tether. Treatment of $[RhCl(\eta^2\text{-alkene})_2]$ (alkene = C_8H_{14}, C_2H_4) with $C_6H_5CH_2CH_2P^tBu_2$ gives the five-coordinate hydrido(aryl) complex **46**, which is a distorted trigonal bipyramid with axial phosphorus atoms.[49] The metalation is reversed when **46** reacts with CO and phenylacetylene to give, respectively, the planar rhodium(I) carbonyl and phenylvinylidene complexes of the P-coordinated ligand (Scheme 9); dihydrogen cleaves the Rh–C σ-bond forming the P-coordinated dihydrido complex **47**. Treatment of **46** with $AgPF_6$ also reverses the metalation leading to the salt **48**, which, like **41**, contains one η^6-arene-coordinated tethered ligand and one monodentate, P-coordinated ligand.

Although the iridium(I) analogue of **44** can be made similarly from $[Ir(acetone)_2(\eta^2\text{-}C_8H_{14})_2]PF_6$, a more useful precursor is the dimer $[Ir(\mu\text{-OMe})(1,5\text{-COD})]_2$.[50] This reacts with the phosphonium salt $[PhCH_2CH_2P^iPr_2H]BF_4$ in acetone to give a P-coordinated salt $[Ir(acetone)(1,5\text{-COD})(^iPr_2PCH_2CH_2Ph)]BF_4$, which, under dihydrogen, gives the dihydridoiridium(III) salt **49** containing the tethered phosphine (Scheme 10). A similar procedure can be used to prepare the analogue of **49** containing $^tBu_2PCH_2CH_2OPh$, the intermediate in this case being the ether-coordinated salt $[Ir(^tBu_2PCH_2CH_2OPh\text{-}\chi^2 O,P)(1,5\text{-COD})]BF_4$. The

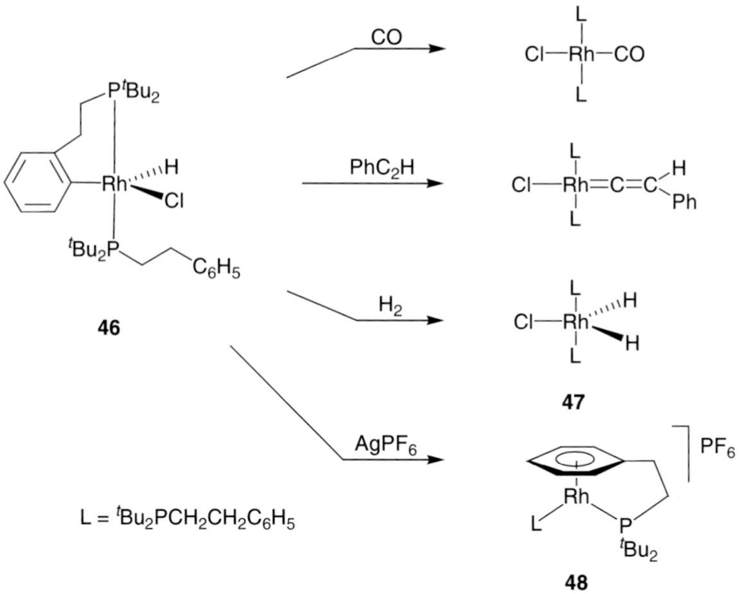

46

CO → $Cl{-}Rh(L)_2{-}CO$

PhC_2H → $Cl{-}Rh(L)_2{=}C{=}C(H)Ph$

H_2 → $ClRh(L)_2H_2$ **47**

$AgPF_6$ → **48** PF_6

L = $^tBu_2PCH_2CH_2C_6H_5$

Scheme 9.

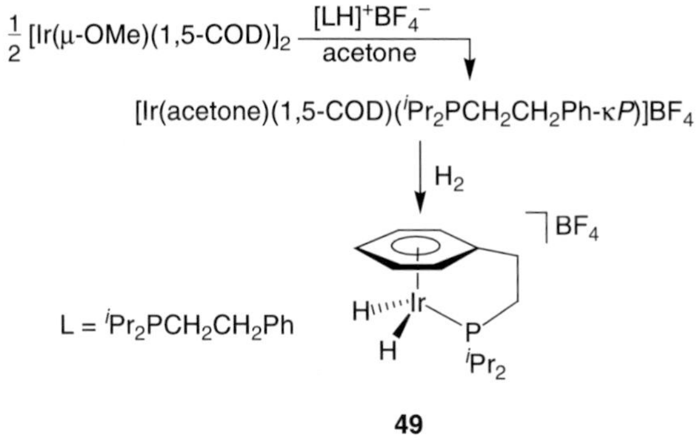

$\frac{1}{2}$ [Ir(μ-OMe)(1,5-COD)]$_2$ $\xrightarrow[\text{acetone}]{[LH]^+BF_4^-}$ [Ir(acetone)(1,5-COD)(^{i}Pr$_2$PCH$_2$CH$_2$Ph-κ*P*)]BF$_4$ $\xrightarrow{H_2}$ **49** BF_4

L = $^iPr_2PCH_2CH_2Ph$

Scheme 10.

dihydrido complexes stoichiometrically hydrogenate ethene or propene to give the corresponding iridium(I)–alkene complexes, e.g., **50a**, **50b**, the reaction being reversed under hydrogen; a similar reaction occurs with diphenylacetylene to give the analogous alkyne complex **51**, the *Z*-stilbene complex **52** being detectable as an intermediate (Scheme 11). The X-ray structure of **51** resembles that of **45a** in that the arene has a slightly inverse boat conformation, the *ipso*-carbon (and, to a lesser extent, the carbon atom *para* to it) being bent towards the metal atom. The coordinated arene in the dihydridoiridium(III) salt **49** is displaced readily by acetonitrile to give *fac*-[IrH$_2$(MeCN)$_3$(^{i}Pr$_2$PCH$_2$CH$_2$Ph-χ^2*P*)]BF$_4$, whereas when the propene complex **50b** is treated with acetonitrile the arene ring undergoes metalation forming

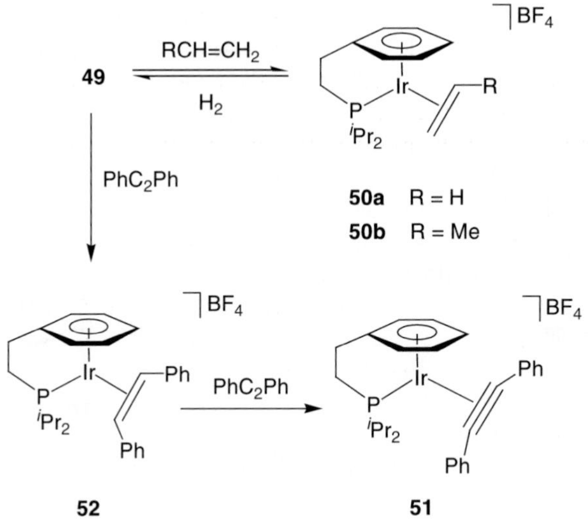

SCHEME 11.

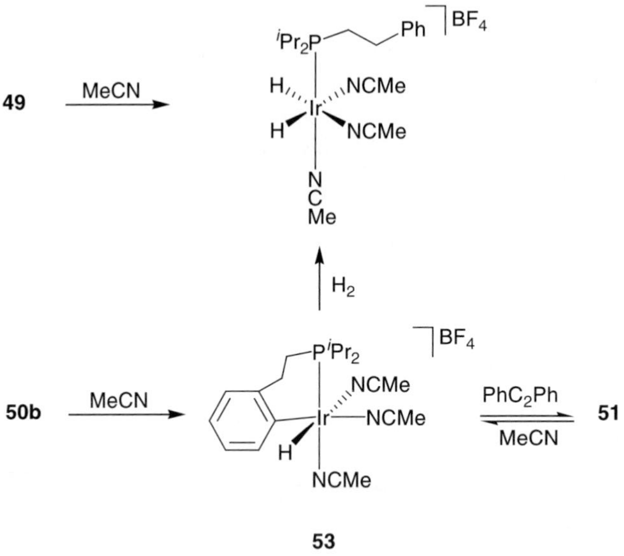

SCHEME 12.

the hydrido(aryl) salt **53** (Scheme 12). The metal-carbon bond in **53** is cleaved by dihydrogen and the metalation can be partly reversed by addition of diphenylacetylene, which re-forms complex **51**.[50]

A large number of strapped arene complexes of ruthenium(II) have been prepared by treatment of $[RuCl_2(\eta^6\text{-arene})]_2$ complexes with tertiary phosphines of the type $R_2P{\sim}Ar$ in which an aromatic group is separated from the phosphorus atom by a two- or three-atom chain. In the first step, which usually occurs at or just above room temperature, the chloride bridges of the dimer are broken to give a half-sandwich, P-coordinated complex [$RuCl_2(\eta^6$-arene)($R_2P{\sim}Ar$)]. Subsequent heating causes displacement of the original arene by the pendant arene, as shown in Scheme 1b. The conditions under which the second step occurs in reasonable yield depend markedly on the nature of the arene in the original dimer, the number of atoms in the tether, and especially on the steric bulk of the substituents on phosphorus.

A readily available and commonly used precursor is the *p*-cymene (cym, 1,4-$MeC_6H_4CHMe_2$) complex [$RuCl_2(\eta^6$-cym)]$_2$, which reacts with arene–tertiary phosphines having two connecting atoms, such as $C_6H_5CH_2CH_2PR_2$ (R = Et, Ph, Cy, ^{t}Bu), to give initially the P-coordinated complexes [$RuCl_2$(η^6-cym)($P{\sim}C_6H_5$)]; on heating in chlorobenzene at 130 °C these are converted into the tethered complexes **54**–**57** in yields of 60–95%.[51–53] The corresponding complex **58** containing $C_6H_5SiMe_2CH_2PPh_2$ can be made similarly from the *p*-cymene P-donor precursor by heating the latter in CH_2Cl_2/THF at 120 °C for 168 h.[54] On the other hand, Therrien *et al.*[55] obtained less than 5% of the tethered complex of 1,2-$C_6H_4(CH_2OH)(CH_2CH_2PPh_2)$ by heating or UV-irradiation of its *p*-cymene P-donor complex.

When there are three atoms in the tether, the *p*-cymene complex is generally a satisfactory precursor only when the substituents on phosphorus are bulky, e.g., cyclohexyl, isopropyl, or *t*-butyl. For example, the ligand $C_6H_5(CH_2)_3PCy_2$ forms the tethered complex **59a** in 91% isolated yield by heating the initially formed *p*-cymene complex [$RuCl_2(\eta^6$-cym){$Cy_2P(CH_2)_3C_6H_5$-κ*P*}] in chlorobenzene at 130 °C for 18 h,[56,57] or in ca. 80% isolated yield in CH_2Cl_2 at 120 °C for 48 h.[54] The methyl-substituted derivatives **59b**, **59c** have also been made in good yield by the first procedure and the $C_6H_5(CH_2)_3P^iPr_2$ tethered complex **60** by the second procedure (42 h heating, 90% yield).[54]

The use of refluxing chlorobenzene to bring about the second step (chelation) of Scheme 1b for arene–ruthenium(II) complexes was introduced by Smith and Wright,[58] who reported the isolation, in 50% yield, of the first tethered arene–phosphine complex of ruthenium(II), [$RuCl_2${η^6-$C_6H_5(CH_2)_3PPh_2$-κ*P*}] (**61**) from [$RuCl_2(\eta^6$-cym){$C_6H_5(CH_2)_3PPh_2$-κ*P*}] by this method. Unfortunately, this result has proved to be irreproducible.[59,60] According to NMR studies, *p*-cymene is displaced when [$RuCl_2(\eta^6$-cym){$C_6H_5(CH_2)_3PPh_2$-κ*P*}] is heated in C_6D_5Cl at 130 °C, but little or no **61** can be detected. Attempts to form it by heating the precursor in CH_2Cl_2 or CH_2Cl_2/THF also failed. Smith and Wright[58] have reported an alternative, but experimentally less convenient, preparation of **61** in which a solution of [$RuCl_2(\eta^6$-cym){$C_6H_5(CH_2)_3PPh_2$-κ*P*}] is electrolysed at +1.5 V (*vs.* SCE), thus presumably generating a labile ruthenium(III) species that readily loses *p*-cymene. Reduction at +1.0 V (*vs.* SCE) then forms **61** in 75% yield.

To prepare complexes containing either two or three atoms in the tether, the best precursor is the dichloro(η^6-arene)ruthenium(II) complex of an aromatic ester such as ethyl benzoate[55] or methyl *o*-toluate.[61] When coordinated to ruthenium(II), these ligands are more labile than benzene or *p*-cymene.[12] As in the case of *p*-cymene, the first product is the P-coordinated complex, although this does not need to be isolated. For example, ethyl benzoate is displaced from [$RuCl_2(\eta^6$-$C_6H_5CO_2Et$){1,2-$C_6H_4(CH_2OH)(CH_2CH_2PPh_2)$-κ*P*}] in CH_2Cl_2 at 120 °C to give, after 24 h, the tethered complex [$RuCl_2\{\eta^6$-1,2-$C_6H_4(CH_2OH)(CH_2CH_2PPh_2)$-κ$P\}$] in 97% yield. The corresponding 1,3-$C_6H_4(CH_2OH)(CH_2CH_2PPh_2)$ derivative is prepared similarly; its structure has been confirmed by X-ray crystallography.[55] The corresponding tethered complexes derived from $C_6H_5CH_2CH_2PR_2$ [R = Ph (**55**), Cy (**56**)] and $(C_6H_5)_2CHCH_2PR_2$ (R = Ph, Cy) are formed from [$RuCl_2(\eta^6$-$C_6H_5CO_2Et)]_2$ in high yield even in refluxing chloroform.[62] The tethered complex **58** derived from $C_6H_5SiMe_2CH_2PPh_2$ is obtained from its methyl *o*-toluate P-donor precursor in CH_2Cl_2/THF at 120 °C after 36 h in 71% yield (cf. 168 h for the *p*-cymene precursor).[54,60] The tethered complexes derived from $C_6H_5(CH_2)_3PMe_2$ (**62**), $C_6H_5(CH_2)_3PPh_2$ (**61**), $C_6H_5(CH_2)_3P^iPr_2$ (**60**), $C_6H_5(CH_2)_3P^tBu_2$ (**63**), and $C_6H_5(CH_2)_3PCy_2$ (**59a**) are formed in similarly good yields from the P-donor adducts with [$RuCl_2(\eta^6$-1,2-$MeC_6H_4CO_2Me)]_2$ in CH_2Cl_2; the presence of THF serves to shorten reaction times.[54,59,60] This procedure has been extended to the synthesis of tethered complexes in which the entering arene is substituted with methyl groups but the yields are much poorer. Thus, the pentamethylphenyl compound **64** is formed in only 7% yield when the precursor [$RuCl_2(\eta^6$-1,2-$MeC_6H_4CO_2Me)$ $\{Ph_2P(CH_2)_3C_6Me_5$-κ$P\}$] is heated in CH_2Cl_2/THF at 120 °C for 24 h, but the yield can be increased to 35% by use of di-*n*-butyl ether as solvent (140 °C, 16 h). The yield of the 2,4,6-trimethylphenyl compound **65** is only 18% in CH_2Cl_2/THF at 120 °C, and the yields in CH_2Cl_2 alone or in di-*n*-butyl ether are even poorer.[54]

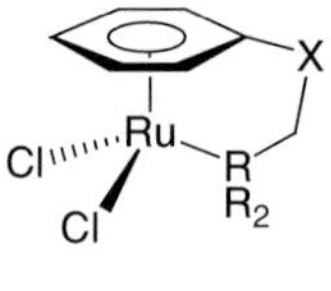

54 X = CH_2, R = Et
55 X = CH_2, R = Ph
56 X = CH_2, R = Cy
57 X = CH_2, R = ^{t}Bu
58 X = $SiMe_2$, R = Ph

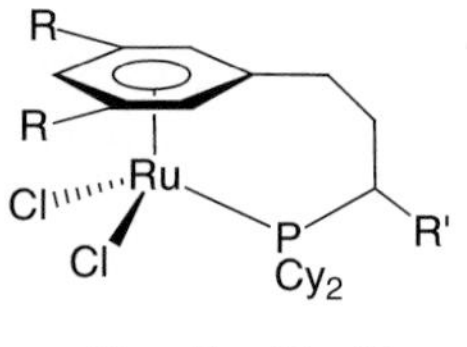

59a R = R' = H
59b R = Me, R' = H
59c R = H, R' = Me,

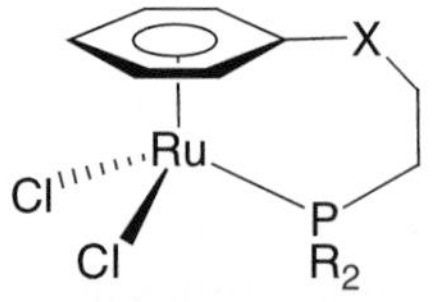

60 X = CH_2, R = ^{i}Pr
61 X = CH_2, R = Ph
62 X = CH_2, R = Me
63 X = CH_2; R = ^{t}Bu
66 X = O; R = ^{t}Bu

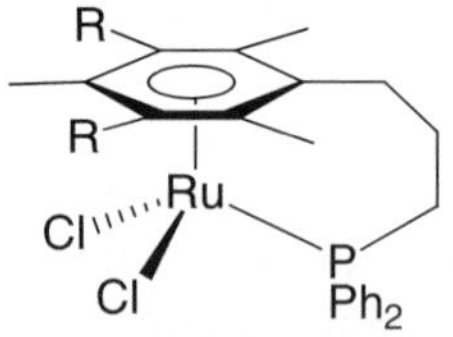

64 R = Me
65 R = H

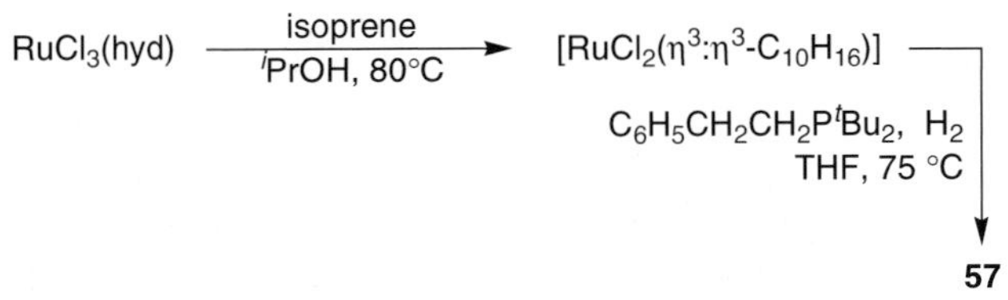

$RuCl_3(hyd)$ —isoprene, ^iPrOH, 80°C→ $[RuCl_2(\eta^3:\eta^3\text{-}C_{10}H_{16})]$ —$C_6H_5CH_2CH_2P^tBu_2$, H_2, THF, 75 °C→ **57**

SCHEME 13.

The formation of tethered arene–phosphine complexes of ruthenium(II) is clearly favoured by the presence of bulky substituents on phosphorus. This effect is so marked for tert-butyl that the $^tBu_2P$ complex **57** can be made directly from hydrated $RuCl_3$ by heating the latter with isoprene in 2-propanol and then heating the intermediate $[RuCl_2(\eta^3:\eta^3\text{-}C_{10}H_{16})]$ *in situ* with $C_6H_5CH_2CH_2P^tBu_2$ in THF under an atmosphere of dihydrogen at 75 °C (Scheme 13).[53] The tethered complex **66** derived from $C_6H_5OCH_2CH_2P^tBu_2$ can be made similarly. This effect of bulky substituents is reminiscent of the effect of P^tBu_2 in facilitating the cyclometalation of transition metal complexes, which, by analogy with the Thorpe–Ingold *gem*-dimethyl effect in organic chemistry, has been attributed to a combination of entropic and enthalpic effects, viz., a reduction in the loss of rotational entropy and in destabilizing gauche-interactions attendant on cyclization.[63,64] For further discussion, see Section III.

Tethered arene complexes of ruthenium(II) are also formed readily when the conformational freedom of the tether is restricted, for which purpose the biphenyl unit has proved ideal. Thus, (2-dicyclohexylphosphino)biphenyl and its derivatives react with $[RuCl_2(\eta^6\text{-}C_6H_6)]_2$ in DMF at 100 °C to give directly the tethered complexes **67**–**69** in yields of 31–96% without detection or isolation of the presumed P-coordinated intermediates.[65]

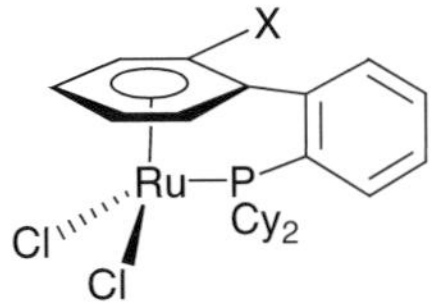

X; Ru—P; Cy_2; Cl; Cl

67 X = H
68 X = Me
69 X = NMe_2

Similar complexes are formed by acid-promoted P–C bond cleavage in the bis(acetato)ruthenium(II) complexes of atropisomeric, substituted biphenyl (Biphep) and binaphthyl (Binap) ditertiary phosphines that have found wide application as homogeneous catalysts for various enantioselective organic reactions, particularly hydrogenation. The topic has been reviewed.[66] For example, reaction of the MeO–Biphep complexes **70a**–**70c** with two equiv. of HBF_4 gives the tethered arene complexes **71a**–**71c** as a consequence of P–C bond cleavage and subsequent hydrolysis [Eq. (8)].[67] These complexes contain a novel phosphinite anion, $[R_2POBF_3(OH)]^-$, which is stabilized by hydrogen-bonding to BF_4^-. To avoid the

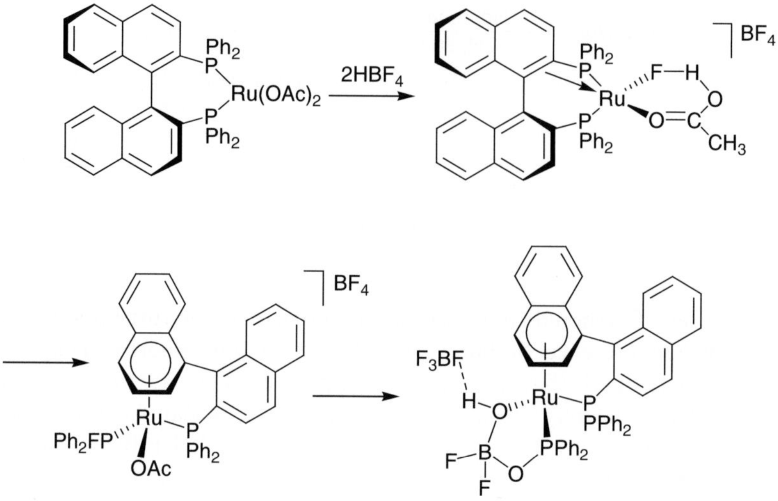

SCHEME 14.

complication of BF_4^- hydrolysis, complex **70b** has been heated with wet trifluoromethane sulfonic acid, which gives the (triflato)(hydroxydiphenylphosphine) complex **72** [Eq. (9)].[68] Formally, acetate is removed by protonation and water adds across the P–C bond, the process being assisted by pre-coordination of one of the formal double bonds adjacent to PPh_2 and the C–C single bond. Similar reactions occur in the Binap system and the sequence of events in the reaction of HBF_4 with $[Ru(OAc)_2(Binap)]$ revealed by NMR spectroscopy is as shown in Scheme 14. In both Binap and Biphep series, the PPh_2OH complexes undergo phenyl migration from P to Ru on reaction with alcohols (ROH) to give P(OH)(OR)Ph complexes as just one diastereomer [Eq. (10)];[69] with tert-butanol/water/THF mixtures the dimeric species **73** containing μ-$PO(OH)_2$ is obtained.[70] Triflato complexes analogous to **72** in the Binap series with PCy_2 and P^iPr_2 in place of PPh_2 have also been reported; these complexes react slowly with methanol to give the corresponding cationic hydrido compounds.[71]

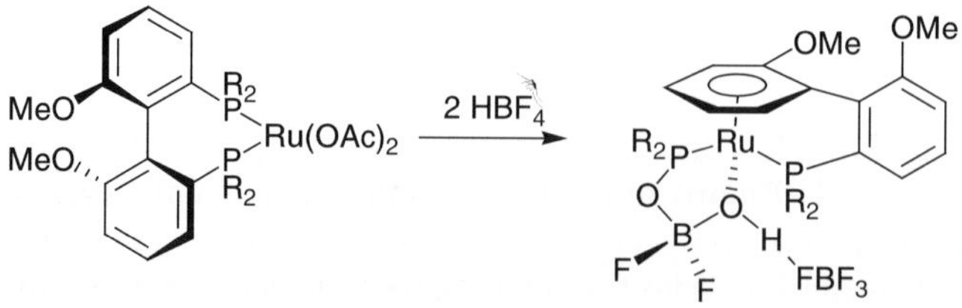

(8)

70a $R = 3,5\text{-}C_6H_3{}^tBu_2$
70b $R = C_6H_5$
70c $R = 4\text{-}C_6H_4CH_3$

71a $R = 3,5\text{-}C_6H_3{}^tBu_2$
71b $R = C_6H_5$
71c $R = 4\text{-}C_6H_4CH_3$

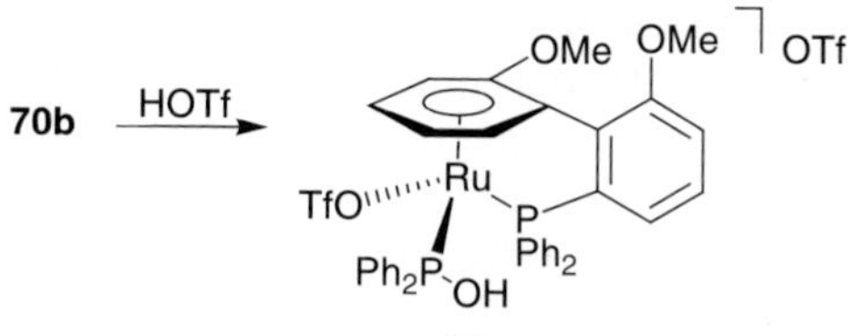

(9)

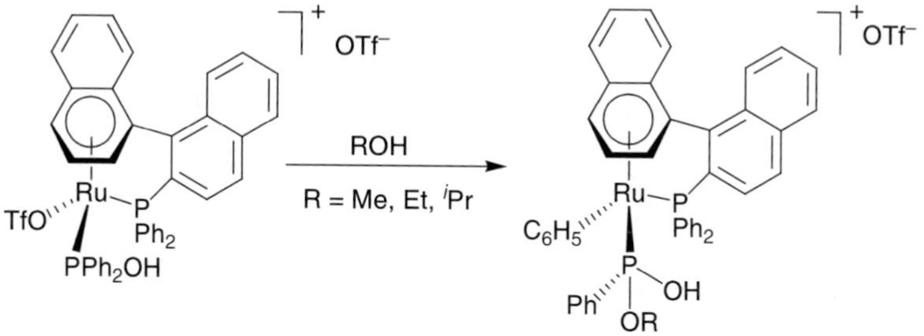

(10)

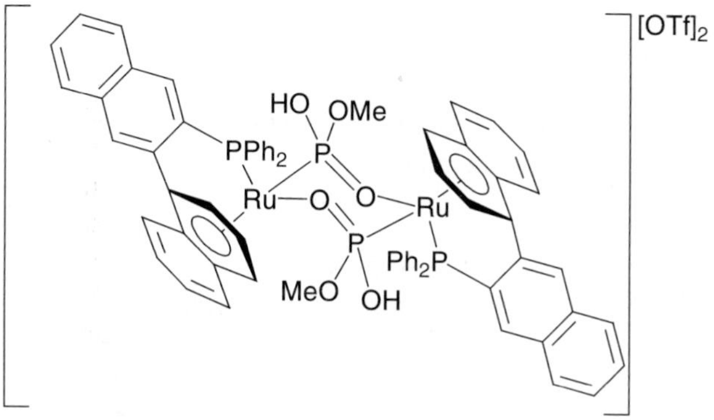

73

Tethered arene complexes can also be generated in the MeO–Biphep series in reactions that are not acid-promoted. Treatment of the bis(acetato) complex **70a** with three equivalents of alkyllithium reagents gives the bis(alkyl) complexes **74a**, **74b** by cleavage of one of the Ru–P bonds:[72]

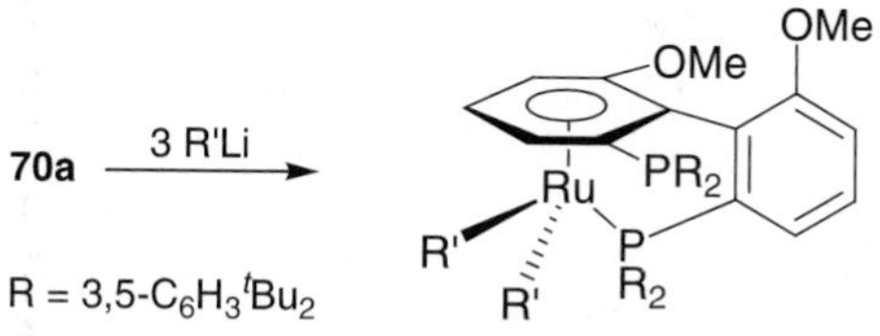

(11)

74a R' = Me
74b R' = CH_2SiMe_3

Displacement of a coordinated arene by a pendant arene (Scheme 1b) has been used to prepare tethered arene complexes in which the auxiliary ligand is a carbene.

When $[RuCl_2(\eta^6\text{-cym})]_2$ is heated in toluene at 100 °C with the carbenes **75** and **76**, *p*-cymene is displaced and the tethered complexes **77** and **78** are formed in high yield [Eq. (12)]. Complex **77** can also be made by heating **75** with $[RuCl_2(\eta^6\text{-}C_6Me_6)]_2$ in xylenes at 140 °C; in this case, use of toluene at 100 °C gives an non-tethered intermediate containing C_6Me_6 and the coordinated carbene.[73]

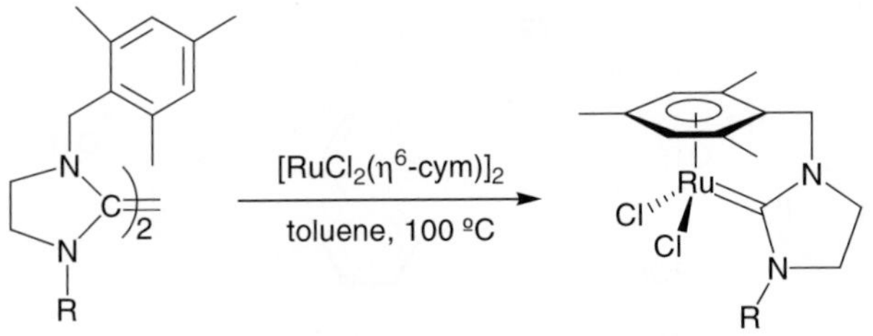

[RuCl₂(η⁶-cym)]₂, toluene, 100 ºC (12)

75 R = CH_2CH_2OMe
76 R = CH_2-2,4,6-$C_6H_2Me_3$

77 R = CH_2CH_2OMe
78 R = CH_2-2,4,6-$C_6H_2Me_3$

An example of a tethered arene complex of ruthenium(II) in which the auxiliary ligand is a σ-bonded carbon atom is complex **80**, which is formed by the action of $AgBF_4$ and $P(OMe)_3$ on the η^4-tetraphenylcyclobutadiene η^3-allyl complex **79** [Eq. (13)]. In the proposed mechanism, the allyl group migrates to the four-membered ring, which opens to generate an intermediate cation **81**, the pendant arene of which coordinates to the metal.[74]

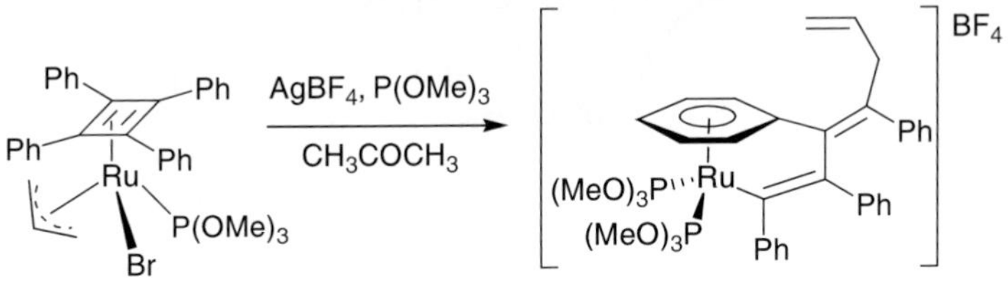

$AgBF_4$, $P(OMe)_3$, CH_3COCH_3 (13)

79 **80**

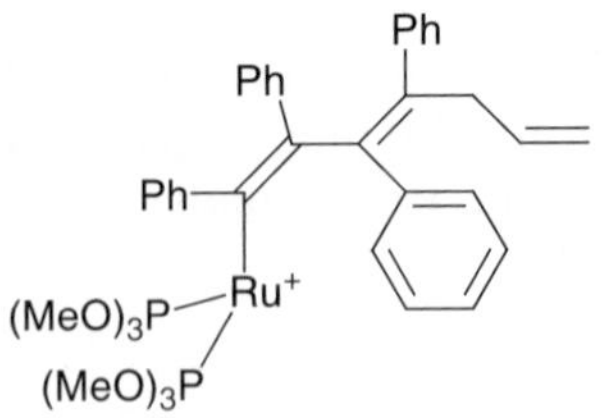

81

A series of cationic, strapped, η^6:η^1:η^1-arene complexes of ruthenium(II) (**82**) containing both carbon and phosphorus as the auxiliary atoms has been obtained by a coupling reaction of the iminophosphorane complexes **83** with an excess of 1, 1-diphenyl-2-propyn-1-ol, the *p*-cymene in **83** being displaced by one of the phenyl groups of the propargylic alcohol [Eq. (14)].[75] When the mesitylene analogue of **83** ($R_F = p$-C_5F_4N) is employed, an intermediate allenylidene complex **84** can be isolated in which the electrophilic NR_F group of **83** has dissociated from the metal; this then couples slowly with the allenylidene residue.

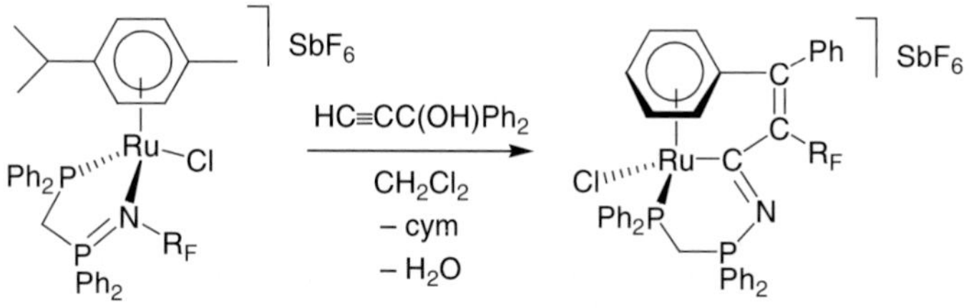

$$\mathbf{83}\ \xrightarrow[\text{CH}_2\text{Cl}_2,\ -\text{cym},\ -\text{H}_2\text{O}]{\text{HC}\equiv\text{CC(OH)Ph}_2}\ \mathbf{82} \qquad (14)$$

$R_F = p$-C_5F_4N, p-C_6F_4CN, p-C_6F_4CHO, C_6-1,2,5-F_3-3,4-$(CN)_2$

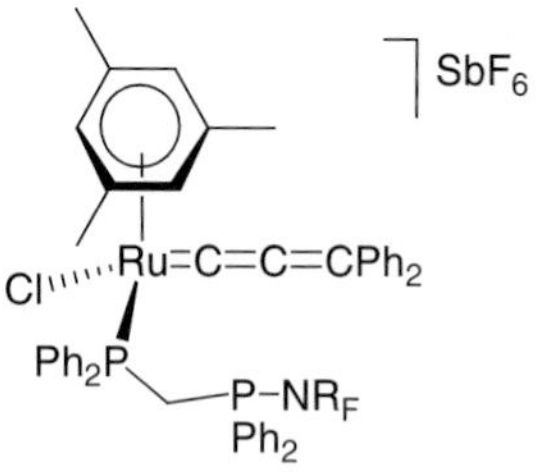

84

Unsuccessful attempts have been made to generate tethered arene derivatives of ruthenium carbonyl clusters by use of $C_6H_5(CH_2)_3PPh_2(L)$ to displace CO from appropriate precursors. The complexes $[Ru_6C(CO)_{16}L]$ and $[Ru_3(CO)_9L_3]$ contain the intact cluster core and the ligands are only P-bonded. In $[Ru_6C(CO)_{13}L]$, however, the ligand bridges two adjacent metal atoms *via* phosphorus and one of the P–Ph groups.[76,77]

D. *Syntheses via Intramolecular Construction of the Strap (Scheme 1c)*

Several examples are known of tethered arene ruthenium(II) complexes having either sulfur or phosphorus as the auxiliary donor atom in which the tether results from an intramolecular condensation between a substituent on the arene and a functional group on the donor atom. The dication **85**, which results from the reaction of $[RuCl_2(\eta^6\text{-}C_6Me_6)]_2$ with $AgPF_6$/acetone in the presence of the macrocyclic thioether 1,4,7-trithiacyclononane (9-aneS_3), undergoes three successive deprotonations on treatment with KO^tBu (>3 equivalents) (Scheme 15).[78,79] The first two lead to C–S bond cleavage and formation of the isolable intermediate **86**. In the last step, one of the methyl groups of η^6-C_6Me_6 is deprotonated and the

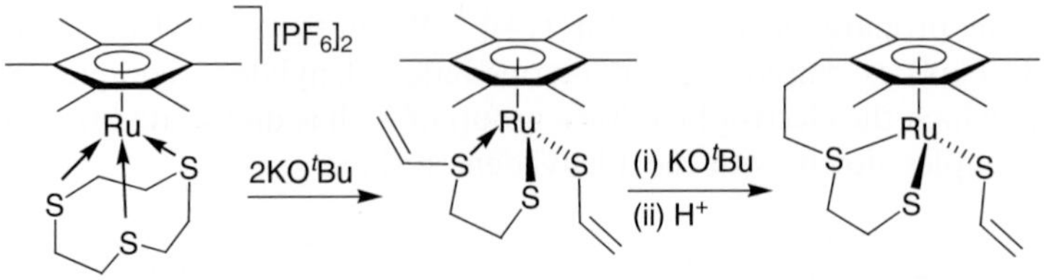

SCHEME 15.

resulting carbanion undergoes Michael addition to the vinyl group of the thioether–thiolate ligand $CH_2 = CHSCH_2CH_2S^-$ to give the tethered complex **87**.

A similar reaction sequence occurs when the corresponding dication containing the macrocycle 9-aneNS_2 is treated with an excess of KO^{t}Bu or KOH [Eq. (15)]. Even a suitably designed non-macrocyclic precursor can generate a tethered arene, as shown in Eq. (16).[80]

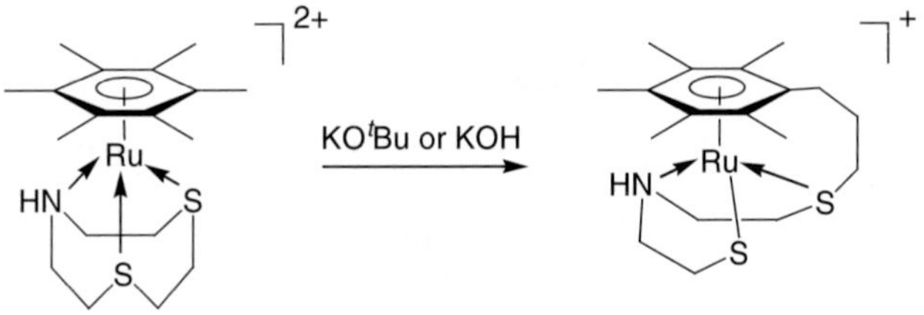

(15)

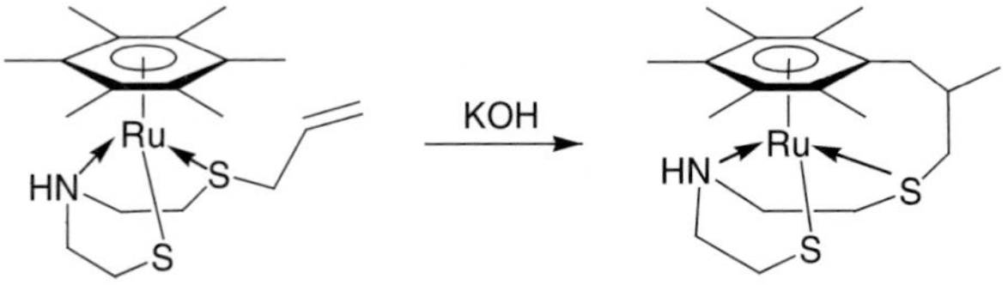

(16)

The same approach has been used to prepare tethered arene–phosphine and –arsine complexes of ruthenium(II) containing three carbon atoms in the strap. Treatment of the vinyldiphenylphosphine derivatives **88a**–**88e** of various methylarene ruthenium(II) complexes with KO^{t}Bu (1.0 mol per Ru) in refluxing acetonitrile for 48 h gives the corresponding tethered complexes **61**, **64**, and **89**–**91** [Eq. (17)].[81] Reported yields range from 48 to 70%. The precursors **88a**–**88e** can be generated *in situ* and do not need to be isolated. In the case of the *p*-cymene precursor, only the methyl group on the arene is deprotonated. This procedure provides an alternative approach to the tethered complex **61** of $C_6H_5(CH_2)_3PPh_2$ and it would appear to be the most convenient procedure for the synthesis of the (pentamethyl)phenyl complex **64** (see Section II.C). In the tertiary arsine series, however, yields are only ca. 20%.[82]

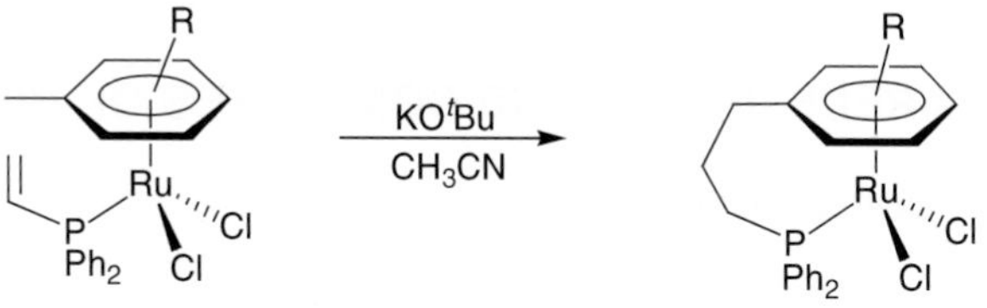

(17)

R			R	
R = 3,5-Me$_2$	**88a**		R = 3,5-Me$_2$	**89**
4-^{i}Pr	**88b**		4-^{i}Pr	**90**
Me$_5$	**88c**		Me$_5$	**64**
4-Me	**88d**		4-Me	**91**
H	**88e**		H	**61**

In a reaction that similarly relies on the acidity of benzylic protons in arene–ruthenium(II) complexes, the η^6-mesitylene complex **92** containing bidentate $(C_6F_5)_2PCH_2CH_2P(C_6F_5)_2$ loses two molecules of HF on treatment with proton sponge to give the di-strapped salt **93** in which each tether contains three carbon atoms [Eq. (18)]).[83]

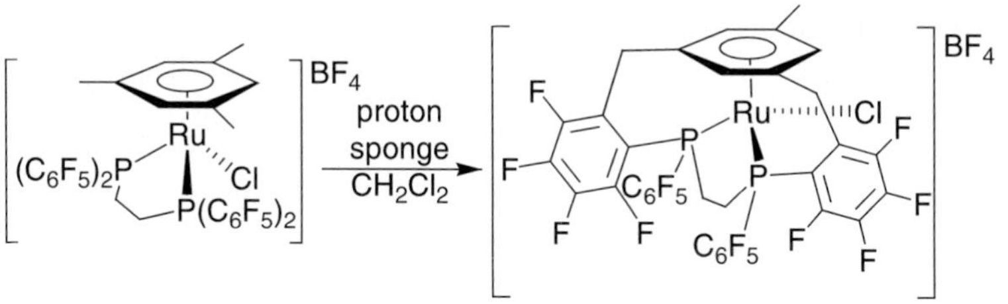

(18)

92 **93**

E. *Syntheses via Alkyne Condensation (Scheme 1d)*

So far as we know, this approach has been used only once, although it should be capable of extension. The octahedrally coordinated rhodiacyclopentadiene complex **94**, which results from the reaction of $[RhCl(PPh_3)_3]$ with 2 mol of (2-phenylethynyl)diphenylphosphine, reacts with diphenylacetylene to give the di-strapped arene–rhodium(I) complex **95**, from which the *p*-terphenyl-based bis(diphenylphosphine) can be released by heating with NaCN [Eq. (19)].[84]

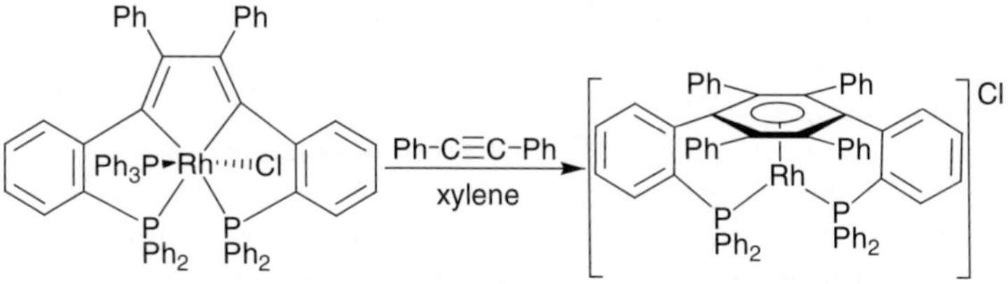

(19)

94 **95**

III

TETHERED ARENE–PHOSPHINE COMPLEXES OF RUTHENIUM(II)

A. *General Considerations*

It is clear from the preceding section that the field of tethered arene–metal complexes is dominated by ruthenium and by arene–phosphines as ligands. In part, this situation has arisen because of the current surge of interest in the catalytic properties of ruthenium complexes in organic synthesis.[85,86] Moreover, the tethered arene complexes are usually air-stable, crystalline solids with a well-defined, half-sandwich molecular geometry that, in principle, can lock the configuration at the metal centre. These compounds should, therefore, be ideal both for the study of the stereospecificity of reactions at the metal centre and for stereospecific catalysis.

B. *Structural Features*

The geometries observed from the numerous X-ray structural determinations are very similar to those of analogous non-tethered complexes, the inter-bond angles of the tripodal set of ligands being close to 90°. The centroid of the η^6-arene ring frequently does not lie directly above the metal centre but displacements are generally small (ca. 0.01–0.08 Å). In complexes containing a three atom-strap, the benzylic carbon atom is usually close to coplanar with the carbon atoms of the attached arene, without distortion of the bond lengths and angles in the tether. This may not be the case for complexes having two connecting groups in the tether, and the bond adjacent to the η^6-arene may be bent away from the mean plane of the η^6-arene ring, e.g., in [$RuCl_2(\eta^6$-$C_6H_5SiMe_2CH_2PPh_2$-κP)] (**58**), the angle for the Si–C(C_6H_5) bond is about 14°.[60]

The Ru–P bond lengths in tethered arene–phosphine complexes of ruthenium(II) generally fall in the range 2.30–2.35 Å and are significantly shorter, by as much as 0.07 Å, than those of η^6-arene complexes containing the same, non-tethered ligand. For example, the Ru–P bond lengths in [$RuCl_2$(η-cym){$C_6H_5(CH_2)_3P^iPr_2$-κP}] and [$RuCl_2${η^6-$C_6H_5(CH_2)_3P^iPr_2$-κP}] are 2.393(3) and 2.338(2) Å, respectively.[54] Bulky substituents on phosphorus tend to give rise to longer Ru–P bonds, the longest being in the tert-butyl derivatives [$RuCl_2(\eta^6$-$C_6H_5CH_2CH_2P^tBu_2$-κP)] [2.3976(13) Å][52] and [$RuCl_2${η^6-$C_6H_5(CH_2)_3P^tBu_2$-κP}] [2.413(9) Å].[54] A correlation between Ru–P bond length and the size of substituents on phosphorus has been observed in complexes of the type [$RuCl(\eta^5$-$Cp^*)(PR_3)_2$]; the Ru–P bond lengths increase in the order $R_3 = Me_3 < Me_2Ph < MePh_2 < Ph_3$.[87]

The complexed arene rings in tethered complexes of ruthenium(II) are close to planar, though the *ipso*-carbon atom is often pulled slightly towards the metal atom. In the phosphine complexes, the Ru–C(arene) distances *trans* to the P-donor (2.22–2.29 Å) are significantly greater than those *trans* to the Ru–Cl bonds (2.15–2.25 Å). This feature is also evident in non-tethered complexes of the type [$RuCl_2(\eta^6$-arene)(PR_3)] and can be attributed to the higher *trans*-influence of PR_3 relative to that of Cl^-.[88]

The most significant structural difference between tethered and non-tethered arene–phosphine complexes that contain the same phosphine lies in the conformation of the tether. In non-tethered complexes containing three methylene units between the arene and phosphorus, the five connected atoms P(1)–C(1)–C(2)–C(3)–C(4) adopt a conformation similar to the enthalpy-preferred *anti–anti* conformation of *n*-pentane.[89,90] For example, in crystalline [$RuCl_2$(η^6-cym){$Me_2P(CH_2)_3C_6H_5$-κ*P*}], the torsion angles P(1)–C(1)–C(2)–C(3) and C(1)–C(2)–C(3)–C(4) lie in the range 165.1–176.9°[53] and thus are close to the "ideal" 180°,[86,87] which minimizes non-bonded hydrogen–hydrogen repulsions. By contrast, in the tethered complex [$RuCl_2${η^6-$C_6H_5(CH_2)_3PMe_2$-κ*P*}] (**62**),[60] rotation about both the C(1)–C(2) and C(2)–C(3) bonds causes the torsion angles P(1)–C(1)–C(2)–C(3) and C(1)–C(2)–C(3)–C(4) to be only 73(1)° and 67(1)°, respectively. In **62** the phosphorus and C(3) atoms lie on the same side of the C(1)–C(2) bond, and the C(1) and C(4) atoms lie on the same side of the C(2)–C(3) bond, so that the tether adopts a conformation similar to the (gauche–gauche)$^{\pm}$ conformation of *n*-pentane. The higher energy of this conformation, which amounts to 13.8 kJ mol^{-1} in the case of *n*-pentane,[87] must be overcome by the greater stability of the tethered complex (chelate effect) relative to that of its non-tethered counterpart. Bulky substituents such as ^{t}Bu on phosphorus may favour formation of the tethered complex by reducing the energy difference between the two conformations. This could arise in part from an increase in the energy of the *anti–anti* conformation in the non-tethered complex. There is also a noticeable increase in the torsion angle P(1)–C(1)–C(2)–C(3) in the tethered complexes with increasing bulk of the substituent on phosphorus, resulting in a greater difference between the two torsion angles P(1)–C(1)–C(2)–C(3) and C(1)–C(2)–C(3)–C(4). The differences are 6°, 13°, 18°, 21°, and 19° for the Me_2P (**62**), Ph_2P (**61**), $^iPr_2P$ (**60**), Cy_2P (**59a**), and $^tBu_2P$ (**63**) complexes, respectively.[54,60] Since there are no significant differences in the bond angles in the tether in this series, the non-bonded H–H repulsions in the tether may be reduced significantly by the bulky substituents on phosphorus.

The chelate rings in the tethered complexes containing methylene-connecting groups are undoubtedly inverting in solution. This has been demonstrated by a variable-temperature ^{1}H NMR study of [$RuCl_2$(η^6-$C_6H_5SiMe_2CH_2PPh_2$-κ*P*)] (**58**) in CH_2Cl_2. The equivalent Si–Me groups at room temperature become non-equivalent at −110 °C as a result of the slowing down of inversion between the two envelope conformations of the chelate ring arene–Si–C–P–Ru, the phosphorus atom lying either above or below the plane containing this chain of atoms.[54]

C. *Reactivity*

Although few comparative studies have been carried out, tethering appears to increase the stability of the Ru–arene bond, as expected. The tethered complexes usually show parent-ion peaks in their EI- or FAB-mass spectra, whereas non-tethered complexes show only peaks arising from loss of arene. Thermogravimetric analysis establishes that the non-tethered complexes [$RuCl_2$(η^6-cym){$Cy_2P(CH_2)_3Ar$-κ*P*}] (Ar = C_6H_5, 3,5-$Me_2C_6H_3$) lose *p*-cymene at ca. 185 °C, whereas the tethered complexes lose the strapped arene only at ca. 300 °C.[56,57]

The η^6-arene in the tethered complexes $[RuCl_2\{\eta^6\text{-}C_6H_5(CH_2)_3PR_2\text{-}\kappa P\}]$ [R = Ph (**61**), ^{i}Pr (**60**), Cy (**59a**)] is displaced in boiling acetonitrile, but the reactions are much slower than those for the corresponding non-tethered complexes $[RuCl_2(\eta^6\text{-}C_6H_6)\{R_2P(CH_2)_3C_6H_5\text{-}\kappa P\}$.[54] For example, the reaction of **61** with boiling acetonitrile requires 264 h for completion, whereas the reaction is completed in 48 h for the non-tethered benzene complex. The product is the same in both cases, viz., $[RuCl(NCMe)_4\{R_2P(CH_2)_3C_6H_5\text{-}\kappa P\}]Cl$, as a mixture of *trans*- and *cis*-isomers containing octahedrally coordinated ruthenium(II); for R = Ph, these occur in a 5:1 ratio. The product having R = Ph is identical with that formulated incorrectly in Ref. 58 as $[RuCl_2(NCMe)_3\{Ph_2P(CH_2)_3C_6H_5\text{-}\kappa P\}]$. When there are only two connecting groups in the strap, the tethered arene is replaced more easily by boiling acetonitrile. Thus, both $[RuCl_2(\eta^6\text{-}C_6H_6)\{Ph_2PCH_2SiMe_2C_6H_5\text{-}\kappa P\}]$ and $[RuCl_2\{\eta^6\text{-}C_6H_5SiMe_2CH_2PPh_2\text{-}\kappa P\}]$ lose the coordinated arene completely after 48 h under these conditions.[54]

The mono-strapped arene complexes readily undergo replacement of one or both chloride ligands in the tripod. While such reactions are generally similar to those of non-tethered arene–ruthenium(II) complexes, the products are sometimes more stable and there is less tendency to lose arene during the reaction. The complexes $[RuCl_2(\eta^6\text{-}C_6H_5CH_2CH_2PPh_2\text{-}\kappa P)]$ (**55**) and $[RuCl_2(\eta^6\text{-}C_6H_5CHPhCH_2PPh_2\text{-}\kappa P)]$ are converted into the corresponding di-iodo complexes on treatment with sodium iodide in acetone at room temperature.[61] The complexes $[RuCl_2\{\eta^6\text{-}C_6H_5(CH_2)_3PR_2\text{-}\kappa P\}]$ [R = Ph (**61**), Me (**62**)] react with an excess of silver acetate to give the corresponding bis(κO-acetato) complexes. Similar reactions with sodium acetylacetonate (acac) and dimethyldithiocarbamate (dtc) in the presence of $AgPF_6$ give the salts **96a**, **96b**, **97a**, and **97b** (Scheme 16).[54] Complex **61** also reacts with neutral ligands in the presence of NH_4PF_6 to give salts $[RuCl(L)\{\eta^6\text{-}C_6H_5(CH_2)_3PPh_2\text{-}\kappa P\}]PF_6$ [L = PMe_3, PPh_3, $P(OMe)_3$, $P(OPh)_3$, py].[91] The tethered phosphite complex **16b** reacts with CO in the presence of $AgPF_6$ to form a labile mono-carbonyl salt $[RuCl(CO)\{\eta^6\text{-}C_6H_5(CH_2)_3OP^iPr_2\text{-}\kappa P\}]PF_6$,[24] and the tethered complexes **89** and **90** [Eq. (17)] react similarly with phenyl isocyanide in the presence of $NaPF_6$ to give the salts $[RuCl(CNPh)\{\eta^6\text{-}Ar(CH_2)_3PPh_2\text{-}\kappa P\}]PF_6$ (Ar = 3,5-$Me_2C_6H_3$, 4-$^iPrC_6H_4$).[92]

Solvento-cations are generated by treatment of $[RuCl_2\{\eta^6\text{-}C_6H_5CH_2CH_2P^tBu_2\text{-}\kappa P\}]$ (**57**) with $AgPF_6$ in acetone or acetonitrile.[53] The acetonitrile salt $[RuCl(NCMe)(\eta^6\text{-}C_6H_5CH_2CH_2P^tBu_2\text{-}\kappa P)]PF_6$ can be isolated, but attempted isolation of the acetone salt leads to the PF_6 salt of a di-μ-chloro dication, $[Ru_2(\mu\text{-}Cl)_2(\eta^6\text{-}C_6H_5CH_2CH_2P^tBu_2)_2](PF_6)_2$. The labile coordinated solvents are replaced by PMe_3 to give the stable salt $[RuCl(PMe_3)(\eta^6\text{-}C_6H_5CH_2CH_2P^tBu_2\text{-}\kappa P)]PF_6$. If the preparation of **57** (Scheme 13) is modified in the second step by addition of Et_3N, the product is the hydrido(chloro) derivative of **57**, viz., $[RuHCl(\eta^6\text{-}C_6H_5CH_2CH_2P^tBu_2\text{-}\kappa P)]$. The corresponding $C_6H_5OCH_2CH_2P^tBu_2$ system behaves similarly.[53]

The mono-strapped complex **98**, which is prepared by the ethyl benzoate displacement method (Section II.C), has planar chirality owing to the presence of different 1,3-substituents on the η^6-arene and can be separated into its enantiomers by chiral HPLC. Each enantiomer reacts with $AgOTf/H_2O$ to give the corresponding di-strapped aqua cation in which a second chiral centre is present because there are now three different groups in the tripodal set. However, each

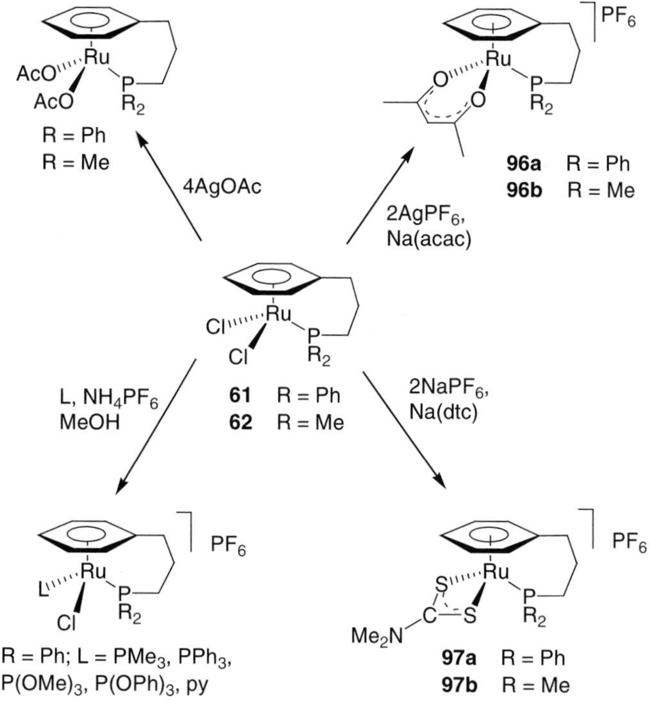

SCHEME 16.

enantiomer gives just a single diastereomer, (R_{Ru}, S) and (S_{Ru}, R), **99a** and **99b**, respectively [Eq. (20)], instead of the expected pair. Thus the configuration about the metal centre is locked and stereochemical integrity is retained, even in the presence of coordinating solvents such as DMSO, benzaldehyde, water, ethanol, and acetonitrile.[93] If a similar sequence is carried out with the enantiopure auxiliary camphorpyrazole instead of 3,5-bis(trifluoromethyl)pyrazole, the corresponding $RuCl_2$ complex is formed as a 1:1 mixture of diastereomers, (R, R_P) and (R, S_P), which can be separated by silica-gel chromatography. Treatment of each diastereomer with $TlOTf/H_2O$ then forms the aqua dication with coordinated pyrazole, the process again being diastereoselective. The resulting diastereomers (R, R_P, S_{Ru}) and (R, S_P, R_{Ru}) have been structurally characterized.[94]

CF3 N N Cl Ru P Ph2 Cl CF3 —AgOTf / H2O→ CF3 N N Ru P Ph2 H2O CF3]2+ (20)

(*S*) **98** [(*R*) **98**] (*R*Ru , *S*) **99a** [(*S*Ru , *R*) **99b**]

The planar chiral tethered complex $[RuCl_2(\eta^6\text{-}2'\text{-}Me_2NC_6H_4C_6H_4\text{-}2\text{-}PCy_2\text{-}\kappa P)]$ (**68**) reacts with ligands (L) in the presence of $AgSbF_6$ to give salts $[RuCl(L)(\eta^6\text{-}2'\text{-}Me_2NC_6H_4C_6H_4\text{-}2\text{-}PCy_2\text{-}\kappa P)]SbF_6$, in which the metal centre is also chiral.[65] If L is itself chiral and non-racemic (L*), this can generate a maximum of four diastereomers: one pair with L* *anti* to NMe_2, the other with L* *syn* to NMe_2. These are indeed observed with L* = (*R*)-(+)-α-methylbenzylamine or (*R*)-(+)-methyl-*p*-tolyl sulfoxide but the diastereomers could not be separated; hence, **68** could not be resolved in this way. If L is achiral, and not prochiral, a maximum of two diastereomers, *anti* and *syn*, **100a** and **100b**, each of which exists as a pair of enantiomers, is possible in principle.

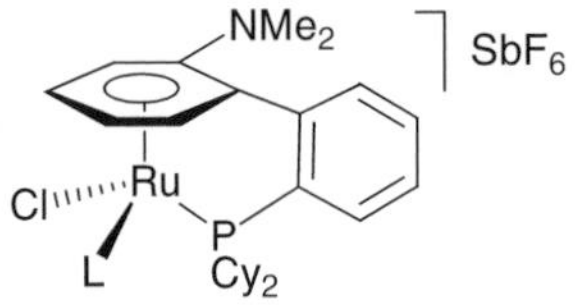

NMe2 SbF6 Cl Ru L P Cy2

100a *anti*

(enantiomeric pair)

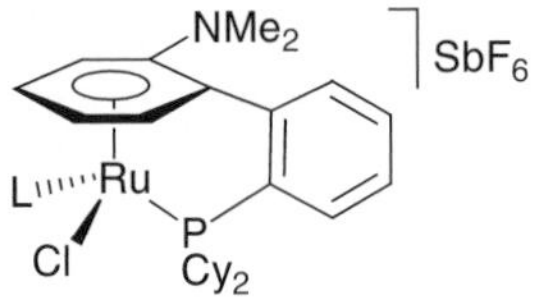

NMe2 SbF6 L Ru Cl P Cy2

100b *syn*

(enantiomeric pair)

In the case of $L = PPh_3$, however, the reaction is almost diastereoselective (>96% formation of the *anti*-diastereomer), apparently because the incoming PPh_3 is directed to a particular coordination site by steric interaction with the NMe_2 substituent. For $L = PMePh_2$, PMe_2Ph, and PMe_3, the diastereoselectivities are ca. 90%, although the *syn*-diastereomer for $L = PMe_2Ph$ can be isolated by repeated recrystallization. No *syn–anti* interconversion occurs over long periods; hence, this system, like those of Therrien *et al.*,[93,94] retains its stereochemical integrity. For $L = PPh_3$, the isolated *anti*-diastereomer undergoes spontaneous resolution on crystallization. The manually separated enantiomers show the expected mirror-image CD spectra, and X-ray crystallographic study establishes that (−)-$[RuCl(PPh_3)\{\eta^6\text{-}2'\text{-}Me_2NC_6H_4C_6H_4\text{-}2\text{-}PCy_2\}]SbF_6$ (**100a**: $L = PPh_3$) is the (R_{Ru}, *R*) enantiomer. When L is a chiral tertiary phosphine, only two diastereomers in ca. 1:1 ratio are obtained rather than the expected four, but these could not be separated. The formally 16-electron dication $[Ru(PPh_3)(\eta^6\text{-}2'\text{-}Me_2NC_6H_4C_6H_4\text{-}2\text{-}PCy_2)]^{2+}$, which is generated from enantiopure **100a** ($L = PPh_3$) by treatment with another equivalent of $AgSbF_6$, catalyses the asymmetric Diels–Alder reaction of methacrolein with cyclopentadiene at −24 °C with high conversion (95%) and diastereoselectivity (96%) but the enantioselectivity to the (−)-*exo*-adduct is disappointingly poor (19%).[65]

Good diastereoselectivities (de = 82–90%) are also found in the reactions of $[RuCl_2\{\eta^6\text{-}(R)\text{-}C_6H_5CH(Me)CH_2CH_2PPh_2\text{-}\kappa P\}]_2$ with primary and secondary amines (L) in the presence of $NaBF_4$ to form the salts $[RuCl(L)\{\eta^6\text{-}(R)\text{-}C_6H_5CH(Me)CH_2CH_2PPh_2\text{-}\kappa P\}]BF_4$. The major diastereomers of the aniline and piperidine complexes were shown by X-ray structural analysis to have the (S_{Ru}, R_c) configuration; for the aniline complex this is retained for at least a week at room

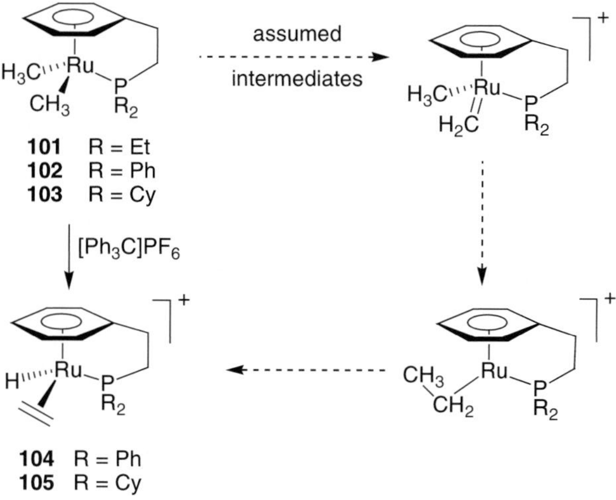

SCHEME 17.

temperature. The diastereoselectivity is much poorer (de = 14%) for L = imidazole.[94a]

The complexes $[RuCl_2(\eta^6\text{-}C_6H_5CH_2CH_2PR_2\text{-}\kappa P)]$ [R = Et (**54**), Ph (**55**), Cy (**56**)] react with an excess (>5 equiv.) of methyllithium to give the corresponding dimethyl complexes **101**–**103** in ca. 35% isolated yield.[51] Except for **101**, they appear to be thermally more stable than comparable non-tethered complexes such as $[RuMe_2(\eta^6\text{-}C_6H_6)(PPh_3)]$.[12] Treatment of **102** and **103** with $[Ph_3C]PF_6$ abstracts hydride from one of the methyl groups giving the hydrido(ethylene) salts **104** and **105** by the sequence shown in Scheme 17.[95] Although there is no evidence from the X-ray structure of **105** for an agostic interaction between Ru–H and C_2H_4, 2D EXSY measurements show that reversible insertion–deinsertion does occur. Similar chemistry has been established in the non-tethered $Ru(\eta^6\text{-}C_6Me_6)(PPh_3)$ series.[96]

One of the Ru–CH_3 bonds in **102** and **103** is cleaved by the acid $[Et_2OH]^+$ $[B\{C_6H_3(CF_3)_2\text{-}3,5\}_4]$, {the anion is abbreviated as $[B(Ar_F)_4]^-$}, in the presence of ligands (L), as shown in Scheme 18.[97] For L = CO, the isolable (methyl)(carbonyl) cations **106** and **107** are formed; whereas for L = C_2H_4, the corresponding (methyl)(ethene) cation (R = Ph) is unstable and decomposes in CH_2Cl_2 to give the di-μ-chloro dication **108**. Again, similar chemistry has been demonstrated in the non-tethered $Ru(\eta^6\text{-}C_6Me_6)(PPh_3)$ series except that, in this case, the intermediate ethene cation decomposes to give the cation $[RuH(\eta^6\text{-}C_6Me_6)(\eta^2\text{-}2\text{-}CH_2{=}CHC_6H_4PPh_2\text{-}\kappa P)]^+$, presumably *via ortho*-metalation of PPh_3.[96] Evidently, the presence of the tether inhibits this process. For L = C_2H_2, the isolated products are polyacetylene and the cyclic η^3-allyl cations **109** and **110**; the latter are believed to arise by a *trans*-insertion of acetylene into the Ru–CH_3 bond and two subsequent *cis*-insertions followed by ring closure. When L = norbornene, the product is ring-opened poly(norbornene), presumably formed *via* Ru–methylidene intermediates; $B(C_6F_5)_3$ and $[Ph_3C]^+BF_4^-$ can also be used as activators for this ring-opening metathesis polymerization (ROMP) process.[97]

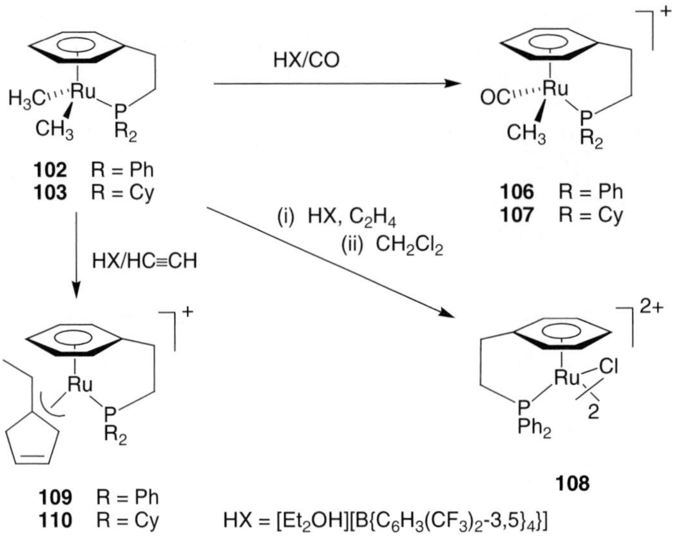

SCHEME 18.

Both non-tethered complexes [RuCl$_2$(η-cym){Cy$_2$P(CH$_2$)$_3$Ar-κ*P*}] (Ar = C$_6$H$_5$, 3,5-Me$_2$C$_6$H$_3$) and their tethered counterparts **59a**–**59c** show comparable reactivity and stereoselectivity at 60 °C in the cyclopropanation of styrene by decomposition of ethyl diazoacetate. In the temperature range 40–50 °C, however, the non-tethered complexes are far more active in promoting the decomposition of ethyl diazoacetate. Apparently, different Ru–carbene species are generated from the tethered and non-tethered precursors; in the latter the arene is probably displaced during the reaction. The tethered complexes, activated either by EtO$_2$CCHN$_2$ or Me$_3$SiCHN$_2$ (TMSD, trimethylsilyldiazomethane), are also poor catalysts for ROMP of cyclooctene, and, again, arene must be lost to generate the active species.[56,57] The two atom-tethered complexes **55** and **56** and their analogues, [RuCl$_2${η^6-C$_6$H$_5$CHPhCH$_2$PR$_2$-κ*P*}] (R = Ph, Cy), can be activated for ROMP of norbornene in various ways. The chloride ligands can be abstracted with AgBF$_4$, AgSbF$_6$, or Na[B(Ar$_F$)$_4$] in methanol, generating a presumably solvated dication that retains the tethered arene; its catalytic activity is strongly anion-dependent, increasing in the order $BF_4^- < SbF_6^- < B(Ar_F)_4^-$, reflecting the decreasing tendency of these anions to interact with the metal centre.[62] The catalytically active species formed from AgBF$_4$ with subsequent addition of TMSD apparently has no coordinated arene, and addition of TMSD alone also generates an active species that contains no coordinated arene.

The allenylidene salt **111** has been prepared by treatment of **59a** first with AgOTf in CH$_2$Cl$_2$ and then with 1,1-diphenyl-2-propyn-1-ol.[52] Non-tethered versions of such compounds are easily synthesized alternatives to the Grubbs complexes for RCM reactions, but, although **111** is catalytically active, it is not longer-lived than the non-tethered complexes under the reaction conditions (toluene, 80 °C). In the reaction of the tethered complex [RuCl$_2${η^6-1,2,4-Me$_3$C$_6$H$_2$-5-(CH$_2$)$_3$PPh$_2$-κ*P*}] with 1,1-diphenylpropynol in methanol/CH_2Cl_2 in the presence of $NaPF_6$, the product is the alkoxycarbene cation **112**, which presumably is derived by addition of methanol to the α,β-carbon atoms of a putative allenylidene intermediate. Alkoxycarbene cations containing the groups =C(OMe)Me and =C(OMe)CH_2Ph are obtained by reaction of the tethered complex with Me_3SiC_2H and PhC_2H, respectively, under similar conditions. The tethered dichloro complex catalyses the hydration of phenylacetylene to acetophenone in refluxing 95% ethanol over 24 h, but its reactivity is comparable with that of its non-tethered analogues.[92]

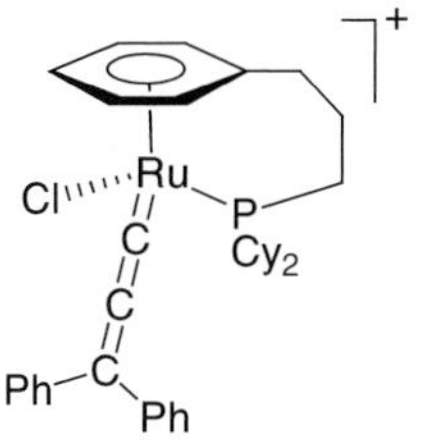

111 **112**

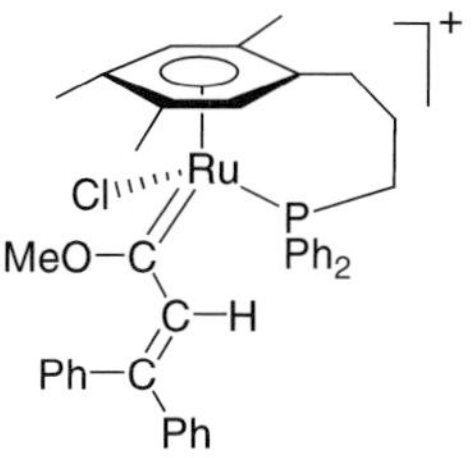

D. *Redox Behaviour*

The tethered arene complexes of ruthenium(II), [$RuCl_2(\eta^6$-Ar~PR_2-κ*P*)], show reversible one-electron oxidation in CH_2Cl_2 at room temperature by cyclic voltammetry, the electrode potentials $E_{1/2}$ being in the range +0.5–0.79 V (*vs.* FcH/FcH^+). A selection of values is given in Table I. A comparison of the CV and ACV (alternating current voltammogram) traces of complex **61** and ferrocene in CH_2Cl_2 shows the reversibility of the two couples to be comparable.[54] The magnitude of $E_{1/2}$ is affected by the electron-donating or electron-withdrawing abilities of the ligands in the expected way.[60,81] For example, complex **61** has a higher $E_{1/2}$-value than **64** as a consequence of the electron-donating methyl groups in the latter. The CV behaviour of the tethered complexes is generally similar to that of their non-tethered analogues.[98]

The changes occurring on electro-oxidation of both tethered and non-tethered complexes have been studied by spectroelectrochemistry at ca. 228 K in CH_2Cl_2 in the presence of 0.3 M [^{n}$Bu_4N]BF_4$.[54,99] For example, when a potential of +0.65 V (*vs.* FcH/FcH^+) (ca. 100 mV greater than the $E_{1/2}$-value) is applied to a solution of the $C_6Me_5(CH_2)_3PPh_2$ complex **64** under these conditions, the broad band at 27,400 cm^{-1} due to **64** is gradually replaced by new bands at 29,100 cm^{-1}, 24,600 cm^{-1}, and 19,500 cm^{-1}. When the exhaustive oxidation is complete, the original spectrum is regenerated by applying a potential of −0.55 V (*vs.* FcH/FcH^+). The electronic spectrum of the electrogenerated Ru(III) species is stable indefinitely at 228 K but begins to change at 263 K; at 283 K it reverts slowly to the spectrum of its precursor **64**. The ESR spectrum of the anodic oxidation product in a frozen glass shows three *g*-values (2.32, 2.19, 2.00) similar to those of pseudo-octahedral Ru^{III} complexes. The evidence therefore points to the electro-generation of an arene–ruthenium(III) cation, $[\mathbf{64}]^+$, which is long-lived at 228 K and has some

TABLE I
REDOX POTENTIALS (VS. FcH/FcH$^+$) OF SELECTED TETHERED ARENE–PHOSPHINE COMPLEXES OF RUTHENIUM(II)

Complex	$E_{1/2}$ (V)	Ref.
[$RuCl_2\{\eta^6$-$C_6H_5(CH_2)_3PPh_2$-$\kappa P\}$] (**61**)	+0.77[a]	54, 60
	+0.74[b]	81
	+1.34[c]	58
[$RuCl_2\{\eta^6$-$C_6H_5(CH_2)_3PMe_2$-$\kappa P\}$] (**62**)	+0.71[a]	54, 60
[$RuCl_2\{\eta^6$-$C_6H_5(CH_2)_3P^iPr_2$-$\kappa P\}$] (**60**)	+0.78[a]	54
[$RuCl_2\{\eta^6$-$C_6H_5(CH_2)_3PCy_2$-$\kappa P\}$] (**59a**)	+0.76[a]	54
[$RuCl_2\{\eta^6$-$C_6H_5(CH_2)_3P^tBu_2$-$\kappa P\}$] (**63**)	+0.82[a]	54
[$RuCl_2\{\eta^6$-4-$MeC_6H_4(CH_2)_3PPh_2$-$\kappa P\}$] (**91**)	+0.67[b]	81
[$RuCl_2\{\eta^6$-4-$^iPrC_6H_4(CH_2)_3PPh_2$-$\kappa P\}$] (**90**)	+0.66[b]	81
[$RuCl_2\{\eta^6$-3,5-$Me_2C_6H_3(CH_2)_3PPh_2$-$\kappa P\}$] (**89**)	+0.61[b]	81
[$RuCl_2\{\eta^6$-2,4,6-$Me_3C_6H_2(CH_2)_3PPh_2$-$\kappa P\}$] (**65**)	+0.65[a]	54, 60
[$RuCl_2\{\eta^6$-$C_6Me_5(CH_2)_3PPh_2$-$\kappa P\}$] (**64**)	+0.55[a]	54, 60
	+0.47[b]	81
[Ru(acac)$\{\eta^6$-$C_6H_5(CH_2)_3PPh_2$-$\kappa P\}$]PF_6 (**96a**)	+1.10[a]	54
[Ru(dtc)$\{\eta^6$-$C_6H_5(CH_2)_3PPh_2$-$\kappa P\}$]PF_6 (**97a**)	+0.94[a]	54
[Ru(dtc)$\{\eta^6$-$C_6H_5(CH_2)_3PMe_2$-$\kappa P\}$]PF_6 (**97b**)	+0.88[a]	54

[a]Scan rate of 100 mV s^{-1}, 0.5 M [$^nBu_4N$]PF_6 in CH_2Cl_2 at 293 K, Ag/AgCl/CH_2Cl_2 reference electrode [$E_{1/2}$ (FcH/FcH$^+$) = +0.55 V].
[b]Scan rate of 250 mV s^{-1}, 0.1 M [$^nBu_4N$]PF_6 in CH_2Cl_2, Ag/AgCl (aq.) reference electrode [$E_{1/2}$ (FcH/FcH$^+$) = +0.51 V].
[c]vs. SCE.

stability even at room temperature. The spectroelectrochemical behaviour of complexes **59a**, **60**, **63**, **65**, and **89** is generally similar. For the complexes containing $Me_2P(CH_2)_3C_6H_5$ (**62**) and $Ph_2PCH_2SiMe_2C_6H_5$ (**58**), the Ru^{III} species can be electrogenerated at 228 K but decompose irreversibly above ca. 263 K, presumably with displacement of the tethered arene. The spectroelectrochemical behaviour of the complex of $C_6H_5(CH_2)_3PPh_2$ (**61**) is not reversible, even at 228 K. One-electron oxidation products of the non-tethered complexes can be generated similarly but they are generally less stable.[54]

It is clear from the $E_{1/2}$-values that a powerful one-electron oxidant is required to generate a ruthenium(III) complex of a tethered arene–phosphine. From the reaction of **64** with the aminium salt [N(4-$BrC_6H_4)_3$]$SbCl_6$ (E_o = +0.70 V *vs.* FcH/FcH$^+$ in CH_2Cl_2), the salt [**64**]$^+$ [$SbCl_6$]$^-$ has been isolated and structurally characterized.[54] The Ru–C (arene) bond distances in the salt are in the range 2.228–2.336(5) Å, i.e., ca. 0.1 Å longer than those in the precursor **64**.[60] Both features are consistent with the expected weakening of the metal–arene interaction in going from the $4d^6$ to the $4d^5$ electron configuration. Although the tether probably contributes to the stability of [**64**]$^+$, the most important factor is the presence of methyl groups on the arene ring, since, by the same technique, it is possible to prepare the non-tethered analogue [$RuCl_2(\eta^6$-$C_6Me_6)(PPh_3)$]$SbCl_6$ from [$RuCl_2(\eta^6$-$C_6Me_6)(PPh_3)$].[54] Attempts to isolate either tethered or non-tethered arene–ruthenium(III) salts having fewer methyl substituents by use of $[N(4\text{-}BrC_6H_4)_3]SbCl_6$ as oxidant have not been successful, perhaps because the oxidation potentials of the Ru(II) precursors are too high.

IV
TETHERED NON-BENZENOID AROMATIC COMPLEXES

The preparation of tethered complexes in which $C_5H_5^-$ or C_6H_6 is replaced by the isoelectronic, seven-membered ring tropylium(cycloheptatrienylium) ion, $C_7H_7^+$, has been described by Tamm *et al.*,[100–105] whose primary motivation has been to compare the C_5R_5Ru and C_7H_7Mo synthons in organometallic and catalytic chemistry. The published work concerns complexes in which oxygen and phosphorus are the atoms tethered to the seven-membered ring.

In the oxygen series, the starting material is (2-hydroxy-3,5-dimethylphenyl)tropylium ion (**113**), which is converted in four steps *via* the iodo-complexes **114** and **115** into the tethered aryloxomolybdenum(0) complex **116** (Scheme 19).[100] Protonation of **116** with HBF_4 gives cation **117**, the labile Mo–O bond of which is replaced by xylyl isocyanide to give the non-tethered complex **118**. Although one carbonyl ligand of complex **114** is replaced by tertiary phosphines, the reaction with dppe takes an unexpected course. As shown in Scheme 20, elimination of HI occurs without displacement of CO to give the tethered complex **119**, which contains η^3-cycloheptatrienyl; the complex is pseudo-octahedral if the seven-membered ring is assumed to occupy just one coordination site. The ^{31}P nuclei of **119** are equivalent at room temperature and display the expected inequivalence only at $-30\,^{\circ}C$ owing to a rapid interconversion of enantiomers and a migration (probably *via* successive 1,2-shifts) of the metal centre about the seven-membered ring. The η^7-coordination of the ring is restored when the CO groups are removed by UV-irradiation, giving the air-sensitive, tethered Mo(0) complex **120**. The latter is readily oxidized by $[(FcH)_2]PF_6$ to a stable, 17-electron Mo(I) cation $[\mathbf{120}]^+$.[100]

In the phosphorus series, the starting materials are the (cycloheptatrienyl) phosphines, $1\text{-}C_7H_6\text{-}2\text{-}C_6H_4PR_2$ (R = Ph, ^{i}Pr), which react with $Mo(CO)_6$ to give the tethered η^6-cycloheptatriene complexes **121a**, **121b**. These are converted in four steps to the 17-electron, tethered dibromo (η^7-tropylium)molybdenum(I) complexes **122a**, **122b** (Scheme 21).[101–104] The bromide ligands of these complexes are replaced successively by (trimethylsilyl)methyl on treatment of **122a**, **122b** with Me_3SiCH_2MgBr and, in the presence of the Grignard reagent, both the dibromides and the bis(alkyls) catalyse ROMP of norbornene. Reduction of **122b** with $NaBH_4$ gives the η^2-BH_4 complex **123**, which is a useful precursor (Scheme 22). The BH_4 ligand is surprisingly stable to hydrolysis but is converted into the monohydride **124** by ethanolic PPh_3. The BH_4 ligand is removed from **123** by the mild acid $[PhNHMe_2][BPh_4]^-$ in the presence of ligands, such as 2,6-$Me_2C_6H_3NC$, NBD, PhC_2H, *t*-BuC_2H, and Ph_2C_2, to give the tethered Mo(0) complexes **125–129** (Scheme 22). The phenylacetylene complex catalyses the cyclotrimerization of phenylacetylene at 80 °C for 12 h to a ca. 2:1 mixture of 1,3,5- and 1,2,4-triphenylbenzene; some linear oligomerization also occurs.[105]

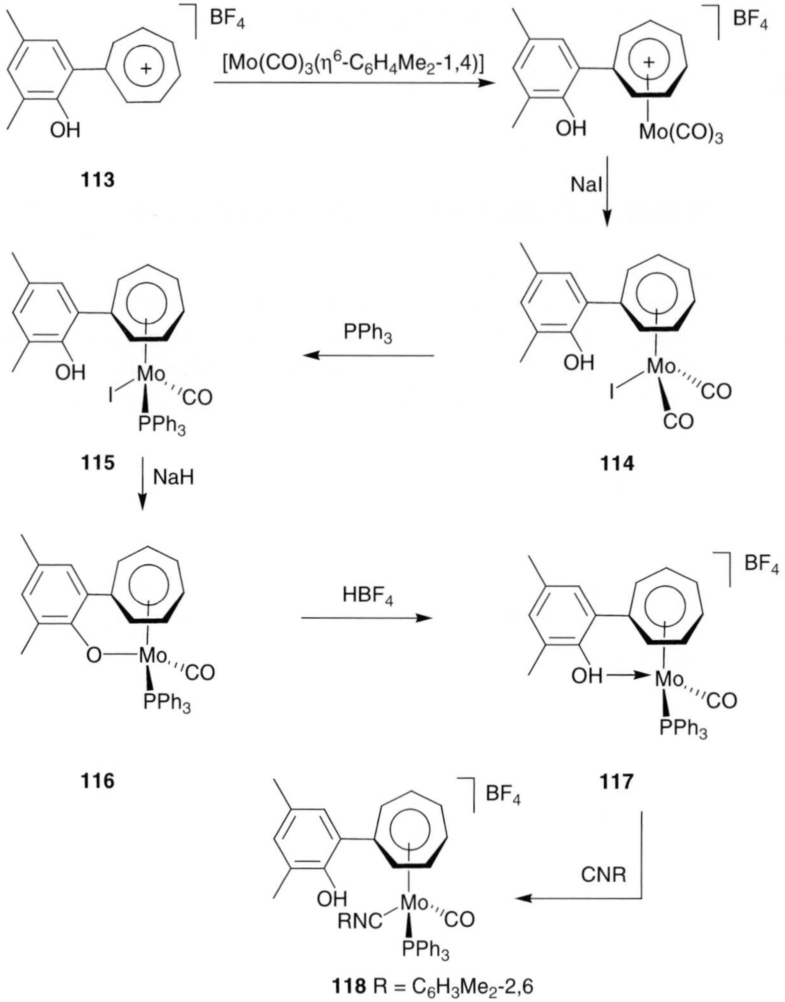

SCHEME 19.

Treatment of the acetato complex [Mo(OAc-κ^2O)(η^7-C_7H_7-2-$C_6H_4P^iPr_2$-κP)] with diazoalkanes $RCHN_2$ (R = Ph, $SiMe_3$) in the presence of Me_3SiCl gives alkylidene complexes **130a** and **130b**, whereas $Ph_2C=N_2$ gives an N-donor complex of diphenyldiazomethane, without elimination of N_2.[105] An active ROMP-catalyst is generated on adding $TlPF_6$ to **130b**. The vinylcarbene complex **131** is obtained by reaction of dichloromethane (acting as a source of CH_2) with the tolane cation **129**.[105]

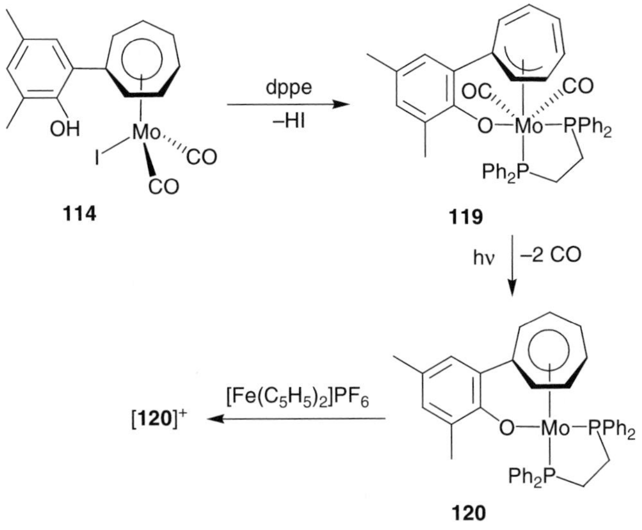

SCHEME 20.

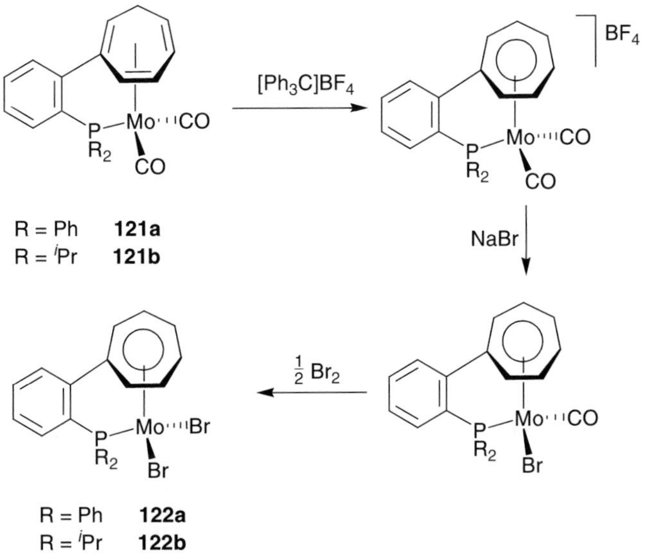

SCHEME 21.

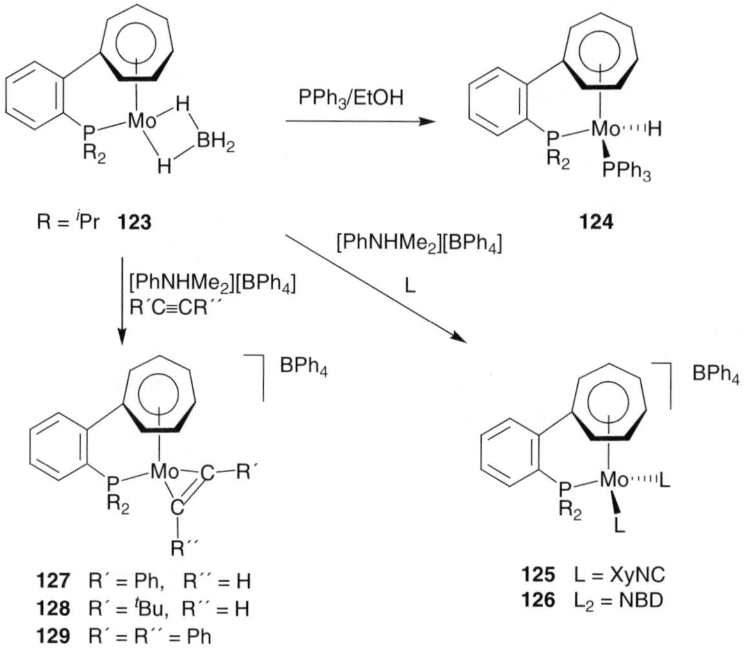

SCHEME 22.

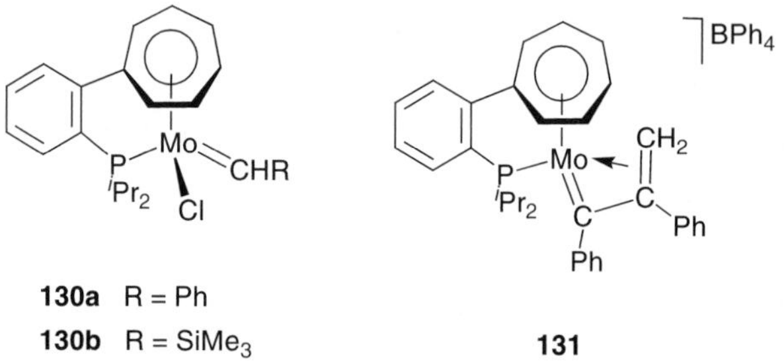

ACKNOWLEDGMENTS

Joanne Adams acknowledges the receipt of an Australian Postgraduate Award. We thank Mrs Rosemary Enge for her invaluable help in preparing the manuscript.

REFERENCES

(1) Jeffrey, J. C.; Rauchfuss, T. B. *Inorg. Chem.* **1979**, *18*, 2658.
(2) Bader, A.; Lindner, E. *Coord. Chem. Rev.* **1991**, *108*, 27.
(3) Braunstein, P. *J. Organomet. Chem.* **2004**, *689*, 3953.
(4) Okuda, J. *Comments Inorg. Chem.* **1994**, *16*, 185.
(5) Jutzi, P.; Dahlhaus, L. *Coord. Chem. Rev.* **1994**, *137*, 179.
(6) Jutzi, P.; Siemeling, U. *J. Organomet. Chem.* **1995**, *500*, 175.

(7) Wang, B.; Deng, D. *New J. Chem.* **1995**, *19*, 515.
(8) Jutzi, P.; Redeker, D. *Eur. J. Inorg. Chem.* **1998**, 63.
(9) Okuda, J.; Eberle, T. (A. Togni, R. Halterman, Eds.) *Metallocenes*, Vol. 1, Weinheim, Wiley-VCH, **1998**, Chapter 7.
(10) Siemeling, U. *Chem. Rev.* **2000**, *100*, 1495.
(11) Butenschön, H. *Chem. Rev.* **2000**, *100*, 1527.
(12) Bennett, M. A.; Smith, A. K. *J. Chem. Soc., Dalton Trans.* **1974**, 233.
(13) Kasahara, A.; Izumi, T.; Tanaka, A. *Bull. Chem. Soc. Jpn.* **1967**, *40*, 699.
(14) Abdul-Rahman, S.; Houlton, A.; Roberts, R. M. G.; Silver, J. *J. Organomet. Chem.* **1989**, *359*, 331.
(15) Nesmeyanov, A. N.; Rybinskaya, M. I.; Krivykh, V. V.; Kaganovich, V. S. *J. Organomet. Chem.* **1975**, *93*, C8.
(16) Nesmeyanov, A. N.; Krivykh, V. V.; Petrovskii, P. V.; Rybinskaya, M. I. *Dokl. Akad. Nauk SSSR* **1976**, *231*, 110.
(17) Nesmeyanov, A. N.; Krivykh, V. V.; Rybinskaya, M. I. *J. Organomet. Chem.* **1979**, *164*, 159.
(18) Nesmeyanov, A. N.; Krivykh, V. V.; Panosyan, G. A.; Petrovskii, P. V.; Rybinskaya, M. I. *J. Organomet. Chem.* **1979**, *164*, 167.
(19) Merlic, C. A.; Miller, M. M. *Organometallics* **2001**, *20*, 373.
(20) Ohnishi, T.; Miyaki, Y.; Asano, H.; Kurosawa, H. *Chem. Lett.* **1999**, 809.
(21) Miyaki, Y.; Onishi, T.; Kurosawa, H. *Inorg. Chim. Acta* **2000**, *300–302*, 369.
(22) Miyaki, Y.; Onishi, T.; Ogoshi, S.; Kurosawa, H. *J. Organomet. Chem.* **2000**, *616*, 135.
(23) Miyaki, Y.; Onishi, T.; Kurosawa, H. *Chem. Lett.* **2000**, 1334.
(24) Kitaura, R.; Miyaki, Y.; Onishi, T.; Kurosawa, H. *Inorg. Chim. Acta* **2002**, *334*, 142.
(25) Hannedouche, J.; Clarkson, G. J.; Wills, M. *J. Am. Chem. Soc.* **2004**, *126*, 986.
(26) Mahaffy, C. A.; Pauson, P. L. *J. Chem. Res., Synop.* **1979**, 126; *J. Chem. Res., Miniprint* **1979**, 1752.
(27) Zimmerman, C. L.; Shaner, S. L.; Roth, S. A.; Willeford, B. R. *J. Chem. Res., Synop.* **1980**, 108; *J. Chem. Res., Miniprint* **1980**, 197.
(28) Muetterties, E. L.; Bleeke, J. R.; Sievert, A. C. *J. Organomet. Chem.* **1979**, *178*, 197.
(29) Traylor, T. G.; Stewart, K. J.; Goldberg, M. J. *J. Am. Chem. Soc.* **1984**, *106*, 4445.
(30) Howell, J. A. S.; Dixon, D. T.; Kola, J. C.; Ashford, N. F. *J. Organomet. Chem.* **1985**, *294*, C1.
(31) Traylor, T. G.; Goldberg, M. J. *J. Am. Chem. Soc.* **1987**, *109*, 3968.
(32) Thorn, M. G.; Etheridge, Z. C.; Fanwick, P. E.; Rothwell, I. P. *Organometallics* **1998**, *17*, 3636.
(33) Thorn, M. G.; Etheridge, Z. C.; Fanwick, P. E.; Rothwell, I. P. *J. Organomet. Chem.* **1999**, *591*, 148.
(34) Robertson, G. B.; Whimp, P. O., Colton, R.; Rix, C. J. *Chem. Commun.* **1971**, 573.
(35) Colton, R.; Rix, C. J. *Aust. J. Chem.* **1971**, *24*, 2461.
(36) Chatt, J.; Watson, H. R. *J. Chem. Soc.* **1961**, 4980.
(37) Kerschner, J. L.; Torres, E. M.; Fanwick, P. E.; Rothwell, I. P.; Huffman, J. C. *Organometallics* **1989**, *8*, 1424.
(38) Lockwood, M. A.; Fanwick, P. E.; Rothwell, I. P. *Polyhedron* **1995**, *14*, 3363.
(39) Lentz, M. R.; Fanwick, P. E.; Rothwell, I. P. *Organometallics* **1997**, *16*, 3574.
(40) Lentz, M. R.; Fanwick, P. E.; Rothwell, I. P. *Organometallics* **2003**, *22*, 2259.
(41) Bishop, P. T.; Dilworth, J. R.; Zubieta, J. A. *J. Chem. Soc., Chem. Commun.* **1985**, 257.
(42) Dilworth, J. R.; Zheng, Y.; Lu, S.; Wu, Q. *Inorg. Chim. Acta* **1992**, *194*, 99.
(43) Hartshorn, C. M.; Steel, P. J. *Angew. Chem., Int. Ed. Engl.* **1996**, *35*, 2655.
(44) Singewald, E. T.; Mirkin, C. A.; Levy, A. D.; Stern, C. L. *Angew. Chem., Int. Ed. Engl.* **1994**, *33*, 2473.
(45) Singewald, E. T.; Shi, X.; Mirkin, C. A.; Schofer, S. J.; Stern, C. L. *Organometallics* **1996**, *15*, 3062.
(46) Sassano, C. A.; Mirkin, C. A. *J. Am. Chem. Soc.* **1995**, *117*, 11379.
(47) Slone, C. S.; Mirkin, C. A.; Yap, G. P. A.; Guzei, I. A.; Rheingold, A. L. *J. Am. Chem. Soc.* **1997**, *119*, 10743.
(48) Werner, H.; Canepa, G.; Ilg, K.; Wolf, J. *Organometallics* **2000**, *19*, 4756.
(49) Canepa, G.; Brandt, C. D.; Werner, H. *Organometallics* **2001**, *20*, 604.
(50) Canepa, G.; Sola, E.; Martín, M.; Lahoz, F. J.; Oro, L. A.; Werner, H. *Organometallics* **2003**, *22*, 2151.

(51) Umezawa-Vizzini, K.; Guzman-Jimenez, I. Y.; Whitmire, K. H.; Lee, T. R. *Organometallics* **2003**, *22*, 3059.
(52) Fürstner, A.; Liebl, M.; Lehmann, C. W.; Picquet, M.; Kunz, R.; Bruneau, C.; Touchard, D.; Dixneuf, P. H. *Chem. Eur. J.* **2000**, *6*, 1847.
(53) Jung, S.; Ilg, K.; Brandt, C. D.; Wolf, J.; Werner, H. *J. Chem. Soc., Dalton Trans.* **2002**, 318.
(54) Adams, J. R. PhD Thesis, Australian National University, Canberra, **2003**.
(55) Therrien, B.; Ward, T. R.; Pilkington, M.; Hoffmann, C.; Gilardoni, F.; Weber, J. *Organometallics* **1998**, *17*, 330.
(56) Simal, F.; Jan, D.; Demonceau, A.; Noels, A. F. *Tetrahedron Lett* **1999**, *40*, 1653.
(57) Jen, D.; Delaude, L.; Simal, F.; Demonceau, A.; Noels, A. F. *J. Organomet. Chem.* **2000**, *606*, 55.
(58) Smith, P. D.; Wright, A. H. *J. Organomet. Chem.* **1998**, *559*, 141.
(59) Bennett, M. A.; Harper, J. R. (G. J. Leigh, N. Winterton, Eds.), *Modern Coordination Chemistry: The Legacy of Joseph Chatt*, Royal Society of Chemistry, Cambridge, **2002**, pp. 163–168.
(60) Bennett, M. A.; Edwards, A. J.; Harper, J. R.; Khimyak, T.; Willis, A. C. *J. Organomet. Chem.* **2001**, *629*, 7.
(61) Pertici, P.; Salvadori, P.; Biasci, G.; Vitulli, G.; Bennett, M. A.; Kane-Maguire, L. A. P. *J. Chem. Soc., Dalton Trans.* **1988**, 315.
(62) Abele, A.; Wursche, R.; Klinga, M.; Rieger, B. *J. Mol. Catal. A* **2000**, *160*, 23.
(63) Shaw, B. L. *J. Am. Chem. Soc.* **1975**, *97*, 3856.
(64) Shaw, B. L. *J. Organomet. Chem.* **1980**, *200*, 307.
(65) Faller, J. W.; D'Alliessi, D. G. *Organometallics* **2003**, *22*, 2749.
(66) Geldbach, T. J.; Pregosin, P. S. *Eur. J. Inorg. Chem.* **2002**, 1907.
(67) den Reijer, C. J.; Rüegger, H.; Pregosin, P. S. *Organometallics* **1998**, *17*, 5213.
(68) den Reijer, C. J.; Wörle, M.; Pregosin, P. S. *Organometallics* **2000**, *19*, 309.
(69) Geldbach, T. J.; Drago, D.; Pregosin, P. S. *Chem. Commun.* **2000**, 1629.
(70) Geldbach, T. J.; Pregosin, P. S.; Albinati, A.; Rominger, F. *Organometallics* **2001**, *20*, 1932.
(71) Geldbach, T. J.; Pregosin, P. S.; Albinati, A. *Organometallics* **2003**, *22*, 1443.
(72) Feiken, N.; Pregosin, P. S.; Trabesinger, G. *Organometallics* **1997**, *16*, 3735.
(73) Çetinkaya, B.; Demir, S.; Özdemir, I.; Toupet, L.; Sémeril, D.; Bruneau, C.; Dixneuf, P. H. *New J. Chem.* **2001**, *25*, 519.
(74) Crocker, M.; Green, M.; Nagle, K. R.; Williams, D. J. *J. Chem. Soc., Dalton Trans.* **1990**, 2571.
(75) Cadierno, V.; Díez, J.; García-Álvarez, J.; Gimeno, J. *Chem. Commun.* **2004**, 1820.
(76) Phillips, G.; Hermans, S.; Adams, J. R.; Johnson, B. F. G. *Inorg. Chim. Acta* **2003**, *352*, 110.
(77) Johnson, B. F. G.; Khimyak, T.; Wansel, F. W.; Phillips, G.; Hermans, S.; Adams, J. R. *J. Cluster Sci.* **2004**, *15*, 315.
(78) Bennett, M. A.; Goh, L. Y.; Willis, A. C. *J. Chem. Soc., Chem. Commun.* **1992**, 1180.
(79) Bennett, M. A.; Goh, L. Y.; Willis, A. C. *J. Am. Chem. Soc.* **1996**, *118*, 4984.
(80) Shin, R. Y. C.; Tan, G. K.; Koh, L. L.; Goh, L. Y. *Organometallics* **2004**, *23*, 6293.
(81) Ghebreyessus, K. Y.; Nelson, J. H. *Organometallics* **2000**, *19*, 3387.
(82) Nelson, J. H.; Ghebreyessus, K. Y.; Cook, V. C.; Edwards, A. J.; Wielandt, W.; Wild, S. B.; Willis, A. C. *Organometallics* **2002**, *21*, 1727.
(83) Bellabarba, R. M.; Saunders, G. C.; Scott, S. *Inorg. Chem. Commun.* **2002**, *5*, 15.
(84) Winter, W. *Angew Chem., Int. Ed. Engl.* **1976**, *15*, 241.
(85) Bruneau, C., Dixneuf, P. H., Eds., *Ruthenium Catalysts and Fine Chemistry (Topics in Organometallic Chemistry, No. 11)*, Springer Verlag, Berlin and Heidelberg, 2004.
(86) Murahashi, S.-I., Ed., *Ruthenium in Organic Synthesis*, Wiley-VCH, Weinheim, 2004.
(87) Smith, D. C.; Haar, C. M.; Luo, L.; Li, C.; Cucullu, M. E.; Mahler, C. H.; Nolan, S. P.; Marshall, W. J.; Jones, N. L.; Fagan, P. J. *Organometallics* **1999**, *18*, 2357.
(88) Bennett, M. A.; Robertson, G. B.; Smith, A. K. *J. Organomet. Chem.* **1972**, *43*, C41.
(89) Dale, J. *Stereochemistry and Conformational Analysis*, Universitetsforlaget, Oslo, 1978, Verlag Chemie, New York, Weinheim, 1978, pp. 95–98.
(90) Eliel, E. L.; Wilen, S. H.; Mander, L. N. *Stereochemistry of Organic Compounds*, Wiley, New York, 1994, pp. 604–605.
(91) Smith, P. D.; Gelbrich, T.; Hursthouse, M. B. *J. Organomet. Chem.* **2002**, *659*, 1.
(92) Ghebreyessus, K. Y.; Nelson, J. H. *Inorg. Chim. Acta* **2003**, *350*, 12.

(93) Therrien, B.; Ward, T. R. *Angew. Chem., Int. Ed.* **1999**, *38*, 405.
(94) Therrien, B.; König, A.; Ward, T. R. *Organometallics* **1999**, *18*, 1565.
(94a) Pinto, P.; Marconi, G.; Heinemann, F. W.; Zenneck, U. *Organometallics* **2004**, *23*, 374.
(95) Umezawa-Vizzini, K.; Lee, T. R. *Organometallics* **2004**, *23*, 1448.
(96) Werner, H.; Kletzin, H.; Höhn, A.; Paul, W.; Knaup, W.; Ziegler, M. L.; Serhadli, O. *J. Organomet. Chem.* **1986**, *306*, 227.
(97) Umezawa-Vizzini, K.; Lee, T. R. *Organometallics* **2003**, *22*, 3066.
(98) Devanne, D.; Dixneuf, P. H. *J. Organomet. Chem.* **1990**, *390*, 371.
(99) Adams, J. R.; Bennett, M. A.; Yellowlees, L. J., unpublished work.
(100) Tamm, M.; Bannenberg, T.; Dressel, B.; Fröhlich, R.; Holst, C. *Inorg. Chem.* **2002**, *41*, 47.
(101) Tamm, M.; Dressel, B.; Urban, V.; Lügger, T. *Inorg. Chem. Commun.* **2002**, *5*, 837.
(102) Tamm, M.; Baum, K.; Lügger, T.; Fröhlich, R.; Bergander, K. *Eur. J. Inorg. Chem.* **2002**, 918.
(103) Tamm, M.; Dressel, B.; Lügger, T.; Fröhlich, R.; Grimme, S. *Eur. J. Inorg. Chem.* **2003**, 1088.
(104) Tamm, M.; Dressel, B.; Baum, K.; Lügger, T.; Pape, T. *J. Organomet. Chem.* **2003**, *677*, 1.
(105) Tamm, M.; Dressel, B.; Lügger, T. *J. Organomet. Chem.* **2003**, *684*, 322.

Index

D

E

F

G

R

S

T

U

V

X

Z

Cumulative List of Contributors for Volumes 1–36

Cumulative Index for Volumes 37–54